Information Sources in the

Earth Sciences

Guides to Information Sources

A series under the General Editorship of
D. J. Foskett, MA, FLA
and
M. W. Hill, MA, BSc, MRIC

This series was known previously as 'Butterworths Guides to Information Sources'.

Other titles available are:

Information Sources in Polymers and Plastics
   edited by R. J. Adkins

Information Sources in Energy Technology
   edited by L. J. Anthony

Information Sources in the Life Sciences
   edited by H. V. Wyatt

Information Sources in Physics (Second edition)
   edited by Dennis F. Shaw

Information Sources in Law
   edited by R. G. Logan

Information Sources in Management and Business
(Second edition)
   edited by K. D. C. Vernon

Information Sources in Politics and Political Science: a survey
worldwide
   edited by Dermot Englefield and Gavin Drewry

Information Sources in Engineering (Second edition)
   edited by L. J. Anthony

Information Sources in Economics (Second edition)
   edited by John Fletcher

Information Sources in the Medical Sciences (Third edition)
   edited by L. T. Morton and S. Godbolt

Information Sources in the
Earth Sciences
Second Edition

Editors
David N. Wood,
Joan E. Hardy and
Anthony P. Harvey

BOWKER-SAUR
London • Edinburgh • Munich • New York
Singapore • Sydney • Toronto • Wellington

British Library Cataloguing in Publication Data

Information sources in the earth sciences.—
2nd ed.
1. Earth sciences. Information sources
I. Hardy, Joan E. (Joan Evelyn), *1932–*
II. Wood, D. N. (David Norris)
III. Harvey, Anthony P. (Anthony Peter),
1940– IV. Wood, D. N. (David Norris)
Use of earth sciences literature. V. Series
550.'7

ISBN 0-408-01406-7

Library of Congress Cataloging-in-Publication Data
available on request.

*Bowker-Saur is the library and information division of Butterworths,
Borough Green, Sevenoaks, Kent TN15 8PH*

Cover design by Calverts Press
Printed on acid-free paper
Printed and bound in Great Britain by
Biddles Ltd, Guildford and King's Lynn

Series Editors' Foreword

Daniel Bell has made it clear in his book *The Post-Industrial Society* that we now live in an age in which information has succeeded raw materials and energy as the primary commodity. We have also seen in recent years the growth of a new discipline, information science. This is in spite of the fact that skill in acquiring and using information has always been one of the distinguishing features of the educated person. As Dr Johnson observed, 'Knowledge is of two kinds. We know a subject ourselves, or we know where we can find information upon it.'

But a new problem faces the modern educated person. We now have an excess of information, and even an excess of sources of information. This is often called the 'information explosion', though it might be more accurately called the 'publication explosion'. Yet it is of a deeper nature than either. The totality of knowledge itself, let alone of theories and opinions about knowledge, seems to have increased to an unbelievable extent, so that the pieces one seeks in order to solve any problem appear to be but a relatively few small straws in a very large haystack. That analogy, however, implies that we are indeed seeking but a few straws. In fact, when information arrives on our desks, we often find those few straws are actually far too big and far too numerous for one person to grasp and use easily. In the jargon used in the information world, efficient retrieval of relevant information often results in information overkill.

Ever since writing was invented, it has been a common practice for men to record and store information; not only fact and figures, but also theories and opinions. The rate of recording accelerated after the invention of printing and moveable type, not because that in itself could increase the amount of recording but because, by making it easy to publish multiple copies of a document and sell them at a profit, recording and distributing information became very lucrative and hence attractive to more people. On the other hand, men and women in whose lives the discovery of the handling of information plays a large part usually devise ways of getting what they want from other people rather than from books in their efforts to avoid information overkill. Conferences, briefings, committee meetings are one means of this; personal contacts through the 'invisible college' and members of one's club are another. While such people do read, some of them voraciously, the reading of published literature, including in this category newspapers as well as books and journals and even watching television, may provide little more than 10% of the total information that they use.

Computers have increased the opportunities, not merely by acting as more efficient stores and providers of certain kinds of information than libraries, but also by manipulating the data they contain in order to synthesize new information. To give a simple illustration, a computer which holds data on commodity prices in the various trading capitals of the world, and also data on currency exchange rates, can be programmed to indicate comparative costs in different places in one single currency. Computerized data bases, i.e. stores of bibliographic information, are now well established and quite widely available for anyone to use. Also increasing are the number of data banks, i.e. stores of factual information, which are now generally accessible. Anyone who buys a suitable terminal may be able to arrange to draw information directly from these computer systems for their own purposes; the systems are normally linked to the subscriber by means of the telephone network. Equally, an alternative is now being provided by information supply services such as libraries, more and more of which are introducing terminals as part of their regular services.

The number of sources of information on any topic can therefore be very extensive indeed; publications (in the widest sense), people (experts), specialist organizations from research associations to chambers of commerce, and computer stores. The number of channels by which one can have access to these vast collections of information are also very numerous, ranging from professional literature searchers, via computer intermediaries, to Citizens' Advice Bureaux, information marketing services and information brokers.

The aim of the Guides to Information Sources (formerly Butterworths Guides to Information Sources) is to bring all these sources and channels together in a single convenient form and to present a picture of the international scene as it exists in each of the disciplines we plan to cover. Consideration is also being given to volumes that will cover major interdisciplinary areas of what are now sometimes called 'mission-oriented' fields of knowledge. The first stage of the whole project will give greater emphasis to publications and their exploitation, partly because they are so numerous, and partly because more detail is needed to guide them adequately. But it may be that in due course the balance will change, and certainly the balance in each volume will be that which is appropriate to its subject at the time.

The editor of each volume is a person of high standing, with substantial experience of the discipline and of the sources of information in it. With a team of authors of whom each one is a specialist in one aspect of the field, the total volume provides an integrated and highly expert account of the current sources, of all types, in its subject.

D. J. Foskett
Michael Hill

Preface to the Second Edition

The first edition of this book was published in 1973 under the title *Use of Earth Sciences Literature*. The second edition has been completely revised to include many of the new publications and information services now available. Since 1973 almost one million journal articles and several thousand textbooks have been published around the world in the earth sciences, and when one considers that much of interest to the earth scientist is also to be found in the literature of chemistry, physics, engineering, biology etc the amount of potentially useful information available is truly enormous.

The purpose of this book is, like its predecessor, to advise the reader on the most appropriate sources of information by clarifying the structure of earth sciences information and by citing particular sources which, based on personal experience, the individual contributors feel able to recommend.

The editors are indebted to the contributors for their patience, understanding and enthusiasim in preparing their revised chapters and are particularly thankful to Mr Michael O'Donoghue, Mr Duncan McKay, Mr Jim Colman and Professor Richard Selley who agreed to write their contributions at short notice.

Thanks are due also to staff at the British Library Document Supply Centre who checked many of the journal titles and dates, and also to librarians at the Geological Society, Imperial College and King's College London, who gave assistance with references. Dr Eric Robinson of University College London read much of the manuscript and offered valuable advice, for which the editors are grateful.

The following organizations gave permission for parts of their publications to be reproduced: Mineralogical Society (*Mineralogical Abstracts*); Institute for Scientific Information (*Science Citation Index*); Chemical Abstracts Service (*Chemical Abstracts*); American Geological Institute (*Bibliography and Index of Geology*).

Contributors

Douglas A Bassett PhD
Director, National Museum of
Wales, Cardiff

Keith M Clayton CBE PhD FGS
Dean, School of Environmental
Sciences, University of East
Anglia, Norwich

Jim B Colman MSc MIMM
British Geological Survey,
Keyworth, Nottingham

Stephen C Francis PhD FGS
formerly Department of
Geology, Chelsea College,
London

Anthony Hall PhD FGS
Department of Geology, Royal
Holloway and Bedford New
College, Egham, Surrey

John Hopson BSc Dip Inf Sc
British Library Information
Sciences Services, London

D A Jenkins PhD
D B Johnson
W I Kelso
School of Biochemistry and Soil

Science, University College of
North Wales, Bangor

Mary Lynette Larsgaard
Map and Imagery Laboratory,
Library, National Cartographic
Information Centre, University
of California, Santa Barbara

Ann Lum BSc Dip Lib ALA
Palaeontology Library, British
Museum (Natural History),
London

Duncan J McKay MA Dip Lib
Earth Sciences Information
Officer, Britoil plc, Glasgow

John S Milsom PhD FGS
Department of Geological
Sciences, University College,
London

Michael J O'Donoghue FGS
British Library Science
Reference and Information
Service, London

R A S Ratcliffe
formerly Meteorological
Office, Bracknell

Richard C Selley PhD DIC FGS
Head, Department of Geology,
Imperial College of Science,
Technology and Medicine,
London

P W G Tanner PhD DIC FGS
Department of Geology,
University of Glasgow

**Jennifer M Todd BA Dip Lib
ALA**
formerly Librarian, The
Kennedy Library, Department
of Earth Sciences, University of
Leeds

John N Walsh PhD FGS
Department of Geology, Royal
Holloway and Bedford New
College, Egham, Surrey

Contents

CHAPTER ONE

Introduction

DAVID N WOOD and JOAN E HARDY

It has been said that 'any science is a body of public, organized knowledge' and it is a desirable feature of scientific life that scientists should communicate their results: new scientific information is incorporated into the body of scientific knowledge only when it is communicated, usually in a published form. In this way unnecessary duplication of scientific effort may be avoided, and stimulation of thought may lead to the development of new ideas.

The longest established means of interchanging ideas is by private communication, either orally or by letter. The scientific conference of the present-day is an acknowledged vehicle for this type of dissemination of topical research and knowledge. However, it is impossible for any one scientist to have personal contact with all researchers whose individual work may impinge upon his own field of interest: to keep track of new and relevant information the scientist must turn to the literature. It is necessary, therefore, for the scientist to understand the nature of this communication system. Although the term 'information explosion' has been taken to refer to the considerable increase in scientific activity since 1950, nevertheless the scientific journals, *Journal des Sçavans* and the *Philosophical Transactions of the Royal Society*, were first published in 1665, with the purpose of broadcasting new scientific discoveries. It is interesting to note the first paragraph of the first issue of the *Philosophical Transactions*: "There is nothing more necessary for promoting the improvement of philosophical matters than the communicating to such as apply their studies and endeavours that way, such things as are discovered or put in practice by others. It is therefore thought fit to employ the press as the most proper way to gratify those whose engagement in such studies and delight in the advancement of learning and profitable discoveries doth entitle them to the knowledge of what this Kingdom or other parts of the world do from time to time afford." Since the seventeenth century the journal has become the primary means of published communication of new discoveries by the increasing number of research workers, and the number of scientific and technological journals has increased to accommodate the burgeoning body of work. The growth of the literature has itself been the subject of many papers.

Although the bulk of papers relevant to a specific field of work will appear in a limited number of journals, say about 20, the geologist must be aware of relevant information in a wider selection of journals in 'fringe' disciplines. The concentration of major works in a small number of geological journals is supported by the application of a refereeing system. Periodical publications, of course, are not the only source of scientific information. Other types of literature exist and, like the periodical, have increased in number at a rate which reflects the general growth of scientific activity. These other types of publication include additional primary sources such as theses, reports, patents, standards, and maps, as well as secondary literature such as review journals, textbooks, monographs, handbooks, encyclopaedias, and many conference proceedings.

The information contained in this body of literature may be of use to the scientist in a number of tasks: keeping up-to-date with current work; locating specific facts such as the optical properties of a mineral, the chemical composition of a certain rock, or the manufacturer of a particular instrument; and carrying out an exhaustive search for information on a particular subject. Whatever the reason for using it, the literature is probably the most expensive tool that the scientist has at his disposal. Despite its cost and potential value, few scientists are taught about the structure and use of their subject literature. There is a tendency in most universities to assume that a knowledge of scientific literature is gained intuitively, and most students obtain degrees, and even higher degrees, without really discovering how to use this valuable research tool effectively. On the other hand, considerable time and effort is devoted to teaching students how to use relatively inexpensive equipment such as a petrological microscope, an X-ray fluorescence spectrograph, or a flame photometer.

Earth scientists in particular have been slow to appreciate the problems posed by the literature explosion. This is partially understandable in view of the fact that the growth of literature has not been quite as spectacular in the earth sciences as in other subject fields such as chemistry and physics. It has been increasing nevertheless, and for this reason, and because of certain characteristics displayed by the literature, the earth scientist cannot afford to remain indifferent to the information problem. The characteristics in question include the interdisciplinary nature of much of the geological literature, the fact that it is more international in scope than the literature of any other subject, and that it remains current for a much greater period than the literature of other scientific disciplines.

The first edition of this book was prepared with a view to its being used as the basis for a course on the literature of the earth sciences and also a guide to students and practising geologists wishing to become familiar with the literature. The second edition has endeavoured to outline developments since 1973, for example online databases and databanks, as well as updating the earlier text. The book is not intended to be a comprehensive reference tool; the more important literature is of course mentioned, but for details of additional publications the reader will find himself referred to other reference works.

The first published attempt to introduce geologists to the various forms of scientific publication was *Guide to geologic literature*, by R M Pearl (McGraw-Hill, 1951), an eminently readable book, but most literature guides since then have been mainly reference works: this work, like its first edition, endeavours to introduce the whole subject of searching the earth sciences literature. Since both types of work will be of use to the practising geologist, details of such publications are given below.

MacKay, J W (1974) Sources of information for the literature of geology; an introductory guide. 2nd ed. London: Geological Society of London
Van Balen, J *compiler* (1978) Geography and earth sciences publications, 1973–1975; an author, title and subject guide to books reviewed, and an index to the reviews. Pierian Press
Ward, D C *et al*(1981) Geologic reference sources. 2nd ed. Metuchen (N.J.) & London: Scarecrow Press
Wood, D N *editor* (1973) Use of earth sciences literature. London: Butterworths

Articles on specific aspects of geological literature appear from time to time in periodical publications in the librarianship and information science fields, in particular the publications of the Geoscience Information Society:

Geoscience Information Society. *Proceedings*. Vol.1 1970–

CHAPTER TWO

Earth science libraries and their use

JENNIFER M TODD

Earth science libraries are by no means a recent development. The oldest geological library in the world was founded in 1809 by members of the Geological Society of London and contained not only books donated by members, but also a considerable fossil and rock collection. Only three years later in 1812, the Society found it necessary to appoint an officer to take charge of the library and mineral collection, and to act as draughtsman to the Society. This officer was given the title of Keeper of the Museum, and devoted his time to the above duties. As the collection continued to expand, the emphasis on the various duties of the Keeper began to change as there developed a demand for the skills associated with the duties of a librarian. Thus, in 1842 the Committee of the Geological Society were able to report their ". . . great satisfaction in bringing under the notice of the Council the judicious manner in which the library has been arranged . . . The whole of the volumes have been ticketed, lettered and numbered, and to each of these a reference has been made in the Catalogue".

This was a very early attempt to introduce bibliographical control not only by means of a catalogue, but also by the production of an elementary classification scheme. Under this scheme, the literature was to be divided into various subject groupings and a letter of the Roman alphabet assigned to each group. Thus:

1. Transactions	A,B,C,D
2. Proceedings	E
3. Periodicals	F,G
4. General Treatises on Geology	H
5. Practical Geology	I,J

6. Mineralogy	K
7. Crystallography, Mineralogy etc	L
8. Topography	M
9. Chemistry	N
10. Physics etc	P
11. Natural History	R

Readers could then establish which letter had been assigned to the subject that they required, and so find the literature on a shelf labelled accordingly.

Interest in geology as an independent subject continued to develop in various countries throughout the nineteenth century as national and local geological societies and surveys came into being, often producing their own literature, and it has continued to expand throughout this century. New branches of the subject have been developed and studies of existing ones have been extended, or combined with previously separate branches. These developments covered a much wider field of interest and embraced all aspects of the earth sciences.

The 1950s and 1960s saw an "information explosion" with a rapid increase in the amount of documented information available, and this has continued to be the case as more and more information is made available in a variety of forms (journal, book, thesis, microfilm and microfiche). At the same time, geological literature has differed from the other sciences in that it has often remained valid for a longer period. This mass of information has therefore had to be stored in such a way that information seekers can obtain ready access to it. To this end, earth sciences libraries have developed and expanded together with their subject.

Many libraries accepted the idea that information should be on open access and readily available to readers, and so it became essential that they be given every assistance in finding it. To do this, it was helpful to have related subjects brought together, and the need to find a method of systematically arranging an ever increasing volume of literature led to the invention of a number of classification schemes. Some are general schemes which are widely used by a variety of libraries, others are specialized subject schemes and therefore less frequently encountered, and, more rarely still, a library may use an "in-house", self-devised scheme.

The three general schemes most widely used in Europe and North America are the Decimal Classification, the Universal Decimal Classification and the Library of Congress schemes. A fourth, the Colon Classification scheme, devised by S R Ranganathan, has made an impact on library science ideas in the west, but it is rarely found in use outside of the Asian continent, or more especially, India.

The first of the three schemes to gain widespread acceptance was one devised by an American, Melvil Dewey, and first published in 1876. It became known as the Decimal Classification scheme (DC) as Dewey chose Arabic numerals for his notation. He then divided all knowledge into ten main classes each of which was divided into ten main divisions, and these in turn were divided into ten sections. Dewey was not the first librarian to introduce arrangement by subject, but what he did was to introduce the idea of applying the notation to the books rather than to the shelves, thus allowing a new book to be inserted at the most suitable point in the sequence, rather than just be added at the end of the subject collection. The scheme has a somewhat North American bias and is therefore widely used there, but it has also been widely adopted by public and general libraries in Britain, and is currently into the nineteenth edition.

However, it does not always meet successfully the special requirements of science and technology libraries, where the subject matter has greatly expanded during this century. It was precisely for this reason that the Universal Decimal Classification scheme (UDC) was devised by two Belgians, Paul Otlet and Henri La Fontaine. They obtained permission from Dewey to adapt the fifth edition of DC to make it not only more applicable to European libraries, but also to expand its coverage of science and technology in order to accommodate all new developments. Although the notation is similar to that of DC in that it retains the use of Arabic numerals and uses the same major divisions, it has not retained the three digit minimum, e.g. Earth Science is 550 in DC, but 55 in UDC. Despite the many similarities, the two schemes have inevitably tended to draw apart as UDC has been further revised and expanded to cater for new developments in science and technology, where its schedules are at their most detailed. It is for this reason that UDC has been adopted internationally and is used by many academic and subject libraries in Europe and North America, and has also gained widespread acceptance in the USSR.

The third scheme in general use, especially in North America, is the Library of Congress scheme (LC) which was first devised to meet the needs of the rapidly expanding Library of Congress, Washington, at the beginning of this century. As it was created to fit a library collection already in existence, and not based on abstract theory, it does have certain advantages over other schemes. Many US libraries now use LC cards which provide both cataloguing and classification information. The ready accessibility of LC cards, and the ready availability of catalogue records on the LCMARC database, has prompted some academic libraries in the US to make the change from DC to LC.

LC differs from DC and UDC not only in the breakdown of its major classes, but also in its use of the Roman alphabet to symbolize them. Main classes are denoted by a single capital letter and, in most of them, the major sections are denoted by a second capital letter. Arabic numerals are then used to denote the divisions.

When using a library it is generally helpful for the reader to have some understanding of the pattern of which the individual classification symbols are a part. The arrangement of the major classes in the three sections can be seen thus:

DC

000	General works
100	Philosophy
200	Religion
300	Social Sciences
400	Languages
500	Pure Sciences
600	Technology
700	The Arts
800	Literature
900	History

UDC

0	Generalities
1	Philosophy/Ethics/Psychology
2	Religion/Theology
3	Social Sciences/Law
4	Philology/Linguistics
5	Pure Sciences
6	Applied Sciences
7	The Arts
8	Literature
9	Geography/History

LC

A	General works
B	Philosophy/Religion
D	History-general and Old World
E–F	History
G	Geography
H–P	(see LC schedules)
Q	Science
T	Technology
U	Military Science
V	Naval Science
Z	Bibliography

Earth science is located in the pure science class in DC (500) and UDC (5), and the science class in LC (Q). The science class in each scheme is then divided into ten (DC and UDC), or twelve sections (LC).

DC		UDC	
500	Pure Sciences	5/50	Exact sciences in general
510	Mathematics	51	Mathematics
520	Astronomy	52	Astronomy
530	Physics	53	Physics
540	Chemistry	54	Chemistry
550	Earth Sciences	55	Earth Sciences. Geology.
560	Paleontology		Meteorology
570	Life Sciences	56	Paleontology
580	Botanical Sciences	57	Biological Sciences
590	Zoological Sciences	58	Botany
		59	Zoology

LC

Q	Science (general)
QA	Mathematics
QB	Astronomy
QC	Physics
QD	Chemistry
QE	Geology
QH	Natural History
QK	Botany
QL	Zoology
QM	Human Anatomy
QP	Physiology
QR	Bacteriology

There is a great deal of similarity between the schemes at this point as each has attempted a logical arrangement of related subjects. The Earth Science/Geology sections are further divided into subject areas:

DC

550	Sciences of the Earth (and other worlds)
551	Geology, meteorology, general hydrology
552	Petrology
553	Economic geology
554–599	Treatment by continent etc.:
	554 Europe
	555 Asia
	556 Africa
	557 North America
	558 South America
	559 Other parts of the world and extra-terrestrial worlds

UDC

55	Earth sciences, geology, geophysics
550	Ancillary sciences, geophysics, seismology, electricity, magnetism, geochemistry, geobiology
551	General geology, meteorology, climatology, historical geology, stratigraphy, paleogeography
552	Petrology, petrography
553	Economic geology, mineral deposits
554	Hydrosphere, hydrology

Palaeontology follows in a separate section in both DC (at 560) and UDC (at 56).

LC

QE	Geology
1–350	General
351–399	Mineralogy
420–499	Petrology
500–625	Dynamic and structural geology
651–700	Stratigraphic geology
701–996	Paleontology, paleozoology, paleobotany

The notation can sometimes be still further expanded or synthesized to cover various subject aspects of a single publication. However, few libraries can afford, or have the space, to duplicate stock to cover every relevant point in the scheme, and a publication can usually only occupy one place on the shelves, e.g. a book on coal deposits in Wales can be shelved either in the economic geology section, or under Wales in the area section.

In addition to printed literature, a library may hold a microfilm or microfiche collection which, because of its format, cannot be shelved with the books, even though classified in the same way. This also applies to oversize books and to reprint or pamphlet collections. Other material frequently shelved outside the main book collection includes journals, which now tend to be accommodated in a separate section and arranged by alphabetical order of title, rather than classified by subject. Current periodicals are often separated from the main journal collection and displayed on a special rack. Other material of interest to the geologist such as maps and rocks are occasionally found in the library, but are more usually housed as completely separate collections.

The key to the shelf arrangement is the catalogue which enables the user to find the required literature through a number of approaches, even though there is only one physical shelf location. An alphabetical index of some sort is always needed as a point of entry to the classification scheme, in order to enable the user to find the notation. The purpose of the catalogue was defined by C A Cutter as long ago as 1876, and as a broad summary it is still true today:

(1) to enable a person to find a document of which either a) the author, or b) the title, or c) the subject is known.
(2) to show what the library has a) by a given author, b) on a given subject (plus related subjects), and c) in a given form of literature.
(3) to assist in the choice of document through description in the catalogue of a) the edition, and b) the content.

The physical form of the index system or catalogue varies considerably in different types of library. For many years the most popular form of catalogue has been the card catalogue, with standard size cards filed in drawers. It is extremely convenient for adding new entries and achieving direct entry to a required access point. Other forms of catalogue are the sheaf catalogue consisting of thin slips of paper held in binders, and printed catalogues in book form. However, both these types of catalogue have the disadvantage of inflexibility with regard to adding new entries.

The last ten years have brought new developments in catalogue format, and library holdings may now also be indexed in microfilm or microfiche. The greatest development has been the use, usually by larger libraries, of computers to store the cataloguing information, and from this database an up-to-date print-out or microfiche version of the catalogue may be produced; current developments are aimed at giving direct access to the catalogue from public terminals. This form of cataloguing is likely to gain widespread acceptance by libraries with the financial and staff resources to operate it, but smaller or isolated libraries will not always be able to adopt computerized systems.

Whatever the physical format of the catalogue, each entry contains standard information regarding author (can be a person, institution, etc.) and classification, with some additional detail regarding publisher, date of publication, library accession number etc.,

e.g.   LEEDER, Michael R
                   Sedimentology: process and product.
                   London: Allen and Unwin, 1982.
              82–72   QE 471

There are various rules governing the way that the catalogue is compiled, and these vary from "in-house" rules used by only one library, to those that are internationally recognized such as the Anglo-American cataloguing rules (AACR) or the Library of Congress rules for descriptive cataloguing. The Anglo-American rules are widely used in both Britain and North America, and therefore provide a certain uniformity of practice. However, it cannot be complete uniformity as there are separate British and American texts which take account of national preferences. The preface to the code states that it was originally devised for use in "larger libraries of a scholarly character", but it is now used by many types of library, both large and small. The Library of Congress rules, which are based on their own interpretation of AACR, have gained in popularity along with the classification scheme, especially in the USA.

In Britain, and Europe generally, the most popular type of catalogue is the classified catalogue, while in the USA it is the dictionary catalogue. A classified catalogue consists of a systematic arrangement of related topics, using the notation as the key. It is therefore not possible to combine the subject and author entries in one sequence and there has to be at least two, and sometimes three, separate parts to a classified catalogue. These are (1) an author or combined author/title alphabetical section, (2) the classified sequence including extra entries for books covering more than one subject, and (3) an alphabetical subject index which acts as a guide to the classified sequence. Sometimes (1) and (3) are combined in a single alphabetical sequence in the style of a dictionary catalogue.

The dictionary catalogue consists of subject headings interfiled alphabetically with author/title entries. Cross references to related subjects are achieved by use of the terms "see" and "see also". There are two major methods of filing entries in an alphabetical catalogue and these are "word by word" and "letter by letter", e.g. mineral petrology would be followed by mineralogy in a "word by word" sequence, but would be preceded by it in a "letter by letter" arrangement.

However efficiently constructed, a library catalogue and shelf arrangement can only be a reflection of what is actually available in that library, and no single library can hope to maintain a comprehensive collection of all the available literature. The enormous growth in scientific literature in the past decade has stimulated what could be termed an "information revolution", with the increasing use of new techniques by libraries to keep pace with the volume of literature produced. The concept of a computerized database emerged in the early 1960s, with an enormous growth in the use of machine-readable databases taking place from the mid 1970s.

As more libraries acquire the necessary hardware, i.e. computers of varying degrees of sophistication, they are able to offer regular, current awareness services. A Selective Dissemination of Information service (SDI) will retrieve references on a specific topic from the latest computerized database of citations to the literature (books, journals, reports, etc.) based on a match of keywords. These databases are computerized versions of printed indices such as *Bibliography and Index of Geology* (the printed index of GeoRef), and are usually updated on a monthly basis. They contain descriptive information about the publications that they list, i.e. author, title, volume, number, pagination, date and sometimes an abstract.

There are a number of earth science databases which have either international or national coverage, or are confined to a specific subject area, and they provide services such as SDI, retrospective searches, bibliographies, etc. They include systems such as Geological Reference File (GeoRef), GeoArchive, Geobase and Canadian Center for Geoscience Data, of which the first two are perhaps the most widely used and provide the most extensive coverage. Further reference to these services may be found in Chapter 6.

However, a great many libraries still do not have the resources to enable their readers to take advantage of these services, and those that do often find that they are unable to provide all the literature cited. Indeed it is obvious that no single library can possess all the material needed to satisfy its readers, and that discovering that certain information exists may then necessitate an interlibrary loan request or even a personal visit to another library.

There are a number of different types of earth science libraries, or earth science collections within a larger library, and the services that they provide depend very much upon the function that they are intended to fulfil. In national collections such as the British Library Science Reference and Information Service and the British Museum (Natural History) Library situated in London, or the Library of Congress in Washington, the extensive geology collections are part of a much larger science collection. In public libraries the geology stock is also part of a larger collection, but in this case it is part of a general collection and is likely to consist of a fairly small number of books of the type most likely to be needed by an interested amateur.

In-depth collections in the UK can be found in the academic sector, i.e. as subject collections within a university, polytechnic, or college library, or as special libraries within an earth science department in one of these institutions. Other special libraries include those serving the British Geological Survey with the main survey collection based in Keyworth, Nottinghamshire, and a small number of regional branches, including London. There are also various geological societies scattered throughout the country and many of them have sizeable library collections, although none so extensive as that of the Geological Society of London. Earth science libraries have also been developed by various branches of industry to serve their own particular interests.

Individual academic, survey, society and industrial libraries all have their own rules and regulations regarding use by non-members, and usually prefer readers to contact them before actually visiting the library. Useful lists of British geology libraries can be found in a number of sources, including E Robinson's *Guide to the Geology Libraries in London*, and the *Geological Directory of the British Isles*, edited by J A Diment: as is the case with many directories, the latter cannot claim comprehensive coverage as not all libraries answered the questionnaire. However, it is a most useful list and gives details of all types of geology library, including opening hours, size and content of stock, special collection details, and classification systems used.

Although many of these libraries are available as potential suppliers of information to those unable to find the material that they require in their own library, most readers prefer, in the first instance, to try and acquire what they need from the British

Library Document Supply Centre. This library is based at Boston Spa in Yorkshire, and is unique in the library world as a centralized, national lending library, offering rapid loan and photocopy services. A public reading room is available for use by visitors, but access to the interloan system is through the borrower's own library organization.

The USA library scene follows a similar pattern in many respects, with in-depth collections maintained by university and college libraries, or by individual earth science departments. The United States Geological Survey is based in Washington and has a number of branches with libraries scattered throughout the country. Individual states also have their own geological surveys which concentrate on various aspects of the geology of that state, and they maintain state survey libraries, often based in university departments. There are also museum, society, and industrial libraries. Details of such libraries and their stock can be found in two useful publications: (1) *Directory of Geoscience Libraries: US and Canada*, published in 1974 by the Geoscience Information Society, and (2) *A Directory of Information Resources in the United States: geosciences and oceanography*, published in 1981 by the National Referral Center. The National Referral Center is based in the Library of Congress in Washington, and is a free service which directs seekers of information to those libraries or organizations able to supply it. These information sources, together with a description of their stock and services offered, are listed in a subject-indexed, computerized file, which is maintained by professional analysts and used primarily by referral specialists at the Center. From this file the referral specialists are able to provide enquirers with details of appropriate information sources, including names, addresses and telephone numbers, together with some details of what each can provide.

At the end of this chapter reference is made to a number of publications both for Great Britain and the USA which list sources of geological information in those countries, and a useful world-wide list of the most important sources of geological information listed country by country was published in the *Encyclopedia of Library and Information Science* in 1973 (Lea *et al*). Some libraries also publish their own library catalogues, although it is usually only the larger organizations which have the resources to do this, e.g. the British Museum's *General Catalogue of Printed Books* with Supplements, the Library of Congress' *National Union Catalog* and the United States Geological Survey's *Catalog of the USGS Library*.

Whatever the size or type of library, it is important to remember that the staff are there primarily to help the reader obtain maximum benefit from it. Just as the libraries themselves have grown and developed since the founding of the first one in the early nineteenth century, so librarians have extended the scope of their original duties to play a much more active part in the dissemination of information process. There are now a number of associations formed specifically for earth science librarians, such as the GIS (Geoscience Information Society — a member society of the American Geological Institute, and an associate society of the Geological Society of America), and the GIG (Geological Information Group of the Geological Society of London). These organizations were formed for the purpose of exchanging information and ideas, not only in order to enrich the professional careers of their members, but also to create the best service possible for those seeking information in any aspect of the earth sciences.

References

Ball, H W (1979) The evolution of a national collection. *Special Papers in Palaeontology*, **22**, 49–56

Bartlett, L A (1982) A method for core selection of geoscience journals for libraries with limited resources. *In*: Kidd, C M (ed), *Second International Conference on Geological Information, May 23–27, Colorado School of Mines, Golden, Colorado. Proceedings*: Volume **2**, 356–370

British Library (Science Reference and Information Service) (1986) *Guide to government departments and other libraries*. London: The British Library Board

Burk, C F Jr. (1982) A worldwide list of course and reference databases in the geosciences. *Database*, **5**, (2), 11–21

Burkett, J (ed) (1974) *Government and related library and information services*. 3rd edn. London: The Library Association

Dewey Decimal Classification and Relative Index. (1979) 19th edn. Albany (New York): Forest Press

Diment, J A (ed) (1978) *Geological directory of the British Isles: a guide to information sources. (Geological Society of London Miscellaneous Paper No. 10)*. London: The Geological Society of London

Foskett, A C (1971) *The subject approach to information*. 2nd edn. London: Clive Bingley

Geological Society of London. Annual Report (1842). *In:Proceedings of the Geological Society of London*, **III** (II, No. 86), 616–617

Gorman, M and Winkler, P W (1978, 1982) *Anglo-American cataloguing rules*. 2nd edn. London: The Library Association

Griffiths, J-M (1982) Main trends in information technology. *Unesco Journal of Information Science*, **IV**, (4), 230–238

Hardy, J E (1976) *Libraries for the geologist in and around London*, revised by S Dibley. 4th edn. London: Imperial College Lyon Playfair Library

Harvey, A P (1982) Geological information, past, present and future. *In*: Kidd, C M (ed) *Second International Conference on Geological Information, May 23–27, Colorado School of Mines, Golden, Colorado. Proceedings*: Volume **1**, 10–18

Harvey, A P (1978) A history of geological serial publication in the United Kingdom. *Earth and Life Science Editing*, **6**, 7–12

Henton, T (1985) *The Geologist's Directory*. 3rd edn. London: The Institution of Geologists

Kesner, R M and Jones, C H (1984) *Microcomputer applications in libraries: a management tool for the 1980s and beyond*. London: Aldwych Press.

Lea, G, Briers, P M and Harvey, A P (1973) Geological libraries and collections. *Encyclopedia of Information Science*, **9**, 283–309

LenRoot-Ernt, L and Darnay, B T (1982) *In: Subject directory of special libraries and information centers*. 7th edn. Volume **5**: *Science and technology libraries*. Detroit (Michigan): Gale Research Company

Line, M B (1977) National libraries. *In*: Whatley, H A (ed) *British librarianship and information science 1971–1975*. 132–145. London:The Library Association

Masterson, A R (1982) The role of the information specialist in US state geological surveys. *In*: Kidd, C M (ed) *Second International Conference on Geological Information, May 23–27, Colorado School of Mines, Golden, Colorado. Proceedings*: Volume **2**, 258–274

Merrill, G P (1924) *The first one hundred years of American geology*. New Haven (Connecticut): Yale University Press

Nag, D (1980) Earth science information systems and services. *Annals of Library Science and Documentation*, **27**, 102–105

National Referral Center (1981) *A directory of information resources in the United States:geosciences and oceanography*. Washington: Library of Congress

Needham, C D (1971) *Organizing knowledge in libraries: an introduction to information retrieval*. 2nd edn. London:Andre Deutsch

Price, J F (1981) A directory of information resources in the United States: geosciences and oceanography. *Geoscience Information Society Proceedings*. **11**, 81–88

Robinson, E (comp) (1982) *A guide to the geology libraries in London*. 3rd edn. London:University of London

Thurston, N (1980) Collection development in geoscience libraries:serials. *Geoscience Information Society Proceedings*, **10**, 17–34

Tseng, S C (comp) (1984) *Library of Congress rule interpretations of AACR 2, 1978–1984*. Cumulated edn. Metuchen (New Jersey) & London:The Scarecrow Press

United States Library of Congress (1950, reprinted 1967) *Classification. Class Q. Science*. 5th edn. Washington: USGPO

Universal Decimal Classification (1982). London: British Standards Institution

Walker, R D and Parker D (1974) *Directory of geoscience libraries: US and Canada*. 2nd edn. Washington:Geoscience Information Society

Woodward, H B (1907) *The history of the Geological Society of London*. London:The Geological Society of London

CHAPTER THREE

Primary literature

JOHN HOPSON

Scientific literature is usually divided into two categories, primary and secondary. Primary literature is normally defined as containing new information such as the first reports of laboratory studies and field investigations, details of new hypotheses, descriptions of new equipment, and so on, or a new interpretation of previously known information. Secondary literature by contrast seeks to organize and distill primary information, place it in context and arrange it so that it is more easily accessible such as in abstracting journals, encyclopedias, data compilations and textbooks. There are various types of primary literature, including periodicals, reports, theses, government and international publications, and conference proceedings. Each of these types is described in this chapter. The reader should be aware, however, that there are other types of primary literature which are not included here as they are of little interest to, and are little used by, earth scientists, e.g. patents and standards. Information about these primary sources is given in a number of publications, such as *Use of Chemical Literature*, edited by R T Bottle (3rd edn, Butterworths, 1979). It is also worth noting that maps, although they cannot be considered 'literature' as such, can be valuable as sources of primary information. This is of course especially true in the earth sciences, where geological maps are of particular importance. These are dealt with in Chapter 7.

In reality the distinction between primary and secondary literature is somewhat arbitrary. Periodicals are, for instance, essentially a type of primary literature, but they can carry previously known information, such as research reviews. Books and monographs are basically secondary sources but sometimes

contain original ideas or hypotheses. Notwithstanding this in-
distinct boundary, as a generalization the division between
primary and secondary literature is valid. This chapter describes
some types of primary literature together with foreign language
literature and translations. Secondary literature is covered in the
two following chapters.

Periodicals

Periodicals, here including journals, bulletins, serials, etc., are the
pre-eminent vehicle for scientific communication and are the
principal means by which scientists report new information to one
another. The characteristic features of primary periodicals are that
they are issued, regularly or irregularly, as part of a numbered
series and that they contain new and original information.

The first scholarly periodical, *Journal des Sçavans*, was published in Paris in January 1665. It was quickly followed by others,
most notably the *Philosophical Transactions of the Royal Society*,
the first issue of which appeared in March 1665. Since then
Philosophical Transactions has been published continuously and it
is the oldest scientific periodical in existence. The number of
periodicals has grown ever since: by 1800 there were about 100
titles, by 1850 around 1,000 and by the end of the nineteenth
century the number had grown to 5,000 or so. At the time of
writing it is estimated that there are over 50,000 periodicals
currently being published world-wide.

Hand-in-hand with this continuous overall growth has been the
trend towards fragmentation by subject and ever increasing
specialization. The earliest periodicals covered all sciences and the
'useful arts' as well. However, as science itself became more
organized, and divided into the various disciplines recognizable
today, so scientific literature has become more specialized and
periodicals devoted to particular subjects have developed.
Although there are still many general science periodicals (e.g.
Nature, Science, Journal of the Royal Society of New Zealand, and
Proceedings of the Royal Irish Academy) most are devoted to
narrower subject areas.

The first periodical covering only a single branch of the earth
sciences was *Journal des Mines* (1795–1815) and throughout the
nineteenth century new earth sciences periodicals regularly
appeared, e.g. *Quarterly Journal of the Geological Society* (1845),
Proceedings of the Geologists' Association (1859), *Geological
Magazine* (1864), and the *Journal of Geology* (1893). In the
twentieth century the trend towards further specialization has

continued, in science itself and hence in scientific literature. This is well illustrated by the titles of some new periodicals which have appeared over the last 10 years or so: *Journal of Structural Geology* (1979), *Precambrian Research* (1974), *Journal of Metamorphic Geology* (1983), and the *Journal of Vertebrate Paleontology* (1981).

The earliest periodicals were almost exclusively issued by learned societies, but there is now a variety of different types of organizations involved in periodical publication. Nevertheless, societies are still responsible for a large number of periodicals, including many of the more prestigious titles. A characteristic feature of most society-originated periodicals is their strict refereeing process, whereby all submitted papers are passed to an independent, anonymous referee for comment and criticism. By this means a society is able to maintain the quality of its periodicals and this in turn maintains the reputation of the society. Periodicals of this type include *Bulletin of the Geological Society of America* (1890–), *Proceedings of the Ussher Society* (1962–) and *Mineralogical Magazine* (1876–) published by the Mineralogical Society of Great Britain.

Over the last twenty or thirty years commercial publishers have shown an increasing interest in scientific periodical publishing. Among the most active of such publishers are: Elsevier, with titles such as *Tectonophysics* (1964–), *Chemical Geology* (1966–) and the *Journal of Volcanology and Geothermal Research* (1976–); Pergamon Press with *Geochimica et Cosmochimica Acta* (1951–), *Organic Geochemistry* (1977–) and *Geothermics* (1972–); and Academic Press who publish *Cretaceous Research* (1980–), and *Quarternary Research* (1970–). In addition, many periodicals are now published by commercial publishers on behalf of learned societies: examples include *Journal of the Geological Society*, published by Blackwell Scientific Publications, *Earth Surface Processes and Landforms*, published by John Wiley and Sons for the British Geomorphological Research Group, and the recently launched *Geology Today* (1985–), aimed at "encouraging and spreading the pleasures of geology to a wider audience", published by Blackwell Scientific Publications in association with the Geological Society and the Geologists' Association.

Also produced by commercial organizations are trade journals, which help to keep scientists, technologists and business men in touch with current developments in their particular industry. Characteristic features of trade journals include a large amount of space devoted to news items, typically shorter and less technical articles, and a higher proportion of advertisements than is found in learned journals. For many trade journals advertising revenue accounts for a large part of their total income and helps to keep the price down. In the earth sciences most of the trade journals are concentrated in the applied subject areas, particularly the extractive industries: examples include *Mining Magazine* and *Mines and Quarry*.

Government publications form a large and very important part of the primary literature of the earth sciences due, at least in part, to the essentially regional nature of the subject. Most countries of the world have a government-funded 'geological survey' type of organization and most of these publish results of their research in memoirs, bulletins, etc., many of which are referred to in Chapters 7 and 8. In the United Kingdom there is the British Geological Survey (BGS), while in the United States is the US Geological Survey. BGS publishes through Her Majesty's Stationery Office (HMSO) and new publications are announced in the various HMSO guides, e.g. *Daily List of Government Publications*, which appears on Mondays to Fridays inclusive, and the annually produced *Government Publications*. In addition to these general lists HMSO also produces a number of Sectional Lists, one of which, List 45, gives details of BGS publications: others include List 37, Meteorological Office. Details of publications produced by the US Geological Survey are given in their monthly *New Publications of the Geological Survey*.

Periodicals are also produced by independent research institutes, such as the Scott Polar Research Institute which publishes *Polar Record* (1931–). Educational establishments too produce many periodicals, e.g. *Bulletin of the Geological Institutions of the University of Uppsala* (1910–).

Most periodicals are issued with volume or annual indexes which are useful for locating particular items, by subject or by author, and can greatly ease the 'browsing' of a long run of volumes. In some cases indexes covering a much longer period have been produced. Examples of periodicals with cumulated indexes of this type include the *Journal of the Geological Society* (formerly the *Quarterly Journal . . .*) with indexes for volumes 1–50, 51–90 and 91–118, *Geological Magazine*, volumes 9–100 and *Proceedings of the Geologists' Association*, volumes 41–50, 51–60 and 61–70.

Before moving on from this general area of periodicals it is worth making some mention of the pattern of their use. Several citation studies have been carried out on the use of earth sciences literature, Gross and Woodford (1931), Craig, J E G (1969), Woodford (1969), Wood (1973) and Garfield (1974), and a number of generalizations can be made from their results. Periodicals are much the most heavily used type of literature; a relatively small number of periodicals account for most of the use; at least for British and US earth scientists, little use is made of foreign (particularly foreign language) material; and the useful life of earth sciences literature is generally longer than for other sciences. Craig, G Y (1969) gives estimates of the length of useful life of the literature of several geological subjects.

Geographic origin of periodicals

The world-wide nature of science and technology is reflected by the fact that scientific and technical periodicals are published in practically every country in the world. This is particularly true of sciences like geology, stratigraphy, palaeontology and meteorology for which there is a strongly regional flavour to much of the research that is carried out. Not surprisingly, however, relatively few countries are responsible for most of the periodicals that are produced. A recent paper by Carpenter and Narin (1980) gives an analysis of nearly 25,000 periodicals received by the British Library Lending Division during 1973. Each title was assigned to one of nine broad subject groups, one of which was 'earth and space science' and then analysed for country of origin. As one might expect, the United States produces the largest number of periodicals and, together with the USSR, the United Kingdom, Germany, France, Japan and Canada, is responsible for almost 60% of the titles in the earth and space sciences. A full country-by-country breakdown of the periodicals covered by the survey is given in Table 3.1 below. The dominance of western countries, although likely to continue for the foreseeable future, will probably decline as developing countries increase their output of scientific and technical literature.

Lists of periodicals

One of the best known lists of periodicals is *Ulrich's International Periodicals Directory* (annual, 2 vols, Bowker) the 1986–87 edition of which lists over 68,000 current periodicals arranged under 534 subject headings. For each entry it gives the title of the periodical, name and address of the publisher, frequency and Dewey Decimal classification number. Additional information, such as the ISSN (International Standard Serial Number), subscription price, language and year first published, is given if available. There is a companion volume to Ulrich's, entitled *Irregular Serials and Annuals: an international directory* (annual, Bowker). This publication covers items such as conference proceedings, annual reports, miscellaneous publications etc., in fact anything produced irregularly or less frequently than twice a year. The 1986–87,

Table 3.1. Number and % of periodicals in earth & space science, and all fields, from each country (after Carpenter and Narin, 1980)

Subject Country	Earth & Space Science Number	%	All Fields Number	%
United States	513	17.2	4,987	20.1
United Kingdom	229	7.7	3,136	12.6
Germany (East & West)	214	7.2	1,983	8.0
France	158	5.3	1,278	5.2
USSR	361	12.1	2,175	8.8
Japan	130	4.4	1,521	6.1
Australia	80	2.7	521	2.1
Canada	139	4.7	587	2.4
India	37	1.2	464	1.9
Israel	10	0.3	57	0.2
Italy	79	2.6	833	3.4
Sweden	52	1.7	420	1.7
New Zealand	26	0.9	138	0.6
South Africa	37	1.2	189	0.8
Rest of East Europe	218	7.3	1,740	7.0
Rest of West Europe	400	13.4	2,989	12.1
Rest of Asia	79	2.6	446	1.8
Rest of Africa	59	2.0	248	1.0
Rest of Near East & North Africa	20	0.7	105	0.4
Central & South America	141	4.7	984	4.0
	2,982		24,801	

edition contains some 35,500 entries under 466 subject headings. Between editions these two publications are updated by *The Bowker International Serials Database Update*, formerly *Ulrich's Quarterly* (quarterly, Bowker) which gives details of new titles, cessations, changes of title, etc. The same publisher also produces *Sources of Serials* (2nd edn, 1981) which lists approximately 100,000 titles, arranged first by country, then by publisher and/or corporate author: over 63,000 publishers and corporate authors are listed. This 'family' of publications from Bowker is also available as an online database through DIALOG Information Services, as a CD-ROM, and on microfiche. Although entitled simply *Ulrich's International Periodicals Directory* it includes *Irregular Serials and Annuals, Ulrich's Quarterly* and *Source of Serials* as well.

Other useful publications include *The Standard Periodical Directory* (8th edn, Oxbridge Communications Inc., 1985) which gives details of more than 60,000 titles, and nationally-based lists such as the *Directory of Japanese Scientific Periodicals* (revised edn, National Diet Library, 1979) which gives details of 8,901 titles. In the United Kingdom there is *Current British Journals: a bibliographical guide*, edited by D P Woodworth and C M Goodair (4th edn, British Library Document Supply Centre, 1986). This work is published in association with the UK Serials Group and was first issued in 1970 as *Guide to Current British Journals*. It is arranged in a classified order, using the Universal Decimal Classification and contains 7,499 entries.

Union lists, and lists of holdings issued by individual libraries, are particularly useful in that they not only give bibliographic information, but also provide a location for the periodical in question. One of the largest union lists, out-of-date but still useful for locating older titles, is the *World List of Scientific Periodicals* (4th edn, 3 vols, Butterworths, 1963–65) which gives details of about 60,000 scientific and technical periodicals published between 1900 and 1960. In addition to the title, publisher, start date, etc., it also gives the holdings of a large number of British libraries and the recommended abbreviation for the periodical. Similar to the *World List*, though of wider scope, is the *British Union-Catalogue of Periodicals* (*BUCOP*), published in four volumes by Butterworths between 1955 and 1958. *BUCOP* originally covered the period up to 1955 and a supplement was issued in 1962 covering the period up to 1960. Throughout the 1960s and 1970s further supplements and cumulations were issued bearing a secondary title 'New Periodical Titles'. Also during the 1960s *BUCOP* combined with the *World List* and adopted the sub-title 'Incorporating World List of Scientific Periodicals'. *BUCOP* has itself now been replaced by *Serials in the British Library: together with locations and holdings of other British and Irish libraries*. This is published quarterly by British Library Bibliographic Services and annual cumulations are issued on microfiche.

In North America there is the *Union List of Serials in Libraries of the United States and Canada* (2nd edn, 4 vols, H W Wilson, 1943). Supplements covering the period 1941–43 and 1944–49 were published in 1945 and 1953 respectively. Since then it has been updated by the Library of Congress publication *New Serial Titles*, issued monthly with annual, quinquennial and occasional larger cumulations. Also of value is the *Union List of Scientific Serials in Canadian Libraries* (8th edn, 2 vols, Canada Institute for Scientific and Technical Information, 1980) which lists some 48,000 titles together with the holdings of 250 libraries.

Among the lists of periodical holdings of individual libraries are: *Serial Publications in the British Museum (Natural History) Library* (3rd edn, 3 vols, British Museum (Natural History), 1980) which lists about 17,000 serials and includes abbreviations; the

Science Museum's *Periodicals on Open Access* issued three times a year on microfiche; and *Current Serials Received* (annual, British Library Document Supply Centre). The April 1985 edition of the last publication listed about 54,000 current titles and reported that a further 2,500 were on order. It is arranged alphabetically, with two supplementary sequences listing transliterated cyrillic titles and cover-to-cover translations. The British Library Document Supply Centre also produces *Keyword Index to Serial Titles (KIST)* on microfiche, with a replacement set being issued quarterly. *KIST* is a keyword-out-of-context listing, arranged alphabetically by significant title words and, unlike *Current Serials Received*, includes defunct as well as current titles. Originally *KIST* was restricted to the holdings of the British Library Document Supply Centre (formerly the British Library Lending Division) but it now includes all the periodicals held by the British Library Science Reference and Information Service (formerly the Science Reference Library). Records from other British Library departments are being added, as are the records from some other libraries, including the Science Museum Library and the Cambridge University Library. At the time of writing *KIST* contains details of over 300,000 titles, including cross references.

Subject oriented lists of periodicals and lists of periodicals held by specialized libraries can be particularly useful. Earth sciences publications of this type include: *Periodicals on Geology held by the Science Reference Library*, compiled by O Bradley and S Bird (British Library, Science Reference Library, 1978); the *List of Serials held in the Geological Survey of Canada Library: 1982*, compiled by W Stark and R Pleasant (Geological Survey of Canada, 1983) (*Geological Survey of Canada Paper* **83–17**); *Union Catalogue of Serials in the Geological Survey of India Libraries* (Indian National Scientific Documentation Centre, 1970); and the *List of Serial Publications held in the Library of the Geological Society, London* (Geological Society, 1978). The last publication lists some 2,700 and, in addition to the full title, gives abbreviations in accordance with British Standard BS 4148: 1985. Also still of use, although now rather out-of-date, is *Soupis Periodik Geologickych Ved,* edited by J Lomsky (Nakladatelstvi Ceskoslovenske Akademie Ved, 1959), which is particularly useful for Eastern European material.

Many of the producers of abstracting and indexing publications, or bibliographic databases, publish lists of source periodicals. Among these, and of particular interest to earth scientists, is the *GeoRef Serials List and KWOC Index* (annual, American Geological Institute) which is available on microfiche or on paper: the 1982 edition lists over 7,700 serials. A similar product was issued by Geosystems, producers of *Geotitles* and its associated database GeoArchive, entitled *GeoSources*.

As has been mentioned earlier in this chapter, the useful life of earth sciences literature is generally longer than for other sciences, therefore the most current union lists are not necessarily the only relevant source: the lists so far mentioned have mostly covered current or twentieth-century periodicals, and tracing earlier titles through these publications, especially of journals which have ceased publication, may be difficult. Two lists which are helpful in this respect are the *Catalogue of Scientific and Technical Periodicals 1665–1895*, by H C Bolton (2nd edn, Smithsonian Institution, 1897) (*Smithsonian Miscellaneous Collections*, Vol. **40**) and *Catalogue of Scientific Serials of all Countries . . . 1633–1876*, by S H Scudder (Harvard University, 1879). This last publication was reprinted in 1965 by Krauss Reprints.

As stated earlier, periodicals are the pre-eminent vehicle for scientific communication, preferred by both authors and readers alike. Despite their manifest success, perhaps to some extent because of it, periodicals are faced with a number of difficulties and are posing problems for their publishers, their authors and their readers. Doubts have been expressed recently as to whether their pre-eminent status can be maintained in the future.

The most serious problem facing periodicals concerns the proliferation of scientific literature; the number of articles produced each year is vast, 60,000 in the earth sciences alone would be a conservative estimate, and the total number of current periodicals is always increasing. This increase in the volume of literature being produced is in part the result of the increased number of scientists working today, a fact which is to be welcomed. Much less welcome is the well known 'publish or perish' syndrome whereby, in the interests of furthering their careers, some scientists feel compelled to publish the results of their research, even if they do not merit publication. Furthermore, it often appears that the quantity of articles, rather than their quality, is the criterion used to judge the abilities of an individual. Thus it is sometimes thought desirable, at least from the author's point of view, to publish the results of a piece of research in a series of papers rather than produce a single, well-rounded document. It follows from this that new information is more thinly spread and hence it is more difficult to locate.

This ever increasing volume of published literature has imposed a considerable strain on libraries, particularly on their finances, and over the last decade or so this has coincided with a reduction in real terms of the funds available to many libraries for the purchase of new stock. The same period has also seen a dramatic rise in the production costs of publishing, and periodical prices have outstripped the rate of inflation. The inevitable result has been a fall in the circulation of many periodicals which in turn has led to higher unit costs, increased prices, and yet fewer subscriptions. A further problem for libraries is that of storage and conservation. Many libraries are now over-crowded with books and periodicals and do not possess adequate shelf-space to house them, or sufficient funds for binding and preservation.

Another serious problem facing periodicals is the delay in publication. Formal publication of an article, involving as it does in most cases a strict refereeing process, is seldom rapid. Furthermore some of the major, prestigious periodicals have many more papers submitted to them than do those of a lesser status. In consequence these prestige periodicals sometimes develop a backlog of papers which have been accepted and are awaiting publication: in such circumstances delays are unavoidable.

Despite all this, scientific periodicals continue to exist in much the same form as they have done for over a hundred years. To some extent their problems have been eased by the introduction of new technology which has assisted with more rapid typesetting and printing, better retrieval through computerized databases, and more widespread use of microforms. However, various alternatives to periodicals have been suggested, and some tried, some of which are briefly described below:

(1) Synopsis journals — Under this system periodicals consist of synopses, of perhaps two pages in length, with full papers made available in one of a variety of ways, e.g. on microfiche, from the publisher, from the author, etc.

(2) Distribution of papers as separates — This system leaves the periodical article essentially unchanged, but alters the means of distribution. Instead of receiving a periodical with perhaps only one article of interest in each issue, a scientist obtains, and pays for, only those articles he or she requests.

(3) Depositories for additional material — Here lengthy compilations of statistical data or background information are omitted from articles, but are stored and made available from a central depository.

(4) Electronic journals — Still largely experimental, the electronic journal is currently the subject of much research but some publishers (e.g. McGraw-Hill) have launched lists of journal titles which are available in this form.

(5) Reports — These already form a substantial part of the scientific literature, but there is a suggestion that they should replace periodicals as the main source of primary information.

An interesting example of what happens when a conventionally produced journal attempts to find an alternative format is shown by the *Bulletin of the Geological Society of America* (1890–). This is a highly prestigious journal but, faced with increasing production costs and a backlog of papers awaiting publication, the Society decided, in the late 1970s, to publish the *Bulletin* in a two-stage format. Part 1 consisted of synopses; Part 2, on microfiche, contained full papers. The two stage *Bulletin* commenced in January 1979 but was not a success. Authors were reluctant to submit papers and as a result the *Bulletin* decreased in both size and quality. The experiment was abandoned in January 1982, when the *Bulletin* returned to its previous printed format.

None of the suggested alternatives has yet 'caught on'. Scientists are essentially conservative in their reading and writing habits and for the foreseeable future the continued dominance of periodicals seems assured. Even so it is hard to believe that some sort of electronic journal will not eventually replace today's conventionally-produced periodical.

Lastly in this section, there is an entirely different sort of problem associated with periodicals; it arises from the common practice of using abbreviated periodical titles in the list of references appended to articles. Some abbreviations are more or less self-evident, e.g. *Geol. Mag.* (*Geological Magazine*); others considerably less so, e.g. *Min. Metal.* (*Mineraçao e Metalurgia*). A great deal of time can be wasted, not only by readers searching a library catalogue, but also by diligent authors seeking the correct abbreviation — there are a number of different systems currently used, e.g. the short forms used in the *World List* and the recommendations given in British Standard BS4148: 1985. A few periodicals, for example, *Bulletin of the Geological Society of America* and *Cretaceous Research*, now insist on full titles being given in a list of references: it can only be hoped that this practice will become more widespread.

Reports

A second important source of primary information is reports, the largest type of non-conventional literature, which has come to be known as 'grey literature'. Typically reports are issued in series, invariably at irregular intervals, and they usually carry an alphanumeric identification code. These codes were originally devised and used for security purposes as a means by which a document could be cited or described without reference either to its title or its author. Reports are generally submitted directly to the person or organization which commissioned the research and they are often progress reports rather than reports of completed research. Many reports are therefore issued more for administrative reasons, to keep the funding body or department informed, rather than the desire to communicate any worthwhile scientific information.

The majority of reports are made available to the general public soon after they are produced. Some however, those which contain information not thought suitable for unlimited access, such as defence reports, are 'restricted'. These documents are usually referred to as classified reports and may be restricted for only a short time, or in some cases, almost indefinitely. Reports are generally cheaply produced and issued in unpublished or semi-published form. They are rarely printed and often consist of pages of duplicated typescript: many are available only on microfiche. Although they are produced and issued in most countries of the world, by far the greatest number emanate from the United States, principally from US government departments and agencies. The total number of reports issued each year is somewhat uncertain; estimates vary from 100,000 to around 500,000.

The fact that most reports are submitted direct, without any formal refereeing process, together with their typically 'rough and ready' format, means that they can be produced more quickly than would be the case if they were published in periodicals. This is the principal advantage of reports as a source of information. They are also usually much longer than periodical articles and contain somewhat more detailed information. However, reports are frequently criticized by both scientists and librarians: the major criticism being that many of them are of doubtful scientific merit. Without the formal refereeing process there is no adequate mechanism of quality control and it is frequently said of reports that they all too often contain faulty interpretation of data and poorly supported conclusions.

Reports have always posed problems for the librarian, particularly in their collection and bibliographic control. They are often issued out of numerical order (some report numbers never appear at all) and some are classified and therefore unavailable. Together these factors can make it difficult to establish whether all the available reports in a particular series are held or not. Another problem is the continuing practice of referring to a report by its code number alone. These codes rarely contain any element of classification and seldom give any clue to the likely subject matter. Furthermore, many reports have more than one number. For instance the numbers AD-A130 115/9, AFGL-TR-82–0396 and AFGL-ERP-816 all refer to the same document. A useful guide to report codes is the *Dictionary of Report Series Codes*, by L E Godfrey and H F Redman (2nd edn, Special Libraries Association, 1973) which lists over 20,000.

The production format of reports also attracts some criticism. Most are unattractive, an irritating though not an important problem, but occasionally some pages are almost unreadable, a feature which cannot be so easily ignored. Reports available only on microfiche do not enhance their appeal to users, who have a well-documented aversion to microforms of any kind.

Many countries have established clearing-houses with responsibility for collecting, issuing and providing information about reports. In the United States the main source of information about reports, and of reports themselves is the National Technical Information Service (NTIS), formerly the Clearinghouse for Federal Scientific and Technical Information. The NTIS collection now exceeds one million reports, and to this about 70,000 are added each year. NTIS publishes *Government Reports Announcements and Index (GRA&I)* which is the most comprehensive source of information about new reports and which combines the functions of two earlier publications, *Government Reports Announcements* and *Government Reports Index*. *GRA&I* is issued semi-monthly and is arranged by subject, using the COSATI (Committee on Scientific and Technical Information) classification. This has 22 broad subject categories divided up into 178 sub-categories. Category 4, Atmospheric Sciences and Category 8, Earth Sciences and Oceanography are the two categories most likely to be of interest to readers of this book. Each issue of *GRA&I* has five indexes: a keyword index; a personal author index; a corporate author index; a contract/grant number index; and a NTIS order/report number index. Copies of reports listed in *GRA&I* are available from NTIS, either as hard copy or on microfiche.

A second major source of information about reports is *Scientific and Technical Aerospace Reports (STAR)*, issued semi-monthly by the National Aeronautics and Space Administration (NASA). *STAR* has a world-wide coverage of all report series of interest in the field of space research and in addition it also includes NASA – owned patents, translations and dissertations. It is arranged under 10 major subject divisions divided into 74 specific subject categories. Individual issues have subject, personal author, corporate source, contract number and report/accession number indexes: cumulated indexes are issued annually.

Also of value is *Energy Research Abstracts (ERA)* published semi-monthly by the United States Department of Energy. *ERA* is a comprehensive abstract journal covering reports issued worldwide. It also covers all literature, including periodical articles, patents, theses, etc., originated by the US Department of Energy. It is arranged by subject with 40 subject categories and 294 sub-categories.

In the United Kingdom the principal collection of report literature is held by the British Library Document Supply Centre (BLDSC). BLDSC, like its predecessors the British Library Lending Division and National Lending Library of Science and Technology, attempts to collect report literature comprehensively and, in addition to British reports, acquires all NTIS products on microfiche. Information about new British reports is given in their monthly publication, *British Reports, Translations and Theses* (*BRTT*), formerly *BLLD Announcement Bulletin*. *BRTT* is arranged by subject using modified COSATI categories. Category 04, Atmospheric Sciences is incorporated into Category 08, Earth Sciences and Oceanography and there are no sub-categories. A keyterm index is included with each monthly part and cumulated author and keyterm indexes are provided on microfiche at quarterly intervals. Until recently a second major UK collection of reports was held by the Department of Industry's Technology Reports Centre (TRC). However in 1982 TRC's functions (and its collections) were transferred to BLDSC and its publication *R&D Abstracts* was discontinued.

Further information about reports and guides to report literature can be found in *Use of Reports Literature*, edited by C P Auger (Butterworths, 1976).

Theses

Another useful source of primary information is theses, sometimes called dissertations, which are usually produced in partial fulfilment of the requirements for higher degrees, such as PhDs. Theses are researched and written-up under the supervision of a member of the academic staff of the degree awarding institution and before acceptance they are 'refereed' by an external examiner, who is a recognized authority in the subject field of the thesis, and the candidate is examined on his thesis by both an internal and an external examiner. A thesis is generally required to show evidence of some original research and should be 'worthy of publication'. A large proportion of the theses that are produced each year subsequently form the basis of one or more periodical articles: these are often co-authored by the candidate who presented the thesis and his or her supervisor. The process of formal publication in a periodical, however, can take considerable time and it is not uncommon for an article to appear well over a year after the presentation of the thesis on which it is based. This fact serves to emphasise the timeliness, and hence the value of, thesis literature as a source of new information. Another valuable characteristic of theses is that they are generally required to contain a detailed literature survey. These literature surveys can be extremely useful as bibliographies, particularly when a scientist is starting research in a hitherto unfamiliar field.

In the United Kingdom candidates for higher degrees are usually required to submit two or three typewritten copies of their theses. After acceptance, copies are usually housed in both the main library of the degree-awarding body and in the department where the research was undertaken. Other countries, however, have adopted different practices and in some parts of the world candidates are expected to have theses printed and published.

The quality of theses is very variable and depends not only on the abilities of the candidate but also on those of his or her supervisor and the external examiner. However, the main problems associated with the use of thesis literature as a source of primary information are those of discovery and availability. The difficulties involved, first in finding out about a potentially useful thesis and then in obtaining a copy, have not encouraged their use. Many scientists are of the opinion that in view of their variable quality and the problems encountered in obtaining them, theses are not worth the effort.

Recent years have seen an enormous improvement in guides to thesis literature and many degree-awarding institutions have now relaxed their restrictions on making theses available to outside individuals and organizations. In some cases, however, it may still be necessary to get permission from the department before one can obtain a copy. Recent years have also seen an enormous growth in the number of theses that are produced annually and this fact, together with their timeliness compared with more formally produced periodical articles, make it impossible to ignore them as a source of primary information.

The most comprehensive source of information about theses is probably *Dissertation Abstracts International*, published since 1969 by University Microfilms International (UMI). UMI is a commercial organization which has an agreement with most United States and Canadian universities whereby it microfilms, and makes available, the theses that they produce. Until 1976 *Dissertation Abstracts International* was issued in two parts: *Section A, The Humanities* and *Section B, The Sciences and Engineering*, both of which appear monthly. In 1976 a third part, *Section C, European Abstracts* (quarterly), was added, reflecting UMI's increasing interest in European theses. Each of the three sections is arranged by subject with author and keyword indexes. Other US sources of information about theses include: *American Doctoral Dissertations*, (annual, University Microfilms International); *Masters Abstracts: a catalog of selected Masters theses on microfilm*, (quarterly, University Microfilms International); and *Masters Theses in Pure and Applied Sciences by Colleges and Universities of the United States and Canada*, (annual, Plenum Press). The 1982 edition of this last publication lists over 11,000 theses from 242 contributing organizations.

Details of older US theses can be traced using *Comprehensive Dissertation Index, 1861–1972*, which was published in 1973 by Xerox University Microfilms (now University Microfilms International). This work consists of 37 volumes, 33 of which are devoted to particular subjects, with a four-volume author index. Volume 16 covers the earth sciences, and the whole work contains details of more than 417,000 dissertations. Supplements, consisting of five volumes each and listing about 35,000 theses, are issued annually. There is also a ten-year supplement covering the period 1973–1982 in 38 volumes.

UMI's files are now computerized and their Comprehensive Dissertation Index database can be searched online through host systems, e.g. DIALOG Information Services.

British universities and other degree-awarding institutions have, until recently, been somewhat more reluctant to make their theses generally available than their overseas counterparts. Since 1970 the British Library Document Supply Centre (BLDSC) has had a policy of collecting as many British theses as possible, by borrowing and microfilming them. After some initial reluctance, most universities and the Council for National Academic Awards now participate in this scheme. Recent additions to the BLDSC collection are listed in their monthly publication *British Reports, Translations and Theses (BRTT)*, to which reference has already been made on page 29.

Another British organization which actively promotes the use of theses as sources of information is Aslib (formerly the Association of Special Libraries and Information Bureaux). Aslib gives details of new British theses in its publication *Index to Theses accepted for Higher Degrees by the Universities of Great Britain and Ireland and the Council for National Academic Awards*, which was first published in 1950. From volume **35** (1986) abstracts are included and publication is quarterly.

Older British theses can be traced using the *Retrospective Index to Theses of Great Britain and Ireland, 1716–1950*, edited by R R Bilboul and F L Kent (ABC Clio, 1976). This work consists of five volumes: Volume **1**, *Social Sciences and Humanities*; Volume **2**, *Applied Sciences and Technology*; Volume **3**, *Life Sciences*; Volume **4**, *Physical Sciences*; and Volume **5**, *Chemical Sciences*.

In addition to the aforementioned publications which list theses on all subjects and from a number of degree-awarding institutions, there are also lists from particular institutions, and others devoted to particular subjects. Examples of the former type include the University of London publication *Theses and Dissertations accepted for the Degrees of M.Phil. and Ph.D.*, and *Titles of Dissertations approved for Ph.D., M.Sc. and M.Litt. Degrees in the University of Cambridge during the Academic Year . . .* , issued annually. The last printed edition of the former was for 1979–80; a microfiche version covers 1981–85.

Of more specific interest to earth scientists are the subject oriented lists of theses, such as the *Bibliography of Theses written for Advanced Degrees in Geology and Related Sciences at Universities and Colleges in the United States and Canada through 1957*, compiled by B J Chronic and H Chronic (Pruett Press, 1958). This publication gives details of over 11,000 theses and has been supplemented by a series of guides: *Bibliography of Theses in Geology, 1958–1963*, compiled by B J Chronic and H Chronic (American Geological Institute, 1965); *Bibliography of Theses in Geology, 1964*, by D C Ward (*Geoscience Abstracts*, Vol.7, no.12, pt 1, 1965); *Bibliography of Theses in Geology, 1965–1966*, edited by D C Ward and T C O'Callaghan (American Geological Institute, 1969); and *Bibliography of Theses in Geology, 1967–1970*, edited by D C Ward (*Geological Society of America Special Paper* **143**, 1973). Since 1971 North American theses on geology have been covered by the *Bibliography and Index of Geology*, published monthly by the American Geological Institute, and the corresponding computerized database GeoRef.

Other specialized bibliographies of earth science theses include: *Titles of Research Theses, 1960–1975: geology of the British Isles and offshore areas*, compiled by A V Hodgson and D J C Laming (2nd edn, Bibliographic Press, 1976) which lists about 1,350 titles, and *Theses on Scottish Geology 1960–68*, by W D I Rolfe (*Scottish Journal of Geology*, Vol.6, no.4, 1970). This latter publication is regularly updated in the same journal. Occasionally individual universities issue lists of theses produced on specific subjects. An example is the *Bibliography of Geology Theses and Dissertations at Syracuse University 1879–1979*, compiled by D F Merriam (*Syracuse University Geology Contribution*, **7**, 1980).

Details of other bibliographies of theses can be found in *A Guide to Theses and Dissertations: an annotated international bibliography of bibliographies*, by M M Reynolds (Gale, 1975), which contains details of over 2,000 such publications. Bibliographies of theses can also be traced using abstracting and indexing services such as the *Bibliography and Index of Geology* (monthly, American Geological Institute) and *Geotitles* (monthly, Geosystems) or their associated databases, GeoRef and GeoArchive respectively.

Conference publications

Conferences, known variously as symposia, colloquia, congresses, seminars and meetings are a long established means by which scientists can meet and exchange information on topics of mutual and current interest. Furthermore, through conference publications this information can also be communicated to scientists who did not attend. Conferences are usually organized by learned or professional societies; very often more than one society is involved. In size they range from small local meetings to huge international gatherings: in subject scope they may deal only with a small specialized topic or they may cover all aspects of a particular subject.

Critics of conferences frequently point out that much of the information presented by speakers is already available in the literature and that some papers are of a low calibre: nevertheless there are good papers published in conference proceedings and the problem is to identify these. Other criticisms of conferences concern the standard of presentation of the papers and the practical organization of the conference. Nevertheless, in many cases they do report the results of original research for the first time. They are unquestionably popular with participants and the number of such gatherings increases each year: the chief benefit to be derived from attending a conference is thought by many to be in the informal contacts made, rather than in the formally presented papers, which may be of low calibre, report nothing new, or announce results of research prematurely.

Most conferences generate at least some literature and this can be divided into two main types: pre-conference literature and post-conference literature. Pre-conference literature typically consists of a detailed programme of the sessions together with abstracts of the papers which are to be presented. In many cases it also includes brief biographical details of the speakers and a complete list of all the conference attendees. It is rarely published in the conventional sense: usually it is produced in duplicated form and in most cases it is distributed only to registered participants of the conference. Collection and bibliographic control of pre-conference literature is fraught with difficulties. A great deal of it is of an essentially ephemeral nature; however its ephemeral nature does not stop it being cited (or asked for) and some of it can be valuable.

Pre-conference literature is most easily located by using one of the published lists of forthcoming conferences, e.g. *Forthcoming International Scientific and Technical Conferences* (quarterly, Aslib); *World Meetings* (quarterly, Macmillan) and then by writing to the organizers direct. Many scientific journals also publish information about forthcoming conferences. In the earth sciences field, for instance, lists of forthcoming conferences are regularly featured in *Geotimes* (monthly, American Geological Institute) and *British Geologist* (quarterly, Institution of Professional Geologists).

Post-conference literature, generally referred to as conference proceedings, is usually published either in the form of individual books or within established periodicals. Sometimes however, conference proceedings become periodicals in themselves, particularly where the conference is numbered or held at regular, say yearly, intervals. For details of conference proceedings probably the best guide is *Index of Conference Proceedings Received*, published monthly by the British Library Document Supply Centre, with annual, 5-year and 10-year cumulations. There is also an 18-year cumulation on microfiche which covers the period 1964–81. *Index of Conference Proceedings Received* is particularly useful in that it not only gives bibliographic details but also provides locations for the item in question. It is also available online as Conference Proceedings Index, through BLAISE-LINE. Other guides include Interdok's *Directory of Published Proceedings*, issued monthly with annual cumulations, and *Proceedings in Print*, issued bi-monthly and also with annual cumulations.

Conference proceedings which are published in the form of books can be traced through general bibliographic tools such as the *British National Bibliography* (weekly, British Library Bibliographic Services) and subject oriented publications, e.g. *Books in the Earth Sciences and related topics* (quarterly, Bibliographic Press Ltd) which has a separate index for conference proceedings.

The guides so far mentioned give details of conference proceedings, but it is often useful to be able to trace individual conference papers. Most subject-based abstracting and indexing services and their related databases attempt to cover conference papers, but there are also publications devoted solely to this type of literature. One such guide is *Conference Papers Index*, formerly *Current Programs*, which is published monthly by Cambridge Scientific

Abstracts and includes about 100,000 papers annually. It should be noted however that this is a list of papers presented, rather than papers published. For published papers there is the *Index to Scientific and Technical Proceedings* (*ISTP*), issued monthly with annual cumulations by the Institute for Scientific Information. First published in 1978, *ISTP* indexes nearly 90,000 items annually, taken from over 3,000 proceedings which, according to the publishers, represents almost half of all proceedings published.

Foreign language literature and translations

A survey by Hawkes (1967) gives a language breakdown of the 1961 primary literature of the earth sciences. The survey included over 30,000 items and the results showed that 27% of the articles were English, while Russian, the largest single language represented, accounted for some 30%. A more recent survey by Wood (1979), covering the 1977 literature, suggests that English, with 55.5%, has now overtaken Russian, with 23.5%, as the dominant language. Further evidence to this effect has been provided in a survey by Connor and Manheim (1982). Their survey, also of the 1977 literature, included nearly 90,000 items and estimated that of these 48% were in English and 41.2% were in Russian. A comparison of the results of the three above-mentioned surveys is given in Table 3.2. Further evidence of the increasing use of English is provided by the fact that many periodicals which were hitherto published in other languages now state that the 'preferred' language is English. In some cases even the title has changed, e.g. *Contributions to Mineralogy and Petrology* (monthly, Springer) which was formerly published under the title *Beitrage zur Mineralogie und Petrographie*. Despite this trend towards publication in English there still exists a considerable

Table 3.2. Breakdown by language of earth sciences literature (after Hawkes, 1967; Wood, 1979*; Connor and Manheim, 1982**)

Year	1961	1977*	1977**
Language			
English	27.0%	55.5%	48.0%
Russian	30.0%	23.5%	41.2%
French	11.0%	8.5%	3.0%
German	11.0%	5.0%	3.0%
Other languages	21.0%	7.3%	5.0%
	100.0%	99.8%	100.2%

volume of literature, amounting to tens of thousands of articles each year, which is published in other languages. Wood (1979) has drawn attention to the apparent low level of use of foreign language material and suggests this can be attributed to three main causes: 1) much geological information is of purely local interest, 2) many earth scientists are unfamiliar with the bibliographic tools that are available and simply fail to discover the existence of foreign language publications of potential interest, 3) many earth scientists cannot cope with foreign language literature and when faced with the 'language barrier' simply give up. Ellen (1979) reports the results of a questionnaire survey of the foreign language problems facing British scientists. The author herself describes the results of the survey as 'rather depressing'. They show that whereas most scientists felt that they could cope with French language literature, few of them could cope with German, and fewer still with Russian. The survey also shows that a high percentage of respondents had been faced with a foreign language problem within the month prior to receiving the questionnaire.

One attempt to minimise the problem of literature written in a foreign language is the 'journal in translation'. There are three main types:

(1) Cover-to-cover translations, which give a full translation of all the articles appearing in a particular foreign language periodical.
(2) Selective single-source translations, which give full or partial translations, from a single foreign language periodical, of those articles which are considered worthwhile.
(3) Selective multi-source translations, which contain full or partial translations of articles from two or more foreign language periodicals.

Some publishers specialize in publishing translation journals. Two of the biggest of these are Consultants Bureau, a subsidiary of Plenum Press, which publishes around 100 titles, and Allerton Press which has around 40. Not surprisingly, in view of the intellectual effort involved, many translation journals have very high subscription prices, often amounting to many times the cost of the original periodical. Most of the translations journals that are currently available translate from Russian into English. However there are some which translate from Russian into other languages, and from other languages into English.

One of the most useful guides to this type of periodical is *Journals in Translation* (4th edn, British Library Document Supply Centre/International Translations Centre, 1988), which gives bibliographic details of over 1,000 single- and multi-source translations journals. It is arranged alphabetically by translated title, has keyword and original title indexes and includes both current and dead titles. Cover-to-cover translations of cyrillic serials are also listed, in a separate sequence, in *Current Serials Received*, the April 1986 edition of which lists nearly 200 titles. Another useful publication is *Journals with Translations held by the Science Reference Library*, by B A Alexander, published in 1985, which contains over 500 entries and includes both live and dead titles. It merges and updates two earlier publications from the Science Reference Library (now the British Library Science Reference and Information Service): *Holdings of Journals in Translation (Aids to Readers*; no. **1**, rev. edn, 1981) and *Translations Series (Aids to Readers*; no. **19**, rev. edn, 1982). A list of cover-to-cover translations of earth sciences periodicals is given in the paper by Wood (1979).

Over the last twenty years or so the number of translation journals has steadily increased. However they are able to cover only a relatively small percentage of the total number of non-English language publications that are produced each year and many articles are translated individually. Most of these individual translations are either unpublished or semi-published; some are issued within a report series. Finding out about them and then locating a copy can be difficult and many countries have 'clearing-houses' responsible for collecting and disseminating information about such translations.

On an international level there is the International Translations Centre, formerly the European Translations Centre, which is based at Delft in the Netherlands. It publishes, jointly with the Centre National de la Recherche Scientifique, *World Transindex*, which commenced in 1978 and replaced three earlier publications, namely *World Index to Scientific Translations*; *Translations Bulletin*; and *Bulletin des Traductions*. *World Transindex* is arranged under COSATI subject headings and has annual indexes. It is also available as an online database through the ESA-IRS host system. With effect from 1987 this publication became *World Translations Index* both as a printed bibliography and an online database.

The *World Translations Index* has subsumed the *Translations Register-Index* which was published from 1967 in the US by the National Translations Center, based, until its demise in 1988, at the John Crerar Library in Chicago. Older translations can be traced by using the *Consolidated Index of Translations into English* (Special Libraries Association, 1969) which covers the period from 1953 to 1966.

In the United Kingdom the British Library Document Supply

Centre has a long-standing policy of collecting amd making available individual translations. At the time of writing nearly 500,000 are held, with an annual intake of around 11,000. Details of newly acquired translations are given in the monthly publication *British Reports, Translations and Theses (BRTT)*, described earlier in this chapter.

As stated in an earlier section of this chapter, many translations, particularly those which are produced by or for government departments are issued in a report series. These can be traced using the various guides that are available, e.g. *Government Reports Announcements and Index*, or by using one of the subject-based secondary services such as the *Bibliography and Index of Geology*.

If it has not been possible to locate an existing translation through the various sources of information indicated above, it may be thought necessary to have a translation specially prepared. In certain cases it may be possible to arrange an informal translation, by a university colleague in a language department for instance. However, such translations may be of doubtful scientific merit and a full, professional translation is sometimes required. Not surprisingly, having a translation specially prepared is usually an expensive process. To assist in finding a translator there are a number of directories available: one example is the *Index of Members of the Translators' Guild*, published in loose-leaf format, which in addition to the main listing has subject and less-common languages indexes and a geographical listing of members. Another example is *Translation and Translators: an international directory and guide*, by S Congrat-Butler (Bowker, 1979). As part of its services to its members Aslib maintains an 'Index of Approved Translators'.

Some libraries and institutions offer a translation service whereby a member of staff, competent in the language concerned, will give an oral rendering in English of the main points in an article in a foreign language journal. An example of this is the Linguistic Aid Service offered at the British Library Science Reference and Information Service.

References

Carpenter, M P and Narin, F (1980) The subject composition of the world's scientific journals. *Scientometrics*, **2**, (1), 53–63

Connor, M and Manheim, F T (1982) World geoscience literature and translation. *in*: Kidd C M (ed) *Second International Conference on Geological Information, May 23–27, Colorado School of Mines, Golden, Colorado. Proceedings*: Volume 2 404–409. (*Oklahoma Geological Survey, Special Publication* **82–4**)

Craig, G Y (1969) Communication in geology. *Scottish Journal of Geology*, **5**,(4), 305–321

Craig, J E G (1969) Characteristics of the use of geology literature. *College and Research Libraries*, **30**,(3), 230–236

Ellen, S R (1979) Survey of foreign language problems facing the research worker. *Interlending Review*, **7**,(2), 31–41

Garfield, E (1974) Journal citation studies. X. Geology and geophysics. *Current Contents*, **30**, July 24 1974, 5–9

Gross, P L K and Woodford, A O (1931) Serial literature used by American geologists. *Science*, **73**, 660–664

Hawkes, H G (1967) Geology. *Library Trends*, **15**,(4), 816–828

Wood, D N (1973) Primary literature. *in*: Wood, D N (ed). *Use of Earth Sciences Literature*, 17–44. London: Butterworths

Wood, D N (1979) Earth science and the foreign language problem. *in*: Harvey, A P and Diment, J A (eds) *Geoscience Information; a state-of-the-art review*, 226–246. (Proceedings of the First International Conference on Geological Information, London, April 10–12 1978)

Woodford, A O (1969) Serial literature used by American geologists, 1967. *Journal of Geological Education*, **17**,(3), 87–90

CHAPTER FOUR

Secondary literature: reference and review publications

JOHN HOPSON

The primary literature is the medium for reporting results of original research and by definition it should contain new information which has yet to be absorbed into the body of scientific knowledge. However, by its very nature, the primary literature is a disconnected and unorganized collection of information. Locating specific facts can be a difficult, time-consuming, and often frustrating, process. There has, therefore, grown up a body of secondary literature. Secondary publications are characterized by the fact that they contain no new information. They are compiled from primary sources and their purpose is to organize and rearrange the information into a more easily accessible form.

This chapter introduces some types of secondary literature, including encyclopedias, dictionaries, handbooks and directories. Examples of each type are given and, where appropriate, sources of further information are indicated. The subject chapters which appear later in the book will mention those secondary publications which are of particular importance to the various branches of the earth sciences. In addition readers are also referred to three basic sources of information on reference publications. The first two are general in coverage: *Guide to Reference Books*, edited by E P Sheehy (10th edn, American Library Association, 1986) and *Walford's Guide to Reference Material*, Volume 1, *Science and Technology*, edited by A J Walford (4th edn, Library Association, 1980). The third is of particular interest to geologists, *Geologic Reference Sources*, edited by D C Ward *et al* (2nd edn, Scarecrow Press, 1981). It is arranged into general, subject, and regional sections and lists over 4,000 items. According to the preface it is "intended mainly as a 'ready-reference' guide; therefore annotations are minimal except in the general section where some of the principal information sources are discussed".

Encyclopedias

Encyclopedias are probably the most familiar and most commonly used of all the reference books that are available. Their purpose is to summarize existing knowledge and to present it to the user in a readily-accessible form. Primarily a tool of first resort, encyclopedias are most frequently used to provide introductory information, particularly when the subject is an unfamiliar one.

In subject scope encyclopedias range from very general works which attempt to cover all human knowledge, to narrowly specialized publications confined to a single subject area. In size they range from large multi-volume works to comparatively slim, single volumes. Most encyclopedias employ an alphabetical arrangement of subject entries, usually accompanied by a detailed index. Some, however, use a systematic arrangement with the information organized under a number of broad subject headings, rather like a text-book or treatise. The publishers of these systematic encyclopedias claim that theirs is the more logical approach, since similar subjects are grouped together and not arbitrarily separated as is the case in those that are arranged alphabetically. The index is particularly important in those encyclopedias which adopt a systematic arrangement: without a good index much of the information can be almost irretrievably lost and information, no matter how accurate and comprehensive, is of no value whatever if it cannot be found.

Although the compilation of an encyclopedia is usually controlled by an editorial panel, individual contributions are made by specialists recognized as experts in their own particular field. In many cases these individual contributions are signed and in most of the better encyclopedias they are usually accompanied by a list of references, to guide the non-specialist in the subject to more in-depth information.

Encyclopedias, particularly those which aim to cover a wide range of subjects, are sometimes criticized by non-contributing specialists who claim that topics within their field of interest have been over-simplified. This is not, perhaps, altogether untrue but neither is it entirely fair. A contribution written by one expert in a particular field is not really meant for another expert in the same field, who would anyway already possess the information. Rather, it is written for someone who is searching for information peripheral to, or outside, his own speciality, and bearing in mind the fact that the information is, of necessity, highly condensed, some superficiality is to be expected.

A more serious problem for encyclopedias, and one where criticism is perhaps more valid, is that of currency. Knowledge, particularly scientific knowledge, does not stand still so an encyclopedia is, to a greater or lesser extent, out-of-date even before it is published. Thereafter the information 'decays' until eventually it ceases to be of value other than to historians of the subject. The rate at which this decay occurs is largely dependent on the subject. It is likely, for instance, that an encyclopedia of computer science will become out-dated more quickly than an encyclopedia of palaeontology. Even within individual specialized encyclopedias the rate of decay will not be constant; an encyclopedia of geology published in the late 1950s might still provide useful information on some subjects, but it would be distinctly unreliable on others, such as plate tectonics or lunar geology. One way for publishers to ensure that the information contained in an encyclopedia is as up-to-date as possible is the rapid publication of a completely new edition. However, compilation of an encyclopedia is a lengthy process and the production of a new edition every year or so is not a practical proposition, particularly in the case of large multi-volume works. In an effort to circumvent this problem the publishers of some of the larger encyclopedias have adopted a policy of continuous revision. Under this system, those entries in most need of revision are rewritten and, with reprinting taking place every one or two years, new information can be incorporated into the encyclopedia. An alternative method for keeping up-to-date is for publishers to issue supplements which describe the progress made in various subjects since the publication of the current edition.

There are many general encyclopedias covering all subjects but the best known, and probably the most widely available, is *The New Encyclopaedia Britannica*, published by Encyclopaedia Britannica Inc. First published as a 3-volume work in 1768, *Britannica* now consists of 32 volumes, divided into 4 sections. Volumes 1–12 form the "Micropaedia" which contains alphabetically arranged short entries intended as a 'ready reference'. The "Macropaedia", volumes 13–29, is also arranged alphabetically but has far fewer, though very much longer, articles for 'knowledge in depth'. The other 3 volumes are made up of the "Propaedia" (one volume), which contains the 'outline of knowledge and guide to *Britannica*', and a 2-volume index of about 400,000 entries. *Britannica* is updated by means of continuous revision and reprinting within the edition: the current 32-volume format dates from the 1985 revision of the 15th edition, first issued in 1974. (A lengthy review of this revision appeared in *Reference Books Bulletin 1985/6*, published by the American Library Association in 1986). *Britannica* is also supplemented by *Britannica Book of the Year* and the *Yearbook of Science and the Future*, both published by Encyclopaedia Britannica Inc. Further details of *Britannica* and other general encyclopedias can be found in *Encyclopedia Buying Guide: a consumer guide to general encyclopedias*, by K F Lister (3rd edn, Bowker, 1981) which, in over 500 pages, evaluates all the non-specialized encyclopedias available in the United States. Another guide to general encyclopedias is *Encyclopedia Ratings*, by J P Walsh (6th edn, Reference Books Research Publications, 1983), a fold-out sheet which describes 19 general encyclopedias.

Of the English language encyclopedias which cover the whole range of science and technology the *McGraw-Hill Encyclopedia of Science and Technology* (6th edn, McGraw-Hill, 1987) is by some considerable margin the largest and most comprehensive. It consists of 20 volumes, one of which is an index, and contains some 7,700 entries from about 4,000 contributors. First published in 1960 new editions have appeared every 6 or 7 years, and since 1962 it has been updated annually by means of the *McGraw-Hill Yearbook of Science and Technology*. This, like the encyclopedia, employs an alphabetical arrangement of entries and, in addition, each volume contains a number of review articles on topics of current interest.

Also covering science and technology as a whole, and in some depth, is *Van Nostrand's Scientific Encyclopedia*, edited by D M Considine (6th edn, Van Nostrand Reinhold, 1983) which is available either as a single volume, or as a two-volume set. There are 7,300 entries from 200 authors and there are nearly 9,600 cross-references, which, according to the preface, 'make the encyclopedia virtually self-indexing': there is no separate index. First published in 1938, the first four editions had a serious weakness in that they contained no bibliographies. With the appearance of the 5th edition in 1976, this defect was rectified and there are now several thousand references listed throughout the encyclopedia. Perhaps the only valid criticism which can be levelled at these two general scientific encyclopedias is that they concentrate too much on US information and do not fully reflect the international nature of science and technology. Nevertheless, both of them contain a wealth of information which is both reliable and authoritative, and either can be consulted with confidence.

On a somewhat smaller scale, and at a correspondingly lower price, is the *Phaidon Encyclopedia of Science and Technology*, edited by J D Yule (Phaidon, 1978) which contains some 5,500 entries in nearly 600 pages. This encyclopedia is directed primarily towards the layman and one-third of the space is devoted to diagrams and photographs, mainly in colour. According to the preface an additional feature is that "there are more than 1,000 brief biographical notices of the men and women who have contributed most to the development of modern science and technology". Earth scientists represented include James Hutton, Sir Charles Lyell and William Smith. Also of interest is the *McGraw-Hill Concise Encyclopedia of Science and Technology*, editor-in-chief S P Parker (McGraw-Hill, 1984), which has over 7,000 text entries and a 30,000 entry index.

A more specialized encyclopedia, covering the earth sciences, is the *McGraw-Hill Encyclopedia of the Geological Sciences*, editor-in-chief D N Lapedes (McGraw-Hill, 1978). According to the publishers the articles "either have been taken from the *McGraw-Hill Encyclopedia of Science and Technology* or have been written especially for this volume". There are 560 entries and the arrangement is alphabetical, unlike *The Cambridge Encyclopedia of Earth Sciences*, edited by D G Smith (Cambridge University Press, 1982), which adopts a systematic arrangement with the information divided into 27 sections. Two other encyclopedias which use systematic rather than alphabetical arrangement are: *Planet Earth: an encyclopedia of geology*, edited by A Hallam (Elsevier-Phaidon, 1977) and *The New Larousse Encyclopedia of the Earth*, by L Bertin (Hamlyn, 1972).

A series of earth science encyclopedias is being produced under the general heading *The Encyclopedia of Earth Sciences Series*, originally published by Reinhold and now published by Hutchinson Ross. This is a series of single volume, autonomous encyclopedias, each of which deals with a separate branch of the earth sciences. Originally eight volumes were planned but the project has expanded considerably since the publication of Volume **I**, *The Encyclopedia of Oceanography*, edited by R W Fairbridge in 1966, so that now 24 volumes are scheduled. Each volume is arranged alphabetically with a separate index and has cross-references not only within the volume but to entries in other volumes as well. One of the most recent, Volume **IVB**, *The Encyclopedia of Mineralogy*, edited by K Frye and published in 1981, has, in addition to the main alphabetical sequence, a mineral glossary which forms a second, also alphabetical, sequence of roughly 3,000 entries in a little over 200 pages. Another feature of this encyclopedia is that the preface contains a section on the literature of mineralogy and provides a list of nearly 200 journals likely to be of interest to the mineralogist. A list of all the volumes published to date in *The Encyclopedia of Earth Sciences Series* is given as Appendix 1 of this chapter.

The branch of the earth sciences which probably has the largest number of specialized encyclopedias is that which deals with minerals and gemstones. Many of these volumes are beautifully illustrated and some concentrate on the photogenic aspects of the subject almost to the exclusion of written information. Two of the more substantial works are: *The Encyclopedia of Minerals*, by W L Roberts *et al* (Van Nostrand Reinhold, 1974) and *The Encyclopedia of Minerals and Gemstones*, edited by M O'Donoghue (Orbis, 1976). The former is arranged alphabetically and for each mineral gives a 'best reference in English': the latter employs a systematic arrangement.

Three encyclopedias with a slightly unusual subject coverage are: *Standard Encyclopedia of the World's Mountains*, edited by A J Huxley (Weidenfeld and Nicolson, 1962); the *Standard Encyclopedia of the World's Oceans and Islands*, edited by A J Huxley (Weidenfeld and Nicolson, 1962); and the *Standard Encyclopedia of the World's Rivers and Lakes*, edited by R K Gresswell and A J Huxley (Weidenfeld and Nicolson, 1965). Each has a main A-Z sequence of entries on the world's major rivers, lakes, mountains, etc., and a gazetteer section which gives briefer details for the remainder.

Other encyclopedias of interest to earth scientists include: the *McGraw-Hill Encyclopedia of Ocean and Atmospheric Sciences*, editor-in-chief S P Parker (McGraw-Hill, 1980); *The Encyclopedia of Prehistoric Life*, edited by R Steel and A P Harvey (Mitchell Beazley, 1979); and *Grzimek's Encyclopedia of Evolution*, edited by G Heberer and H Wendt (Van Nostrand Reinhold, 1976). This last work is a translation of the original German language edition, published in Zurich in 1972. The information is organized in 23 sections or chapters with headings such as: 'The Origin of Life'; 'Invertebrate Evolution in the Paleozoic Era'; and 'The Tertiary Period — The Age of Mammals'.

Lastly, still of value although now perhaps a little out-of-date, is the *International Dictionary of Geophysics*, edited by S K Runcorn *et al* (2 vols + folder of maps, Pergamon Press, 1967). Despite its misleading title this publication is unquestionably an encyclopedia and has articles up to 10 pages in length, in many cases accompanied by lengthy bibliographies.

Dictionaries

Monolingual dictionaries

A monolingual dictionary is essentially an alphabetical list of terms each of which is defined. The definitions are often supplemented by additional information, for instance, pronunciation, origin, usage, synonyms and variant spellings. Frequently a dictionary will also contain an appendix, or set of appendices, giving information not considered appropriate to the main body of entries. Commonly these appendices take the form of 'useful data' such as conversion tables, chemical tables, physical constants and mathematical symbols.

There are a number of similarities between dictionaries and encyclopedias. Both range between the very general and the highly specialized, both can be large multi-volume works or small single volumes and both, typically, employ an alphabetical arrangement. Furthermore, just as human knowledge is always expanding, necessitating frequent editions of encyclopedias, so the vocabulary expands in order to accommodate new developments and dictionaries too require regular updating. It is not always clear whether a particular publication is a dictionary or should more properly be termed an encyclopedia. Certainly the title of a publication does not always accurately reflect its contents, as we have seen in the case of the *International Dictionary of Geophysics* (see above), which is without doubt an encyclopedia. As a general guide a dictionary's prime concern is the terminology and its main function is to define the terms rather than to explain them.

Large general dictionaries such as *Webster's Third New International Dictionary* (3rd edn, Merriam, 1976) and the *Oxford English Dictionary* (13 vols, Clarendon Press, 1933; Supplement I 1972; Supplement II 1976), contain definitions of some scientific terms, as do most of the smaller, more readily available volumes, such as *Chambers Twentieth Century Dictionary* (New edn, Chambers, 1983). However, a scientist will often need a dictionary prepared especially for his or her needs. With well over 100,000 definitions, the *McGraw-Hill Dictionary of Scientific and Technical Terms*, editor-in-chief S P Parker (3rd edn, McGraw-Hill, 1984) is the most extensive of those dictionaries which cover the whole field of science and technology. An alternative is the *Dictionary of Science and Technology*, edited by T C Collocott and A B Dobson (revised edn, W & R Chambers, 1974), which contains over 50,000 entries in 1,328 pages though being rather old it is short of many modern words. There are many other dictionaries covering this general field of science and technology, mostly for students and non-specialists. A popular dictionary in this category is *The Penguin Dictionary of Science*, by B Uvarov and A Isaacs (6th edn, Penguin Books, 1986).

Most scientific disciplines are covered by one or more single-subject dictionaries. The most comprehensive of those in the earth sciences area is the *Glossary of Geology*, edited by R L Bates and J A Jackson (2nd edn, American Geological Institute, 1980).

Widely accepted as the standard dictionary in its field it contains some 36,000 brief definitions, many with references, and it is well cross-referenced. The first edition, published in 1972 and edited by M Gary *et al*, was itself the successor to the earlier *Glossary of Geology and Related Sciences with Supplement* (2nd edn, American Geological Institute, 1960). As an illustration of the extent to which the vocabulary of geology has expanded it is interesting to note that the 1960 edition contained 18,000 definitions whilst that of 1980 has twice as many, 36,000. An abridged paperback version of the *Glossary* has been published as the *Dictionary of Geological Terms* (3rd edn, Doubleday, 1984): intended for students and laymen this contains approximately 8,500 entries.

Another well-known and long established dictionary in the earth sciences area is *Challinor's Dictionary of Geology*, edited by A Wyatt (6th edn, University of Wales Press, 1986). This has considerably fewer entries than those listed above, only some 1,500, but these are generally longer, averaging about twenty lines each. Great emphasis is placed on the origin and usage of the terms and the publishers claim ". . . a special feature being the copious quotations and references. . .". They also suggest that it is intended more as a "useful companion" than for occasional reference. A newer addition to the number of earth science dictionaries is the *McGraw-Hill Dictionary of Earth Sciences*, edited by S P Parker (McGraw-Hill, 1984) which has definitions of more than 15,000 terms in over 800 pages.

Other earth science dictionaries, mostly aimed at the layman or student include: *The Penguin Dictionary of Geology*, by D G A Whitten and J R V Brooks (Penguin Books, 1972); *The Penguin Dictionary of Geography*, by W G Moore (6th edn, Penguin Books, 1981); *A Dictionary of Earth Sciences*, edited by S E Stiegeler (Macmillan, 1976); *Earth Sciences: a dictionary of terms and concepts*, by D Dineley *et al* (Arrow Books, 1976); and *A Dictionary of the Natural Environment*, by F J Monkhouse and J Small (Edward Arnold, 1978). This last work represents a revision of the physical geography terms taken from the earlier *Dictionary of Geography*, by F J Monkhouse (2nd edn, Edward Arnold, 1970). Most of those mentioned above contain illustrations in the form of line drawings.

In common with other scientific subjects, the earth sciences possess a large number of rather more specialized dictionaries, each of which covers a particular branch of the subject. One example of this type, *A Dictionary of Mining, Mineral and Related Terms*, compiled by P W Thrush (US Department of the Interior, Bureau of Mines, 1968), is particularly worthy of note because of its size and comprehensive treatment of the subject. It has 1,269 pages with 150,000 definitions in about 55,000 individual entry terms. Also worth noting is the *International Tectonic Dictionary: English terminology*, edited by J G Dennis (American Association of Petroleum Geologists, 1967) (*AAPG Memoir* 7). In addition to defining each term, this dictionary gives details of the derivation, history and usage, synonyms and related terms. It also has a bibliography and a separate index.

There are many other specialized dictionaries of this type, including: the *Dictionary of Gemmology*, by P G Read (Butterworth Scientific, 1982); *Encyclopedic Dictionary of Exploration Geophysics*, by R E Sheriff (2nd edn, Society of Exploration Geophysicists, 1984); *Dictionary of Applied Geology, Mining and Civil Engineering*, by A Nelson and K D Nelson (George Newnes, 1967); *The Petroleum Dictionary*, by D F Tver and R W Berry (Van Nostrand Reinhold, 1980); *Meteorological Glossary*, by D H McIntosh (5th Edn, HMSO, 1972); *Glossary of Meteorology*, edited by R E Huschke (American Meteorological Society, 1959); *The New Dinosaur Dictionary*, by D F Glut (Citadel Press, 1982); and the *Ocean and Marine Dictionary*, by D F Tver (Cornell Maritime Press, 1979).

One branch of the earth sciences where the standard use of names is of particular importance is mineralogy. *The Glossary of Mineral Species*, by M Fleischer (3rd edn, Mineralogical Record, 1980) has nearly 3,200 entries in the main table. Arranged alphabetically, each mineral is defined in terms of its chemical formula, crystal system, colour (where distinctive), and any relations to other minerals: references to the literature are also given. A different aproach is adopted by *An Index of Mineral Species and Varieties arranged Chemically*, by M H Hey (2nd edn, British Museum (Natural History), 1955; Appendix 1963; 2nd Appendix 1974). Commonly known by its cover title, *Chemical Index*, this publication arranges the minerals by chemical formula but also has an alphabetical index of accepted mineral names (see also the chapter on mineralogy). Another dictionary of interest is *Mineral Names: what do they mean?*, by R S Mitchell (Van Nostrand Reinhold, 1979) which includes derivations for mineral names currently used.

Interlingual dictionaries

In addition to monolingual dictionaries, most scientists will occasionally have to use bilingual or multilingual (polyglot) dictionaries to assist with translating foreign language literature. There are many thousands of interlingual dictionaries available and as with the monolingual type, these can be general, general scientific, or specialized. The choice of which dictionary to use is likely to be dictated by local availability, but a select list of foreign language-English dictionaries is given as Appendix 2 of this chapter.

Details of other scientific dictionaries, both monolingual and interlingual, can be located in the following bibliographies —

Bibliography of Interlingual Scientific and Technical Dictionaries. 5th edn, UNESCO, 1969

Bibliography of Mono- and Multilingual Dictionaries and Glossaries used in Geography as well as in related Natural and Social Sciences, compiled by E Meynen. Franz Steiner, 1974

Dictionaries, Encyclopedias and other Word-Related Books, edited by A M Brewer. 2nd edn, 2 vols, Gale Research, 1979

International Bibliography of Specialized Dictionaries. 6th edn, K G Saur, 1979 (*Handbook of International Documentation and Information*; vol.4)

Russian-English Dictionaries with aids for Translators: a selected bibliography, by W Zalewski. 2nd edn, Russica Publishers Inc, 1981

Dictionaries of abbreviations

Over the last twenty years or so, there has been an enormous increase in the use of the various types of abbreviations, including contractions, initialisms and acronyms, particularly in science and technology. There are now many dictionaries concerned solely with abbreviations and one of these, *Acronyms, Initialisms and Abbreviations Dictionary* (annual, 3 vols, Gale Research) amply demonstrates this proliferation. The first edition, which was published in 1960, contained 12,000 terms whereas the eleventh, the 1987 edition, has over 375,000.

Other dictionaries of abbreviations include:

Anglo-American and German Abbreviations in Science and Technology, by P Wennrich. 3 vols, K G Saur, 1976–78; Supplement 1980. (*Handbook of International Documentation and Information*, vol.14)

Glossary of Russian Abbreviations and Acronyms. Library of Congress, 1967

Ocran's Acronyms: a dictionary of abbreviations and acronyms used in scientific and technical writing, by E B Ocran. Routledge and Kegan Paul, 1978

Abbreviations Dictionary, by R De Sola. 7th edn, Elsevier, 1986

Handbooks

To the working scientist handbooks are possibly the most useful of all the various reference publications that are available. Typically they consist of a miscellaneous collection of tables, charts, data, mathematical formulae and physical constants and they are most commonly used to provide answers to the sort of 'everyday' problems that require straightforward factual information. Originally intended as compact, 'handy' volumes of essential facts and figures, some handbooks have expanded considerably to become large multi-volume works. Many handbooks consist of page after page of tables interrupted by only the briefest explanatory text and can be difficult to use. Indeed, many require an in-depth knowledge of the subject if they are to be used effectively.

Some information, such as mathematical tables and physical constants is of value to scientists working in all subjects and there are many general handbooks containing this sort of information. Examples include: the *CRC Handbook of Chemistry and Physics* (annual, CRC Press) first published in 1914; *Tables of Physical and Chemical Constants*, edited by G W C Kaye and T H Laby (14th rev.edn, Longmans, 1986) first published in 1911; and *Smithsonian Physical Tables*, compiled by W E Forsyth (9th edn, Smithsonian Institution, 1954) (*Smithsonian Miscellaneous Collections*, **120**).

One of the larger collections is the *International Critical Tables of Numerical Data, Physics, Chemistry and Technology*, edited E W Washburn (7 volumes plus index, McGraw-Hill, 1926–33), which covers the primary literature up to 1924 and gives full bibliographic references so that a scientist can form his own opinion of the accuracy and status of the data that are included.

Another multi-volume collection is Landolt-Bornstein's *Zahlenwerte und Funktionen aus Physik, Chemie, Astronomie, Geophysik und Technik* (6th edn, Springer, 1950–59). The first edition of this now massive work was published in 1883 and consisted of 281 pages in a single volume. Since 1961 a 'new series' of separate volumes has been produced under the title *Zahlenwerte und Funktionen aus Naturwissenschaften und Technik — Neue Serie* (Springer, 1961-) each of which covers a specialized topic within one of the following six basic groups:

(1) Nuclear Physics and Technology
(2) Atomic and Molecular Physics
(3) Crystal and Solid State Physics
(4) Macroscopic and Technical Properties of Matter
(5) Geophysics and Space Research
(6) Astronomy, Astrophysics and Space Research

In the 'new series' the tables of contents and introductions are in both German and English.

All the handbooks mentioned above are general items of potential interest to most scientists, but there are a number of more specialized items covering the earth sciences. The following is a select list of some of the more important titles:

AGI Data Sheets. American Geological Institute, 1975
Chemical and Determinative Tables of Mineralogy (without the silicates), by R M Pierrot. Masson, 1979
CRC Handbook of Physical Properties of Rocks, edited by R S Carmichael. CRC Press, 1982
Gemstone and Mineral Data Book, by J Sinkankas. Winchester Press, 1972
Guide to Classification in Geology, by J W Murray. Ellis Horwood, 1981
Handbook of Geochemistry, edited by K H Wedepohl. 2 vols, Springer, 1969–1978. Volume 2 consists of 5 loose-leaf binders.
Handbook of Geophysics and Space Environments, edited by S L Valley. Revised edn, US Air Force, Cambridge Research Laboratories, 1965
Handbook of Meteorological Instruments. 2nd edn, 8 vols, HMSO, 1980–
Handbook of Physical Constants, edited by S P Clark jnr. Revised edn, Geological Society of America, 1966 (*Geological Society of America, Memoir* **97**)
Smithsonian Meteorological Tables, edited by R J List. 6th edn, Smithsonian Institution, 1949 (*Smithsonian Miscellaneous Collections*, **114**)

A useful guide to this type of publication is *Handbooks and Tables in Science and Technology*, edited by R H Powell (2nd edn, Library Association, 1983) which gives details of nearly 3,500 handbooks.

Yearbooks

A 'yearbook' is a periodical publication, usually appearing annually, which describes the events of the previous year. This is achieved, either by reviewing progress and describing major developments in a particular field or, where appropriate, by presenting a statistical summary of a particular industry. Some yearbooks combine both review and statistical information in a single volume. Most yearbooks in the earth sciences subject area are concerned with figures of production, consumption, imports and exports in the mining and petroleum industries.

The following is a select list of some of the more important titles:

Annual Statistical Bulletin — OPEC. This publication gives a statistical summary of the world's petroleum industry, including figures for production, consumption, imports, exports etc.
Australian Mineral Industry — Annual Review, published by the Australian Government Publishing Service for the Bureau of Mineral Resources, Geology and Geophysics. Gives reviews and statistical data for 64 individual commodities.
Canadian Minerals Yearbook, published by Energy, Mines and Resources Canada, and covers the metals, minerals and fuel industries. First published in 1886.
Metal Bulletin Handbook, published annually by Metal Bulletin Books. First issued in 1914 as *Quin's Metal Handbook and Statistics*; from 1982 it has split into 2 volumes: volume **1**,*Prices*; and volume **2**,*Statistics and Memoranda*.
Metal Statistics, published annually by Fairchild Publications.

Metals Week Price Handbook, published annually by *Metals Week*. This year-book gives price details for 46 metals.

Mineral Commodity Summaries, published annually by the US Department of the Interior, Bureau of Mines. Gives an up-to-date summary of 86 non-fuel minerals.

Mineral Facts and Problems, published every 5 years by the US Department of the Interior, Bureau of Mines. Gives reviews and statistical information on all the principal minerals.

Minerals Yearbook, produced by the US Department of the Interior, Bureau of Mines, in 3 volumes. Volume 1 deals with metals and minerals, volume 2 covers domestic area reports, and volume 3 international area reports.

Ocean Yearbook, published every other year by the University of Chicago Press.

United Kingdom Mineral Statistics, published annually for the British Geological Survey by HMSO. First published in 1973, it gives production, consumption, import and export data.

World Mineral Statistics, published annually for the Institute of Geological Sciences by HMSO.

In addition to those mentioned above, there are a number of more general yearbooks. Two of these, described in an earlier section of this chapter, are the *Britannica Book of the Year* (annual, Encyclopaedia Britannica Inc) and the *McGraw-Hill Yearbook of Science and Technology* (annual, McGraw-Hill) which update their respective encyclopedias by giving an account of the major events of the preceding year.

Other general yearbooks include, *The Statesman's Yearbook* (annual, Macmillan) and the *Europa Yearbook* (annual, 2 vols, Europa Publications). Both are arranged country-by-country and both include a limited amount of statistical information on mining, etc., for each country. Also of interest is the *UK Annual Abstract of Statistics* (annual, HMSO) which contains statistical information on all aspects of life in the United Kingdom, including meteorological statistics. Each edition contains figures for the previous 10 years.

Directories

A directory is essentially a list of names and addresses, of organizations or of individuals, linked together by some common feature or features. In addition to this basic information, directories of organizations frequently give details of the history, purpose, membership totals, publications, products and officers of the organization, while directories of individuals may give their qualifications, research interests and date of birth.

In geographic coverage directories may be international, national or local, and in size they can range from large general works to small specialized volumes. Like encyclopedias and dictionaries, directories are of little value if the information they contain is out-of-date, and although some are issued as one-off publications, most are revised at regular intervals, often annually. The word 'directory' may appear in the title, but this is not always the case and a directory may be issued as a 'handbook', 'yearbook', 'who's who', or simply 'list of . . .'.

Directories can be categorized into three main groups: directories of commercial organizations — trade directories; directories of non-commercial organizations — institutional directories; and directories of individuals — personal directories.

Trade directories

Numerically the largest group are trade directories and thousands are produced every year. Sometimes known as 'buyers' guides', they can be used either to find out which companies can supply particular products or services, or to obtain specific information about an individual company. Trade directories range from specialized volumes covering a single industry, such as the *USA Oil Industry Directory* (annual, PennWell Publishing), to nationally oriented general works such as *Kompass United Kingdom*, published annually by Kompass Publishers, and the *Thomas Register of American Manufacturers and Thomas Register Catalog File* (annual, Thomas Publishing) which in 1986 was in 21 volumes. Most of the trade directories of interest to earth scientists are concerned with the minerals and fuels industries, e.g. *The Geophysical Directory* (annual, Geophysical Directory Inc), *European Petroleum Directory* (annual, PennWell Publishing), *Industrial Minerals Directory*, edited by B Coope (2nd edn, Metal Bulletin Books, 1982), the *American Mining, Minerals and Oil Contact Directory* (David Ell Press, 1981), and the *Directory of Mines and Quarries*, compiled by P M Harris *et al* (British Geological Survey, 1984).

Sometimes trade directories are published in trade journals, e.g. *World Mining Yearbook: catalog, survey and directory* which appears in *World Mining* (monthly, Miller Freeman). A valuable guide to this type of directory is *Trade Directory Information in Journals* (6th edn, British Library Science Reference and Information Service, 1986). Details of other trade directories can be found by consulting *Trade Directories of the World*, published in a loose-leaf format by Croner Publications and updated by supplements.

Institutional directories

There are many institutional directories with a world-wide coverage. Prominent among these is *The World of Learning*, published annually by Europa Publications, and arranged country by country. For each of over 150 countries listed it gives the learned societies, professional associations, research institutes, libraries, museums, universities, polytechnics and colleges, and there is a combined organization index. Also covering the world is the *Guide to World Science* (2nd edn, Francis Hodgson, 1974–76) which consists of 24 volumes, each of which is devoted to a particular geographic region or country. Volume **1** covers the United Kingdom; Volumes **22** and **23** deal with the United States. The *Guide to World Science* is being updated by a number of volumes in the series *Longman Guide to World Science and Technology*. The first volume to appear was *Science and Technology in the Middle East*, by Z Sardar (Longman, 1982). Other volumes produced to date cover Latin America (1983), Japan (1984), China (1984), and USA (1986).

Also international in scope are the *World Guide to Scientific Associations*, managing editor B Verrel (4th edn, K G Saur, 1984) (*Handbook of International Documentation and Information*, vol. **13**) which lists more than 22,000 national and international organizations; *World Guide to Trade Associations*, editor B Verrel (3rd edn, K G Saur, 1985) (*Handbook of International Documentation and Information*, vol.**12**) which includes about 31,000 trade associations; and the *Yearbook of International Organisations* (annual, 3 vols, K G Saur).

More restricted in terms of geographic coverage are the *Directory of European Associations*, edited by I G Anderson (3rd edn, 2 vols, CBD Research, 1981), *European Research Centres* (6th edn, 2 vols, Longman, 1985), and *The Scientific Institutions of Latin America* (California Institute of International Studies, 1970).

At a national level British directories include: *Directory of British Associations and Associations of Ireland*, edited by G P Henderson and S P Henderson (9th edn, CBD Research, 1988), and *Trade Associations and Professional Bodies of the United Kingdom*, edited by P Millard (6th edn, Pergamon Press, 1979). Similar publications for the United States include: the *Encyclopedia of Associations* (annual, 4 vols, Gale Research), the 1984 edition of which lists 17,644 organizations, *National Trade and Professional Associations of the United States* (annual, Colombia Books), and the *Research Centers Directory* (annual, 2 vols, Gale Research). This last publication is supplemented between editions by *New Research Centers*. Other national directories of this sort include: for Australia, the *Directory of Australian Associations*, edited by B Chan (2nd edn, Australian Reference Research Publications, 1981) which lists nearly 8,000 associations; for Japan there is *Scientific Research Institutes* (Japan Society for the Promotion of Science, 1980).

There are several directories which cover universities; giving world-wide coverage are: the *World Guide to Universities* (2nd edn, 4 vols, K G Saur, 1976) (*Handbook of International Documentation and Information*, vol.**10**) and the *International Handbook of Universities* (9th edn, Macmillan, 1983), while for the Commonwealth there is the *Commonwealth Universities Yearbook* (annual, 4 vols, Association of Commonwealth Universities). This last publication, first issued in 1914, is arranged country-by-country and can be particularly useful in that it gives a full list of the teaching staff for all the universities covered.

The institutional directories dealt with so far have all been general items in terms of their subject coverage. However there are many directories concerned solely with earth sciences organizations. For British geologists probably the best directory available is *The Geologist's Directory*, edited by T Henton (3rd edn, Institution of Geologists, 1985). First published in 1980, this directory is divided into 10 sections. Section 3 'Geology in Education' lists all the university and polytechnic departments which run courses in geology, while Section 8 'Geological Information Services' gives lists of libraries, museums, geological societies, current British periodicals on geology and related subjects, and publishers of geological literature. In addition to these non-commercial organizations, there is also a substantial amount of trade information. Section 4 'Geology in Industry' includes a list of oil companies, while Section 6 'Specialist Services' contains lists of drilling companies, mud logging companies and geophysical contractors.

Of international coverage is *Earth and Astronomical Sciences Research Centres: a world directory of organisations and programmes*, consultant editor J M Fitch (Longman, 1984) which lists about 3,500 organizations from 131 countries and has subject and title of establishment indexes. Two other international directories are: the *Worldwide Directory of National Earth Science Agencies and Related International Organisations*, compiled by W E Bergquist *et al* (US Geological Survey, 1981) (*US Geological Survey Circular* **834**) which gives details of geological surveys and similar bodies in a country-by-country arrangement, and 'A directory of societies in earth science' published annually in *Geotimes* (monthly, American Geological Institute), the most recent in the October 1986 issue. As in the case of general works, many earth science directories are devoted to particular types of institutions. For educational establishments there is the *Directory of Geoscience Departments, United States and Canada* published annually by the American Geological Institute. Societies are covered by *A Directory of Natural History and Related Societies in*

Britain and Ireland, compiled by A Meenan (British Museum (Natural History), 1983), while libraries are dealt with in the *Directory of Geoscience Libraries, United States and Canada*, edited by R D Walker and D Parker (2nd edn, Geoscience Information Society, 1978) and the *Geological Directory of the British Isles: a guide to information sources*, edited by J A Diment (Geological Society (London), Geological Information Group, 1978) (*Geological Society Miscellaneous Paper*, **10**).

Personal directories

Probably the most common type of personal directory, and the simplest, is in the form of a membership list produced by a learned society or professional association. Typically a society's list of members will contain information such as full name, address, qualifications, status within the society (fellow, member, associate, etc.), and the date he or she was elected. Some of these lists are issued separately, e.g. *The Mineralogical Society List of Members*, the *Association of Engineering Geologists Directory*, the *Directory of Members of the British Micropalaeontological Society*, and the *Geological Society List of Members*, while others are published within one of the society's regular publications, e.g. 'AAPG membership directory and annual report' which is issued every year in the *Bulletin of the American Association of Petroleum Geologists*.

Similarly, many research establishments issue lists of their staff members, often in their annual reports. For example an up-to-date staff directory is published in the *British Geological Survey Annual Report* every year. Lists of university staff can be found in most of the annually produced university calendars.

Another type of personal directory tries to list all the specialists working in a particular field. Some of these are national in coverage, e.g. *US Directory of Marine Scientists* (National Academy of Science, 1982), while others cover the whole world, e.g. *Directory of Palaeontologists of the World* (3rd edn, International Palaeontological Association, 1976), the *International Directory of Marine Scientists* (Food and Agriculture Organisation, 1977) and the *World Directory of Mineralogists*, edited by F Cesbron (3rd edn, International Mineralogical Association, 1985).

Easily the largest personal directory in terms of individuals included is the *Current Bibliographic Directory of the Arts and Sciences*, formerly *Who is Publishing in Science*, produced annually by the Institute for Scientific Information as a spin-off from their various databases. The publishers claim that this directory annually includes 480,000 entries, giving name, organizational address and brief bibliographical details of the articles published by each author during the year of coverage.

For more detailed information about an individual one can consult one of the many biographical directories that are available. In Britain there is *Who's Who*, published annually by A & C Black since 1897. Entries are removed on the death of the individual and are subsequently included in *Who Was Who*: to date seven volumes have been produced, covering the periods 1897–1915, 1916–28, 1929–40, 1941–50, 1951–60, 1961–70, 1971–80 and there is also a cumulated index for the whole period 1897–1980. Most other countries have their own 'who's who' type publication, e.g. *Who's Who in America* (annual, 2 vols, Marquis-Who's Who).

The above publications are general in coverage and include entries for eminent people from all walks of life. However, scientists have their own biographical directories, e.g. *American Men and Women of Science, Physical and Biological Sciences* (15th edn, 7 vols, Bowker, 1979) which contains well over 100,000 entries, *Who's Who of British Scientists* (3rd edn, Simon Books, 1980), and *Who's Who in Science in Europe* (4th edn, 3 vols, Longman, 1984). There are also a number of more specialized works, including *Who's Who in Ocean and Freshwater Science*, edited by A Varley (Francis Hodgson, 1978) and *Financial Times Who's Who in World Oil and Gas* (7th edn, Longman, 1982).

Although its main function is to provide a national register of research, *Current Research in Britain*, published annually by the British Library, can also serve as a personal directory. Formerly published as *Research in British Universities, Polytechnics and Colleges (RBUPC)*, it now appears in 4 volumes: Physical Sciences, Biological Sciences, Social Sciences, and the Humanities. Each volume is organized under a number of broad subject headings. The Physical Sciences volume of the 1986 edition contains sections on: physical geography; geology and mineral technology; and geophysics, meteorology and oceanography. Each volume also has its own name and subject indexes. *Current Research in Britain* is now available online through Pergamon Infoline.

As stated earlier, there are literally thousands of directories published every year and those mentioned above are only a selection of some of the more important works. Additional titles can be traced by using one of the following guides to directories:

Current British Directories, by C A P Henderson. 10th edn, CBD Research, 1985
Current European Directories, by G P Henderson. 2nd edn, CBD Research, 1981

The Directory of Directories, C A Marlow and R C Thomas, editors. 4th edn, Gale, 1986; supplemented between editions by *The Directory Information Service*
Directory of Technical and Scientific Directories. 5th edn, Longman, 1988
Guide to American Directories, edited by B Klein. 11th edn, B Klein Publications, 1982
International Bibliography of Special Directories, edited by H Lengenfelder. 7th edn, K G Saur, 1983 (*Handbook of International Documentation and Information*, vol.5)
The Top 3,000 Directories and Annuals. 7th edn, compiled & edited by M Rasdall, Alan Armstrong and Associates, 1986
Trade Directories of the World. Croner Publications, 1969–

Textbooks and monographs

Commonly grouped together under the simple collective term 'books', there are differences between textbooks and monographs, specifically in their respective functions. However the division between them is not always clear cut and there is an almost imperceptible grading of the one into the other. Furthermore the division between books and other reference publications, such as encyclopedias, is not always immediately apparent.

A textbook will usually contain somewhat more detailed information than will an encyclopedia but it is still mainly concerned with the principles of a subject. It is primarily a teaching instrument and its role is one of selection and simplification. A monograph on the other hand is usually concerned with a narrowly defined single topic and within that topic it attempts to be comprehensive. Textbooks and monographs do not usually contain any original information but reorganize information previously published in the primary literature and arrange it into a more convenient form.

Identifying books of potential interest is the major problem facing the scientist. There are literally millions of books in existence and to this number tens of thousands of new publications are added every year. Although only a percentage of these cover science and technology, the problem is nevertheless a formidable one and searching for books by browsing the shelves of one's local library, or by searching through their catalogue, is something of a hit-and-miss affair. As mentioned in Chapter 2, some libraries do have very large comprehensive collections and in many cases their catalogues have been published. One of the largest of these is the *General Catalogue of Printed Books* issued by the British Museum (the library departments of which became part of the British Library in 1973). This covers the literature up to 1955 and supplements have been issued for the periods 1956–65, 1966–70, 1971–75 and 1976–82 (this last supplement is on microfiche). A new cumulation covering the literature up to 1975 is being published by K G Saur (1979–), who have also published a hard copy supplement for the period 1976–82. Another department of the British Library whose catalogue is published is the Science Reference and Information Service, which issues *SCICAT*. This consists of author/title and classified catalogues on microfiche and covers the period 1975 onwards: it is revised and reissued quarterly. A more specialized library catalogue, and one of particular interest to earth scientists, is the *Catalog of the US Geological Survey* (25 vols, with supplements, G K Hall, 1964–) listing the stock of the largest geological library in the world.

Library catalogues do not contain the most up-to-date information on recently published and about-to-be published items, but there are a number of publications available to assist in finding out about new books. Among these are general, often nationally based, bibliographies, e.g. *British National Bibliography* (weekly with cumulations, British Library, Bibliographic Services) first published in 1950 and the *Cumulative Book Index* (monthly with cumulations, H W Wilson) first published in 1898. A more specialized bibliography of interest to earth scientists is *Books in the Earth Sciences and Related Topics* (quarterly, Bibliographic Press), first published in 1977. Also of value are the lists of new accessions published by many libraries, e.g. *New Books & Serials*, issued monthly by the British Geological Survey, Library Services. Other sources of current bibliographic information include: *Books in Print* (annual, 7 vols, Bowker), its sister publication *Subject Guide to Books in Print* (annual, 4 vols, Bowker), and its British equivalent, *British Books in Print* (annual, 4 vols, Whitaker), all of which are available also as microfiche and on CD-ROM (Compact Disc-Read Only Memory).

Most scientific journals carry book reviews from time to time and many do so on a regular basis, e.g. *Nature* (weekly, MacMillan), *New Scientist* (weekly, IPC), and *Geotimes* (monthly, American Geological Institute). Book reviews are particularly useful in that they not only give bibliographic details but also usually provide an expert assessment of the merit, or otherwise, of a publication. They frequently offer a comparison between the book being reviewed and earlier publications on the same subject and this can be of great assistance when deciding which books to read or buy.

Electronic publication of secondary sources

For many years some publishers have applied computer technology to assist with the compilation, production, and exploitation of secondary publications: these databases have been made available for online searching through 'host' services, as described in Chapter 6. The more recent application of optical disc technology, particularly the CD-ROM (Compact Disc-Read Only Memory), has aroused increasing interest. The ability to hold large databases on CD-ROMs, and make these directly available for access by the end-user of the information on suitable hardware in the library, is an attractive library service. Reference works such as *Books in Print* and *McGraw-Hill Encyclopedia of Science and Technology*, and abstracting services such as *Dissertation Abstracts* and others in the fields of medicine and education, are already in use in many libraries.

Whether this trend to make secondary publications available in electronic form leads to the demise of the 'hard copy' versions of dictionaries, encyclopedias, etc., is, for the time being at least, a matter of conjecture.

Review articles

The continued growth in the primary literature of science has rendered the problem of keeping up-to-date with recent developments in one's own particular field, let alone developments outside one's immediate area of interest, almost impossible. Review articles play an important part in trying to solve this problem, by giving a critical assessment of the primary literature published on a particular subject over a given period of time. They are usually written by acknowledged experts, and are often commissioned by a publisher or editor, rather than simply submitted as is the case with primary research articles.

Woodward (1977) drew attention to the fact that reviews are written and used for a variety of purposes. He suggests that peer evaluation is probably their single most important function, but also lists a number of others including: "collation of information from different sources"; "compaction of existing knowledge"; "identification of emerging specialities"; "initial orientation in a new field"; and "current awareness in related fields". Because of the delays inherent in producing reviews, they are not suitable for current awareness within one's own speciality, but they can be useful as a kind of 'safety net' to ensure that no significant papers have been missed. In addition, review articles can be a useful guide to the relevant literature on a subject and can save a considerable amount of time in conducting a literature survey.

The value and popularity of reviews was emphasized by Craig (1969) who reported the results of a small survey on the reading habits of 30 geologists. Each was asked to list the three most important papers in geology since 1950 and about ⅓ of these were review articles, a figure in marked contrast to the number of reviews in the total number of geology papers produced annually. Craig concludes that "In other words, review articles are held to be more important (33⅓%) than their abundance (2–5%) would seem to indicate".

In fact the usefulness of review articles has been accepted for some considerable time and the earliest scientific journals, such as the *Philosophical Transactions of the Royal Society*, often contained reviews. The first journal devoted entirely to reviews, *Berlinischer Jahrbuch für die Pharmacie*, started publication in 1795, and during the nineteenth century a number of review journals (all German) appeared which covered earth sciences subjects, including *Geographisches Jahrbuch* (1866–1942) and *Jahresbericht uber die Fortschritte der Chemie und Mineralogie* (1822–51).

Most earth science journals carry review articles from time to time, and there are now a number of journals devoted exclusively to reviews. These latter journals often have titles such as *Advances in . . .*, *Progress in . . .*, *Developments in . . .*, and *Annual Review of. . . .* Guides to review serials of this type include *Directory of Review Serials in Science and Technology, 1970–1973: a guide to regular or quasi-regular publications containing critical, state-of-the-art or literature reviews*, by A M Woodward (Aslib, 1974) and the UNESCO publication *List of Annual Reviews of Progress in Science and Technology* (2nd edn, 1969). A list of review serials in the earth sciences is given in Appendix 3 of this chapter.

Identifying review articles in abstracting or indexing periodicals can present difficulties. Those appearing in journals devoted exclusively to reviews are relatively easy to spot but many review articles are published in primary research journals and few of these include the word 'review' in the title. A useful bibliographic source is the *Index to Scientific Reviews* (semi-annual, Institute for Scientific Information). This publication covers all scientific subjects and the publishers claim that 26,000 reviews are indexed every year from over 3,000 journals.

References

Craig, G Y (1969) Communication in geology. *Scottish Journal of Geology*, **5**,(4), 305–321

Woodward, A M (1977) The roles of reviews in information transfer. *Journal of the American Society for Information Science*, **28**,(3), 175–180

APPENDIX 1: The encyclopedia of earth sciences series
Volumes published to date

Vol. 1 *The Encyclopedia of Oceanography*, edited by R W Fairbridge (1966)
Vol. 2 *The Encyclopedia of Atmospheric Sciences and Astrogeology*, edited by R W Fairbridge (1967)
Vol. 3 *The Encyclopedia of Geomorphology*, edited by R W Fairbridge (1968)
Vol. 4A *The Encyclopedia of Geochemistry and Environmental Sciences*, edited by R W Fairbridge (1972)
Vol. 4B *The Encyclopedia of Mineralogy*, edited by K Frye (1981)
Vol. 6 *The Encyclopedia of Sedimentology*, edited by R W Fairbridge and J Bourgeois (1978)
Vol. 7 *The Encyclopedia of Paleontology*, edited by R W Fairbridge and D Jablonski (1979)
Vol. 8 *The Encyclopedia of World Regional Geology*, Part 1: *Western Hemisphere*, edited by R W Fairbridge (1975)
Vol. 10 *The Encyclopedia of Structural Geology and Plate Tectonics*, edited by C K Seyfert (1987)
Vol. 12 *The Encyclopedia of Soil Science*, Part 1: *Physics, Chemistry, Biology, Fertility and Technology*, edited by R W Fairbridge and C W Finkl jnr. (1979)
Vol. 13 *The Encyclopedia of Applied Geology*, edited by C W Finkl jnr. (1984)
Vol. 15 *The Encyclopedia of Beaches and Coastal Environments*, edited by M L Schwartz (1982)

APPENDIX 2: A select list of foreign-language/English general technical dictionaries

Modern Chinese-English Technical and General Dictionary. 3 vols, McGraw-Hill, 1963
Alford, M H T and Alford, V L *Russian-English, English-Russian Scientific and Technical Dictionary.* 2 vols, Pergamon, 1970
Collazo, J L *English-Spanish, Spanish-English Encyclopedic Dictionary of Technical Terms.* 3 vols, McGraw-Hill, 1979
De Vries, L P *French-English Science and Technology Dictionary.* 4th edn, McGraw-Hill, 1976
De Vries, L P and Jacolev, L *German-English Science Dictionary.* 4th edn, McGraw-Hill, 1978
Dorian, A F *Dictionary of Science and Technology, English-German, German-English.* 2nd edn, Elsevier, 1978, 1981
Gullberg, I F *Swedish-English Dictionary of Technical Terms.* 2nd edn, Norsedt, 1977
Walther, R *Polytechnical Dictionary, English-German, German-English.* Pergamon, 1968

Earth sciences dictionaries

Bargilliot, A *Vocabulaire Pratique Anglais-Français et Français-Anglais des Termes Techniques concernant la Cartographie.* Institut Géographique National, 1944

Bradley, J E S and Barnes, A C *Chinese-English Glossary of Mineral Names*. Consultants Bureau, 1963

Brazol, D *Dictionary of Meteorological and Related Terms: English-Spanish, Spanish-English*. Hachette, 1955

Cagnacci Schwicker, A *International Dictionary of Metallurgy, Mineralogy, Geology, Mining and Oil Industries (English-French-German-Italian)*. Technoprint International, 1968

Jackson, I *Nordic Glossary of Hydrology (English-Danish-Finnish-Icelandic-Norwegian-Swedish)*. Almqvist and Wiksell, 1984

Michel, S P and Fairbridge, R W *Dictionary of Earth Science: English-French, French-English*. Masson, 1980

Pfannkuch, H O *Elsevier's Dictionary of Hydrogeology (English-French-German)*. Elsevier, 1969

Proulx, G J *Standard Dictionary of Meteorological Sciences: English-French, French-English*. McGill-Queen's University Press, 1971

Rogoyski, D A *Glossary of Polish-English Meteorological Terms*. NTIS, 1968

Singer, L *Russian-English-German-French Hydrological Dictionary*. Scientific Information Consultants, 1967

Sofiano, T A, Lebedev, A P and Khain, V E *Russian-English Geological Dictionary*. Fizmargiz, 1960

Telberg, V G *Russian-English Dictionary of Paleontological Terms*. Telberg Book Corp., 1966

Telberg, V G and Deruguine, T *Basic Russian-English Geological Dictionary*. Telberg Book Corp., 1960

UNESCO *International Glossary of Hydrology (English-French-Russian-Spanish)*. WMO, 1974

Visser, W A *Geological Nomenclature (English-Dutch-French-German-Spanish)*. 3rd edn, Nijhoff, 1980

Watznauer, A *Dictionary of Geosciences: German-English, English-German*. 2nd edn, 2 vols, Elsevier Scientific, 1982

Zyłka, R *Geological Dictionary (English-Polish-Russian-French-German)*. Wydawnictwa Geologiczne, 1970

APPENDIX 3: Some current review publications in the earth sciences

Advances in Geophysics (1952–) Academic Press, annual

Advances in Hydroscience (1964–) Academic Press, annual

Annual Review of Earth and Planetary Sciences (1973–) Annual Reviews Inc, annual

Developments in Sedimentology (1964–) Elsevier, irregular

Earth Science Reviews (1966–) Elsevier, bi-monthly

Physics and Chemistry of the Earth (1956–) Pergamon, irregular

Progress in Physical Geography (1977–) Edward Arnold, quarterly

Reviews in Engineering Geology (1962–) Geological Society of America, irregular

Reviews of Geophysics and Space Physics (1963–) American Geophysical Union, quarterly

CHAPTER FIVE

Secondary literature: bibliographies, abstracts and indexes

MICHAEL O'DONOGHUE

No reader can be expected to read sets of journals or individual monographs from end to end in the hope that something relevant to his enquiry will turn up. Though almost all journals have their own indexes, annual and cumulative in most cases, they all suffer from the limited range of keywords that space allows. There is little scope for linking topics, and editorial idiosyncrasies can also greatly vitiate the usefulness of an index. *Abstracting journals* are more useful but suffer to some extent from the same defects and often cost substantially more than the journals from which they take their material. However, since all major abstracting journals cover a wide range of primary material their use is obvious and most are compiled by trained abstractors. *Bibliographies*, the third class of secondary literature, can be much more than lists of titles with varying provision of bibliographic detail; some are established subject guides and some are as current as the abstracting journals.

Bibliographies

From a simple reading list for undergraduates to a highly organized guide to all types of publication on a small specific topic, the bibliography remains a list of written materials. At one time concerned mainly with monographs, the coverage of bibliographies now extends over journals, conference proceedings, theses, official governmental and intergovernmental publications and patents. Some bibliographies confine themselves to particular date ranges within their topics, others are compiled with country or language as their prime basis. Still others may be concerned with certain types of material only, such as geological survey publications. There are as many different ways of arranging the entries as there are types of bibliography but most have author and subject indexes, however capricious the mode of entry may be.

Earth science is one of the few scientific subjects to retain a multi-disciplinary content reminiscent of the seventeenth and eighteenth centuries. For this reason and because the older literature retains some value, the early bibliographies are still useful, covering a good deal of general science as well as more specific topics. The growth of information in science as a whole led D J Reuss to compile *Repertorium Commentationum a Societatibus Litteraris Editarum* (Göttingen, 1801–1821). This bibliography of sixteen volumes is a guide to articles found in the journals of scientific societies, with a starting date of 1665 — the date from which *Philosophical Transactions of the Royal Society* and *Journal des Sçavans* begin (see Chapter 3). The Royal Society continued the work with its nineteen-volume *Catalogue of Scientific Papers*, in which nineteenth-century literature is covered. This author index appeared in four series between 1866 and 1925. Series 1 (1800–1863) comprises vols 1–6; series 2 (1864–1873) vols 7–8; series 3 (1874–1883) vols 9–11, and 1800 to 1883 is covered by a supplementary volume. The years 1884 to 1900 are covered by series 4 in vols 13–19. Only the mathematics, mechanics and physics section of a projected *Subject Index to the Catalogue* ever materialized.

With the onset of the twentieth century this work was superseded by a new Royal Society publication *International Catalogue of Scientific Literature* in which books and journal articles were listed annually for the years 1901–1914, but delays to publication resulted in the 1914 volume being completed only in 1921. There are seventeen subject divisions, and author and subject indexes: meteorology, mineralogy, geology, geography and palaeontology — sections F,G,H,J and K — are the ones most likely to be consulted by earth scientists.

Useful geological material can also be found in *Bibliographia Zoologiae et Geologicae*, compiled by Louis Agassiz and revised by H E Strickland. This was published by the Ray Society in 1848, and reprinted in 1968 by Johnson in their series *Sources of Science*. Bibliographic entries are in author order, and periodical titles, arranged by place of publication, are listed in volume 1.

Many monographic bibliographies deal with countries or topics falling within definite limits: a good example is K M Clayton's *A Bibliography of British Geomorphology* (1964). The *KWIC Index to Rock Mechanics Literature* was compiled by the Rock Mechanics Section at Imperial College, London and published in 1969 by the American Institute of Mining, Metallurgical and Petroleum Engineers. More recently there have appeared N C Rosen's *Bibliography of Geology of Iran* (1969, printed 1973), *Bibliography of Jamaican Geology*, edited by M Kinghorn (1977), and R A Smith's *Bibliography of the Geology and Geomorphology of Cumbria* (1974).

Some bibliographies form part of an essentially non-bibliographical series. Examples include *Bibliography of Continental Drift and Plate Tectonics*, published as *Special papers* **142** and **164** of the Geological Society of America in 1973–75, and *Bibliography of Palaeontology in Japan 1961–1975*, published as *Special papers* **22** of the Palaeontological Society of Japan. Other bibliographies, like the *Science Museum Library Bibliographic Series*, form series of their own, with no material other than that held in stock in the library included. Many libraries, on identifying a bibliography while scanning issues of serials on receipt, will give them analytical entries in their catalogues: this has always been the practice of the Science Reference and Information Service of the British Library.

It is by no means always the case that major monographs will have extensive, or even adequate, bibliographies. Some authors feel that with the coming of universal access to journal article databases, only other monographs need be listed. A recent honourable exception is in *Emerald and other Beryls*, by John Sinkankas (Chilton, 1981).

Good bibliographies can also be found in F J Pettijohn's *Sedimentary Rocks*, (3rd edn, Harper & Row, 1976) and in W G McGinnies *et al*, *Deserts of the World*, (University of Arizona, 1968). Palaeontological and stratigraphic bibliographies are listed, with about 1,000 other works, in B Kummel and D Raup's *Handbook of Paleontological Techniques* (Freeman, 1965). The recently-produced second edition of H G Reading's *Sedimentary Environments and Facies* (Blackwell Scientific, 1986) also contains an extensive bibliography.

First time enquirers can be helped by some encyclopedia articles which may contain short lists of references. Broad-based encyclopaedias such as the *McGraw-Hill Encyclopedia of Science and Technology* (see Chapter 4) include bibliographies, e.g. *McGraw-Hill Basic Bibliography of Science and Technology* and *A Selective Bibliography in Science and Engineering*, which are well-worth consulting. The *Selective Bibliography* contains entries in Dewey Decimal Classification order. More extensive lists, although somewhat dated, can be located in Besterman's *A World Bibliography of Bibliographies* (1965–66). This lists 117,187 separate bibliographies under 15,829 subject headings. The list of geological bibliographies, arranged by country, occupies 21 pages. Local information can often be found in the journals of natural history societies, the compilers of which often go to considerable lengths to give adequate references to their papers.

The list of geological bibliographies planned by the Geoscience Information Society has not yet materialized; this was intended to update de Margerie's *Catalogue des bibliographies géologiques* (Gauthier-Villars, 1896), which gives details of 4,000 bibliographies published between 1726 and 1895: it is especially strong in biographical material. The planned list of bibliographies would also have updated *Catalogue of Published Bibliographies in Geology 1896–1920*, be E B Mathews (1923): this formed volume **6**, part 5, no. 36 of *Bulletin of the National Research Council* and contains 3,699 titles. H P Little's *List of Manuscript Bibliographies in Geology and Geography* (1922) is no. 27 of *National Research Council Reprint and Circular Series*. On a smaller scale and covering a more limited field is H K Long's *Bibliography of Bibliographies on the Geology of the States of the United States*, forming volume **7**, part 7, of *Geoscience Abstracts* (1965). Geology and its related subjects can also be found in bibliographies dealing generally with a particular country or group of countries: examples are A E Gropp's *A Bibliography of Latin American Bibliographies* (Scarecrow Press, 1968) and the New Zealand Library Association's *A Bibliography of New Zealand Bibliographies* (1967). Enquirers would also do well to consult G Bridson and A P Harvey's 'A checklist of natural history bibliographies and bibliographical scholarship, 1966–1970', published in *Journal of the Society for the Bibliography of Natural History*, vol. **5**, part 6, 1971. More general is H W Wilson Company's twice-yearly *Bibliographic Index* with annual and larger cumulations.

The GeoArchive database has sections with the numbers 913000 (bibliographies and reviews), 913100 (subject bibliographies) and 914100 (regional bibliographies) and these codes can be conveniently searched online; such a search will locate many bibliographic items not referred to above.

Abstracting and indexing publications

Clearly the monographic bibliography takes a long time to compile and get published; even when it is available it inevitably suffers from slippage of currency, and may not include references to papers in journals. To take care of both these aspects abstracting and indexing publications have been introduced, usually as periodicals; in many cases they are also mounted as databases. Generally the list of items included in an indexing journal will be arranged by subject, author, journal title or keyword. An abstracting journal carries this basic reference information supplemented with abstracts; abstracts may be by the author of the paper himself (this is almost *de rigueur* with papers in remoter languages) but are more usually compiled by a professional or expert abstracter. There should be enough data given in the abstract to enable the reader to decide whether or not he wishes to see the original paper. Some abstracts attempt to go further and give the principal arguments and main data (the 'informative abstract') but this incurs extra cost and loss of currency of the information.

It is interesting to compare the treatment given by different abstract compilers and to this end examples of five different abstracts of the same paper are given in *Figure 5.1*. The most detail is given by *Mineralogical Abstracts* and the least by *Science Citation Index*; in the latter case little more than the title and source is provided.

(1)   **Sulfate-reducing bacteria and silica solubility: a possible mechanism for evaporite diagenesis and silica precipitation in banded iron formations**[1]

STUART J. BIRNBAUM

Division of Earth and Physical Sciences. The University of Texas, San Antonio, TX 78285, U.S.A.

AND

JOHN W. WIREMAN

Biosan Laboratories, 10657 Galazie, Ferndale, MI 48220, U.S.A. and Department of Biological Sciences, Wayne State University, Detroit, MI 48202, U.S.A.

Received December 10, 1984
Revision accepted May 22, 1985

Selective replacement of sulfate-evaporite minerals by silica and the precipitation of silica in association with sulfide mineral phases in banded iron formations may be mediated by the metabolic activities of sulfate-reducing bacteria. Hydrogen sulfide is known to be a product of this metabolism and is often called upon as a source of sulfur for metallic sulfides in sedimentary rocks. We report here on the influence that chemical changes induced by bacterial sulfate reduction have on silica solubility.

Controlled in vitro growth experiments with *Desulfovibrio desulfuricans* and silica show (1) this organism can grow in silica concentrations as great as 400 ppm with no inhibition and (2) growth in the presence of silica yields a decrease in dissolved silica.

Growth experiments with 80 ppm silica produced a lowering in dissolved silica from 80 ppm to 60 ppm, a 25% decrease, in just 30 h. Control experiments in the absence of cells resulted in no effective decrease in dissolved silica. The ability of sulfate-reducing bacteria to remove silica from solution may be related to local changes in pH and to hydrogen bonding of amorphous silica followed by polymerization to higher weight molecules.

––––––––––

Le remplacement sélectif de minéraux d'évaporite de la classe des sulfates et la précipitation de la silice associée aux phases minérales de sulfures dans les

Fig.5.1

Fig. 5.1 *continued*

formations de fer rubanées résultent possiblement de l'intervention d'activites métaboliques des bactéries réductrices des sulfates. On sait que ce métabolisme produit de l'hydrogène sulfuré auquel on réfère souvent comme étant la source de soufre nécessaire à la formation des sulfures métalliques dans les roches sédimentaires. Ici nous décrivons l'influence des changements chimiques provoqués par la réduction bactérienne des sulfates sur la solubilité de la silice.

Des études de croissance contrôlées in vitro avec *desulfovibrio desulfuricans* en présence de silice révèlent que (1) cet organisme peut croître sans inhibition même dans des concentrations aussi fortes que 400 ppm et (2) la croissance en présence de silice entraîne une diminution de la silice en solution.

Les expériences de croissance avec 80 ppm de silice provoquent une diminution de la silice soluble de 80 ppm à 60 ppm, soit une baisse de 25%, en seulement 30 h. Des expériences-témoins en absence de cellules ont montré qu'il ne se produit pas de diminution de la solubilité de la silice. Cette capacité que possèdent les bactéries réductrices des sulfates de diminuer la silice en solution peut être reliée aux variations locales de pH et à la formation d'une liaison chimique de l'hydrogène avec la silice amorphe suivie d'une polymér-isation en molécules de poids moléculaires plus élevés.

[Traduit par le journal]
Can. J. Earth Sci. **22**, 1904–1909 (1985)

(2) 86M/4555   Sulfate-reducing bacteria and silica solubility: a possible mechanism for evaporite diagenesis and silica precipitation in banded iron formations. S.J. Birnbaum & J.W. Wireman, *Canadian Journal of Earth Sciences*, **22(12)**, 1985, pp 1904–1909.

Selective replacement of sulphate-evaporite minerals by silica and the precipitation of silica in association with sulphide mineral phases in banded iron formations may be mediated by the metabolic activities of sulphate-reducing bacteria. Hydrogen sulphide is known to be a product of this metabolism and is often called upon as a source of S for metallic sulphides in sedimentary rocks. The influence that chemical changes induced by bacterial sulphate reduction have on silica solubility is reported. *In-vitro* growth experiments with *Desulfovibrio desulfuricans* and silica show that this organism can grow in silica concentrations $\leqslant 400$ ppm, and that growth in the presence of silica yields a decrease in dissolved silica. Growth experiments with 80 ppm silica produced a lowering of dissolved silica from 80 to 60 ppm in 30 hr. Control experiments in the absence of cells resulted in no effective decrease is dissolved silica. The ability of sulphate-reducing bacteria to remove silica from solution may be related to local changes in ph and to hydrogen bonding of amorphous silica followed by polymerization to higher weight molecules.   p.Br.

(3)   104; **152755d Sulfate-reducing bacteria and silica solubility: a possible mechanism for evaporite diagenesis and silica precipitation in banded Iron Formations**. Birnbaum, Stuart J.; Wireman, John W. (Div. Earth Phys. Sci., Univ. Texas, San Antonio, TX78285 USA). *Can. J. Earth Sci. 1985, 22(12), 1904–9 (Eng)*. *Selective replacement of sulphate-evaporite minerals by SiO_2 and the pptn. of SiO_2 in assocn. with sulfide mineral phases in banded Fe formations may be mediated by the metabolic activities of sulfate-reducing bacteria. H_2S is a product of this metab. and is often called upon as a source for S for metallic sulfides in sedimentary rocks. The influence that chem. changes induced by*

Fig. 5.1 *continued overleaf*

Fig. 5.1 *continued*

bacterial sulfate redn. *have on SiO$_2$ soly. is discussed. Controlled in vitro growth expts. with Desulfovibrio desulfuricans* and SiO$_2$ show: (1) this organism can grow in SiO$_2$ concns. ≤400 ppm with no inhibition and (2) growth in the presence of SiO$_2$ yields a decrease in dissolved SiO$_2$. Growth expts. with 80 ppm SiO$_2$ produced a lowering in dissolved SiO$_2$ from 80 to 60 ppm, a 25% decrease in just 30 h. Control expts. in the absence of cells resulted in no effective decrease in dissolved SiO$_2$. The ability of sulfate-reducing bacteria to remove SiO$_2$ from soln. may be related to local changes in pH and to H bonding of amorphous SiO$_2$ followed by polymn. to higher wt. mols.

(4)**Birnbaum, Stuart J.** (Univ. Tex., Div. Earth and Phys. Sci., San Antonio, TX, United States); and **Wireman, John W.** Sulfate-reducing bacteria and silica solubility; a possible mechanism for evaporite diagenesis and silica precipitation in banded iron formations: *in* Role of organisms and organic matter in ore deposition — Le rôle des organismes et de la matière organique dans la formation des gisements métallifères (Macqueen, R. W., editor; *et al*), Canadian Journal of Earth Sciences = Journal Canadien des Sciences de la Terre, 22(12), p. 1904–1909 (French sum.), illus., 45 ref., December 1985. *Meeting:* May 16, 1984, London, ON, Canada.

(5)   8 В78. Восстанавливающие сульфат бактерии и растворимость кремнезема; возможный механизм диагенеза эвапоритов и осаждения кремнезема в формациях железистых кварцитов. Sulfate-reducing bacteria and silica solubility: a possible mechanism for evaporite diagenesis and silica precipitation in banded iron formations. B i r n b a u m S t u a r t J., Wireman John W. «Can. J. Earth Sci.», 1985, 22, № 12, 1904—1909 (англ.; рез. фр.). ISSN 0008—4077 CA

Экспериментальные исследования выполнялись с анаэробными бактериями, восстанавливающими сульфат, Desulfovibrio desulfuricans в интервале рН от 6,9 до 9,2. Кремнезем не оказывает явного подавляющего влияния на рост клеток. Рост бактерий приводит к снижению рН и к осаждению кремнезема из р-ров. Механизм такого влияния жизнедеятельности бактерий до конца не понят. Возможным механизмом, позволяющим восстанавливающим сульфат бактериям удалять из р-ров кремнезем, является локальное изменение рН и связывание водорода аморфным кремнеземом, после чего происходила полимеризация молекул. Полученные результаты могут быть использованы для объяснения генезиса железистых кварцитов.   В. И. Баженов

(6) ● **Birnbaum S J Wireman**————————————
J W SULFATE-REDUCING BACTERIA AND SILICA SOLUBILITY. A POSSIBLE MECHANISM FOR EVAPORITE DIAGENESIS AND SILICA PRECIPITATION IN BANDED IRON FORMATIONS
CAN J EARTH   22(12):1904–1909   85   45R
UNIV TEXAS, DIV EARTH & PHYS. SCI, SAN ANTONIO, TX 78285, USA.

(1) Summary of original article: *Canadian Journal of Earth Sciences*, **22**, 1985
(2) *Mineralogical Abstracts*
(3) *Chemical Abstracts*
(4) *Bibliography & Index of Geology*
(5) *Referativnyi Zhurnal*
(6) *Science Citation Index — Source Index*

Abstracting journals go back at least as far as the eighteenth century; Kronick's *A History of Scientific and Technical Periodicals* (1962) quotes C G Hoffmann's *Aufrichtige und unpartheyische Gedancken über die Journale Extracte und Monaths-Schriften, Worrinnen dieselben extrahiret, wann es nutzlich suppliret oder wo es nothig, emediret werden; nebst einer Vorrede von der Anneemlichkeit, Nutzen und Fehlern gedachter Schrifften.* This work began in 1714 and finished in 1717. There were only two volumes, each consisting of 12 issues. Nearly 40 journals were scanned, including *Journal des Sçavans* (one of the first two scientific periodicals to be issued). The first abstracting journals seem quickly to have exhausted their mainly German compilers; only 14 lasted longer than 10 years during the period 1714–1790. The world's first real scientific abstracting service, *Pharmaceutisches Centralblatt* (later *Chemisches Zentralblatt*) began in 1830 and ceased publication in 1969.

The first regular guide to earth science literature is probably *List of Geological Literature added to the Geological Society's Library during the year. . . .* This began in 1895 and appeared annually up to 1936, covering the literature up to 1934. Subject indexes were compiled and the index was arranged alphabetically by author and title.

With a date of first issue slightly earlier than the *List*, the *Bibliography of North American Geology* first saw the light in 1886, the publisher being the US Geological Survey. It is an index to material covering the North American continent, Greenland, the West Indies and the various islands under US influence such as Guam. Papers by American authors in foreign journals are included if they deal with North America or are general; papers on North America by foreign authors are included regardless of place of publication, and papers of a general nature are included if they appear in North American journals. This makes the coverage somewhat restricted compared to the *List* but the *Bibliography* is still in existence. Hard copy is issued in the *Bulletin* series of the US Geological Survey and it is now available online. It appears more or less annually, with cumulations.

Complementary to the *Bibliography of North American Geology* is *Bibliography and Index of Geology exclusive of North America*. The Geological Society of America began to publish this work in 1934 (covering the literature for 1933) and publication continued annually under this title until vol.**32** (1968). From then on the current title *Bibliography and Index of Geology* was chosen to indicate the world-wide coverage of the publication and most entries have short abstracts. Cumulative indexes are available; the 'Cumulative index' is a subject index, and the 'Cumulative bibliography' lists the citations in author order.

From 1901 to 1942 *Geologisches Zentralblatt* covered a good deal of European literature. From 1931 onwards it appeared as Abteilung A. *Geologie* (issued fortnightly) and Abteilung B. *Palaeontologie* (also called *Palaeontologisches Zentralblatt* and issued monthly): each volume has an author and a subject index. Also in German, and beginning life in 1925, is *Neues Jahrbuch für Mineralogie, Geologie und Paläontologie*: Abteilung A. *Mineralogie und Petrographie, Referate*; Abteilung B. *Geologie und Paläontologie, Referate*. Before 1925 it had several predecessors with various titles whose details can be found in *BUCOP*. From 1926 a title modification made it *Neues Jahrbuch für Mineralogie, Geologie und Paläontologie, Referate*. Between 1926 and 1927 it was issued in two parts: Abteilung A. *Mineralogie, Petrographie* and Abteilung B. *Geologie, Paläontologie*. Between 1927 and 1942 it was issued in three parts with six issues per volume: Teil 1. *Kristallographie, Mineralogie*; Teil 2. *Allgemeine, Geologie, Petrographie, Lagerstattenlehre*; Teil 3. *Historische und Regionale Geologie, Paläontologie*: publication ceased in 1942.

In 1943 the *Neues Jahrbuch für Mineralogie, Geologie und Paläontologie, Referate* and *Geologisches Zentralblatt* were merged in *Zentralblatt für Mineralogie, Geologie und Paläontologie*, at this date changing status from a primary to a secondary publication. Between 1943 and 1949 it appeared in three parts. Teil 1 dealt with mineralogy and crystallography; Teil 2 with petrography and geochemistry, and Teil 3 with palaeontology and general geology. In 1950 it split into two separate titles, *Zentralblatt für Geologie und Paläontologie* and *Zentralblatt für Mineralogie*.

A French equivalent was published under the title *Bibliographie des Sciences Géologiques* from 1923, being published by the Société géologique de France. Until 1930 it appeared quarterly and became annual after that date. Abstracts were not given, entries being arranged first by country and then by journal title. The annual version appeared in 18 subject sections, each with subsections and each volume had an author index. Between 1947 and 1960, when publication ceased, references were taken from *Bulletin Analytique* and, after 1955, from *Bulletin Signalétique*.

Bulletin Analytique began in 1940 and was produced by the Centre de Documentation du Centre National de la Recherche Scientifique. It appeared monthly and contained abstracts of the world's scientific and technical literature. Earth science material was represented only by mineralogical and geochemical entries between 1940 and 1945 but a section *Sciences de la Terre* began in 1946. At first this was divided into three sections, *Minéralogie, pétrographie, géologie* but, in 1947, a section *Paléontologie*, and in

1952, *Physique du globe*, were added. In 1956 the title became *Bulletin Signalétique* (see below).

Bulletin Signalétique, taking over from *Bulletin Analytique* in 1956, was issued in 31 separately available sections and appeared monthly. It has a somewhat complicated bibliographical history and was replaced by the title *Informascience (Documentation centre), Bulletin Signalétique* in 1977. The sections likely to be of most interest to earth scientists are 161, *Cristallographie*; 210, *Minéralogie, Géochimie, Géologie Extraterrestre, Pétrographie*; 214, *Géologie Appliquée, Formations Superficielles*; and section 216, *Géologie paléontologie*. Over 7000 journals are covered, plus reports and dissertations. Abstracts are usually in French but titles are given in their original language; an in-house classification scheme is used.

From 1968 *Bibliographie des Sciences de la Terre* was published by the Bureau de Recherches Géologiques et Minières, Orléans (BRGM). It was issued monthly in eight subject sections, most of which are further subdivided. Each division has its own author and subject indexes and there are also geographical and other indexes where relevant. The *Bibliographie* ran until 1971 when it was replaced by *Bulletin Signalétique*; at that time the *Bibliographie* consisted of: cahier A, covering *Minéralogie et géologie* which was replaced by section 220 of the *Bulletin Signalétique*; cahier B, covering *Gitologie et economie minière* (replaced by section 221); cahier C, covering *Roches cristallines* (222); cahier D, *Roches sédimentaires* (223); cahier E, *Stratigraphie et géologie régionale* (224); cahier F, *Tectonique et géophysique* (225); cahier G, *Hydro-géologie, géologie de l'ingenieur et formations superficielles* (226) and cahier H, *Paléontologie* (227).

IBZ (Internationale Bibliographie der Zeitschriftenliteratur aus allen Gebieten des Wissens) began in 1911 but in 1965 started to appear in a new series of volumes. Though covering up to 10,000 journals, publication is traditionally slow and subjects are very wide-ranging. Subject indexes give keywords in German, French and English but bibliographic information is only given at the German entry: cross-references for French and English equivalents are provided.

One of the most comprehensive abstracting services in existence is the USSR Academy of Science's *Referativnyi Zhurnal* which was first published in 1954. Over 20,000 journals are abstracted as well as patents, standards, dissertations, maps and monographs. Sections likely to interest the earth scientist include geodesy and aerial surveying, geophysics, geography, geology, mining, chemistry and soil science and agricultural chemistry. All sections are issued monthly and have monthly indexes. Author indexes are arranged in Roman and Cyrillic alphabetical sequences. There is a formula and a patent index for the chemistry section, and a geographical index for the geography section. Abstracts are in Cyrillic but full bibliographic details are given in the language of the paper. Subject indexes are in Cyrillic. English-language contents lists can be found in some of the sections (in most Russian bibliographies in monographs, English-language works are listed last in a group and enable the reader to get some idea of the subject of the book). The subject sequence of the abstracts is the same from issue to issue and this is helpful to those whose Russian is slight. In 1970 the then National Reference Library of Science and Invention (now the Science Reference and Information Service of the British Library) published *A Guide to Referativnyi Zhurnal*, by E J Copley.

All sections, apart from *Geodeziya* and *Pochvovedenie* (Soil science), are available in sub-series. Geology is divided into general geology; stratigraphy and palaeontology; geochemistry, mineralogy, petrography; Quaternary period, geomorphology of land and ocean bottom; geological and geochemical prospecting; hydrogeology, engineering geology, geocryology; ore deposits; non-metallic useful minerals; deposits of useful combustible minerals; techniques of geological prospecting. The publication *Abstracting Services* (International Federation for Documentation, 1969) gives details of sub-series in other sections.

Although there are some long-established abstracting journals in the earth sciences, as described above, others have lasted only a short time, for example *Geoscience Abstracts* which covered only literature published in, or about, North America.

Among other useful abstracting publications to which the earth scientist may occasionally have recourse are:

Annotated Bibliography of Economic Geology, 1928–65
Bibliography of Seismology, 1966–
British Geological Literature, 1964–68
Geographical Abstracts, 1972–
Geophysical Abstracts, (Washington) 1929–71
Mineralogical Abstracts, 1922–

The practice of citation is important both for the subject under consideration and for the author who wants to know how frequently his work is referred to, and by whom. The best-known citation index is *Science Citation Index* (*SCI*) which is published commercially from Philadelphia by the Institute for Scientific Information. Two main indexes, published bi-monthly, the *Citation Index* and the *Source Index*, allow identification of cited papers by a particular author, on particular subjects, and the names of journals in which the citations have appeared, with the citing authors' names. In the *Source Index* is a list of items arranged alphabetically by author, with full bibliographic details, which have appeared in the journals scanned: in the early issues of *SCI*, British geological journals were poorly represented. In the *Citation Index* references are listed in alphabetical order of first author, and under each entry is a list of the authors who cited it in the period covered by the index. Titles of the citing papers have to be obtained by referral to the *Source Index*. In order to save space there is heavy use of abbreviation both of names and publication titles.

A reader will usually go straight to the *Citation Index* with the knowledge of some papers in his field, perhaps written some time ago. He will be led from these items to the current citing papers listed beneath whereupon reference to the *Source Index* gives full bibliographic details. These papers in turn can be obtained and the same process can be initiated: this process has been called "cycling". In 1967 the *Permuterm Subject Index* appeared as the third element of *SCI*. This index contains the words used in the titles of the papers listed in the *Source Index*. All the words in the titles are arranged in alphabetical order and under each appears a list of the words which occur with it in any of the titles. In *Figure* 5.2 a paper on mineral layering in granite can be found by looking under "layering". In the list both "mineral" and "granite" appear and alongside each is the name of the author (Claxton) of a paper on that subject. Bibliographic details can then be found in the *Source Index*. Useful guides to *SCI* can be found in papers by Cawkell 'Search strategies using the *Science Citation Index*' in *Computer based information retrieval systems* (1968) and by Malin 'The *Science Citation Index*' in *Library Trends*, **16**(3), 1968.

SCI has its own bibliography which appears in the publication itself. In 1982 the Institute for Scientific Information published *GeoSciTech Citation Index* which ran only from 1981–1982, two annual issues only. The work is subtitled '*An international interdisciplinary index to the literature of geology, earth science, mineralogy, atmospheric sciences and other disciplines*'. For both years there are the three volumes; *Source Index, Citation Index* and *Permuterm Subject Index*. The *Source Index* volume includes a corporate index which appears first in order of country (including state), and then again alphabetically, with a world-wide coverage.

The coverage of the various abstracting services may appear confusing to enquirers and there is, of course, a certain amount of overlap. Each service is aimed at a different clientele; this may be language- or subject-oriented. No abstracting service can cover its field in its entirety so, for a full bibliographic search, it is necessary

Permuterm index　　　　　Citation index

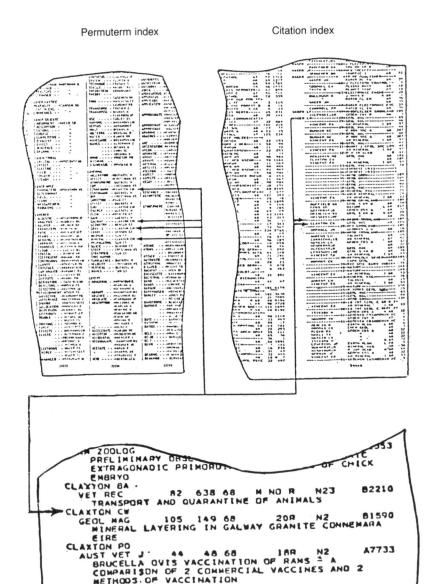

Source index

Figure 5.2 Science Citation Index

to scan several relevant services. For the first edition of this book the author made a survey of the extent of coverage, and coverage overlap, of abstracting services in a number of fields by checking, in appropriate services, bibliographies which he had compiled in different subject fields. The general conclusions from the findings in 1973 are probably still valid but the application of computerized online searching services should lessen the concern then expressed about omission and overlap.

The time-lag, referred to earlier, between publication of the original paper and of its abstracts, encouraged some publishers to produce so-called 'current awareness' publications. These publications usually give only the titles of papers and precede the abstracting and indexing journals.

The best known of the current awareness publications is *Current Contents*, which began life in 1969, and is published by the Institute of Scientific Information in Philadelphia. Various subject editions are produced and the publication appears weekly. The issues are made up of reproductions of the contents pages of recently published journals in the particular field covered: the contents pages are arranged in broad subject groups and the journals represented are listed on the cover. Many journal publishers send their contents pages at proof stage so that the issues of *Current Contents* appear as fast as possible, and distribution arrangements try to maintain the speed.

Geotitles (formerly *Geotitles Weekly*) was published by the British company Geosystems and it endeavoured to draw attention to publications, news items, and events, very rapidly.

CHAPTER SIX

Computerized information services and geological databases/databanks

DUNCAN J McKAY AND MICHAEL O'DONOGHUE

The output of published primary literature in the earth sciences has been estimated to have grown to 90,000 items per annum and for many years publishers have used computers to assist in the compilation of the large abstracting and indexing journals mentioned in the previous chapter, and secondary sources referred to in Chapter 4, e.g. directories such as *American Men and Women of Science, Marquis Who's Who*, bibliographies such as *Books in Print*, and encyclopedias such as the *Encyclopaedia Britannica*. An important advantage of this method of accessing information is that it usually improves the currency of the information in the reference tools represented. Since the first edition of this book (1973) the developments in the mechanized information field have been enormous and have led in great part to the growth of information science in general: the development of sophisticated online information retrieval systems, and advances in both computer technology and telecommunications have ensured that this information explosion has been managed to the benefit of the earth scientist.

Modern technology makes it possible to access large bibliographic databases via a 'host', e.g. Dialog, BLAISE, etc. The main hosts for earth science files are Dialog, Questel, SDC, ESA-IRS, and Data-Star: others are listed in the Appendix to this chapter under individual databases.

General databases are those such as are accessed through the BLAISE service of the British Library: these include the British Library's catalogues, the *British National Bibliography*, the KIST service (Keyword Index to Serial Titles) which covers the periodical holdings of the Science Reference and Information

Service and the Document Supply Centre of the British Library, the Science Museum Library and Cambridge University Library, and other services. One advantage of using the BLAISE service is that Cataloguing-in-Publication (CIP) entries are available for monographs in course of publication thus offering an up-to-date listing of books.

The best-known databases for geological information are GeoRef and GeoArchive both of which allow literature searches to be undertaken in a fraction of the time required by manual methods, and to a more sophisticated level, but it is essential to be aware that online bibliographic databases are not a complete answer to literature searching. A recent paper (Ahmed, 1986) considers the problems encountered in compiling a full bibliography on a restricted sedimentological topic. The lack of overlap in total coverage, coupled with poor coverage of certain geographical regions, particularly the third world, in the major earth science databases, has been observed by a number of workers.

The computer technology which was first applied to assist in the production of hardcopy abstracting and indexing journals, in particular the indexing of the titles to create keyword-in-context (KWIC) indexes, has been developed with appropriate software to enable fast literature searches to be made, and more recently, a similar development has brought numeric databanks as accessible as bibliographic databases. These parallel developments are now drawing together and earth scientists, together with information scientists, should aim towards an integrated geological information service. Pruett (1986) provides a basis for the format of an 'ideal geoscience information system' which would give such coverage.

The major commercial hosts such as Dialog, Data-Star and ESA-IRS mount a wide range of databases and databanks on very large computers, to which subscribers may have access on the allocation of a user identity code and a password. The key to efficient searching of the host's files is to select the most appropriate files, and determine the keywords (subject word or phrase, synonyms, broader terms, etc.) and their relationship, before logging into the computer; it is possible also to search on an author's name. If an initial response gives too many references, various qualifiers may be used to narrow the search, e.g. publication date, a language limiter, bibliographic type, etc. Thus the very general search for 'geology' may become 'theses, on petroleum geology, published in Germany, between 1969–74'. A geographic limiter may be particularly appropriate in earth science searches, e.g. in the above case the search may be limited to the 'Persian Gulf'. The choice of terms to identify the subject is made easier if a thesaurus has been published for the database being

used. The relationship of selected terms may be defined by using
the usual Boolean operators, 'AND', 'OR', 'AND NOT': these
operators may be used after it has become clear which terms have
produced the most relevant references, when the application of
Boolean logic will draw these out.

Bibliographic information given by the databases can range
from full entries, including abstracts, to short listings of title and
year of publication only: the most usual format chosen gives
author, title, and source details.

Originally searches were done in batch mode and the printed
result sent to the enquirer: modern computer techniques enable
the person searching the database either to print the results online
(though this can be very time consuming and costly if there are
many references and abstracts are included), or to download the
references to disk, disconnect from the computer, and print out
the results offline. The proliferation of desk-top terminals may
mean that the earth scientist will do independent searching of
databases and databanks, as long as the necessary telecom-
munication links are available; however, to use these information
services economically and adequately, training is required and
many libraries have appointed specialist staff to undertake, on
request, online searches for their readers.

Notwithstanding the great advances made in computer searching in recent years, the use of published abstracting services, e.g.
Bibliography and Index of Geology, is by no means a thing of the
past.

Some of the main bibliographic databases in the earth sciences
are described below but a more extensive listing of both databases
and databanks may be found in the Appendix to this chapter.

GEOREF is the database of the American Geological Institute
(AGI) and covers the worldwide technical literature on geology
and geophysics: it corresponds to the six printed AGI publications
   Bibliography and Index of North American Geology, 1961–69
   Bibliography of Theses in Geology, 1965–66
   Geophysical Abstracts, 1966–71
   Bibliography and Index of Geology Exclusive of North America,
   1933–68
   Bibliography and Index of Geology, 1969–
   Bibliography and Index of Micropaleontology, 1972–
The coverage of these publications, and the online service,
includes journals, books and book chapters, conference papers,
government publications, theses, dissertations, reports, maps, and
meetings papers. In March 1981 these records totalled 652,380.

One advantage of the online service is that some additional records, not found in the printed bibliographies, are included, especially references to the *Professional Papers* series of the US Geological Survey since 1978.

GEOARCHIVE was produced by the British firm Geosystems but, at the time of writing, is changing hands: it is advertised as a comprehensive database covering all types of information sources in geoscience as long as they are publicly available and, therefore, small news items may be included as well as longer articles. A *Geosaurus* was published as a guide to the subject indexing of the database.

GEOBASE is a new online database, corresponding to the publications of *Geographical Abstracts* (see Chapter 17). The subjects covered include all the branches of geography, geology, and related aspects of ecology. The service was launched in 1987 with over 250,000 records dating from 1980, and it is updated monthly. As a new service this database has still to be evaluated.

PASCAL is an online service compiled by the Centre de Documentation Scientifique et Technique in France, and is linked with the abstract journal *Bulletin Signalétique* (1973–83) and *Bibliographie Internationale Pascal Sigma: Pascal Thema* (1984–). Worldwide literature of all types but mainly journal articles, theses, reports, and conference proceedings, are entered on the file, normally in the language of origin with a French translation of the title and abstract. After 1982 keywords are given in both English and French and titles translated into English are available also. In 1986 the file size was quoted as about 5,600,000 references, but that covers entries for physics, chemistry, and life sciences, as well as earth sciences.

PETROLEUM ABSTRACTS is an online database for which a licence fee must be paid to the University of Tulsa before it may be accessed. The database began in 1965 and gives world-wide coverage of technical literature and patents in the fields of exploration, development, and production of oil and natural gas. The controlled vocabulary *Exploration and Production Thesaurus* is the listing from which index terms are assigned to the entries in the database: there is also a *Geographic Thesaurus* with its own supplementary descriptors, and a KWOC (Keyword-out-of-Context) list of the *Exploration and Production Thesaurus* and addenda descriptors from the supplementary word list, to assist in the selection of relevant keywords.

Geologists need to be aware also that online services in other disciplines may contain relevant references and searches should be made also in the files of Chemical Abstracts Service, Biological Information Services, and Inspec, to cover such major hard copy abstracting journals as *Chemical Abstracts, Biological Abstracts, Physics Abstracts, Electrical and Electronic Abstracts*, and *Computing and Control Abstracts*. There are many more bibliographic databases on which specialists in library and information units will be able to advise.

During the same period that the bibliographic databases have developed, there has been a large increase in the number of numeric databases, or databanks, which provide access to the basic numeric data required, and gathered, by earth scientists. An examination of the literature during August 1987 traced about 400 databases which provide information of use to earth scientists, and these are listed in the Appendix at the end of this chapter, in two parts. The first details databases considered to be of most significance or interest; the second lists those of more specialized interest.

General directories which identify scientific, geoscientific, and more general databases covering business and financial information, are available in both hardcopy and machine-readable forms: examples include the *Encyclopedia of Information Systems and Services*, 2 vols, (1985–86) and *Database of Databases*. Others are listed in the 'References' section at the end of this chapter.

Since the first edition of this work a number of periodicals have appeared which have carried articles on the application of computer technology in library and information work: some have been short-lived but others have become more established, e.g. *Online* and *Online Review*. In *Database* a paper by Burk (1982) is the only recent work to have covered geoscientific databases in any detail. Other papers in both the annual proceedings of the Geoscience Information Society and the quadrennial International Conference on Geological Information (London, 1978; Denver, 1982; Adelaide, 1986) consider aspects of databases, specifically and in general.

Details of numeric and other source databanks may be similarly located, but with less ease; many of these source databases are of interest only to restricted groups of scientists and tend to receive less publicity than their bibliographic counterparts.

Most easily accessible are the databases listed in the *USGS Circular* **817** (1983) despite some changes which have occurred since it was published. For Western Europe the small guide *Geological Data for Application to the Environment* (British Geological Survey, 1986) provides information on both machine- and non-machine-readable databases. *Geoscience databases in Australia* has been compiled by E P Shelley and published in the *Bureau of Mineral Resources Report* **269** (1985). COGEODATA and COGEODOC'S *Inventory of Automated Data Bases in Europe, Asia and Africa*, appeared in preliminary form in 1988, and should be available on floppy disc in time for the 28th International Geological Congress in 1989.

References

Ahmed, R (1986) Bibliography of lacustrine carbonate sediments and sedimentary carbonate rocks: discussion and a proposal. *Proceedings of the Third International Conference on Geoscience Information, Adelaide*. Volume **1**, 241–4.

British Geological Survey (1986) *Geological Data for Application to the Environment*. Edinburgh: British Geological Survey

Burk, C J (1982) A worldwide list of source and reference databases in the geosciences. *Database*, **5** (June), 11–21.

Codata Directory of Data Sources for Science and Technology *Codata Bulletin* (issues in a number of chapters)

Cuadra Associates Inc. *Cuadra Database*. This corresponds to the online Directory of Online Databases available on SDC and Data-Star

Databanks and databases in geology (1981). (Thematic set of papers) *Journal of the Geological Society of London*, **138**,573–630

Database Directory Service (1986–87). New York:Knowledge Industry

Encyclopedia of Information Systems and Services, 1985–86. 2 vols. New York: Gale

EUSIDIC Database Guide (1983) Oxford:Learned Information

Findlay, M A (1984) Database development in Australia. *Proceedings, National Online Meeting*, 47–55

Gupta, D K (1985) Citation analysis study of sub-disciplines of exploration geophysics. *INICAE*, **4**,(1), 5–13

McKay, D J Unpublished research on the obsolescence of petrological literature

Pruett, N J (1986) State of the art of geoscience libraries and information services. *Proceedings of the Third International Conference on Geoscience Information, Adelaide*. Volume **2**, 15–30

Schmittroth, J and Lucas, A eds (1985–86) *New Information Systems and Services; a periodic supplement to the United States and international volumes of the sixth edition of 'Encyclopedia of Information Systems and Services'*. (Two issues) New York:Gale

Shelley, E P compiler (1985) Directory of government geoscience databases in Australia, 1984. *Bureau of Mineral Resources Report* **269** (Microfiche MF214)

US Geological Survey. Office of the Data Administrator (1983) Scientific and technical, spatial and bibliographic databases and systems of the US Geological Survey. *US Geological Survey Circular* **817**

Williams, M (1989) *Computer-readable Databases*. 5th edn. Detroit: Gale Research

Relevant journals:

Database (1977–   )

Online (1976–   )

Online Review (1976–   )

APPENDIX 1: Database listing

In general this listing of the more significant and/or more easily accessed databases is arranged alphabetically by database name; exceptions are made for those databases not having a specific name which are entered under pseudo-titles. It is divided into two parts: (i) Listing of significant databases (ii) Listing of specialized databases.

The level of information provided is dependent on accessibility during the short time available for compilation (September 1987) and it should be noted that it has not been possible to verify information by direct contact with database producers.

No details of costs are included with the database details as these may vary from host to host and year to year. The vast literature of the online world, including journals such as *Online*, may be consulted for both individual and comparative database reviews.

The minimum entry provides details of database name and producer, the type of database and its coverage: information on the size and frequency of update are provided where possible. An indication of accessibility is also provided, either by listing hosts or by noting availability of public online access; "not available online" indicates that there is no public real-time online access — it should be particularly noted that this type of information is subject to rapid change.

Listing of Significant Databases

ADIGE
  Producer   Center for Stratigraphy & Petrology of the Central Alps
  Coverage. . . .
    Bibliography of Italian geology, plus data relating to geological formations, especially in the Alpine area, and well logs from the Po Plain.
  File Info
    1200 records at 1985.
    Closed file

AESIS
AUSTRALIAN EARTH SCIENCES INFORMATION SYSTEM
  Producer   Australian Mineral Foundation
  Coverage. . . .
    Bibliography of Australian earth science, includes that published both within Australia and without.
  File Info
    36,000 records at 1985 plus 6,500 pa
  Hosts
    Ausinet

AOSI
ALBERTA OIL SANDS INDEX
  Producer Alberta Oil Sands Technology and Research Authority
  Coverage. . . .
    Alberta oil sands and heavy oils
    (see also HERI)
  File Info
    8,100 records at 1985, plus 1,000 pa
  Hosts
    CAN/OLE and QL

APPLIED SCIENCE AND TECHNOLOGY INDEX
  Producer Wilson (HW) & Co
  Coverage. . . .
    Includes energy, petroleum and gas literature, from 1983–
  File Info
    114,000 records at 1985 plus 60,000 pa
  Hosts
    Wilsonline

ARCTIC SCIENCE AND TECHNOLOGY INFORMATION SYSTEM (ASTIS)
  Producer Arctic Institute of North America (Calgary)
  Coverage. . . .
    Includes Arctic earth sciences with an emphasis on North America
  File Info
    13,300 records at 1985
  Hosts
    QL

ASIAN GEOTECHNOLOGY
  Producer Asian Institute of Technology
  Coverage. . . .
    Geotechnical engineering projects relevant to Asia
  File Info
    28,500 at 1986
  Hosts
    ESA/IRS

AUTOMATED MINERALS INFORMATION SYSTEM (AMIS)
  Producer US Bureau of Mines Minerals Information Directorate
  Coverage. . . .
    All aspects of the production and consumption of strategic non-fuel mineral
    commodities
  Hosts
    Not available online

BANQUE DE DONNEES GEOCHIMIQUES I
  Producer Institut Français du Pétrole
  Coverage. . . .
    Petroleum geochemical analyses
  File Info
    27,000 records at 1983
  Hosts
    Not available online

BIBLAT
    Producer   Centro de Informacion Cientifica y Humanistica
    Producer   Universidad Nacional Autonoma de Mexico
    Coverage. . . .
        Multidisciplinary database providing details of papers by Latin American
        authors or on Latin America
    File Info
        27,000 records at 1984
    Hosts
        Questel

BIBLIOGRAPHIC INFORMATION DATABASE
    Producer   British Geological Survey
    Coverage. . . .
        Bibliography of BGS publications
    File Info
        4,100 records at 1986
    Hosts
        Dialog — included in GeoArchive

BIBLIOGRAPHIC INFORMATION ON SOUTH-EAST ASIA
BISA
    Producer   University of Sydney Library
    Coverage. . . .
        Includes earth science and resource information for SE Asia especially
        Indonesia, Malaysia and Singapore.
        Co-operative cataloguing system
    File Info
        20,000 records at 1983
    Hosts
        Ausinet

BIOSIS
    Producer   Biosciences Information Service (Philadelphia, USA)
    Coverage. . . .
        Provides bibliographic coverage of palaeontology, ecology and soils
    File Info
        3.8m at 1986
    Hosts
        Dialog, SDC, BRS, ESA/IRS

CAB ABSTRACTS
    Producer   Commonwealth Agriculture Bureau
    Coverage. . . .
        Primarily concerned with agriculture, CAB Abstracts includes hydrology in
        its coverage of "arid lands"
    File Info
        1.85m at 1986
    Hosts
        Dialog, ESA/IRS, BRS, SDC, Dimdi

CANADIAN SEA ICE INFORMATION SYSTEM
Producer   Center for Cold Ocean Resources Engineering (University of
      Newfoundland)
Coverage. . . .
    Data on historical ice concentration, distribution, morphology, physical and
    mechanical properties for Canadian Arctic and East Coast
Hosts
    Available by direct dial

CANMINDEX — CANADIAN MINERAL OCCURRENCE INDEX
Producer   Geological Survey of Canada (Ottawa)
Coverage. . . .
    Spatial distribution of Canadian minerals and associated bibliographic
    information
Hosts
    Not available online

CENTRAL MEDITERRANEAN DATABASE
Producer   Kingston Polytechnic Geological Services and Earth Sciences Centre
Coverage. . . .
    Bibliographic database covering all aspects of the earth sciences in the Central
    Mediterranean Region
Hosts
    Database may be purchased; not available online

CHEMICAL ABSTRACTS
Producer   Chemical Abstracts (Columbus, Ohio)
Coverage. . . .
    Includes coverage of cosmochemistry, economic geology, fossil fuels,
    geochronology, geothermal resources, mineralogy, petrology, soils and water
    chemistry.
    Minerals are indexed to some 5,800+ chemical substance headings.
File Info
    7.1m in total at 1986
Hosts
    Dialog, ESA/IRS, Data-Star, SDC and others

COAL DATABASE
Producer   IEA Coal Research
Coverage. . . .
    All aspects of coal research and technology
File Info
    48,000 records at 1985 plus 15,000 pa
Hosts
    Inka, Belendis, CAN/OLE and others

COLD
(ANTARCTIC BIBLIOGRAPHY)
Producer   Cold Regions and Engineering Laboratory (Hanover, New Hampshire)
Coverage. . . .
    Coverage includes aspects of the geosciences in Antarctica
File Info
    89,300 records at 1984
Hosts
    SDC

COMPENDEX
(ENGINEERING INDEX)
  Producer   Engineering Information Inc (New York)
  Coverage. . . .
    Includes engineering geology, mining engineering, fuel technology and petroleum engineering
  File Info
    1.5m records at 1986 plus 140,000 pa
  Hosts
    Dialog, SDC, ESA/IRS, Data-Star and others

CONFERENCE PAPERS INDEX
  Producer   Cambridge Scientific Abstracts (Bethesda, USA)
  Coverage. . . .
    Includes conference papers on all aspects of the geosciences
  File Info
    1.3m records at 1986 plus 120,000 pa (totals for all sciences)
  Hosts
    Dialog, ESA/IRS

CONFERENCE PROCEEDINGS INDEX
  Producer   British Library Document Supply Centre (Boston Spa)
  Coverage. . . .
    Includes earth science conference holdings information for BL(DSC)
  Hosts
    Blaise

CURRENT RESEARCH IN BRITAIN
(CRIB)
  Producer   British Library Document Supply Centre (Boston Spa)
  Coverage. . . .
    Coverage includes earth science research at British institutions including universities, polytechnics
  File Info
    65,000 projects; updated annually
  Hosts
    Pergamon

DATA RESOURCES INC SYSTEM
  Producer   Data Resources Inc
  Coverage. . . .
    A number of databanks maintained containing time-series relating to drilling data, energy company activities and other statistical aspects of the energy industries
  File Info
    100,000+ time series at 1985
  Hosts
    Direct dial

DISSERTATION ABSTRACTS
(COMPREHENSIVE DISSERTATION ABSTRACTS)
  Producer   University Microfilms International
  Coverage. . . .
    Multidisciplinary database providing citations to doctoral theses accepted by American and Canadian universities.

File Info
900,000 records at 1985
Hosts
Dialog, BRS, SDC and others

DOCOCEAN

Producer   Centre National pour l'Exploitation des Océans (Brest, France)
Coverage. . . .
Integrated access to ASFA, Ocean Abstracts, Pascal (Oceanology) and CNEXO-BNDO databases; includes coverage of resource exploration
Hosts
BNDO

EARTH SCIENCE INFORMATION DATA SYSTEM
ESIDB

Producer   US Geological Survey Geologic Division
Coverage. . . .
Information about databases held and/or produced by the USGS

EARTHQUAKE DATA FILE

Producer   National Geophysical and Solar-Terrestrial Data Center (USA)
Coverage. . . .
Worldwide data including dates, times, depths of foci for earthquakes
Hosts
Not available online

ECONOMINE

Producer   Bureau du Recherches Géologiques et Minières
Coverage. . . .
Economic aspects of all minerals (fuel and non-fuel) from 1984; for earlier information see PASCAL
File Info
8,000 records at 1987
Hosts
Questel

ELECTRICAL PROPERTIES BIBLIOGRAPHY

Producer   US Geological Survey Geologic Division
Coverage. . . .
Worldwide references regarding the electrical properties of minerals

ENERGY DATABASE
DOE ENERGY/EDB

Producer   US Department of Energy
Coverage. . . .
Coverage includes all aspects of energy; file also includes information from other databases such as INIS
Hosts
Dialog, Inka

ENERGYLINE
  Producer   Environment Information Centre Inc
  Coverage. . . .
    Includes coverage of energy resources and reserves in a general database
    covering scientific, engineering, political and socio-economic aspects of
    energy.
  File Info
    86,000 records at 1986 plus 6000 pa
  Hosts
    Dialog, SDC, ESA/IRS

ESCAP BIBLIOGRAPHIC INFORMATION SYSTEM
  Producer   Economic and Social Commission for Asia and the Pacific
  Coverage. . . .
    Includes mineral and energy resource literature for the ESCAP region (Asia
    and the Pacific)
  File Info
    20,000 records at 1985 plus 9000 pa
  Hosts
    Not available online

FEDERAL RESEARCH IN PROGRESS
  Producer   National Technical Information Center
  Coverage. . . .
    Includes details of federally funded research projects in the earth sciences,
    continues SSIE
  File Info
    71,000 records at 1985
  Hosts
    Dialog

GENERAL SCIENCE INDEX
  Producer   Wilson (HW) & Co
  Coverage. . . .
    Includes earth sciences and oceanography literature
  File Info
    25,000 records at 1985
  Hosts
    Wilsonline

GEOARCHIVE
  Producer   Geosystems
  Coverage. . . .
    Comprehensive coverage of all aspects of the geosciences; with sources
    publicly available and holding information content even where a news item is
    of ephemeral interest
    Printed publications including "Geotitles Weekly" are derived from the
    database.
  File Info
    250,000 records at 1979 plus 60,000 pa
  Hosts
    Dialog

GEOBANQUE
    Producer   Bureau du Recherches Géologiques et Minières
    Coverage. . . .
        All aspects of French underground works (wells, boring, quarries and mines)
    File Info
        400,000 records at 1986
    Hosts
        Questel

GEOBASE
    Producer   Geoabstracts Limited
    Coverage. . . .
        Includes all aspects of geology from 1980– which are included in the *Geographical Abstracts* series
    File Info
        200,000 records at 1987 plus 3,000 pa
    Hosts
        Dialog

GEODX
NATIONAL DATABASE OF STRATIGRAPHICAL NAMES
    Producer   Bureau of Mineral Resources (Canberra)
    Coverage. . . .
        References to Australian geological papers on stratigraphic names
    File Info
        8,000 records
    Hosts
        Not available online

GEOGEN
    Producer   Centre de Recherches Pétrographiques et Geochimiques (Vandoeuvre-les-Nancy, France)
    Coverage. . . .
        Geochemical analyses and related descriptive data
    Hosts
        Not available online

GEOGRAPHIC NAMES INFORMATION SYSTEM
GNIS
    Producer   Defense Mapping Agency. Scientific Data Dept. (Washington DC)
    Producer   US Geological Survey National Cartographic Information Center
    Coverage. . . .
        Maintains machine readable files of world-wide geographic names
    File Info
        4.5m names at 1985
    Hosts
        Not available online

GEOINDEX
    Producer   US Geological Survey
    Coverage. . . .
        Index to original published geological maps — where detail is greater than that shown on published state geologic maps
    Hosts
        Not available online

GEOKART
Producer   Austria. Geological Survey
Coverage. . . .
   Databank of geological maps
File Info
   10,000 entries

GEOLINE
Producer   Bundesanstalt fur Geowissenschaften und Rohstoffe (BGR)
   (Hannover) (presumed)
Coverage. . . .
   All aspects of earth science literature from 1970–. Hydroline database covering
   hydrology and hydrochemistry is included as a sub-set
File Info
   450,000 records at 1986 plus 35,000 pa
Hosts
   Inka

GEOMAGNETISM DATABANK
Producer   British Geological Survey
Coverage. . . .
   World digital centre for geomagnetism
File Info
   0.4Gbyte data

GEOMAPFIL
Producer   US Geological Survey Geologic Division
Coverage. . . .
   Availability catalogue for data types shown on geologic maps; file can be used
   for identifying maps with particular characteristics

GEOMECHANICS ABSTRACTS
Producer   Rock Mechanics Information Service (Imperial College, London)
Coverage. . . .
   All aspects of rock and soil mechanics and engineering geology
File Info
   27,500 records at 1985 plus 1,500 pa
Hosts
   Infoline

GEONAMES
Producer   US Geological Survey Geologic Division
Coverage. . . .
   Rock stratigraphic names for USA
File Info
   17,000 names at 1985
Hosts
   Available as magnetic tape service

GEOREF
Producer   American Geological Institute (Alexandria, VA)
Coverage. . . .
   Bibliographic coverage of all aspects of the geosciences; "the single most
   significant database for geoscientists' literature searching"

File Info
  1m records at 1986 plus 55,000 pa
Hosts
  Dialog, SDC, CAN/OLE

GEOSCAN
  Producer   Geological Survey of Canada (Ottawa)
  Coverage. . . .
    Solid-earth geoscience literature on Canada, or produced therein with
    emphasis on unpublished reports
  File Info
    100,000 records at 1985 plus 1,000 pa

GEOSCITECH
  Producer   Institute for Scientific Information (ISI) (Philadelphia)
  Coverage. . . .
    Bibliographic coverage of all aspects of earth science
  File Info
    300,000 records at 1983 plus 55,000 pa

GLACIER INVENTORY OF CANADA
  Producer   National Hydrology Research Institute (Ottawa)
  Coverage. . . .
    Two files providing details of the physical characteristics of Canadian glaciers
    and related bibliographic information
  File Info
    15,000 glaciers; 3,000 references

GLOBAL SEISMOLOGY DATABANK
  Producer   British Geological Survey
  Coverage. . . .
    Seismic Data for the world
  File Info
    Around 20 Gbyte of data

GPO MONTHLY CATALOGUE/GPO PUBLICATIONS
REFERENCE FILE
  Producer   US Government Printing Office
  Coverage. . . .
    File indexes publicly available documents produced by the legislative and
    executive branches of government. Coverage includes earth sciences
    The reference file details those publications currently in print at the USGPO
  File Info
    154,000 records at 1983
  Hosts
    Dialog, BRS

HAIL BIBLIOGRAPHY
  Producer   Illinois State Water Survey Library
  Coverage. . . .
    Hail and related subjects
  File Info
    1,800 records at 1985
  Hosts
    Not available online

HEAVY OIL/ENHANCED RECOVERY INDEX
(HERI)
    Producer   Alberta Oil Sands Information Center
Coverage. . . .
    Bibliographic coverage of enhanced oil recovery and heavy oils in Alberta
    File Info
        2,500 records in 1985 plus 1,000 pa
    Hosts
        QL, direct dial

IGNEOUS BASE
    Producer   Carnegie Institution (Washington DC)
    Coverage. . . .
    Petrological data for igneous rocks derived from literature worldwide

IMAGERY DATABASE FILE
    Producer   US Geological Survey
    Coverage. . . .
    Catalogue and index of remote sensing imagery
    Hosts
        Not available online

IMMAGE
INFORMATION ON MINING, METALLURGY AND
GEOLOGICAL EXPLORATION
    Producer   Institution of Mining and Metallurgy
        Coverage. . . .
        All aspects of economic geology, mining and extraction of non-ferrous
        minerals
    File Info
        25,000 records at 1987 plus 4,000 pa
    Hosts
        Pergamon

INDIAN NATIONAL GEOSCIENCE INFORMATION
SYSTEM
    Producer   Geological Survey of India (Calcutta)
    Coverage. . . .
    Indian geoscience literature and data
    File Info
        System under development (1986)

INFOIL
    Producer   Norwegian Petroleum Directorate
    Coverage. . . .
    Research and development project directory for the UK and Norway in all
    areas of oil and petroleum activities
    File Info
        1,500 records at 1985
    Hosts
        NSI

INSPEC
  Producer   Institution of Electrical Engineers
  Coverage. . . .
    Physics subfile (a) includes coverage of geophysics and remote sensing
  File Info
    1969–1976 comprises 990,000 records; 1977–1987 file comprises some 1.9m records plus 180,000 pa
  Hosts
    Dialog, ESA/IRS, BRS, Data-Star, SDC

INTERNATIONAL PETROLEUM ABSTRACTS
  Producer   Institute of Petroleum (London)
  Coverage. . . .
    Coverage includes geological and geophysical aspects of oil and gas exploration
  File Info
    1,500 records at 1986 plus 300 pa
  Hosts
    Pergamon

INTERNATIONAL PHOSPHATE RESOURCE DATABASE
  Producer   US Geological Survey
  Coverage. . . .
    All aspects of phosphate resource geology and associated bibliographic references

INTUREGEO
INTERNATIONAL URANIUM GEOLOGY INFO SYSTEM
  Producer   International Atomic Energy Agency
  Coverage. . . .
    Geological and statistical data relating to uranium
  Hosts
    Not available online

JAPAN OCEANOGRAPHIC DATA CENTER
  Coverage. . . .
    Maintains files of oceanographic data for all areas in Japanese territory
  Hosts
    Not available online

JCPDS INTERNATIONAL CENTRE FOR DIFFRACTION DATA
  Producer   JCPDS
  Coverage. . . .
    X-ray diffraction and similar data for minerals, metals and alloys
  File Info
    42,000 records at 1985

JICST SCIENCE AND TECHNOLOGY FILE — EARTH SCIENCES, MINING AND METALLURGY
  Producer   Japan Information Center of Science and Technology (Tokyo)
  Coverage. . . .
    Includes worldwide mining and earth sciences literature
  File Info
    542,000 records at 1985 plus 66,000 pa

LEAD ISOTOPE DATA BANK
(LIDB)
  Producer   US Geological Survey
  Coverage. . . .
    Isotopic data for rocks and ores (worldwide)
  Hosts
    Not available online

LEDA 2
  Producer   European Space Agency
  Coverage. . . .
    Catalogue of images received by Funcino (1975–), Kocina (1978–),
    Maspalosmas (1984–) and US-EROS (1972–1982)
  File Info
    230,000 records at 1985 plus 30,000 pa
  Hosts
    ESA/IRS

MAP CATALOGUE
  Producer   British Geological Survey
  Coverage. . . .
    Map catalogue
  File Info
    120,000 maps at 1986
  Hosts
    Dialog (inc in GeoArchive)

MARINE GEOLOGY DATABASE
  Producer   Bureau du Recherches Géologiques et Minières (presumed)
  Coverage. . . .
    28,880 locations recorded, 21,767 have geological descriptions (at 1986)

MARINE RESOURCE DATA BANK
  Producer   National Geophysical and Solar-Terrestrial Data Center (USA)
  Coverage. . . .
    Analyses of manganese nodules (worldwide)
  Hosts
    Not available online

MEETING AGENDA
  Producer   Commissariat à l'Energie Atomique
  Coverage. . . .
    Includes conference announcements for the earth sciences, which are deleted
    after the meetings have taken place
  File Info
    5,500 records
  Hosts
    Questel

METADEX
METALS ABSTRACTS
  Producer   American Society for Metals
  Producer   Metals Society (London)
  Coverage. . . .
    Coverage includes metal ores
  Hosts
    ESA/IRS, Dialog

METEOROLOGICAL AND GEOPHYSICAL ABSTRACTS
  Producer   American Meteorological Society (Boston, Ma)
  Coverage. . . .
    Subjects covered include meteorology and astrophysics; physical oceano-
    graphy; hydrology and glaciology
  File Info
    57,000 records at 1980 plus 7,200 pa
  Hosts
    Dialog and others

MINERAL RESOURCES DATA SYSTEM
MRDS
  Producer   US Geological Survey
  Producer   University of Oklahoma
  Coverage. . . .
    Comprises the four data banks: CRIB, CRIS, DIST, GEOS which cover all
    aspects of the world's mineral resources
  Hosts
    Available online

MINERALOGICAL ABSTRACTS
  Producer   Mineralogical Society of Great Britain
  Coverage. . . .
    Bibliographic coverage of geochemistry, mineralogy and petrology
  File Info
    plus 5,000 pa

MINERALS AVAILABILITY SYSTEM
  Producer   US Bureau of Mines Minerals Information Directorate
  Coverage. . . .
    Reserve data for 34 US and foreign minerals
  Hosts
    Not available online

MINING TECHNOLOGY
MINTEC
  Producer   Canada Center for Mineral and Energy Technology (CANMET)
  Coverage. . . .
    Bibliographic coverage of mining technology and related topics of relevance to
    Canada

MINSEARCH
   Producer   British Geological Survey
   Coverage. . . .
     Coverage includes geological and exploration aspects of minerals
   File Info
     12,000 records at 1985; closed file?
   Hosts
     Pergamon

MINSYS DATORIUM
   Producer   Geosystems
   Coverage. . . .
     Coverage includes aspects of non-fuel minerals from exploration to production
   File Info
     5m records at 1985 plus 50,000 pa

MOLARS
   Producer   Meteorological Office (Bracknell, UK)
   Coverage. . . .
     Coverage of meteorology, climatology, surface oceanography, planetary
     atmospheres with more limited coverage of fluid mechanics, and computers as
     applied in/to meteorology. Most items available at Meteorological Office
   File Info
     139,000 records at 1986 plus 9,600 pa
   Hosts
     ESA/IRS

NASA
(SCIENTIFIC AND TECHNICAL AEROSPACE REPORTS/
INTERNATIONAL AEROSPACE ABSTRACTS)
   Producer   NASA Scientific and Technical Information Office
   Coverage. . . .
     Coverage of worldwide report literature includes geosciences
   File Info
     1.3m records at 1983 plus 60,000 pa
   Hosts
     ESA/IRS

NATIONAL GEOCHEMICAL DATA BANK
   Producer   British Geological Survey
   Coverage. . . .
     Archival storage of public geochemical data produced or analysed in the UK
   Hosts
     Not available online

NATIONAL GEOTHERMAL INFORMATION RESOURCE
(GRID)
   Producer   University of California
   Coverage. . . .
     Bibliographic coverage of geothermal energy

NATIONAL GRAVITY DATABANK
   Producer   British Geological Survey
   Coverage. . . .
     2x2 km grid gravity data for the UK

NATIONAL OCEANOGRAPHIC DATA BANK
Producer   Marine Information and Advisory Service (Wormley, Surrey)
Coverage. . . .
   UK National Centre for oceanographic data

NATIONAL URANIUM RESOURCE EVALUATION PROGRAM
(NURE)
Producer   US Geological Survey
Coverage. . . .
   Geochemistry and geophysics of uranium
Hosts
   Not available online

NATO PUBLICATION COORDINATION OFFICE INDEX
Coverage. . . .
   References to published results of NATO ASI Series publications, includes
   coverage of geosciences
File Info
   13,000 records at 1987
Hosts
   ESA/IRS

NTIS
Producer   National Technical Information Service
Coverage. . . .
   Published and unpublished material produced as a result of federal or federally
   funded research in the USA; also includes reports received as a result of
   exchange with non-US government agencies.
   Coverage includes natural resources and earth sciences
File Info
   1.15m records at 1982 plus 60,000 pa
Hosts
   Dialog, SDC, ESA/IRS, Data-Star and others

OCEANIC ABSTRACTS
Producer   Cambridge Scientific Abstracts (Bethesda, USA)
Coverage. . . .
   Includes coverage of marine geology, geophysics and geochemistry
File Info
   169,000 records at 1985 plus 9,000 pa
Hosts
   Dialog, ESA/IRS, BNDO

OIL INDEX
Producer   Norwegian Center for Informatics
Producer   Norwegian Petroleum Directorate
Coverage. . . .
   Includes petroleum geology, oil and gas exploration literature published in
   Scandinavia
File Info
   18,500 records at 1985 plus 2,000 pa

250754

P/E NEWS
Producer   American Petroleum Institute
Coverage. . . .
Cover to cover indexing of certain petroleum industry journals includes coverage of exploration (part of the API Central Indexing and Abstracting Service)
File Info
340,000 records at 1984 plus 5,000 pa
Hosts
Dialog, SDC, Data-Star

PASCAL
Producer   Centre National de Recherche Scientifique, Centre de Documentation Scientifique et Technique
Coverage. . . .
Multidisciplinary bibliographic database including coverage of all aspects of the earth sciences
Hosts
ESA/IRS, Questel

PETROCONSULTANTS' DATABANKS
Producer   Petroconsultants SA (Geneva)
Coverage. . . .
A series of databanks covering: exploration and development wells; concession information; oil and gas field data; exploration statistical information and details of geophysical surveys from the world, excluding USA and the Soviet bloc
Hosts
Available by direct dial

PETROLEUM ABSTRACTS
Producer   University of Tulsa Information Services Department
Coverage. . . .
Coverage of earth science articles which contribute to petroleum exploration; thereby excluding igneous geology, geomorphology and most palaeontology
Hosts
SDC (licence fee payable to Tulsa before use)

RARE EARTH INFORMATION CENTER DATABASE
Producer   Energy & Mineral Resources Research Institute, Iowa State University
Coverage. . . .
All aspects including geochemistry of rare earth elements
File Info
25,000 records at 1985
Hosts
Not available online

REGIONAL GEOCHEMICAL DATABASE
Producer   British Geological Survey
Producer   National Geochemical Databank of the UK
Coverage. . . .
Geochemical databank maintained by BGS
File Info
5 Mbyte of data at 1986

RESORS
REMOTE SENSING ONLINE RETRIEVAL SYSTEM
  Producer   Canada Center for Remote Sensing (Ottawa)
  Coverage. . . .
    Bibliographic coverage of technology for and applications of remote sensing
  File Info
    43,000 records at 1985 plus 5,000 pa

ROCK ANALYSES STORAGE SYSTEM
RAAS
  Producer   US Geological Survey Geologic Division
  Coverage. . . .
    Chemical analysis data for rocks
  File Info
    600,000 records at 1982
  Hosts
    GEISCO

ROCK INFORMATION SYSTEM
RKNSYS
  Producer   Carnegie Institution (Washington DC)
  Coverage. . . .
    Published data on the chemical composition of Cenozoic volcanic rocks
  File Info
    16,000 records at 1985
  Hosts
    Not available online

ROSCOPS
REPORTS OF OBSERVATIONS AND SAMPLES COLLECTED BY
OCEANOGRAPHIC PROGRAMS
  Producer   Centre National pour l'Exploitation des Océans (Brest, France)
  Coverage. . . .
    Oceanographic information
  File Info
    8,000 expeditions at 1985 plus 200 pa
  Hosts
    BNDO

SAGEOLIT
  Producer   Geological Survey of South Africa
  Coverage. . . .
    Unpublished geoscience literature on southern Africa; reported in 1982 as
    being extended to include published literature for areas south of 17 degrees
  File Info
    4,100 records at 1982 plus 3,000 pa

SCIENCE CITATION INDEX
  Producer   Institute for Scientific Information (ISI) (Philadelphia)
  Coverage. . . .
    Includes coverage of earth sciences
  Hosts
    Data-Star, Dialog and others

SEISMIC DATA ANALYSIS SYSTEM
(SEDAS)
Producer   US Geological Survey
Coverage. . . .
  Data from earthquake monitoring stations, worldwide
Hosts
  Not available online

SEISMIC RISK DATABANK
Producer   Bundesanstalt fur Geowissenschaften und Rohstoffe (BGR)
(Hannover) (Presumed)
Coverage. . . .
  Factual data on earthquakes in the Federal Republic
File Info
  2,000 records

SELECTED WATER RESOURCE ABSTRACTS
Producer   Water Resources Scientific Information Center (USA)
Coverage. . . .
  Bibliographic coverage of all aspects of water resources
File Info
  173,000 records at 1985 plus 7,000 pa

SIGLE
Coverage. . . .
  Coverage of "grey literature" from within EEC and Sweden, examples of
  grey literature would be translations and internal reports.
  Produced by a consortium of European documentation centres
File Info
  47,000 records at 1984 plus 36,000 pa
Hosts
  Inka, BLAISE

SIRIS
Producer   New Zealand Department of Scientific and Industrial Research
Coverage. . . .
  Includes coverage of earth science and energy information
Hosts
  Not available online

SMITHSONIAN SCIENCE INFORMATION EXCHANGE
SSIE
Producer   Smithsonian Institution
Coverage. . . .
  Coverage of geoscience research and development
  Continued in Federal Research in Progress (qv)
File Info
  439,000 records at 1985; closed file
Hosts
  Dialog, SDC

STATSID
Producer   Institut Français du Pétrole
Producer   Société Nationale Elf Aquitaine
Coverage. . . .
Worldwide oil and gas statistics
File Info
210,000 records at 1980
Hosts
GEISCO

TEKNICKAN AIKAUSLEHTI INDEKSI
TALI
Producer   National Library for Science and Technology (Helsinki)
Coverage. . . .
Includes coverage of Finnish geological literature

UK GEOPHYSICAL SURVEY DATABASE
Producer   British Geological Survey
Coverage. . . .
Details of geophysical surveys undertaken by BGS both on and offshore

UNDERGROUND EXCAVATION AND ROCK PROPERTIES
INFORMATION
(UERPIC)
Producer   Centre for Information and Numerical Data Analysis and Synthesis.
(Illinois)
Coverage. . . .
Includes mechanical, thermophysical, electrical and magnetic properties of
selected rocks and minerals from 1945–
Hosts
Not available online

VINITI
VSESOYUZNYY INSTITUT NAUCHNOY I TEKNICHESKOY
INFORMATSII
Producer   All Union Institute of Scientific & Technical Information (Moscow)
Coverage. . . .
Includes earth science literature of USSR

VOLCANO DATA FILE
Producer   Smithsonian Institution
Coverage. . . .
Location, features and activities of worldwide volcanoes
Hosts
Not available online

WORLD COAL RESOURCES & RESERVES DATABANK
Producer   IEA Coal Research
Coverage. . . .
All aspects of coal resources
Hosts
Not available online

WORLD DATA BANK ON MANGANESE NODULES
Producer   Scripps Institution of Oceanography
Coverage. . . .
Chemical analyses and related bibliographic information for manganese nodules, worldwide
Hosts
Not available online

WORLD DATA CENTER A — GLACIOLOGY (SNOW AND ICE)
Producer   Cooperative Institute for Research in the Environmental Sciences (University of Colorado)
Producer   National Environmental Satellite Data and Information Service (NESDIS)
Coverage. . . .
Maintains an automated system for bibliographic information within its subject areas; in addition it archives data sets on magnetic tape. All aspects of snow and ice are covered. Inventories of ice core data etc are held in machine readable form

WORLD DATA CENTER A — MARINE GEOLOGY AND GEOPHYSICS
Producer   National Geophysical Data Center (National Oceanic and Atmospheric Administration)
Coverage. . .
Magnetic tape services include copies of "underway geophysical data" and a global database of marine geological and geophysical data sets

WORLD DATA CENTER A — METEOROLOGY
Producer   National Climatic Data Center (National Oceanic and Atmospheric Administration)
Coverage. . . .
Center maintains machine readable files relating to all aspects of meteorology

WORLD DATA CENTER A — OCEANOGRAPHY
Producer   National Oceanographic Data Center (National Oceanic and Atmospheric Administration)
Coverage. . . .
Some of the center's holdings of bibliographic and source data are held in machine readable forms

WORLD DATA CENTER A — ROTATION OF THE EARTH
Producer   US Naval Observatory Time Service Division
Coverage. . . .
Observations of variations in latitude and rotation rate of the earth from PZT and astrolabe, together with the time to which the variations are referenced
Hosts
Available by direct dial

WORLD DATA CENTER A — SOLID EARTH GEOPHYSICS
Producer   National Geophysical Data Center (National Oceanic and Atmospheric Administration)
Coverage. . . .
Much of the center's data from all aspects of solid earth geophysics are held on magnetic tape

WORLD GRAVIMETRIC DATA BANK
  Producer   Bureau du Recherches Géologiques et Minières
  Coverage. . .
    Worldwide gravimetric data
  File Info
    3m records
  Hosts
    Not available online

WORLD TRANSINDEX
  Producer   International Translation Center
  Coverage. . . .
    Coverage of translations includes all aspects of the earth sciences
  File Info
    165,000 records at 1985 plus 25,000 pa
  Hosts
    ESA/IRS

Listing of Specialized Databases

ACCU-WEATHER
  Producer   Accu-Weather Inc (PA)
  Coverage. . . .
    Real time updated meteorological and climatological data from the world

ACTIVE WELL DATA FILE
  Producer   Petroleum Information
  Coverage. . . .
    Reports drilling activity from permit to completion
  File Info
    Updated daily; data backup onto Historic Well Data File (qv) at periodic
    intervals
  Hosts
    GEISCO

AEROMAGNETIC SURVEYS DATABASE
  Producer   US Geological Survey Geologic Division
  Coverage. . . .
    Aeromagnetic readings taken from various surveys
  Hosts
    Not available online

AFEE
  Producer   Association Française pour l'Etude des Eaux
  Coverage. . . .
    All aspects of fresh water including hydrologeology and hydrology
  File Info
    70,000 at 1985
  Hosts
    ESA/IRS

AFTERSHOCK STUDIES DATABASE
Producer   US Geological Survey Geologic Division
Coverage. . . .
   Aftershock data collected from temporary networks after major earthquakes
Hosts
   Not available online

AGE PROFILE DATABASE
Producer   US Geological Survey Geologic Division
Coverage. . . .
   Age profile data with coordinates
File Info
   31,640 records at 1983

AIRBORNE GEOPHYSICAL DATABANK
Producer   Finland, Geological Survey
Coverage. . . .
   Magnetic, electromagnetic and radiometric data results
File Info
   Annual increase: 211 values/sec and 500hr flight work

ALASKA SEISMIC STUDIES DATABASE
Producer   US Geological Survey Geologic Division
Coverage. . . .
   Data from South-central Alaska
Hosts
   Not available online

ALASKAN AEROMAGNETIC DATABASE
Producer   US Geological Survey Geologic Division
Coverage. . . .
   Low altitude aeromagnetic data of positions of Alaska

ANTARCTIC METEORITE BIBLIOGRAPHY
Producer   Lunar and Planetary Institute (USA)
Coverage. . . .
   All scientific aspects of Antarctic meteorites
File Info
   750 records at 1985 plus 150 pa
Hosts
   Available by direct dial

ASSESSMENT REPORT INDEX
Producer   British Columbia Branch of Minerals Resources
Coverage. . . .
   Unpublished reports prepared by the mineral industry
Hosts
   Not available online

ATLANTIC MARGIN CORING PROJECT DATABASE
Producer   US Geological Survey Geologic Division
Coverage. . . .
   Series of databases covering aspects of the geochemistry of samples taken on the Atlantic seaboard of the US together with information relating to the cruises themselves

ATLANTIC OUTER CONTINENTAL SHELF DATABASE
  Producer   US Geological Survey Geologic Division
  Coverage. . . .
    Regional grid of CDP seismic data for the offshore eastern seaboard of the
    USA

AUSTRALASIA WELL HISTORY FILE
  Producer   Not known
  Coverage. . . .
    Well information for Australasia
  File Info
    3,200 records at 1985 plus 200 pa

AUSTRALIAN CURRENT OIL WELL PROGRAM REPORTS
  Producer   Lipscombe & Associates
  Producer   Petroleum Information
  Producer   Nielsen (AC)
  Coverage. . . .
    Summary reports on wells currently drilling in Australasia
  File Info
    Updated daily
  Hosts
    Ausinet

AUSTRALIAN OIL EXPLORATION DATABASE
  Producer   Lipscombe & Associates
  Producer   Petroleum Information
  Producer   Nielsen (AC)
  Coverage. . . .
    Factual information on oil and gas exploration in Australia
  File Info
    4,000 records at 1985 plus 500 pa
  Hosts
    Ausinet

B&F NAVIGATION — GEORGES BANK DATABASE
MADATA
  Producer   US Geological Survey Geologic Division
  Coverage. . . .
    Station positions for the first Georges Bank cruise

BALTIMORE CANYON MINERALOGY DATABASE
  Producer   US Geological Survey Geologic Division
  Coverage. . . .
    Mineralogy of the Baltimore canyon; cruise date 1979

BANQUE DE DONNEES PEDOLOGIQUES
  Producer   ORSTOM
  Coverage. . . .
    Soil profiles
  Hosts
    Not available online

BANQUE DES DONNEES DU SOUS-SOL FRANÇAIS
SUBSOIL DATA BANK
Producer   Bureau du Recherches Géologiques et Minières
Coverage. . . .
Geological information on the French sub-soil
File Info
630,000 records at 1983; plus 12,000 pa
Hosts
Not available online

BANQUE GRAVIMETRIC INTERNATIONAL
Producer   Bureau du Recherches Géologiques et Minières
Coverage. . . .
International gravimetric data
File Info
500,000 stations at 1983; 2m records in 1980
Hosts
Not available online

BASE DE DATOS DE SENATES GRAVIMETRICAS
Producer   Institut Geographique National (France)

BASE DE DATOS GEOMAGNETICOS
Producer   Institiuto Geograficio Naccional (Italy)

BEDROCK CHARACTERISTICS DATABANK
Producer   Finland. Geological Survey
Coverage. . . .
Preliminary database covering geological characteristics, eg petrology, of
bedrock samples collected during mapping projects
File Info
2,000 samples at 1986

BEDROCK DATABANK
Producer   Geological Survey of Norway
Coverage. . . .
Machine readable reference archive

BGS/NGDC COMPUTERISED BOREHOLE DATABASE
Producer   British Geological Survey
Coverage. . . .
Data abstracted from BGS/NGDC records archive for England
File Info
5,000 boreholes

BOREHOLE DATABANK
Producer   Geological Survey of Nordrhein-Westfalen
Coverage. . . .
Machine readable borehole log database

BOREHOLE DATABANK OF LOWER SAXONY
Producer   Bundesanstalt für Geowissenschaften und Rohstoffe (BGR)
(Hannover) (presumed)

Coverage. . . .
   Borehole descriptions
File Info
   320,000 records; held offline

BOREHOLE GEOTHERMAL DATABANK
   Producer   British Geological Survey
   Coverage. . . .
      Temperature data for UK onshore boreholes

BOSTON HARBOUR GEOCHEMISTRY DATA
MASX
   Producer   US Geological Survey Geologic Divison
   Coverage. . . .
      Offshore geochemistry project results/data from 1977–1978

BULK MINERAL RESOURCES ASSESSMENT DATA
   Producer   British Geological Survey
   Coverage. . . .
      Borehole reference data, geological data and sample analyses for the UK
   File Info
      150,000 boreholes — not all machine readable

CALIFORNIA REGIONAL GRAVITY DATABASE
   Producer   US Geological Survey Geologic Division
   Coverage. . . .
      Gravity measurements from 64,000 sites in California

CENSUS OF COAL MINES
   Producer   McGraw-Hill Inc
   Coverage. . . .
      Data relating to active coal mines and companies in North America

CENTRAL CALIFORNIA MICROEARTHQUAKE DATABASE
   Producer   US Geological Survey Geologic Division
   Coverage. . . .
      Fundamental parameters from small earthquakes in the central Californian
      Coast Ranges

CHINESE GEOLOGICAL BIBLIOGRAPHIC DATABASE
   Producer   National Geological Library (Beijing, China)
   Coverage. . . .
      Currently at an initial stage of development

CHROMITE DEPOSITS DATABASE
   Producer   US Geological Survey Geologic Division
   Coverage. . . .
      Geological information on US chromite deposits

CHROMITE RESOURCES DATABASE
   Producer   US Geological Survey Geologic Division
   Coverage. . . .
      Chromium deposits and production from the world

CITATN
BIBLIOGRAPHIC REFERENCES DATABASE
  Producer   US Geological Survey Geologic Division
  Coverage. . . .
  Full text database for publications of the Office of Energy and Marine Geology

COAL DATA
  Producer   Alberta Coal Technology Information Centre
  Coverage. . . .
  Chemical analyses of Albertan coals

COAL DATA BANKS
  Producer   US Department of Energy
  Coverage. . . .
  Time series for distribution, production and use of US coal

COAL DEPOSIT DATABANK
  Producer   Geological Survey of Nordrhein-Westfalen
  Coverage. . . .
  Information on hard coal deposits from Nordrhein-Westfalen and Saar
  districts

COAL FILE
  Producer   British Columbia Branch of Minerals Resources
  Coverage. . . .
  Data on British Columbian coal resources
  Hosts
  Not available online

COAL RESEARCH PROJECTS
  Producer   IEA Coal Research
  Coverage. . . .
  Research projects on all aspects of the world-wide coal industry
  Hosts
  Inka, CAN/OLE, Belendis

COAL RESERVES EVALUATION SYSTEM
CORES
  Producer   Geological Survey of South Africa
  Coverage. . . .
  Borehole data relating to coal exploration. Database is used to calculate and
  evaluate coal reserves and resources

COAL-ABS
  Producer   Alberta Coal Technology Information Centre
  Coverage. . . .
  All aspects of coal technology, world-wide, including reserves and resources.
  Chemical analyses of Alberta coals
  File Info
  14,000 records at 1985

COASTLINE DATABASE
Producer   British Geological Survey
Coverage. . . .
Digitised coastline data for Scotland and Norway-Biscay
File Info
1 Gbyte data

COMPUTER INDEX OF HISTORICAL SEISMICITY
Producer   Bureau du Recherches Géologiques et Minières (presumed)
Coverage. . . .
At 1986 dataset comprises some 100,000 observations from 5,000 events since
1981 together with associated bibliographic references

COMPUTERISED LIBRARY OF ANALYSED IGNEOUS ROCKS
CLAIR
Producer
Coverage. . . .
Chemical analyses of igneous rocks, world-wide, information is obtained from
published references
File Info
26,000 records
Hosts
Not available online

COMPUTERISED RESOURCES INFORMATION BANK
CRIB
Producer   US Geological Survey Branch of Resource Analysis
Coverage. . . .
Metallic and non-metallic mineral resources of the world – name and location
of mineral deposits, production, estimated reserves and potential resources.
Hosts
GEISCO

CONDUCTIVITY MEASUREMENT DATABASE
Producer   US Geological Survey Geologic Division
Coverage. . . .
Resistivity and conductivity measurements from the US

CONSORTIUM FOR CONTINENTAL REFLECTION PROFILING
Producer   Cornell University
Coverage. . . .
Data on seismic reflection profiling of the earth's crust and upper mantle in the
USA
Hosts
Not available online

CONTEMPORANEOUS US TERRAIN DIGITAL DATABASE
Producer   US Geological Survey Geologic Division
Coverage. . . .
Terrain elevation data for US

CONTINENTAL SCIENTIFIC DRILLING PROGRAM DATABASE
(CSDP)
  Producer   Lawrence Livermore National Laboratory (Livermore, Ca)
  Coverage. . . .
    Information relating to "scientifically interesting" boreholes drilled in con-
    tinental USA
  File Info
    1,800 records at 1982
  Hosts
    Available online

CONTINENTAL SHELF SUPERFICIAL GEOLOGY DATABASE
  Producer   British Geological Survey
  Coverage. . .
    Data from cruises, including sample descriptions and analyses, from the
    UK continental shelf
  File Info
    12,000 records at 1986

CRUDE OIL ANALYSIS DATA BANK
  Producer   Bartesville Energy Technology Centre (BETC) (Oklahoma)
  Coverage. . . .
    Geochemical data on crude oils, predominantly from fields in the USA
  File Info
    9,000 records at 1985
  Hosts
    Direct dial available

CRYST
  Producer   Cambridge Crystallographic Data Centre (University of Cambridge,
             England)
  Coverage. . . .
    X-ray crystallography search system offering retrieval of plotted structures,
    space groups, unit cell parameters etc
  Hosts
    NIH-EPA

CSATA DATABANK
  Producer   Centro Studi e Applicazioni in Tecnologia Avanzate (Bari)
  Coverage. . . .
    Satellite imagery of Central South Italy

DAAS
DRILLING ANALYSIS DATA
  Producer   Petroleum Information
  Coverage. . . .
    Selected well data from permit to completion; includes success ratios
  File Info
    Historic wells from 1970
  Hosts
    GEISCO

DELFT HYDRO
Producer   Delft Hydraulics Laboratory
Coverage. . . .
All aspects of hydraulic engineering and related subjects
File Info
51,000 records at 1986 plus 7,000 pa
Hosts
QL, ESA/IRS

DGU GROUNDWATER DATABASE
Producer   Groundwater Survey Division Netherlands Organisation for Applied
Scientific Research
Coverage. . . .
Groundwater level data for the Netherlands
Hosts
Not available online

DIAMOND DRILL CORE ARCHIVE
Producer   Finland. Geological Survey
Coverage. . . .
Locational data for boreholes
File Info
5,000 records

DRILLING INFORMATION SERVICES
Producer   Adams and Rountree Technology (Lafayette, USA)
Coverage. . . .
Current and historical drilling information
File Info
250,000 at 1985
Hosts
GEISCO

DST-80
Producer   McAllister & Associates
Coverage. . . .
Drill stem test data from western Canadian petroleum wells
Hosts
Not available online

EARTH SCIENCES INFORMATION SYSTEM
Producer   Soil Survey Institute of the Netherlands
Coverage. . . .
Data taken from borehole samples and mapping
Hosts
Not available online

EASTERN US EARTHQUAKE DATABASE
Producer   US Geological Survey Geologic Division
Coverage. . . .
Earthquake data for USA east of 94 degrees

ECONOMIC MINERALS DATABASE
Producer   British Geological Survey
Coverage. . . .
   Partially machine-readable database of minerals, mining and economic geology; at present only overseas information is machine-readable
File Info
   12,600 machine-readable entries

DEEP GEOLOGY PROGRAMME BOREHOLE DATABASE
Producer   British Geological Survey
Coverage. . . .
   Borehole records and correlation logs for Southern England and onshore UK
File Info
   Around 3,000 records — some not yet computerised

ENERGY AND MINERALS RESOURCES
Producer   Business Publishers Inc
Coverage. . . .
   Full text of energy and minerals resources newsletter covering all aspects of energy from exploration to end user
File Info
   115 records at 1985 plus 50pa
Hosts
   Newsnet

ENERGY BIBLIOGRAPHY AND INDEX
Producer   Gulf Publishing Company
Producer   Texas A&M University (College Station, Texas)
Coverage. . . .
   Provides access to non-journal energy related materials held in Texas A&M University Library
File Info
   20,000 items
Hosts
   SDC

ENERGY DATA — OIL AND GAS PRODUCTION HISTORY
Producer   Dwight's Energy Data Inc
Coverage. . . .
   Oil and gas production information from continental North America
File Info
   749,000 properties at 1985 plus 49,000 pa
Hosts
   GEISCO

ENERGY RESOURCES DATA SYSTEM
(ERDS)
   Producer   Alberta Energy Resources Conservation Board
   Coverage. . . .
      Data from fuel boreholes in Alberta (includes oil, gas, coal and oil-sands)
   Hosts
      Not available online

ENERGYNET
Producer   Environment Information Center Inc
Coverage. . .
Directory of energy related organizations and personal experts
Hosts
Dialog, ESA/IRS

ENGINEERING GEOLOGY DATABANK
Producer   Geological Survey of Nordrhein-Westfalen
Coverage. . . .
Machine readable engineering geology data

ENVIRONMENTAL DATA INDEX
Producer   National Oceanographic Data Center (National Oceanic and Atmos-
pheric Administration)
Coverage. . . .
Includes reference to data sets of geochemical and geological nature

ENVIRONMENTAL GEOLOGY PROJECTS
Producer   British Geological Survey
Coverage. . . .
Hazards, resources, ground stability etc — partly machine readable

ENVIRONMENTAL PROTECTION DATABANK
Producer   Bundesanstalt für Geowissenschaften und Rohstoffe (BGR)
(Hannover) (presumed)
Coverage. . . .
Factual data of relevance to the environment
File Info
8,000 records at 1986

GEOTHERMAL BOREHOLE LOG DATA
Producer   British Geological Survey
Coverage. . . .
Compiled information on geophysical logs
File Info
5,000 boreholes at 1986

EXPLORATORY WELL STATISTICS FILE
Producer   American Petroleum Institute
Producer   American Association of Petroleum Geologists
Coverage. . . .
Drilling statistics for US oil and gas wells using data derived from API well
tickets.
File Info
330,000 wells at 1985
Hosts
GEISCO

GENERALISED SAMPLE DATA SYSTEM
(GSDS)
  Producer   US Geological Survey
  Coverage. . . .
    Data on samples related to oil and gas development; linked to the well history
    control system
  Hosts
    Not available online

GEOCHEMICAL ANALYSES DATA FILE
  Producer   British Geological Survey
  Coverage. . .
    Stream sediment and heavy metals data for the UK continental crust — not all
    in machine readable form

GEOCHEMICAL DATA FILE
  Producer   Finland. Geological Survey
  Coverage. . . .
    Sequential data file of geochemical data for rocks, sediments, peat and water
  File Info
    Annual growth of 40,000 samples

GEOCHEMICAL DATA SYSTEM
  Producer   Saskatchewan Research Council
  Coverage. . . .
    Geochemical analyses and related field data
  Hosts
    Not available online

GEOCHEMICAL DATABANK
  Producer   Geological Survey of Norway
  Coverage. . . .
    Machine readable reference archive

GEOCHEMICAL DATABANK
  Producer   Bundesanstalt für Geowissenschaften und Rohstoffe (BGR)
            (Hannover) (presumed)
  Coverage. . . .
    Multi-element geochemical data for 62 mapping areas
  File Info
    3 million data items

GEOCHEMICAL INFORMATION SYSTEM
  Producer   Centre de Recherches Pétrographiques et Géochimiques
            (Vandoeuvre-les-Nancy, France)
  Coverage. . . .
    Encompasses experimental geochemical data and bibliographic information

GEODAS
  Producer   Osaka City University
  Coverage. . . .
    Geological borehole data
  Hosts
    Not available online

GEODIAL
GEOSCIENCE DATA INDEX FOR ALBERTA
Coverage. . . .
Published and unpublished material for all aspects of Alberta earth sciences
File Info
5,000 records at 1985

GEOELECTRIC SOUNDINGS DATABASES
Producer   Geological Survey of the Netherlands
Coverage. . . .
Data on soundings from the first 10m of the subsurface

GEOIPOD
Producer   Centre National pour l'Exploitation des Océans (Brest, France)
Coverage. . . .
Drilling data from voyages of the Glomar Challenger, 1968–
Hosts
BNDO

GEOLOGICAL AND GEOCHEMICAL ASPECTS OF URANIUM DEPOSITS
Producer   Oak Ridge National Laboratory (USA)
Coverage. . . .
Covers aspects of uranium deposits
File Info
3,200 records at 1980; closed file

GEOLOGICAL DATA CENTER DATA BASE
Producer   Scripps Institution of Oceanography
Coverage. . . .
Includes seismic and magnetic data
Hosts
Not available online

GEOPHYSICAL DATA FILES (DENMARK)
Producer   Denmark. Geological Survey
Coverage. . . .
Machine readable files of geoelectric resistivity measurements and logs from
borehole testing

GEOPHYSICAL DATABANK
Producer   Geological Survey of Norway
Coverage. . . .
Machine readable reference archive

GEOPROJEKT
Producer   Austria. Geological Survey
Coverage. . . .
Information system for geoscientific and geotechnical projects carried out in
Austria

GEOPUNKT
Producer   Austria. Geological Survey
Coverage. . . .
Sample and outcrop data used for analysis and in applied geology

GEOTHERM
GEOTHERMAL RESOURCES FILE
  Producer   US Geological Survey Branch of Resource Analysis
  Coverage. . . .
    Covers geological and hydrological nature of geothermal resources
  Hosts
    Not available online

GESTION ET ETUDE DES INFORMATIONS SPECTROSCOPIQUES
ATMOSPHERIQUES
  Producer   Centre National de Recherches Scientifiques Laboratoire du
             Météorologie Dynamique
  File Info
    251,000 records at 1985 plus 22,850 pa

GRANITES URANIFERES FRANÇAIS
GUF
  Producer   Centre   de   Recherches   Pétrographiques   et   Géochimiques
             (Vandoeuvre-les-Nancy, France)
  Coverage. . . .
    Analyses of French hercynian granites
  Hosts
    Not available online

GRAVIMETRIC DATABASE
  Producer   Bureau du Recherches Géologiques et Minières (presumed)
  Coverage. . . .
    350,000 computerised observations (at 1986) used for gravimetric calculation
    and plotting

GROUND-WATER CHEMISTRY DATABANK
  Producer   Geological Survey of Nordrhein-Westfalen
  Coverage. . . .
    Machine readable databank of ground-water chemistry information

GROUNDWATER ANALYSIS DATABASE
  Producer   Geological Survey of Luxembourg
  Coverage. . . .
    Chemical and bacteriological data on groundwater; primarily drinking water

GROUNDWATER DATA BANK
  Producer   Finland. Geological Survey
  Coverage. . . .
    Sample data from springs, bedrock and wells
  File Info
    6,500 samples

GROUNDWATER DATABANK
  Producer   Geological Survey of Norway
  Coverage. . . .
    Machine readable reference archive

GROUNDWATER RESOURCES DATABANK
Producer   Water Resources Development Dept. (Riyadh, Saudi Arabia)
Coverage. . . .
   Water resources and hydrogeological data for Saudi Arabia
File Info
   Eventually up to 25,000 records

GROUNDWATER SURVEY
Producer   Dienst Grondwaterverkenning (Delft)
Coverage. . . .
   Maintains data files on Dutch groundwater resources
File Info
   6.5m data items

HEAVY METAL ANALYSES DATABANK
Producer   Geological Survey of Nordrhein-Westfalen
Coverage. . . .
   Machine readable databank of heavy metal analyses from soil and rock
   samples

HEAVY METALS IN SEA WATER DATABANK
Producer   Geological Survey of Greenland
Coverage. . . .
   Analyses of dissolved metals and metals in suspension in sea water around
   Greenland; used in pollution monitoring.
File Info
   2,000 analyses at 1986

HISTORIC WELL DATA FILE
Producer   Petroleum Information
Coverage. . . .
   Historical and completed drilling information for USA; active data is held in
   Active Well Data File (qv)
Hosts
   GEISCO

HOTLINE'S WELL INFORMATION NETWORK
Producer   Hotline Energy Reports (Denver, CO)
Coverage. . . .
   Drilling information for Rocky Mountain and West Coast regions of the USA
File Info
   350,000 records at 1985; updated daily
Hosts
   GEISCO, direct dial

HYDROLOGICAL INFORMATION STORAGE AND RETRIEVAL SYSTEM
Producer   Northern Illinois University
Coverage. . . .
   Stream flow and water data
File Info
   1m data items

IHO TIDAL CONSTITUENT BANK
Producer   International Hydrographic Organisation (Monaco)
Coverage. . . .
   Data relating to worldwide tidal harmonic constants
Hosts
   Not available online

INDEX TO BEDROCK GEOLOGICAL MAPPING
Producer   British Columbia Branch of Minerals Resources
Coverage. . . .
   Compilation of geological maps relating to British Columbia
Hosts
   Not available online

INDIAN NATIONAL OCEANOGRAPHIC DATA CENTRE
Coverage. . . .
   Maintains databases relating to all aspects of oceanography in the Indian Ocean
Hosts
   Not available online

INDUSTRIAL MINERAL OCCURRENCE DATA FILE
Producer   Finland. Geological Survey
Coverage. . . .
   Industrial mineral occurrence data
File Info
   350 sites

INDUSTRIAL MINERALS DATABANK
Producer   Geological Survey of Norway
Coverage. . . .
   Machine readable reference archive

INFORMATION INDEX ON GROUND MOVEMENTS
Producer   Bureau du Recherches Géologiques et Minières (presumed)
Coverage. . . .
   Observations and analysis of some 4,000 earth movements, eg. rockfalls, (at 1986)

INSTITUT NATIONAL DE LA RECHERCHE AGRONOMIQUE (MONTPELLIER, FRANCE)
Coverage. . . .
   Series of databases covering all aspects of soil science
File Info
   3,000 records at 1983
Hosts
   Not available online

INSTITUTE OF POLAR STUDIES (OHIO STATE UNIVERSITY)
Coverage. . . .
   Data banks of research results covering all subjects including geophysics and especially glaciological and meteorological data for the polar regions
Hosts
   Not available online

IWARDS
IOWA STATE RESOURCES DATA SYSTEM
    Producer   Iowa State Geological Survey
    Coverage. . . .
        Water resources and related data including information from NAWDEX
    Hosts
        Not available online

LAND SURVEY CARBONIFEROUS DATABASE
    Producer   British Geological Survey
    Coverage. . . .
        Carboniferous for Scotland
    File Info
        0.6 Gbyte of data

MAGNETIC FIELD SURVEY DATA
    Producer   Geological Survey of Canada Earth Physics Branch
    Coverage. . . .
        Component observations, primarily Canadian
    Hosts
        Not available online

MAPPING DATABASE
    Producer   British Geological Survey
    Coverage. . . .
        Digitised geological and thematic mapping

MARINE DATA INDEX
MRNIDX
    Producer   US Geological Survey Geologic Division
    Coverage. . . .
        Indexing of geophysical data collected and/or processed by the Office of
        Energy and Marine Geology

MARINE GEOPHYSICS DATABASE
    Producer   British Geological Survey
    Coverage. . . .
        Gravity, magnetic and positional data for UK continental shelf
    File Info
        About 5 Gbyte of data at 1986

MARINE SEDIMENT GEOCHEMISTRY DATABANKS
    Producer   Geological Survey of Greenland
    Coverage. . . .
        Chemical analyses of sediments — for Zn, Cd, Pb in general and other
        elements for some samples.
    File Info
        1,000 analyses at 1986

MASAGE
    Producer   US Geological Survey Geologic Division
    Coverage. . . .
        Age profile data from DSDP Legs 1–75

MASTER WATER DATA INDEX
  Producer   US Geological Survey
  Coverage. . . .
    Index of sites in USA for which water data are available
  File Info
    480,000 sites at 1984
  Hosts
    Available on direct dial

MASTER WELL FILE
  Producer   American Petroleum Institute
  Coverage. . . .
    Oil and gas well data for USA, includes development and service in addition to
    exploration wells (see also AAPG database)
  File Info
    975,000 wells at 1985

MEDI CATALOGUE
MARINE ENVIRONMENT DATA INFORMATION REFERAL SYSTEM
  Producer   Intergovernmental Oceanographic Commission (UNESCO, Paris)
  Coverage. . . .
    Database providing descriptions of datafiles held by organisations involved in
    aspects of the marine sciences
  Hosts
    Not available online

MEXICAN GEOLOGICAL INFORMATION
  Producer   Instituto de Geologia (Mexico City)
  Coverage. . . .
    Bibliography of Mexican geological information

MINERAL DEPOSIT INVENTORY DATABASE
  Producer   Ontario Geological Survey. Geoscience Data Centre
  Coverage. . . .
    Data on minerals and industrial mineral deposits in Ontario
  File Info
    5,000 records at 1983

MINERAL INDUSTRY LOCATION SUBSYSTEM
  Producer   US Bureau of Mines Minerals Information Directorate
  Coverage. . . .
    Locational information for non-fuel mineral sites
  Hosts
    Not available online

MINERAL OCCURRENCE DATA SYSTEM
  Producer   Newfoundland Government Mineral Development Division
  Coverage. . . .
    Mineral deposit information for Newfoundland and Labrador
  Hosts
    Not available online

MINERAL RECONNAISSANCE PROGRAM GEOCHEMICAL DATA
  Producer   British Geological Survey
  Coverage. . . .
    Data and results from the UK MRP; not all in machine readable format

MINERAL RESOURCES DATABASE
MRDB
  Producer   US Geological Survey Geologic Division
  Coverage. . . .
    Records of mineral deposits and mineral commodities, primarily of the USA

MINES AND GEOLOGICAL PLANS INDEX
  Producer   British Geological Survey
  Coverage. . . .
    Data on mines including abandoned mines for Scotland
  File Info
    1,000 plans

MINES AND QUARRIES INDEX
  Producer   British Geological Survey
  Coverage. . . .
    Active workings and commodities for England and Wales
  File Info
    2,500 listings

MINESEARCH
  Producer   Metals Economic Group (Boulder, CO)
  Coverage. . . .
    Geological, project and economic data for US nonferrous mines, development
    projects and exploration activities; 1980–
  File Info
    69,000 records at 1985 plus 1,000 pa
  Hosts
    Not available online

MINFILE
  Producer   British Columbia Branch of Minerals Resources
  Coverage. . . .
    Data on mineral deposits in British Columbia (originally called MINDEP)
  Hosts
    Not available online

MINSYS
  Producer   Mineral Policy Sector (Canada)
  Coverage. . . .
    Index to the Canadian National Mineral Inventory
  Hosts
    Not available online

MOUNT ST HELENS DATABASE
  Producer   Tectonics Productions (Edmonds, WA, USA)
  Coverage. . . .
    The May 1980 eruption of Mount St Helens and its after effects
  File Info
    1,000 records at 1985
  Hosts
    Not available online

MULTI-CHANNEL SEISMIC DATA INVENTORY
Producer   US Geological Survey
Coverage. . . .
   Inventory of multi-channel seismic data from the east coastal margin

NATIONAL CLIMATIC DATA CENTER
Producer   National Climatic Data Center (Asheville, NC)
Coverage. . . .
   Maintains machine readable data for climatic statistics for the USA
Hosts
   Not available online

NATIONAL COAL RESOURCES DATA SYSTEM
Producer   US Geological Survey
Coverage. . . .
   In 1985 comprised eight databases containing data on coal resources and related geographical and geological information

NATIONAL GEOCHEMICAL DATABASE
Producer   Bureau du Recherches Géologiques et Minières (presumed)
Coverage. . . .
   Coverage of around 100,000 sq km with average of three observations per sq km of soil or stream sediment samples (at 1986)

NATIONAL GEODETIC INFORMATION CENTER
Coverage. . . .
   Machine readable files for all US geodetic data meeting national quality standards
Hosts
   Not available online

NATIONAL GRAVITY DATA BASE
Producer   Geological Survey of Canada Earth Physics Branch
Coverage. . . .
   Station and gravity anomaly data from Canada
Hosts
   Not available online

NATIONAL INDEX OF GROUNDWATER QUALITY
Producer   Bureau du Recherches Géologiques et Minières (presumed)
Coverage. . . .
   Data on the water table together with information on groundwater quality

NEAR SURFACE RAW MATERIALS DATA
Producer   Bundesanstalt für Geowissenschaften und Rohstoffe (BGR) (Hannover) (presumed)
Coverage. . . .
   Factual data on near-surface raw materials

NORSAR
Producer   Norwegian Seismic Array (Kjeller, Norway)
Coverage. . . .
Digital seismic database and associated earthquake catalogues
File Info
50,000 events at 1983
Hosts
Not available online

NORTH EAST ATLANTIC BATHYMETRY DATABASE
Producer   British Geological Survey
Coverage. . . .
Sounding data from "special survey" areas for NE Atlantic and northern
continental shelf
File Info
5 Gbyte of data

OCEANOGRAPHIC LITERATURE REVIEW
Producer   Woods Hole Data Base Inc
Coverage. . . .
Corresponds to part B of *Deep-sea Research* and covers all aspects of
oceanographic research
File Info
31,500 records at 1985 plus 6,500 pa
Hosts
Pergamon

OIL AND GAS JOURNAL ENERGY DATABASE
Producer   PennWell Publishing Inc
Coverage. . . .
Wide variety of time series covering all aspects of the energy industry including
exploration and production
File Info
20,000 time series at 1985
Hosts
GEISCO

OIL AND GAS REPORTS
Producer   Dwight's Energy Data Inc
Coverage. . . .
Provides descriptions of oil/gas wells and leases in the USA and Canada —
includes API numbers

File Info
700,000 plus records at 1985
Hosts
GEISCO

OIL SHALES DATA SYSTEM
OSDS
Producer   US Geological Survey Geologic Division
Coverage. . . .
Fischer assay and saline mineral data for well cores from Colorado and Utah;
much of the information is confidential

OMAN STUDIES ASSOCIATION
    Producer   Centre for Documentation and Research on Oman and the Arabian
            Gulf (Tübingen, West Germany)
    Coverage. . . .
    Machine readable bibliographic indexes to the Oman general bibliography
    which includes coverage of geology
    Hosts
    Not available online

ONTARIO GEOSCIENCE DATA INDEX
    Producer   Ontario Geological Survey. Geoscience Data Centre
    Coverage. . . .
    Reports, maps produced by Ontario Mineral Resources group and those
    submitted by the mining industry in Ontario
    File Info
    15,000 records

ONTARIO MINERAL DEPOSIT FILE
    Producer   Ontario Mineral Resources Group
    Coverage. . . .
    Data on non-fuel mineral deposits in Ontario
    Hosts
    Not available online

ORE DATA FILE
    Producer   Finland. Geological Survey
    Coverage. . . .
    Preliminary file — geological characteristics of metal ore occurrences
    File Info
    250 sites

ORE DATABANK
    Producer   Geological Survey of Norway
    Coverage. . . .
    Machine readable reference archive

ORE INDICATIONS DATA BANK
    Producer   Finland. Geological Survey
    Coverage. . . .
    Data from mineralised glacial boulders and bedrock occurrences
    File Info
    9,000 records at 1986 plus 100 pa

PDS
PETROLEUM DATA SYSTEM
    Producer   Petroleum Information
    Coverage. . . .
    Formerly University of Oklahoma's TOTL data file produced in conjunction
    with US DOE; comprises a series of databases covering aspects of USA
    petroleum activities ranging from well data through crude oil analyses to lease
    information.
    Hosts
    GEISCO

PEAT RESOURCES DATA BANK
 Producer   Finland. Geological Survey
 Coverage. . . .
   Geological and geochemical data from peat samples
 File Info
   7,000 samples (28,000 data items) at 1986

PENN STATE COAL DATABASE
 Producer   Pennsylvania State University. Coal Research Station
 Coverage. . . .
   All aspects of the characteristics and analysis of coal seams
 Hosts
   Not available online

PERMIAN BASIN WELL AND RESERVE FILE
 Producer   US Geological Survey Geologic Division
 Coverage. . . .
   Well information from Permian Basin, USA

PERMITS DATA BASE
 Producer   Petroleum Information
 Coverage. . . .
   All US oil and gas permit details including location, proposed depth etc
 File Info
   Updated daily; permits are deleted after ninety days
 Hosts
   GEISCO

PETBANK
PETROLOGY, CHEMICAL, CHEMICAL ANAL. FILE
 Producer   British Geological Survey
 Coverage. . . .
   Chemical analyses of rocks and minerals

PETROPHYSICS AND REMOTE SENSING BRANCH BIBLIOGRAPHIC DATABASE
 Producer   US Geological Survey Geologic Division
 Coverage. . . .
   Bibliography of all productions from the USGS-Geophysics Branch

PETROS
 Producer   Eastern Washington University
 Coverage. . .
   Petrological data on selected igneous rocks derived from the literature
 Hosts
   Not available online

PRODUCTION DATA FILE
 Producer   Petroleum Information
 Coverage. . . .
   Historical and current production data; and test data where available for US wells
 Hosts
   GEISCO

QHBARK/QHGRUN
Producer   Geological Survey of Sweden (Uppsala)
Coverage. . . .
   Hydrogeology of Sweden; databases include well records and time series for groundwater resources
File Info
   plus 8,000 pa

RADIOMETRIC AGE DATA BASE
RADB
Producer   US Geological Survey
Coverage. . . .
   Published radiometric ages for the USA
Hosts
   Not available online

REAL TIME WEATHER INFORMATION SYSTEM
Producer   Weather Services International Corporation (Bedford, MA)
Coverage. . . .
   Real time and historical meteorological data
File Info
   3,000 stations at 1983

REGIONAL GEOLOGICAL SURVEYS DATABASE
Producer   British Geological Survey
Coverage. . . .
   Multi-topic geological surveys' results which incorporate experimental mapping techniques

ROCK AGGREGATES DATABANK
Producer   Geological Survey of Norway
Coverage. . . .
   Machine readable reference archive

ROCK CHEMICAL DATABASE
PETROCH
Producer   Ontario Geological Survey. Geoscience Data Centre
Coverage. . . .
   Geochemical data on Ontario rock samples — primarily of Precambrian age
File Info
   15,000 records
Hosts
   Not available online

SAND AND GRAVEL DATABANK
Producer   Geological Survey of Norway
Coverage. . . .
   Machine readable reference archive

SASKATCHEWAN COAL DATABASE
(SASCO)
Producer   Saskatchewan Department of Mineral Resources
Coverage. . .
   Borehole and related data used for resource assessment
Hosts
   Not available online

SEABED OBSTRUCTIONS DATABASE
Producer   British Geological Survey
Coverage. . . .
   Submarine topography and obstacles for the northern continental shelf
File Info
   1 Gbyte of data

SEARCH
BIBLIOGRAPHIC REFERENCES DATABASE
Producer   US Geological Survey Geologic Division
Coverage. . . .
   Related to CITATN database; provides bibliographic access thereto

SEDGWICK MUSEUM CATALOGUE
Producer   Sedgwick Museum (Cambridge, England)
Coverage. . . .
   Catalogue of museum's fossil collection
File Info
   200,000 entries at 1983
Hosts
   Not available online

SLOPE STABILITY DATABANK
Producer   Geological Survey of Nordrhein-Westfalen
Coverage. . . .
   Data on slope stability

SOIL SURVEY INFORMATION SYSTEM
Producer   Soil Survey of England and Wales
Coverage. . . .
   Master file for series of soil science databanks
Hosts
   Not available online

SPIN
SEARCHABLE PHYSICS INFORMATION NOTICES
Producer   American Institute of Physics
Coverage. . . .
   Bibliographic coverage includes geophysics and astronomy
File Info
   126,000 records at 1980 plus 24,000 pa
Hosts
   Dialog

STICHTING VOOR BODEMKARTERING
Producer   Soil Survey Institute of the Netherlands
Coverage. . .
   Soil analyses and borehole data for the Netherlands
Hosts
   Not available online

SUBSURFACE DATABASE
Producer   Bureau du Recherches Géologiques et Minières (presumed)
Coverage. . . .
   Site specific dossiers, around half the 360,000 are machine readable the
   balance are being converted; growth of 10,000 pa

SUMMARY RESERVE BASE AND AVERAGE ANALYSIS DATABASE
Producer   US Geological Survey Geologic Division
Producer   US Bureau of Mines
Coverage. . . .
  Estimates of coal reserves on a county basis
  Data not verified by USGS

SURFACE MINING AND ENVIRONMENT INFORMATION SYSTEM (SEAMINFO)
Producer   University of Arizona
Coverage. . . .
  Information on strip mining in the Western USA
Hosts
  Not available online

SURFACE RAW MATERIALS DATABANK
Producer   Denmark. Geological Survey
Coverage. . . .
  Data from special mapping projects (eg brown coal) and from the systematic mapping of raw materials undertaken by regional authorities

TEXAS NATURAL RESOURCES INFORMATION SYSTEM
Coverage. . . .
  General series of databases include coverage of geological, meteorological and hydrologic information
Hosts
  Not available online

TFS
(TECHNICAL & FIELD SERVICES)
Producer   Technical and Field Services
Coverage. . . .
  Bibliographic references and data relevant to Australian exploration activities
File Info
  36,000 mines, 20,000 licences
Hosts
  Not available online

TOTL
Producer   Dwight's Energy Data Inc
Coverage. . . .
  Geological and engineering data on US oil and gas fields and reservoirs from 1968–
  (PI, indicate, 1987, that this data file has become part of PDS; not verified with Dwights)
File Info
  110,000 records at 1985
Hosts
  GEISCO

UMPLIS DATA
Producer   Bundesanstalt für Geowissenschaften und Rohstoffe (BGR) (Hannover) (presumed)
Coverage. . . .
  Data on environmentally connected projects
Hosts
  Not available online — held on magnetic tape

USCHEM
   Producer   US Geological Survey Geologic Division
   Coverage. . . .
      Chemical analyses of coals from the USA

WATER DATA BANK
   Producer   US Department of Agriculture Water Data Laboratory. (Bellsville, MD)
   Coverage. . . .
      Precipitation and run-off data from 300 sites in the USA
   File Info
      7,000 station years (rainfall) and 4,000 station years (runoff) at 1985

WATER DATA SOURCES DIRECTORY
   Producer   US Geological Survey Water Resources Division
   Coverage. . . .
      Directory of organisations that are sources of water data

WATERLIT
   Producer   South African Water Information Center (Pretoria)
   Coverage. . . .
      Includes hydrology
   File Info
      106,000 records at 1986 plus 12,000 pa
   Hosts
      Pergamon, SDC

WATR
STREAMLINE WATER INFORMATION DATABASE
   Coverage. . . .
      Water research in Australia
   File Info
      5,000 records at 1985
   Hosts
      Ausinet

WATSTORE
   Producer   US Geological Survey Water Resources Division
   Coverage. . . .
      Stream, river, reservoir and groundwater information; including chemical analyses
   Hosts
      Not available online

WELL DATABASES
   Producer   Geological Survey of the Netherlands
   Coverage. . . .
      Three databanks, with subsets covering borehole information from deep (>300m), medium and surface (<10m) holes

WELL RECORD DATABANK
   Producer   Denmark. Geological Survey
   Coverage. . . .
      Well log information including location, lithology, biostratigraphy etc
   File Info
      250,000 wells

WORLD WEATHER WATCH
Producer   World Meteorological Organisation (Geneva)
Coverage. . . .
Collects and distributes meteorological, oceanographic and similar data and provides for worldwide interchange of meteorological information
File Info
20,000 stations

XTAL
Producer   Crystal Data Center (US National Bureau of Standards)
Coverage. . . .
Single crystal reduction and search system

YUKON BIBLIOGRAPHY
Producer   Boreal Institute for Northern Studies (Alberta)
Coverage. . . .
Includes climate and earth sciences relating to Yukon
File Info
4,000 records plus 300 pa
Hosts
QL

ZOOLOGICAL RECORD
Producer   Biosis UK (York, England)
Coverage. . . .
Includes coverage of palaeontology and taxonomy
File Info
750,000 records to date.
Hosts
BRS, Dialog

APPENDIX 2: A selection of the main online hosts

Bibliographic Retrieval Services Inc(BRS)
1200 Route 7,
Lathan,
New York NY 12100, USA

Data-Star,
Radio Schweiz AG, Data Star
Laupenstrasse 18a,
CH-3008 Berne, Switzerland

In UK:
Plaza Suite,
114 Jermyn Street,
London SW1Y 6HJ, England

Dialog Information Services Inc.
3460 Hillview Avenue,
Palo Alto, CA 94304, USA

PO Box 188,
Oxford OX1 5AX England

European Space Agency (ESA-IRS)
ESRIN/IRS,
Via Galileo Galilei,
00044 Frascati,
Rome, Italy

IRS Dialtech,
Department of Trade and Industry,
Room 392, Ashdown House,
123 Victoria Street,
London SW1E 6RB, England

Orbit Search Service, (*formerly*
SDC Search Service)
800 Westpark Drive,
Suite 400,
McLean, VA22102

Achilles House,
Western Avenue,
London, W3 0UA,
England

Geisco Ltd,
114–118 Southampton Row,
London WC1B 5AB,
England

BELINDIS (Belgian Information &
Dissemination Service),
Centre du Traitement de
l'Information,
Rue J A de Mot 30,
1040 Brussels, Belgium

INKADAT (STN International)
Fachinformations Zentrum,
Karlsruhe,
D-7514 Eggenstein-Leopoldshafen 2,
FDR

CAN/OLE (Canadian Online
Enquiry),
Canadian Institute for Scientific &
Technical Information,
National Research Council of
Canada,
Montreal Road,
Ottawa, Ontario, Canada K1A 0S2

CHAPTER SEVEN

Geological maps and remote sensing

MARY LYNETTE LARSGAARD

History and development

During the first half of the eighteenth century, geological mapping was just developing; after the mid-1700s it became more and more common, but the earliest true geological maps as we view them date from the beginnings of the nineteenth century, at which time colouring began. Geological map issuance accelerated very rapidly, especially in Great Britain, where engineers planning canals had need of maps showing rock structures and types, and where pure scientific interest in earth structures provided considerable impetus to geological map making. Two landmark geological maps that prepared the way were a geological map of the Paris basin by Cuvier and Brongniart (Paris, 1811) and the map of England and Wales by William Smith in 1812, the latter being the first modern geological map of a large area. These maps served as a pattern for other maps to follow, for example Smith's use of profiles to attempt to give a three-dimensional view.

The first official geological survey was founded in Great Britain in 1835; France had previously assigned two geologists the task of preparing a geological map of the country, but had not constituted a formal survey. Most other official national surveys were established much later (for example, the US Geological Survey was created in 1879), but by the beginning of the twentieth century, small-scale geological maps were available for many countries and some large-scale mapping had begun. This time period also marked the initiation of the first international geological map, the *International Geological Map of Europe* at 1:1,500,000 (now in its third edition), which was begun in 1893 under the aegis of the International Geological Congress, as it still is today.

Although geological mapping of many countries still lags woefully behind topographical mapping — in large part because the former is based on the latter — the twentieth century has witnessed considerable geological map production, primarily of map series and of overall maps, and mainly by national and foreign geological surveys and by various geological societies. Along with the well-established use of aerial photography for compilation is the present use of satellite and radar imagery. (See section on Remote Sensing.)

For accounts of the history and development of geological maps, see *Methods in Geological Surveying*, by E Greenly and H Williams (Murby, 1930); H A Ireland in *Bulletin of the Geological Society of America*, **54**,(9), 1943; *Sur les Cartes Géologiques à l'Occasion du 'Mapoteca Geologica Americana'*, by Jules Marcou (Dodivers, Besançon, 1888); *Early Thematic Mapping in the History of Cartography*, by Arthur Robinson (University of Chicago Press, 1982); and T Sheppard in *Report of the British Association for the Advancement of Science, 1920*. A valuable, but now dated, article on the state of geological mapping is K C Dunham's 'Practical geology and the natural environment of the map — I' in the *Quarterly Journal of the Geological Society of London*, **123**,(1), 1967.

Purpose, preparation, and types

The purpose of geological maps is to identify the landforms, rock structure, and rock types of a given area, and to portray geological units and structure in correct spatial relationship (Lillesand, 1979). The map permits the geologist to see the earth as it would appear with all top-lying materials stripped away, and to observe the differences between rock layers (US Geological Survey, 1978). Geological maps usually differentiate between different formations or units by colour and by letters or symbols. The rock type (or, on detailed maps, the specific formation) is shown by small letters following the age symbol, e.g. KJf stands for the Jurassic Cretaceous Franciscan formation. These colours and symbols are keyed to the legend or explanation, usually in the map margin, but occasionally on a separate sheet. Colours are somewhat standardized, e.g. Quaternary deposits are usually in yellow, Tertiary in orange, Jurassic in green, and so forth. 'Solid' or 'bedrock' maps show bedrock geology; 'drift' or 'surficial' maps depict the surficial geology of Quaternary deposits. (Aune, 1960).

For a geological survey, the geologist uses the basic elements of plane surveying, locating the position of geological features by means of directions and distances, supplemented by earlier maps and records, and other information. Large scale maps are generated from original surveys; smaller scales (e.g. 1:200,000 and smaller) are compiled from larger scale maps, or prepared by means of photogeology. The majority of geological maps (especially detailed maps) are overprinted on a previously prepared base map, preferably a topographical map showing natural and cultural features, with contours and spot heights to depict relief. These reference lines and points enable the geologist to locate, reasonably precisely, the rock outcrops and other geological notations on the map. If the geology of an area is quite complex, uncertain contacts may be shown with a dashed or broken line (Aune, 1960).

Geological maps may be of any size area — the world, a continent, a nation, a province, a quadrangle area. National and state/provincial geological surveys usually select a scale or panoply of scales for mapping, with the most detailed exemplified by the British 1:10,560 series; standard series are usually at larger scales than that, for example the 1:25,000 German series and the 1:24,000 US series.

Bishop (1960), Hageman (1968), Harrison (1963), and Robertson (1956) discuss the reliability or presentation of geological maps. For mapping construction generally see *Thematic Maps: their design and construction*, by David J Cuff and Mark T Mattson (Methuen, 1982) and *Cartographic Design and Production*, by J S Keates (Longman, 1973), while *Geology in the Field*, by R R Compton (Wiley, 1985), *Methods in Geological Surveying*, by E Greenly and H Williams (Murby, 1930), *Geologische Karten*, by H Vossmerbauer (Schweizerbart'sche, 1983), *Geologische Karte, ihre Anfertigung und Ausdeutung*, by H Falke (De Gruyter, 1967), and *Field Geology*, by F J Lahee (McGraw-Hill, 1961) deal more specifically with geological mapping. See also the references at the end of the chapter, Barnes (1981) and Martin (1973).

Works mainly concerned with basic map interpretation and construction of geological sections include *Geological Maps and their Interpretation*, by F G H Blyth (Arnold, 1976), *Advanced Geological Map Interpretation*, by Frank Moseley (Wiley, 1979), *Introduction to Geological Maps and Structures*, by John L Roberts (Pergamon, 1982), *Geological Maps*, by Brian Simpson (Pergamon, 1968), *An Introduction to Geological Maps*, by J A G Thomas (Allen and Unwin, 1977), *Introduction to Geological Structures and Maps*, by G M Bennison (Arnold, 1964), *Coupes et Cartes Géologiques*, by A Foucault and J F Raoult (SEDES, Paris, 1966), and *Simple Geological Structures*, by J I Platt and J Challinor (Murby, 1968).

One way of classifying geological maps is by scale, with reconnaissance maps 1:250,000 or smaller, regional 1;25,000–1;50,000 detailed 1;10,000 or larger (to about 1;1,000), and specialized 1:1,000 or larger (Barnes, 1981). Another way is to classify by subject. First there is the most common, the general geological map that depicts formations in a standard map view. Structural and tectonic maps add to the general geological map such information as faults, depths to basement, major basins, and subsurface position of specific rock units; maps showing the latter are called structure contour maps. Isopach maps describe variations in the thickness of rock units (or rock formations), while lithofacies maps show the variations and limits of rock unit facies.

Maps associated with geological maps, useful to persons working in geology, which yet may not be by strict definition geological maps, are:

geochemical maps: show the relative abundance of chemical elements; used to locate ore deposits and areas unsuitable for agriculture; Unesco has published *Legends for Geohydrochemical Maps* (1975).

geomorphological maps: show land forms, e.g. fault scarps; glacial maps, which depict features of glacial action, are a special type.

geophysical maps: show either magnetic field intensity or variations in gravity; used for oil exploration.

geotechnical maps: usually lithologic maps with physical properties of each unit tabulated; used for military or civil engineering; for a relevant bibliography, see *Zeitschrift fur Angewandte Geologie*, **7**,(3), 1961.

hydrogeological maps: show groundwater resource characteristics, including chemical composition; Unesco has published *International Legend for Hydrogeological Maps* (1970).

palaeoclimatic maps: depict climate of geological ages in the past; see *Descriptive Palaeoclimatology*, by A E M Nairn (Interscience, 1961) and *Climates of the Past: an introduction to paleoclimatology*, by M Schwarzbach (Van Nostrand, 1963).

palaeogeographical maps: depict geography of past geological ages; see A I Levorsen's *Paleogeologic Maps* (Freeman, 1960); more notable examples include Schuchert, C *Atlas of Paleogeographic Maps of North America* (Wiley, 1955) and Wills, L J *A Paleogeographical Atlas of the British Isles and Adjacent Parts of Europe* (Blackie, 1951).

palaeotectonic maps: palaeogeographical maps emphasizing dynamic aspects; for example, the paleotectonic maps of the Jurassic, Triassic, and Permian systems, published by the US

Geological Survey as its *Miscellaneous Geologic Investigations* **I–175**, **I-300**, and **I–515** (1956, 1959, 1967).

palinspastic maps: attempt to show original geographical position of units mapped, by removing effects of faulting, folding, and other factors (Kay, 1945).

palaeocontinental maps: attempt to reconstruct the positions of continents in past times; see two books by Smith A G, *Mesozoic and Cenozoic Paleocontinental Maps* (Cambridge University Press, 1977) and *Phanerozoic Paleocontinental World Maps* (Cambridge University Press, 1981).

soil maps: often reflect the nature of underlying rocks, since they are based on a study of soil (which is derived from underlying rock), delimiting soil associations and series (see also Chapter 19).

Electronic data processing (EDP) has had a substantial and continuing impact on the acquisition, formatting, and use of geological data, most recently in the manipulation of Landsat digital data. Of the many publications on this topic, a few will serve as representative:

Fulton, P (1976) 'Selective geologic map information.' *Photogrammetric Engineering and Remote Sensing*, **52**,(8), 1063–67

Gold, Christopher (1980) Geological mapping by computer. *In: The Computer in Contemporary Cartography*, edited by D Fraser Taylor, pp. 151–190, Wiley.

Guides to geological maps

For current information on geological maps recently published, the most important sources are publishers' catalogues and brochures, *Episodes* (1978–   ; formerly the *Geological Newsletter* of the International Union of Geological Sciences), *Geocarte Information* (BRGM, 1984–   ) and the *Bulletin* of the Commission for the Geological Map of the World (1962–   ). Addresses for geological surveys may be obtained either from the above items, or from:

World Directory of National Earth-science Agencies and Related International Organizations (1981) US Geological Survey, Circular **834**,

Allin, Janet (1982) *Map Source Directory*. 2nd edn. Downsview, Ontario: York Universities Libraries, Map Library

For bibliographical references to maps in printed indexes, the following are most helpful:

1. *Bibliographie Cartographique Internationale*. An annual publication since 1946/47, it ceased with the 1975 (published 1978)

volume. Regional headings in the main text; reference numbers in index to geological maps.

2. *Bibliographie des Sciences de la Terre*, described in Chapter 5. Search under "Cartes" in index.

3. *Bibliography and Index of Geology*. Previous to volume 33, this was entitled *Bibliography and Index of Geology exclusive of North America*; North America was indexed by *Bibliography of North American Geology* (see Chapter 5). The indexing of maps in this bibliography is inclined to change, or was in the past; the user is advised to be cautious and to look under 'geological maps' or 'maps' under regional headings, or under 'engineering geology'. 'Cartography' and 'geological exploration' may also lead to map references. The data base may be searched online by coordinates. Before publication of this bibliography began in 1933, maps were indexed under 'maps, geological' in *List of Geological Literature added to the Geological Society's Library during the year. . .* , published by the Geological Society of London.

4. *Bulletin Signalétique*. From 1969, 'Section 216 Géologie. Paléontologie.' Some maps are indexed under 'cartes' and 'cartographie.'

5. *New Geographical Literature and Maps*. Published semi-annually since 1918, by the Royal Geographical Society; maps and atlases are listed under regional headings.

6. *Referativnyi Zhurnal — Geografiya*. This monthly publication contains a *Kartografiya* sub-section; 1956–

Map librarianship organizations publish newsletters which almost always contain a section on new maps, including geological maps: e.g.

Bulletin of the Association of Canadian Map Libraries (1969–   ) Formerly *Newsletter*.

base line. (1981–   ). Newsletter of the Map and Geography Round Table, American Library Association.

Bulletin of the Special Libraries Association, Geography and Map Division. (1947–   )

Information bulletin. Western Association of Map Libraries. (1969–   )

Publishers' catalogues listing maps include: GeoCenter's *Geo-Katalog* and *GeoKartenbrief* (Stuttgart); Edward Stanford (London); and the Telberg Book Corporation's Geological Map Service (Sag Harbor, New York).

Monographic publications of interest are:

Locke, C B M (1969) *Modern Maps and Atlases: an outline guide to twentieth century map production*. Bingley

Ward, Dederick C (1981) *Geologic Reference Sources: a subject and regional bibliography of publications and maps in the geological sciences*. Metuchen, NJ, & London: Scarecrow Press

In the nonprint arena, computer data bases that may be searched for maps are most obviously GeoRef (the previously mentioned data base of the *Bibliography and Index of Geology*) and GeoArchive. The OCLC data base, a data base intended for cataloguing, acquisition and interlibrary loan by libraries, also contains many maps.

Selected geological maps and map series

The list notes only overall maps and map series intended to cover an entire political area or continent; it is not comprehensive or definitive but is rather a guide to the most important maps issued, mainly by government agencies. Not included are atlases, regional maps, or geophysical maps, previous editions, or historical maps, and indexes listing both books and maps.

The list was compiled from:
1. Martin, E L (1973); this chapter in the previous edition of this book
2. Ward, Dederick C (1981) *Geologic Reference Sources : a subject and regional bibliography of publications and maps in the geological sciences*. Metuchen, N.J., London: Scarecrow Press
3. Publications lists of geological surveys
4. Catalogues of Geological Map Service (Sag Harbor, New York) and GeoCenter (Stuttgart)
5. Maps in the Map Room, Arthur Lakes Library, Colorado School of Mines.

Citations are considerably abbreviated. Publisher is not given if it is the same as author, series author is not given if it is the same as publisher, and only the first author is given. Citations are in the form:

Author. *Title*. Scale; number of sheets. (Publisher, date). (*Series*) Text.

Scales may be approximate. In many cases, citations are not given; the decision as to whether to give a citation or to note publication in text was based on the uniqueness of the title, the alphabet of the title (transliteration systems tend to differ, making it difficult to give a citation acceptable to, let alone recognizable to, all readers), and the completeness of information.

WORLD

Geological — *Atlas Géologique du Monde* = *Geological World Atlas*, 1:10,000,000 (Unesco, 1974–), to total 20 sheets, and tectonic — *Karta Tektoniki Dokembriià Kontinentov*, 1:15,000,000 (Akademiia Nauk SSSR and Glavnoe Upravlenie Geodezii i Kartografii, 1972), 12 sheets, maps have been published. For overall geological maps of the world there are:

  Fahlbusch, Klaus *Geologie der Erde* = *Geology of the Earth* = *Geologia del Mundo*. 1:16,000,000. (Justus Perthes 1969)
  Larson, R L *The Bedrock Geology of the World*. 1:23,230,000. (W H Freeman, 1985)

and for a one-sheet tectonic map:

  Condie, Kent C *Tectonic Map of the World*. 1:40,000,000. Plate 5–4 in: *Plate Tectonics and Crustal Evolution*. (Pergamon, 1976)

AFRICA

The classic geological map of the continent is:

  Commission de la Carte Géologique du Monde. *Carte Géologique Internationale de l'Afrique* = *International Geological Map of Africa*. 1:5,000,000, 9 sheets. (The Commission 1986–)
  Text: *Natural Resources Research*, **3**, 1964 and a new text is expected in 1989

Others have been published at different scales:

  Soviet Union. Vsesoiûznyi Nauchno-Issledovatel'skiĭ Geologicheskii Institut. *Geologicheskaiâ Karta Afriki* = *Geological Map of Africa*. 1:10,000,000, 2 sheets. (Vsesoiûznyi Aerogeologicheskii Trest Ministerstva Geologii SSSR, 1969; 1966 on map). Text.
  Unesco. *Afrique : fond géologique pour servir de base à des synthèses diverses ou à des cartes thématiques* = *Africa : geologic background to serve as a basis for various syntheses or thematic maps*. 1:10,000,000. (Unesco, 1967)

and some are on specific sub-areas of geology:

  Choubert, G *Carte Tectonique de l'Afrique* = *Tectonic Map of Africa*. 1:15,000,000. (Unesco, 1968)   ·
  Choubert, G *Carte Tectonique Internationale de l'Afrique* = *International Tectonic Map of Africa*. 1:5,000,000; 9 sheets. (Unesco, 1968)
  Text: *Sciences de la Terre*, **4**, 1969
  Commission for the Geological Map of the World. Subcommission for the Cartography of the Metamorphic Belts of the World. *Metamorphic Map of Africa* = *Carte Métamorphique de l'Afrique*. 1:10,000,000. (Unesco, 1978)
  Carte Internationale de Quaternaire de l'Afrique = *International Quaternary Map of Africa*. 1:2,500,000; 3 sheets published. (Unesco, 1973–)

Algeria

Geological maps are issued at scales of 1:500,000 (6 sheets; second edition dated 1951–52), 1:200,000, and 1:50,000; the first and second are entitled *Carte Géologique de l'Algérie*, the third, *Carte Géologique Détaillée*.

Angola

The overall geological maps are:

  Portugal. Junta das Missões Geográficas e de Investigações do Ultramar. *Esboço Geológico de Angola*. 1:2,000,000. (Instituto Geográfico e Cadastral, 1954). Text.
  Carvalho, Heitar de. *Geologia de Angola*. 1:1,000,000; 4 sheets. (Laboratório Nacional de Investigacão Científica Tropical, 1980–1982).

There is a 1:250,000 series, entitled *Carta Geológica de Angola*, and also a tectonic map:
   Silva, A T S Ferreira da *Esboço Tectónico de Angola; notícia explicativa.* (Junta da Investigações do Ultramar, n.d.) (*Estudios de Geologia, Anais*, **10**(5))

Benin

France. Bureau de Recherches Géologiques et Minières.
Carte Géologique de la République du Dahomey. 1:1,000,000; 2 sheets. (1960)

Botswana

Geological maps are published at 1:1,000,000:
   Mortimer, C. *Geological map of the Republic of Botswana.* (Geological Survey Department, 1984)
and, at 1:2,000,000, there is the *Provisional Geological Map of Botswana.* (1971). A 1:125,000 series (Quarter degree sheet) is also in existence, beginning in 1963. More recently a geological interpretation of Landsat imagery and air photography has been released and published in Mallick, D I *Photogeological Map of Botswana.* 1:6,000,000. (Botswana Geological Survey, 1978), and *Overseas Geology and Mineral Resources, number* **56**. (British Directorate of Overseas Surveys, map 1979, text 1981) (*D.O.S. (Geol.)* **1218**)

Burundi

The 1:100,000 series, *Carte Géologique du Burundi*, is issued by the Ministère de l'Energie et des Mines.

Cameroon

Accompanying the Direction de Mines et de la Géologie's *Bulletin* **2** is:
   Gazel, J *Carte Géologique du Cameroun.* 1:1,000,000, 2 sheets. (1956)

Cape Verde

A 1:100,000 series, *Carta Geológica do Cabo Verde*, has been published since 1977, by the Portuguese government.

Central African Empire

The French Bureau de Recherches Géologiques et Minières has issued:
   Mestraud, J-L *Carte Géologique de la République Centrafricaine.* 1:1,500,000. (1964)

Chad

   Wolff, J-P *Carte Géologique de la République du Tchad.* 1:1,500,000; 2 sheets. (Bureau de Recherches Géologiques et Minières)

Congo see **Zaire**

Congo (Brazzaville)

A 1:2,000,000 (1951) and a 1:3,000,000 tectonic (1952) maps have been published, along with 1:200,000, 1:100,000, and 1:50,000 map series.
   As for other former French colonies, the Bureau de Recherches Géologiques et Minières has issued a geological map:
   Dadet, P *Carte Géologique de la République du Congo, zone comprise entre les parallèles 2° et 5° sud.* 1:500,000. (1969) (*Mémoires du BRGM* **70**)

Dahomey see **Benin**

Djibouti

Maps at 1:200,000 and 1:100,000 have been published:
Clin, M *Carte Géologique du Territoire Français de Afars et des Issas*. 1:200,000; 4 sheets. (Université de Bordeaux III, Conseil de Gouvernement du T.F.A.I., Centre d'Etudes Géologiques et de Développement, 1970) *Carte Géologique de la République de Djibouti*. 1:100,000; 12 sheets. (Djibouti, Government, 1974–) — previous to 1975, titled *Carte Géologique du Territoire Français des Afars et des Issas*.

Egypt

Recently two geological maps have been published by Hayah al-Misriyah al-Ammah lil-Misahah al-Jiyulujiyah wa-al-Mashruat al-Tadiniyah (Egypt. Geological Survey and Mining Authority).
Geologic Map of Egypt. 1:2,000,000.
Geological Map of Egypt. 1:1,000,000; 6 sheets. (1979)
There are also some 1:500,000, 1:125,000, and 1:10,000 sheets, and a tectonic map:
Siagaev, N A *Tectonic Map of Egypt*. 1:2,000,000. (Geological Survey and Mineral Research Department of Egypt, 1966). Accompanied by the Survey's *Paper* **39**.

Equatorial Guinea

A geological sketch map at 1:500,000 was published by the Spanish Instituto de Estudios Africanos in 1954.

Ethiopia

Besides a few 1:500,000 sheets, there is:
Kazmin, V *Geological Map of Ethiopia and Somalia* = *Carta Geológica dell'Etiopia e della Somalia*. 1:2,000,000. (Ethiopian Geological Survey, 1973)
See also Somalia.

Gabon

Accompanied by the *Mémoire du Bureau de Recherches Géologiques et Minières*, **72**, is:
Hudeley, H *Carte Géologique de la République Gabonaise*. 1:1,00,000. (Bureau de Recherches Géologiques et Minières, 1970)

The Gambia

A 1:500,000 map was issued with the *Bulletin of the Gold Coast Geological Survey*, no. **3**, 1925.
See also Senegal

Ghana

There are 2 series — *Geology of the Ghana Quarter Degree sheets*, 1:62,500, 1958– and *Geology of the Ghana Half Degree sheets*, 1:125,000, 1960–; also a few overall maps:
Ghana. Geological Survey. *Map showing Geology and Field Sheets*. 1:2,000,000. (Geological Survey of Ghana, 197–?)
Bates, D A *Geological Map of Ghana*. 1:1,000,000. (Geological Survey of Ghana, 1958)

Ghana. Geological Survey. *Ghana . . . Geological.* 1:2,000,000. (Geological Survey of Ghana, 1969)

Guinea-Bissau

Besides maps at scales of 1:1,000,000 and 1:2,000,000 published in 1947, there is:
Portugal. Junta da Investigações do Ultramar, *Carta Geológica da Guiné.* 1:500,000. (1968?)

Ivory Coast

Two maps, at two different scales, have been published:
Bararre, E *Carte Géologique de la Côte d'Ivoire.* 1:1,000,000. (Côte d'Ivoire Direction des Mines et de la Géologie, 1965)
and three sheets at 1:200,000 were published in 1973.

Kenya

The *Geological Map of Kenya* at 1:3,000,000 was published in 1969 by the Kenya Mines and Geological Department; there is also a series of 1:125,000 sheets.

Lesotho

A geological map of Basutoland at 1:380,160 was published in 1939.
See also South Africa.

Liberia

United States. Geological Survey. *Geologic Map of the . . . Quadrangle, Liberia.* 1:250,000; 10 sheets. (1977) (*USGS Miscellaneous Investigations Series*, **I-771 D to I-780 D**)
Behrendt, J C *Tectonic Map of Liberia Interpreted from Geophysical and Geological Surveys.* 1:1,000,000. (United States Geological Survey, 1974) (*USGS Professional Paper*, **810**, plate 1)

Libya

Conant, Louis C *Geologic Map of the Kingdom of Libya.* 1:2,000,000. (United States Geological Survey, 1964). (*USGS Miscellaneous Investigations Series*, **I-350A**; reprinted as Plate 2 of *Professional Paper*, **660**)
Goudarzi, G H *Preliminary Structure-Contour Map of the Libyan Arab Republic and Adjacent Areas.* 1:2,000,000. (United States Geological Survey, 1978). (*USGS Miscellaneous Investigations Series*, **I-350 C**)

Madagascar

Series maps *Carte Géologique au 1/100.000*, Carte Géologique au 1/200.000, and *Carte Géologique au 1/500.000*, the last in 8 sheets (1968–72), and an overall geological map, have been issued:
Besairie, Henri *Carte Géologique au 1/1.000.000.* 2nd ed. 1:1,000,000; 3 sheets. (Service Géologique de Madagascar, 1964)
There is also a tectonic map:
Besairie, Henri *Carte Tectonique de Madagascar.* 1:3,000,000. (Service Géologique de Madagascar, 1961)

Malawi

A geological map has been published at two different times: Carter, G S *The Geology and Mineral Resources of Malawi.* (Malawi Geological Survey Department, 1973) (*Bulletin* **6**, rev. ed.; includes provisional black and white 1:1,000,000 map), and *Geological Map of Malawi*. 1:1,000,000. (1966), available from Standford's. There is also a series entitled *Malawi* at 1:100,000.

Mali

The only coverage seems to be sheets in the 1:500,000 series, *Carte Géologique de l'Afrique Occidentale*.

Mauritania

France. Bureau de Recherches Géologiques et Minières. *République Islamique de Mauritanie, Carte Géologique*. 1:1,000,000; 6 sheets. (1968). Text (1975).

Morocco

Both geological (1985) and tectonic (1976) maps have been published at 1:2,000,000, as have various series, at 1:50,000, 1:100,000, and 1:200,000, usually titled *Carte Géologique du Maroc*, and usually part of the *Notes et Mémoires* series of the Service Géologique. There is also:
  Choubert, Georges *Carte Géologique du Maroc*. 1:500,000; 6 sheets. (Service de la Carte Géologique, 1946–55)

Mozambique

In addition to a 1:3,000,000 tectonic map (1957), there is Oberholzer, W F *Republica Popular de Moçambique, Carta Geológica*. 2nd ed. 1:2,000,000. (Direcção dos Serviços de Geologia e Minas, 1976), with a text by Afonso, R S, *A Geologia de Moçambique*. 2nd edn (Imprensa Nacional de Moçambique, 1978)

Namibia

There are both an overall map:
  Miller, R McG *South West Africa/Namibia — Geological Map = Suidwes-Afrika/ Namibie — Géologiese Kaart*. 1:1,000,000; 4 sheets. (Geological Survey of South Africa and South West Africa, 1980)
and 2 map series (both issued by the same agency as the above):
  South West Africa Geological Series 1:250,000 = Suidwes-Afrika Géologiese Reeks 1:250,000. (1977–)
  South West Africa Geological Series 1:125,000. (1938–42)

Niger

  Greigert, J *République du Niger, Carte Géologique*. 1:2,000,000. (Bureau de Recherches Géologiques et Minières, 1967?) Text.

Nigeria

Replacing the map issued in 1965 or 1964 is:
  Nigeria. Geological Survey Department. *Geological Map of Nigeria*. 1:2,000,000. (1974) itself perhaps to be replaced by a new edition in 1983. There are also 2 series, 1:250,000 and 1:100,000.

Portuguese Guinea see **Guinea-Bissau.**

Réunion

A 1:50,000 map, in 4 sheets, was published in 1974.

Rhodesia see **Zimbabwe**

Rwanda

There is a series of maps (*Carte Géologique du Rwanda* 1:100,000, 1964–) and a lithologic map:
Rwanda. Service Géologique. *Carte Lithologique du Rwanda*. 1:250,000. (1981)

Senegal

In addition to the 4-sheet map: France. Bureau de Recherches Géologiques et Minières. *Carte Géologique de la République du Sénégal et de la Gambier.* 1:500,000; 4 sheets. (1962) there is also a series: *Carte Géologique du Sénégal.* 1:200,000; 27 sheets. (Direction des Mines et de la Géologie, 1963–)

Sierra Leone

Sierra Leone. Geological Survey Department. *Geological Map, Sierra Leone.* 1:1,000,000. (Great Britain Directorate of Overseas Surveys, 1960).

Somalia

Geological maps of Somalia are a somewhat complicated situation, because of the former divided colonial status; a 1:400,000 map of French Somaliland was published in 1946, and there is an incomplete 1:25,000 series for the former British Somaliland, as well as Mackay, C *Geological Map of Somaliland Protectorate.* 1:1,000,000. Plate in: Somaliland Oil Exploration Company. *A Geological Reconnaissance of the Sedimentary Deposits of the Protectorate of British Somaliland.* (1954).

In addition, there are for the country as a whole a 1:125,000 series (about 23 sheets out of 68 issued) and also:
Somalia. AGIP Mineraria. *Carta Geologica della Somalia e dell'Ogaden.* 1:500,000; 9 sheets. (AGIP Mineraria, 1957–60)
Consiglio Nazionale delle Ricerche. *Geological Map of Ethiopia and Somalia = Carta Geologica dell'Etiopia e della Somalia = Warqadda Cilmiga Dhulka Ee Itoobiya Iyo Soomaliya.* 1:2,000,000. (1973)

South Africa

Replacing the map published in 1955 is Coertze, F J *et al Géologiese Kaart van die Republiek van Suid-Afrika, Transkei, Bophuthatsuana, Venda en Ciskei en die Koninkryke van Lesotho en Swaziland.* 1:1,000,0000; 4 sheets. (South African Geological Survey, 1984).

Map series are published at 1:125,000 and 1:250,000 with some 1:50,000 sheets of selected areas; all series use the general title: *Republic of South Africa — Geological Survey = Republiek van Suid-Afrika — Géologiese Opname.*

The 1:125,000 series began in 1910, the 1:250,000 series in 1957, and the 1;50,000 series in 1973.

South West Africa see **Namibia.**

Spanish Guinea see **Equatorial Guinea.**

Spanish West Africa see **Western Sahara.**

Sudan

Superseding the 1:4,000,000 map published in 1963 is:
Great Britain. Directorate of Overseas Surveys *Geological Map, the Democratic Republic of the Sudan and Adjacent Areas*. 1:2,000,000; 2 sheets. (1974). (*D.O.S.* **1203A** and **1203B**; see also *Overseas geology and mineral resources*, **49**).
There is also a 1:250,000 series.

Swaziland

Hunter, Donald R *Geological map of Swaziland*. 1:250,000. (Geological Survey and Mines Department, 1966)
Swaziland. Geological Survey and Mines Department *Swaziland — Geological Series 1:50,000*. (British Directorate of Overseas Surveys and the National Survey, 1961–) (*D.O.S.* **1216**; after 1978, *D.O.S. (Geol.)* **1089**)
There is also a 1:25,000 series. (1968–)

Tanzania

The most current map seems to be: Tanzania. Geological Survey. *Geology of Tanzania*. 1:3,000,000. (The Survey, 1967).
There is a 1:125,000 series of quarter degree sheets.

Togo

Maps of Togo are usually included in maps of adjacent countries, for example:
Bates, D A *Geological Map of the Gold Coast and Togoland under British Trusteeship*. 1:1,000,000. (Gold Coast Geological Survey, 1955).

Tunisia

Besides series at 1:50,000 and 1:200,000 there is an overall map:
Castany, G *Carte Géologique de la Tunisie*. 2nd ed. 1:500,000; 2 sheets issued. (Service des Mines, de l'Industrie et de l'Energie, 1951). Text.

Uganda

A geological map at 1:1,250,000 (1961) and series at 1:250,000, 1:125,000 (few sheets issued) and 1:100,000 are available.

Upper Volta

Hottin, G *Carte Géologique de la République de Haute-Volta*. 1:1,000,000. (Direction de la Géologie et des Mines, 1975)

Western Sahara

Besides a 1:50,000 series, there is *Mapa Geológico del Sahara Español y Zonas Limítrofes*. 1:1,500,000. (Instituto Geológico y Minero de España, 1958)

Zaire

There are available an overall map: Zaire. Service Géologique. *Carte Géologique du Zaïre*. 1:2,000,000; 2 sheets. (Musée Royal de l'Afrique Central, 1974) and a series, Zaire. Direction du Service Géologique. *Carte Géologique à l'échelle du 1/200.000*. (197–)

Zambia

Besides 1:75,000 and 1:100,000 series, there is an overall map:
Thieme, J G *Geological Map of the Republic of Zambia*. 1:1,000,000; 4 sheets.
(Zambia Geological Survey, 1977)

Zimbabwe

A provisional geological map at 1:1,000,000 was issued in 1977, and there are
1:100,000 maps for some areas.

THE AMERICAS

LATIN AMERICA

Mexico

Replacing the 1968 geological map of the nation is:
Mexico, Secretaria de Program Ación y Presupuesto. *Carta Geológica de la
República Mexicana*. 1:1,000,000. (1981)
As a part of Mexico's massive and impressive 1:50,000 mapping of the country,
there is:
Mexico. Dirección General de Estudios del Territorio Nacional (DETENAL).
Carta Geológica. 1:50,000. (DETENAL, 1971–)
There are also older series at smaller scales published by the Instituto de Geologia,
Mexico:
Carta Geológica de México, serie de 1:100,000. (1963–)
Carta Geológica de los Estados. 1:500,000. (1964–)
and a tectonic map:
Cserna, Zoltan de *Tectonic Map of Mexico*. 1:2,500,000. (Geological Society of
America, 1961). (*Map and chart series* **MC-2**)

CENTRAL AMERICA

United States. Geological Survey. *Geologic Map of Central America*. 1:1,000,000.
(1957). (*USGS Bulletin* **1034**; Plate 1)

Belize

No map covering the entire nation seems to exist, but there is:
Dixon, C G *Geological Map of Southern British Honduras*. 1:275,000. (Survey
Department, 1955)

Costa Rica

Dondoli, Cesar *Mapa Geológico de Costa Rica*. 1:700,000. (Dirección de
Geología, Minas y Petroleo, 1968)

El Salvador

Both geological maps of this country have been done by a German agency,
Bundesanstalt für Bodenforschung.
*Mapa Geológico General de la República de El Salvador = Geologische
Übersichtskarte der Republik El Salvador*. 1:500,000. (1974)
*Mapa Geológico de la República de El Salvador/América Central = Geologische
Karte der Republik El Salvador/MittelAmerika*. 1:100,000; 6 sheets. (1978–81)

Guatemala

Guatemala. Instituto Geográfico Nacional.
Mapa Geológico de la República de Guatemala. Primera edn, 2a. impression.
1:500,000; 4 sheets. (1975)
Mapa Geológico de Guatemala. 1:250,000. (1966–)
Mapa Geológico de Guatemala. 1:50,000. (1966–)

Honduras

Aceituno, Reniery Elvir *Mapa Geológico de la República de Honduras.*
1:500,000; 4 sheets. (Instituto Geográfico Nacional, 1974)
Honduras. Instituto Geográfico Nacional. *Mapa Geológico de Honduras.*
1:50,000. (1957–)

Nicaragua

Nicaragua. Instituto Geográfico Nacional. *Mapa Geológico, preliminar.*
1:1,000,000. (1974)

Panama (including Canal Zone)

Available are:
Panama. Dirección General de Recursos Minerales.
Mapa Geológico. 1:1,000,000. (1976)
Panama — Mapa Geológico. 1:250,000; 7 sheets. (1969–76)

CARIBBEAN REGION

Geological maps of islands in this region are in the main scattered in books and journals; the following are only a sampling. There is a tectonic map of a portion of the region:
Garrison, Louis E *Preliminary Tectonic Map of the Eastern Greater Antilles Region.* 1:500,000. (United States Geological Survey, 1972). (*USGS Miscellaneous Investigations Series*, I–732)
and a more recent geologic-tectonic map of the entire area:
Case, J E *Geologic-tectonic Map of the Caribbean Region.* 1:2,500,000; 3 sheets. (United States Geological Survey, 1980). (*USGS Miscellaneous Investigations Series*, I–1100)

Cuba

Núñez Jimenez, A *Mapa Geológico de Cuba.* 1:1,000,000. (Instituto Cubano de Recursos Minerales, 1962)
Albear, J F *Mapa Hidrogeológico de Cuba.* 1:1,000,000. (Instituto Cubano de Recursos Minerales, 1965)
Cuba. Academía de Ciencias de Cuba. Comisión Nacional. *Mapa Tectónico de Cuba = Tektonischeskaía Karta Kuby.* 1:1,250,000. (Akademiía Nauk SSSR, Geologicheskii Institut).

Dominican Republic

Zoppis de Sena, R *Atlas Geológico y Mineralógico de la Republica Dominicana.*
1:250,000; 8 sheets. [No publisher given]. (1969)

Guadeloupe

France. Bureau de Recherches Géologiques et Minières. *Carte Géologique du Département de la Guadeloupe.* 1:50,000; 5 sheets. (1962–65).

Haiti

As an example of maps contained in other publications, there is:
   Butterlin, Jacques A *Carte Géologique de la République d'Haiti.* 1:250,000.
   (SCIP, 1952) (*Institut Français d'Haiti, Mémoires* 1; plate)
A map of the same scale was published in 1960 as a part of:
   Institut des Hautes Etudes de l'Amérique Latine. Mémoire, **6**, 1960.

Jamaica

   McFarlane, N *Jamaica — Geology.* 1:250,000. (Jamaica Geological Survey,
   1977)
   Great Britain. Directorate of Overseas Surveys. *Jamaica — Geological Sheets.*
   1:50,000; 30 sheets. (1972–) (*D.O.S.* **1177**)

Martinique

   Grunevald, H *Carte Géologique du Département de la Martinique.* 1:50,000;
   2 sheets. (BRGM, 1961)

Puerto Rico

Puerto Rico has been mapped by the US Geological Survey:
   Briggs, Reginald P *Provisional Geologic Map of Puerto Rico and Adjacent
   Isands.* 1:240,000. (The Survey, 1964) (*USGS Miscellaneous Investigations
   Series,* **I–392**)
There is also a series of maps, at 1:20,000, issued as sheets in USGS's *Geologic
Quadrangle Map* (GQ) series.

Trinidad and Tobago

A geological map of Tobago was published in the *Bulletin of the Geological Society
of America* **59**(8), in 1948. There is also a map of Trinidad, published separately:
   Kugler, H G *Geological Map of Trinidad.* 1:100,000; 2 sheets. (Petroleum
   Association of Trinidad, 1961)

SOUTH AMERICA

Some useful guides to geological maps of South America are *Catalogue of the
Geologic Maps of South America,* by H B Sullivan (American Geographical
Society, 1922), a select bibliography issued by the Commission for the Geological
Map of the World in 1963, and the resource guides being issued by the Pan
American Institute of Geography and History (PAIGH) in association with the
relevant national agencies, as e.g. Peru. National Office for the Evaluation of
Natural Resources (ONERN) *Guide to Cartographic and Natural Resources
Information of Peru.* (PAIGH, 1979). Some of these, like the one for Guatemala
(*Guida Geográfica de Guatemala para Investigadores,* 1978, *Publication* no. **319**),
are issued as a number in PAIGH's *Publication* series.
   The Geological Society of America has been active in the mapping of South
America, replacing its 1945, 1:5,000,000 map, with another one of the same scale in
1964. The Society has also issued:
   Tectonic Map of South America. 1:5,000,000; 2 sheets.
Other geological maps of South America have been issued, for example:
   Goudarzi, G H *Geologic Map of South America.* 1:15,000,000. (US Geological
   Survey, 1977). (*USGS Miscellaneous Field Studies,* **MF-868-A**)

Argentina

Besides a 1:200,000 series, there are geological and hydrogeological maps published by the Servicio Geológico Nacional of Argentina.
Mapa Geológico de la República Argentina. 1:2,500,000; 2 sheets. (1982)
Mapa Hidrogeológico de la República Argentina. 1:5,000,000. (1963) Text.
and also a tectonic map:
Borrello, Angel V *Mapa Geotectónico de la República Argentina.* 1:2,500,000; 2 sheets. (Servicio Geológico Nacional, 1978)

Bolivia

Pareja, Jorge *Mapa Geológico de Bolivia.* 1:1,000,000; 4 sheets. (Yacimientos Petrolíferos Fiscales Bolivianos, 1978). Text.
Botello, Rubin *Carte Tectonique de Bolivie = Mapa Tectónico de Bolivia.* 1:5,000,000. (Universidad Mayor de San Andres, 1973). Text.
Bolivia. Servicio Geológico. *República de Bolivia — Hojas (geológicas).* 1:100,000. (1962–)

Brazil

Besides two series (1:250,000 and 1:1,000,000), both a geological and tectonic map have been published:
Almeida, F F M de *Mapa Geológico do Brasil.* 1:5,000,0000. (Departamento Nacional da Produção Mineral, 1971)
Osório Ferreira, Evaldo *Mapa Tectónico do Brasil.* 1:5,000,000. (Departamento Nacional da Produção Mineral, 1971). Text (*Boletim* **I**, 1972)
Publications have also been done on a state-by-state basis; for São Paulo there is a 1:1,000,000 geological map (1963) and a 1:100,000 series, for Paraná there is a geological map and a 1:50,000 series, for Rio de Janeiro a 1:50,000 series, for Minas Gerais a 1:25,000 series, and for Rio Grande do Sul a 1:1,000,000 geological map (1974).

Chile

Both an overall geological map and several series are published by the Instituto de Investigaciones, Chile. There is the *Mapa Geológico de Chile*, 1:1,000,000; 7 sheets, (1979) and series at scales of 1:50,000, 1:100,000, 1:250,000 (*Preliminares*), and 1:500,000.

Colombia

Besides a 1:100,000 series (*Mapa Geológico del Caudrángulo* . . . , 1969–) and a 1:200,000 *Edición Preliminar*, there is also:
Arango Cálad, J L *Mapa Geológico de Colombia.* 1:1,500,000. (Instituto Nacional de Investigaciones Geológico-Mineras (INGEOMINAS), 1976)
all published by INGEOMINAS, and for a tectonic map, there is:
Colombia. Instituto Geofísico Universidad Javeriana. *Mapa tectónico.* 1:5,000,000. (1975). (*Instituto Geofísico de los Andes Colombianos, Boletín, Serie C, Geología*, **19**).

Ecuador

Series at 1:100,000 and 1:50,000 (*Hojas Geológicas*) and an overall map have been issued:
Granja Ballén, Julio *Mapa Geológico de la República del Ecuador.* 1:1,000,000; 2 sheets. (Servicio Nacional de Geología y Minería, 1969)

Falkland Islands (Malvinas)

Adie, R J *Geological Map of the Falkland Islands.* 1:250,000; 2 sheets. (British Directorate of Overseas Surveys, 1972) (*D.O.S. (Geol.)* **1185 A** and **B**)

The Guianas

Two maps of the three countries have been published, one at 1:2,000,000 by the Geological and Mining Serivce of Surinam, and the other by the French, a 1:2,700,000 sheet in about 1969.

French Guiana

France. Service de la Carte Géologique. *Carte Géologique du Département de la Guyane.* 1:500,000; 2 sheets. (1960)

France. Bureau de Recherches Géologiques et Minières. *Carte Géologique de la France — Département de la Guyane.* 1:100,000. (1956–)

Guyana

British Guiana. Geological Survey. *Provisional Geological Map of British Guiana.* 1:1,000,000. (1962).

See also:

Williams, E *The Folded Precambrian of Northern Guyana Related to the Geology of the Guiana Shield.* Guyana Geological Survey (1967) (*Records*, **5**)

There are also map series by the Geological Survey of British Guiana:

Geological Survey of British Guiana. 1:125,000. (1948–55) and *Geological Survey of British Guiana.* 1:200,000; to total 92 sheets. (1961–)

Surinam

Surinam. Geologisch Mijnbouwkundige Dienst. *Geological Map of Suriname.* 1:500,000; 2 sheets. (1977)

Surinam. Geologisch Mijnbouwkundige Dienst. *Geologische Kaart van Suriname.* 1:100,000. (1954–57)

O'Herne, L *Fotogeologische Kaart van Suriname.* 1:1,000,000. (The Survey, 1966)

Published in: Guiana Geological Conference. *Proceedings.* (1969) was O'Herne, L *Fotogeologische Kaart van Suriname.* 1:500,000; 2 sheets. The Survey (1966). It was described in a periodical article, O'Herne, L (1969) Presentation of the photogeological map of Suriname. *Nederlands Geologisch Mijnbouwkundig Genootschap, Verhandelingen* **27**, 49–52.

Paraguay

Eckel, Edwin B *Geologic Map and Sections of Paraguay = Mapa Geológico y Secciones del Paraguay.* 1:1,000,000. (US Geological Survey, 1959) (*USGS Professional Paper* **327**, plate 1)

Peru

Besides a 1:100,000 series, there is an overall map:

Peru. Instituto de Geología y Minería. *Mapa Geológico del Perú.* 1:1,000,000; 4 sheets. (1975)

Uruguay

In addition to a 1:100,000 series (issued from 1969 on), there are two overall maps:

Caorsi, Juan H *Mapa Geológico de la República Oriental del Uruguay.* 1:500,000; 2 sheets. (Instituto Geológico del Uruguay, División de Geología Económica, 1957)

Carta Geológica de Uruguay. 1:1,000,000. (Dirección de Suelos y Fertilizantes, 1975)

Venezuela

Besides several geological maps and map series at various scales (1:50,000, 1:100,000) there is also a tectonic map:

Mapa Geológico de Venezuela. 1:2,000,000. (COPLANARH, 197–?)

Laforest, R *Mapa Geológico de la República de Venezuela*. 1:2,000,000. Ed. especial. (Dirección de Geología, 1964)

Venezuela. Dirección de Geología. *Mapa Geológica Estructural de Venezuela*. 1:500,000; 30 sheets. (1976)

Bucher, Walter *Geologic-tectonic Map of the United States of Venezuela*. 1:1,000,000. (Geological Society of America, 1950)

Also, on a state basis, and at various scales, is a series published since 1950, *Mapas Geológicos y de Recursos Minerales de los Estados*.

NORTH AMERICA

North American Geologic Map Committee. *Geologic Map of North America*. 1:5,000,000; 2 sheets. (US Geological Survey, 1965)

American Association of Petroleum Geologists. *Basement Map of North America between Latitudes 24° and 60°N*. 1:5,000,000; 2 sheets. (US Geological Survey, 1967)

King, Philip Burke *Tectonic Map of North America*. 1:5,000,000; 2 sheets. (US Geological Survey, 1969)

Canada

The Geological Survey of Canada produces 2 major series, the *"A"* series and a *Preliminary* series, with the former coloured and the latter usually uncoloured. The Geological Surveys or Mines Departments of the provinces, and the Research Council of Alberta, produce geological maps for their respective provinces. The Ontario Ministry of Natural Resources publishes a series, *Index to Published Maps, Geological Series*, which, considering the extensive number of blueline print maps issued in the Ontario Geological Survey's *Preliminary map series*, is just as well. The Geological Survey of Canada issues a series of index maps, *Geological Maps Published by the Geological Survey*; issued since 1960 and updated periodically, these index maps are at 1:1,000,000. The most notable of GSC maps are in the *"A"* series:

Geological Map of Canada. 1:5,000,000. (The Survey, 1969) (*Map* **1250A**)

Tectonic Map of Canada. 1:5,000,000. (The Survey, 1969) (*Map* **1251A**)

Geological Map of Canada. 1:1,000,000; to total 70 sheets. (The Survey, 1965–) (Various numbers)

Greenland

In addition to series:

Geological map of Greenland = Geologisk Kort over Grønland. 1:500,000. Geological Survey of Greenland (1971–)

Quaternary map of Greenland = Kvartaergeologisk Kort over Grønland. 1:500,000. Geological Survey of Greenland (1974–)

Geologisk Kort over Grønland. 1:100,000. Geological Survey of Greenland (1967–) Texts (1973–)

The Geological Survey of Greenland also issues 2 major overall maps:

Escher, A *Tectonic/geological map of Greenland*. 1:2,500,000. (1970)

Weidick, A *Quaternary map of Greenland*. 1:2,500,000. (1971?) Explanation in: Denmark, Grønland Geologiske Undersøgelse. *Rapport* **36**.

United States

The United States Geological Survey and the various state geological surveys (or equivalent agencies) carry out the geological mapping of the nation. Extent of coverage and details of the past year's geological mapping are regularly published (about once a year) by the Survey (USGS) in its yearbook and in Chapter A of *Geological Survey Research* (a subseries of USGS's *Professional Papers series*), and also in the non-agency periodical *Geotimes*.

USGS's major series are:
Geologic atlas of the United States, in folios (1894–1946)
Geologic quadrangle map (GQ)
Miscellaneous investigations series (I); formerly *Miscellaneous geologic investigations*
Oil and gas investigations map (OM)
Oil and gas investigations charts (OC)
Coal investigations map (C)
Geophysical investigations map (GP)
Hydrologic investigations atlas (HA)
Miscellaneous field studies map (MF); mainly black and white preliminary maps
Mineral investigations resource map (MR)
Open file reports (OF)
USGS publishes numerous indexes to geologic mapping, usually on a state-by-state basis:
Geologic map index of [state]: new, computer-generated indexes for all 50 states completed 1983; also indexes non-USGS maps
Geologic and water-supply reports and maps of [state] (USGS publications only)
Publications of the Geological Survey/New publications of the Geological Survey: the former issued in cumulations (1879–1961; 1962–1970; 1970–1980; annual), the latter is monthly
The Survey's *Circular* **1003**, *COGEOMAP: a new era in cooperative geologic mapping* (1987) explains its geologic mapping program.
Non-governmental groups have issued indexes to geological maps also:
Pampe, William R *Maps and Geological Publications of the United States: a layman's guide.* (American Geological Institute, 1978)
Guide to U.S. Government Maps, Geologic and Hydrologic Maps. Documents Index (1975 preliminary ed.; 1978; 1978/79 supplement; 1982).
USGS's most notable maps are:
King, Philip Burke *Geologic Map of the United States (exclusive of Alaska and Hawaii).* 1:2,500,000; 2 sheets + legend sheet. (USGS, 1974). (Text: *USGS Professional paper* **901**)
Bayley, Richard W *Basement Rock Map of the United States, exclusive of Alaska and Hawaii.* 1:2,500,000; 2 sheets. (USGS, 1968)
Cohee, George V *Tectonic Map of the United States, exclusive of Alaska and Hawaii.* 1:2,500,000; 2 sheets. (USGS, 1968)
The American Association of Petroleum Geologists (AAPG) has issued a highway map series:
American Association of Petroleum Geologists. Geological Highway Map Committee. *United States Geological Highway Map Series.* 1:1,875,000; 12 sheets when completed. (AAPG, 1966–)
State Geological Surveys may issue overall maps, map series, and geologic map indexes, but not every Survey necessarily issues all three, although most issue their own map series, usually with titles like *Geologic Map*; overall maps are often issued by USGS. Addresses of State Geological Surveys are given in various places: once a year in *Geotimes*; as an addendum to 'A directory of societies in earth science'; in a list issued by the Association of American State Geologists; sporadically in map

librarianship periodicals, such as the *Special Libraries Association Geography and Map Division Bulletin* (e.g. no. **115**, pp. 41–46, March 1979); and in Janet Allin's *Map Source Directory* (2nd edn. Map Library, York University Library, 1982).

ANTARCTIC REGIONS

Besides the overall maps of the region:
  Institut Français du Pétrole. *Carte Géologique de l'Antarctique.* 1:2,500,000; 7 sheets. (Editions Technip, 1977) Text.
  Soviet Union. Ministerstva Geologii. *Geological Map of Antarctica = Geologicheskaiâ Karta Antarktidi.* 1:5,000,000; 4 sheets. (1978)
Nations doing scientific studies of the continent have issued maps in various series, as for example:
  United States. Geological Survey. *Antarctic Geologic Map A-.* 1:250,000. (1970–) sometimes referred to as *Reconnaissance Geologic Map of Antarctica*
  Japan. National Institute of Polar Research. *Antarctic Geological Map Series.* 1:25,000. (1974–)
There is also a tectonic map:
  Soviet Union. Nauchno-issledovatel'skii Institut Geologii Arktiki. *Tektonicheskaia Karta Poliarnykh Oblastei Zemli = Tectonic Map of Polar Regions of the Earth.* 1:10,000,000; 4 sheets. (Ministerstva Geologii, 1971)
and a metamorphic map:
  Kamenev, E N *Karta Metamorficeskich Facij Antarktidy = Map of Metamorphic Facies of Antarctica.* 1:5,000,000; 4 sheets. (Ministerstva Geologii, 1979). Text (1980)

ARCTIC REGIONS

International Symposium on Arctic Geology, 1st, Calgary. *Geological Map of the Arctic.* 1:7,500,000. (Alberta öciety of Petroleum Geologists, 1960)
Geologicheskiaia Karta Severnoi Poliarnoi Oblasti Zemli. 1:5,000,000; 4 sheets. (Nauchno-issledovatel'skii Institut Geologii Arktiki, 1978)
Tektonicheskaia Karta Severnoi Poliarnoi Oblasti Zemli. 1:5,000,000; 4 sheets. (Nauchno-issledovatel'skii Institut Geologii Arktiki, 1978)

ASIA

The most recent overall geological map of Asia is:
  China. Academy of Geological Science. *Geological Map of Asia.* 1:5,000,000; 20 sheets. (Cartographic Publishing House, 1975)
At the same scale is:
  India. Geological Survey. *Geological Map of Asia and the Far East.* 2nd edn. 1:5,000,000; 4 sheets. (Unesco, 1972) Text: *Earth sciences* **11**.
There is also:
  Commission on the Geological Map of the World. Sub-commission on Mapping Metamorphic Belts of the World. *Metamorphic Map of Asia.* 1:5,000,000; 9 sheets. (Soviet Union, 1978)

Afghanistan

Afghanistan. Geological Survey Department. *Geological Map of Afghanistan.* 1:2,500,000. (Afghan Cartographic Institute, 1969)
Sahab Geographic and Drafting Institute. *Carte Géologique de l'Afghanistan = Geological Map.* 1:2,500,000. (1973?) (*Map no..* **248**)
Shareq, Abdullah *Tectonic Map of Afghanistan.* 1:4,900,000. (Geological Survey of Afghanistan)

Bahrain

Bahrain. Wizaret al-Ashghal wa-al-Kahraba wa-al-Ma. *Bahrin Geology*. 1:50,000; 6 sheets. (Ministry of Works, Power and Water, 1974–1976)
See also Saudi Arabia

Bangladesh see **Pakistan.**

Brunei see **Malaysia.**

Burma

Burma. Earth Sciences Research Division. *Geological Map of the Socialist Republic of the Union of Burma*. 1:1,000,000; 3 sheets. (Myamma Oil Corporation, 1977). Text.

Cambodia see **Vietnam**

Ceylon see **Sri Lanka**

China

Both geological and tectonic maps have been published:
Chung-kuo ti chih k'o hsüeh yen chiu yuan. *Chung-hua Jen Min Kung Ho Kuo Ti Chih T'u*. 1:4,000,000. (Ti T'u Ch'u Pan She, 1976): title may be given alternatively as, *Zhonghua Renmin Gongheguo Dizhi Tu*.
Tectonic Map of the Linear Features of China (by using of the satellite images).1:6,000,000. (Institute of Geology of Mineral Deposits, Chinese Academy of Geological Sciences, 1981)
Terman, Maurice J *Tectonic Map of China and Mongolia*. 1:5,000,000; 2 sheets. (Geological Society of America, 1974). (*Geological Society of America. Map and chart series* **MC-4**).
In the past, sheets at 1:1,000,000, 1:235,000, 1:200,000, 1:100,000 and 1:50,000 were issued, and in the 1970s, or early 1980s, the Tokyo Geographical Society distributed them.

Cyprus

Cyprus. Geological Survey Department. *Geological Map of Cyprus*. 1:250,000. (Department of Lands and Survey, 1979)
Tullstrom, N H O *Hydrogeological Map of Cyprus*. 1:250,000. (Geological Survey Department, 1970)

Hong Kong

Great Britain. Directorate of Overseas Surveys. *Geological Map of Hong Kong, Kowloon and the New Territories*. 1:50,000; 2 sheets. (1972). (*D.O.S.* **1184A**)
Accompanies: Allen, P M *Report on the Geological Survey of Hong Kong*. (The Survey, 1971)

India

Roy, B C *Geological Map of India*. 6th edn. 1:2,000,000; 4 sheets. (Geological Survey of India, 1963). Text.
India. Geological Survey. *Geological Quadrangle Map of India*. 1:253,440. (1977–)
India. Geological Survey. *Geological and Mineral Map of India*. 1:1,000,000; to total 48 sheets. (1976–)

India. Central Ground Water Board. *Hydrogeological Map of India.* 1st edn. 1:5,000,000. (1976). Text.

Eremenko, N A *Tectonic Map of India.* 1:2,000,000; 4 sheets. (Oil and Natural Gas Commission, 1968). Text.

Indonesia

Besides several series (1;1,000,000, 16 sheets, 1975–; 1:250,000, about 300 sheets, 1973–; 1:100,000), there are also geological and tectonic maps:

Indonesia. Direcktorat Geologi. *Geologic Map of Indonesia = Peta Geologi Indonesia.* 1:2,000,000; 2 sheets. (US Geological Survey, 1965) (*USGS Miscellaneous Investigations Series,* **I-414**)

Hamilton, Warren *Tectonic Map of the Indonesian Region.* 1:5,000,000. (US Geological Survey, 1981). (*USGS Miscellaneous Investigations Series,* **I-875-D**)

Iran

A geological map at 1;2,500,000 was published in 1959, and there are a number of series, such as the *Geological Quadrangle Map of Iran* at 1:250,000 (1968–), the series at 1:100,000, *Geological Maps and Sections of Iran* at 1:500,000 (1978), and the 1:1,000,000 map in 6 sheets, the latter two being issued by the National Iranian Oil Company.

Iraq

Besides a 1:50,000 series, there is:

Iraq. Mineral Survey Project. *Geological Map of Iraq.* 1:1,000,000. (Ministry of Development, 1977)

The US Geological Survey's *Professional Paper* **560-G** (1967) has a generalized geological map of south and central Iraq.

Israel

Geological maps at 1:250,000 (1975), and 1:500,000 have been published, and there are 1:100,000 (24 sheets, 1954–1964) and 1:50,000 (1970–) series. A geological photomap at 1:500,000 (1979) has also been published. There is also an index:

Ginzburg, Dov *An Inventory of Geological Maps of Israel.* 2 v. (Geological Survey, Mapping Division, 1976) (*Report MM/IGS,* **1/76**)

Japan

Japan has an extensive geological mapping program, with series at 1:500,000, 1:200,000, and 1:50,000, presently very active. The most prominent of the overall maps are:

Japan. Geological Survey. *Geological Map of Japan.* 4th edn. 1:5,000,000. (The Survey, 1982)

Isomi, Hiroshi *Tectonic Map of Japan.* 1:2,000,000; 3 sheets. (Geological Survey of Japan, 1968)

The Survey has issued indexes to their maps, titled *Index to the Geological Maps of Japan* (vol. 1, 1900–1959; vol. 2, 1960–1969; vol. 3, 1970–1974), and a map index, in 1977, at 1:2,500,000. Titles and legends of maps are often in both Japanese and English.

Jordan

Besides a 5 sheet 1:250,000 map (1954–68), there is also a 1:1,000,000 map published in 1939, a 1:100,000 series (published by the German Bundesanstalt für Bodenforschung in 1974), and a 1:500,000 map and text published in 1975.

Kampuchea see **Vietnam**

Korea

Korea. Research Institute of Energy and Resources. *Taehan Cicil To = Geological Map of Korea.* 1:1,000,000. (1981). Text (1983).
Korea. Research Institute of Energy and Resources. *Geological Map of Korea.* 1:50,000; to total about 250 sheets.

Kuwait

Austria. Geological Survey. *Synoptic Geologic Map of the State of Kuwait.* 1:250,000. (Kuwait Ministry of Commerce and Industry, 1966). Text.
See also Saudi Arabia.

Laos see **Vietnam.**

Lebanon

Besides a 1:50,000 series, there have been geological maps of Lebanon, Syria and neighbouring areas in *Notes et Mémoires sur le Moyen Orient*, **8**, 1966 and in the *Beilage zum Lexique Stratigraphique International*, **3**, fasc. 10, both at 1:1,000,000.
In addition, there are separately published geological and hydrogeological maps:
Dubertret, Louis *Carte Géologique du Liban.* 1:200,000. (Institut Géographique National, 1955)
United Nations. *Carte Hydrogéologique du Liban.* 1:200,000. (United Nations, 1967)

Malaysia

Geological maps of Malaysia tend to be of portions of the area, of Malaya and the Malay Peninsula, or of smaller areas, forming a complex picture. Examples of the sorts of maps extant are listed below:
Malaysia. Jabatan Penyiasatan Kajibumi. *Peta Geologi Semenanjung Malaysia = Geological Map of Peninsular Malaysia.* 8th edn. 1:500,000; 2 sheets. (1985)
Tan, Denis N K *Peta Kajibumi Sarawak dan Sabah, Malaysia = Geological Map of Sarawak and Sabah, Malaysia.* 3rd edn. 1:1,000,000. (Geological Survey of Malaysia, 1985).
There is a series (solid geology) at 1:63,360, and also series at 1:250,000 and 1:125,000.

Nepal

Tater, J M *Geological Map of Nepal.* 1:1,000,000. (Metal Mining Agency of Japan, Department of Mines and Geology, 1978)

Oman see **Saudi Arabia**

Pakistan

A series of 1:253,440/1:250,000 sheets is available for part of west Pakistan.
Bakr, M Abu *Geological Map of Pakistan.* (Geological Survey of Pakistan, 1964, reprinted 1979)

Philippines

Philippines. Geological Survey Division. *Geological Map of the Philippines.* 1:1,000,000; 8 sheets. (1963)
Philippines. Geological Survey Division. *Geology and Mineral Resources Map of Provinces.* 1:250,000. (1974–)

Qatar

Geological maps at 1:200,000 and 1:100,000 have been issued.
See also Saudi Arabia.

Saudi Arabia

There are both nationally produced and US Geological Survey maps of Saudi
Arabia and the Arabian peninsula:
> Saudi Arabia. al-Mudiriyah al-'Ammah lil-Tharwah al-Ma'diniyah. *Kingdom of
> Saudi Arabia, Geologic Map GM*. Scales may vary. (The Kingdom, Ministry
> of Petroleum and Mineral Resources, 1966–)
> United States. Geological Survey. *Geological Map of the Arabian Peninsula*.
> 1:2,000,000. (1963) (*USGS Miscellaneous Investigations Series* **I-270 A**)

USGS has issued a series of 21 1:500,000 sheets of Saudi Arabia (*Miscellaneous
Investigations Series* **I-200 A** through **I-220A**), which are apparently being reissued,
in some cases as GM's, by the Kingdom.

Singapore

A geological map of the Republic was issued in 1976, with 9 sheets at 1:25,000 and 1
at 1:75,000.

Sri Lanka

Besides a 1:253,440 tectonic and a 1:63,360 geological map series (1961–), there is a
1982 geological map at 1:500,000.

Syria

Both geological and tectonic maps have been published:
> V/O Technoexport. *Geological Map of Syria*. 1:1,000,000. (1964)
> Ponikarov, V *Geological Map of Syria*. 1:500,000; 5 sheets. (Department of
> Geological and Mineral Research, 1964)
> Ponikatov, V. *Tectonic Map of Syria*. 1:1,000,000. (Department of Geological
> and Mineral Research, 1964)

In addition, there are geological map series at 1:50,000 (English edn. dating from
1968, English/Arabic from 1978) and at 1:200,000 (19 sheets).

Taiwan

Geological maps have been, or are being, issued at scales of 1:500,000 (1974),
1:300,000 (1953), 1:250,000 (1974) and 1:50,000.

Thailand

Besides a geological map series at 1:250,000 (about 1971–), there is an overall map:
> Javanaphet, Jumchet C *Geological Map of Thailand*. 1:1,000,000; 2 sheets.
> (Royal Thai Survey Department, 1969).

Turkey

Besides 8 sheets of a 1:800,000 geological map (1942–46), there is a 1:500,000 series
(each sheet being accompanied by text), a 1:100,000 series, and a 1:2,500,000 map,
all having the base title *Türkiye Jeoloji Haritasi*.

United Arab Emirates see **Saudi Arabia**

Vietnam, Cambodia, Laos

A 1:2,000,000 geological map was published in 1971, and some 1:500,000 sheets are available, the latter being in 14 sheets and published 1961–64.

Yemens

There is an 8-sheet 1:250,000 scale map.
   Grolier, Maurice J *Geologic Map of the Yemen Arab Republic (San'a')*. 1:500,000. (US Geological Survey, 1978) (*USGS Miscellaneous Investigations Series*, **I-1143B**)
   Beydoun, Z R *Geological Map of Eastern Aden Protectorate*. 1:1,000,000. (British Directorate of Overseas Surveys, 1963) (*D.O.S. (Geol.)* **1148**)
See also Saudi Arabia

EURASIA

   Geological Map of Eurasia = Geologicheskaià Karta Evrazii = Carte Géologique de l'Eurasie. 1:5,000,000; 13 sheets. (Vsesoiuznyi Ordena Lenina Nauchno-Issledovatel'skii Geologicheskii Instutyt, VSEGEI, 1975)
   Akademiia Nauk SSSR. Geologicheskii Institut. *Tektonicheskaia Karta Evrazii = Tectonic Map of Eurasia*. 1:5,000,000; 12 sheets. (Glavnoe Upravlenie Geodezii i Kartografii, Ministerstva Geologii SSSR, 1966)

EUROPE

Various maps have been published under the auspices of the International Geological Congress:
   Carte Géologique Internationale de l'Europe. 2nd edn; 3rd edn. 1:1,500,000. (Bundesanstalt für Bodenforschung; Unesco, 1964–)
   Carte Internationale du Quaternaire de l'Europe = Internationale Quartar-Karte von Europe = International Quaternary Map of Europe. 1:2,500,000; to total 16 sheets. (Bundesanstalt für Bodenforschung, 1967–)
   Carte Tectonique Internationale de l'Europe. 1:2,500,000; 16 sheets. (Académie des Sciences de l'URSS, Comité Géologique d'Etat de l'URSS, 1964)
   International Geological Map of Europe and the Mediterranean region = Carte Géologique Internationale de l'Europe et des Regions Riveraines de la Méditerranée. 1:5,000,000; 2 sheets. (Bundesanstalt für Bodenforschung; Unesco, 1971). Text: *Earth science series* **10**.
As can be seen from the above, Unesco is very active in such mapping, further evidence of which is given by the following:
   International Hydrogeological map of Europe = Carte Hydrogéologique Internationale de l'Europe — Internationale Hydrogeologische Karte von Europe. 1:1,500,000; to total 35 sheets. (Bundesanstalt für Bodenforschung; Unesco, 1970–) Texts.
   Metamorphic Map of Europe = Carte Métamorphique de l'Europe. 1:2,500,000; 17 sheets. (Unesco, 1973)

Albania

   Albania. Ministère d'Industrie et des Mines. Institut de Recherches Industrielles et Minières. *Carte Géologique de l'Albanie*. 1:200,000; 3 sheets. (Ministère d'Industrie et des Mines, 1967)
   Albania. Institut de Recherches Géologiques et Minières. *Carte Tectonique de l'Albanie*. 1:500,000; 2 sheets. (1969)

Austria

Besides 2 series (1:75,000, about 1885–1961; 1:50,000), there are geological, hydrogeological, and tectonic maps:
Vetters, H *Geologische Karte der Republik Österreich und der Nachbargebiete.* 1:500,000; 2 sheets + legend. (Geologische Bundesanstalt, 1968)
Gattinger, T *Hydrogeologische Karte der Republik Österreich.* 1:1,000,000. (Geologische Bundesanstalt, 1969)
Beck-Mannagetta, P *Geologische Übersichtskarte der Republik Österreich, mit Tektonischer Gliederung.* 1:1,000,000. (Geologische Bundesanstalt, 1964). Text (1966)

Azores see **Portugal**

Balearic Islands see **Spain**

Belgium

A 1:1,000,000 geological map was issued in 1950; there are map series at 1:160,000 (complete in 12 sheets in 1945), 1:40,000 (226 sheets), and 1:25,000 (to total 237 sheets; 1958–).

Bulgaria

A geological map (*Bâlgarija — Geoložka Karta*) at 1:1,000,000 was published in 1965, replacing a 1947 map at 1:500,000.

Czechoslovakia

Geological Map of Czechoslovakia (superficial deposits omitted). 2nd edn. 1:1,000,000 (Ustrědní Ustav Geologický, 1982)
Kopecky, Sestavil A *Neotektonicka Mapa CSSR.* 1:1,000,000. (Ustrědní Ustav Geologický, 1973)
Suk, M *Metamorphic Map of Czechoslovakia.* 1:1,000,000. (Ustrědní Ustav Geologický, 1973)
Also there is a 1:200,000 geological series.

Denmark

Several geological series are available — 1:100,000 (1893–), 1:160,000, and 1:250,000 (complete in 4 sheets).
See also Scandinavia.

Faeroe Islands

Denmark. Geologiske Undersøgelse. *Geologisk Kort over Faeroerne = Jardfrødiligt Føroyakort = Geological Map of the Faeroe Islands. Praekvartaeret = Prekvarter = Pre-Quaternary.* 1:50,000; 6 sheets. (Geologiske Undersøgelse)

Finland

There is a 1:1,000,000 geological map:
Finland. Geological Survey. *Suomen Kallioperä = Prequaternary Rocks of Finland; Finlands Berggrund.* 1:1,000,000. (1980)
and 2 series, each issued by the Survey in 2 editions:
Suomen Geologinen Yleiskartta = General Geological Map of Finland. 1:400,000.
1. Kivilajikartta = Pre-Quaternary Rocks. 1903–
2. Maaperäkartta = Surficial Quaternary deposits. 1906–

Suomen Geologinen Kartta = Geological Map of Finland. 1:100,000.
1. Kallioperäkartta = Pre-Quaternary rocks. 1949–
2. Maaperäkartta = Quaternary deposits. 1950–
See also Scandinavia.

France

The Bureau de Recherches Géologiques et Minières (BRGM) is an extremely active agency, mapping not only France but also many French, or previously French, territories. Besides 4 series (1:320,000, complete in 21 sheets, 1892–1962; 1:250,000, complete in 44 sheets; 1:80,000, 1832–1962, 255 sheets; 1:50,000, 1960–, to be complete in 1,060 sheets), the agency also publishes, most notably for our purposes:
    Dottin, O *Carte Geologique de la France et de la Marge Continentale.*
        1:1,500,000. (BRGM, 1980)
    Margat, J *Carte Hydrogéologique de la France, Systèmes Aquifères*. 1:1,500,000.
        (Service Géologique National, 1980)
    A 1:1,000,000 tectonic map in 2 sheets, with text, was issued in 1980.

Germany

Prior to the partition of Germany, much of the country had been covered by various series (1:25,000; 1:50,000; 1:100,000; 1:200,000), issued by the Prussian and other State surveys. Post-partition, most of the States (Länder) have geological map coverage at scales of from 1:300,000 to 1:500,000, and the series at 1:25,000 and 1:100,000 have been continued; hydrogeological maps at various scales are also available.
Overall geological maps are, for the German Democratic Republic (East Germany)
    Schumacher, K H *Geologische Karte der Deutschen Demokratische Republik.*
        1:500,000; 2 sheets. (Geologische Kommission, Forschungsinstitut für die
        Erkundung von Erdöl und Erdgas, 1964)
and for the Federal Republic of Germany (West Germany)
    Germany. Bundesanstalt für Bodenforschung. *Geologische Karte der Bundes-
        republik Deutschland*. 1:1,000,000. (1973)
Various indexes to geological maps have been published:
    Franke, Dietrich (1969) '20 Jahre Geologische Kartierung in der Deutschen
        Demokratischen Republik; ein Überblick über die geologischen Karten-
        werke der DDR'. *Géologie*, **18**, 891–910.
    Schamp, Heinz (1960/61) Die Geologischen Übersichtskarten Deutschlands.
        Geographisches Taschenbuch, 181–91.
    Schamp, Heinz (1961) Ein Jahrhundert amtliche geologische Karten, *Berichte
        zur Deutschen Landeskunde, Sonderheft* **4**.

Greece

A general geological map was published in 1954:
    Renz, Carl *Geologikos Chartes tes Ellados = Geologic Map of Greece*. 1:500,000;
        2 sheets. (Institute for Geology and Subsurface Research, 1954)
and there are 2 major series at scales of 1:50,000 and 1:200,000.

Hungary

A principal series at 1:200,000 (*Magyarország Földtani Térképe*, 35 sheets, 1965–), with texts, is augmented by others at 1:25,000, 1:10,000, and various special sheets. The overall map is:
    Hungary. Földtani Intezet. *Magyarország Földtani Térképe*. 1:300,000; 4 sheets.
        (Földtani Intezet, 1956)

Iceland

A series of 1:250,000 geological sheets, to total 9, is published by Landmaelingar Islands.

Ireland

Before partition, all of Ireland had been covered by a 6-inch survey, and 205 1:63,360 coloured maps (1853–90) had been published, as had a few 1:253,440 sheets. Geological maps recently published include:
Williams, C E *Geological Map of Ireland.* 1:750,000. (Ordnance Survey of Ireland, 1979)
Map series at 1:250,000 and 1:63,360/1:50,000 are published for Northern Ireland.

Italy

Geological series are issued (all with the same base title, *Carta Geologica d'Italia*) at 1:50,000, 1:100,000, and 1:500,000, the last accompanied by texts. A 1:1,000,000 geological map was published in 1969, in 2 sheets. Some 1:25,000 sheets are available for a few areas.

Luxembourg

A 1:25,000 geological series in 13 sheets is published, as is an overall map:
Luxembourg. Service Géologique. *Carte Géologique Générale du Grand Duché de Luxembourg.* 2ième edn 1:100,000. (Service Géologique, 1974)

Netherlands

Map series at 1:200,000 and 1:500,000 are published, as is a 1:600,000 map in 5 sheets:
Netherlands. Rijks Geologische Dienst. *Geologische Overzichtskaarten van Nederland; kaarten, profielen, toelichting.* 1:600,000; 5 sheets. *In:* Zagwijn, W H *Toelichting bij Geologische Overzichtskaarten van Nederland.* (Rijks Geologische Dienst, 1975)

Norway

Several different map series – *Berggrunnskart* 1:50,000, 1972–; *Kvartaergeologisk Kart* 1:50,000, 1972–; *Geologisk Generalkart* 1:250,000, 1917–32; *Kvartaergeologisk Kart* 1:250,000, 1949–60; *Berggrunnskart* 1:100,000, previously *Geologisk Kart*, 1875–; *Berggrunnskart* 1:250,000, 1970– are published, as is an overall map:
Holtedahl, Olaf *Geologisk Kart over Norge, Berggrunnskart = Geological Map of Norway (Bedrock).* 1:1,000,000.
which is accompanied by the same author's work, *Geology of Norway, (Geologiske Undersøkelse Skrifter* **208**, 1960); the map was republished in 1975.
See also Scandinavia.

Poland

The Instytut Geologiczny has published very actively, issuing both separate maps, and maps in a portfolio (but not bound) as an atlas, the latter at scales of 1:1,000,000 (15 sheets, 1955–61), 1:2,000,000 (10 sheets, 1968), 1:2,000,000 (5 sheets, 1956), and 1:3,000,000 (13 volumes, 1959–65). The Instytut issues:
Przegladowa Mapa Geologiczna Polski. 1:300,000; 28 sheets. (1946–55)
also series from 1:10,000 to 1:300,000, and numerous geological maps at 1:500,000 and 1:1,000,000, each of which emphasizes certain formations (titles are given on maps in Polish, Russian and English). These are in English only in the following list:

Rühle, Edward *Geological Map of Poland*. 1:500,000; 4 sheets. (Geological Institute, 1986).
Geological Map of Poland without Cainozoic, Cretaceous and Jurassic formations. 1:500,000; 4 sheets. (1980)
Geological Map of Poland and Adjoining Countries without Cainozoic formations . . . 1:1,000,000; 2 sheets. (1979)
Geological Map of Poland without Cainozoic and Cretaceous formations. 1:500,000; 4 sheets (1978)
Geological Map of Poland without Quaternary formations. 1:500,000; 4 sheets + 2 legend sheets. (1977)
Geological Map of Poland without Cainozoic, Mesozoic and Permian formations. 1:1,000,000. (1972)
Geological Map of Poland without Cainozoic formations. 1:500,000; 4 sheets. (1971/72)
A tectonic map has been issued:
Ksiazkiewicz, Marian *Tectonic Map of Poland*. 1:1,500,000. (Geological Institute, 1974). This map also appeared as a folded plate in: *Geology of Poland*, vol. **IV**, Wydawnictwa Geologiczne.

Portugal

The Portuguese Serviços Geológicos issues a 1:50,000 series (*Carta Geológica de Portugal*, 1944–) and several overall maps:
Carta Geológica de Portugal. 4th edn 1:500,000; 2 sheets. (1972)
Carta Geológica do Quaternario de Portugal. 1:1,000,000 (1969)
Carta Tectónica de Portugal. 1:1,000,000. (1972)
See also Spain.

Romania

Geological series at 1:100,000 (6 sheets), 1:200,000 (about 48 sheets), and 1:500,000 (12 sheets, 1941–64) have been published, as have various overall maps, such as geological (1:1,000,000, 1978), hydrogeological (1:1,000,000), and Quaternary geology (1:1,000,000, 1964).

Scandinavia

The Geological Surveys of Denmark, Norway, Sweden and Finland compiled a geological map in 1933 at 1:1,000,000.

Spain

Various maps of the Iberian peninsula as a whole have been published:
Arangurén Sabas, Félix *Mapa Geológico de la Península Ibérica, Baleares y Canarias*. 5a. edn 1:1,000,000. (Instituto Geológico y Minero de España, 1971). Text.
Riba Arderiú, Oriol *Mapa Litológico de España*. 1:500,000; 5 sheets. (Instituto Geológico y Minero de España, 1969)
Spain. Instituto Geológico y Minero. *Mapa Tectónico de la Península Ibérica y Baleares; contribución al mapa tectónico de Europa de la Comisión Nacional de España* . . . 1:1,000,000. (1972)
The Instituto also issues series at 1:50,000 (*Mapa Geológico Nacional*), 1:100,000 (*Mapas Geológicos*), 1:200,000 (*Mapas Geológicos Provinciales* and *Mapa Geológico Nacional de Síntesis de España*, to total 87 sheets) and 1:400,000 (*Mapa Geológico de España*). There is also a commercially issued map:
Mapa Geológico de España y Portugal. 1:1,250,000. (Falkplan, 1965).

Spitzbergen

A 1:1,000,000 map was issued in 1969.

Svalbard and Jan Mayen Islands

A geological map at 1:1,000,000 and a series at 1:500,000 (1971–), issued by the Norsk Polarinstitutt as part of its *Skrifter* series (nr. **154**), have been published. There is also *Geologic map, Svalbard*. 1:500,000. (Norsk Polarinstitutt, 1984)

Sweden

The Geologiska Undersökning issues 13 series relating to maps:
Aa Combined Quaternary and petrological maps. 1:50,000.
A₁a Petrological maps. 1:200,000.
Ab Combined Quaternary and petrological maps. 1:200,000.
Ac Combined Quaternary and petrological maps. 1:100,000.
Ad Agrogeological maps. 1:20,000.
Ae Quaternary deposits. 1:50,000.
Af Solid rocks and geophysics. 1:50,000.
Ag Hydrogeological maps. 1:50,000, 1:100,000, 1:250,000.
Ba Regional Quaternary and petrological maps.
Bb Special maps.
C Memoirs.
D Peat deposits. 1:100,000.
See also Scandinavia.

Switzerland

All of Switzerland is covered by the 1:100,000 series, which sheets are being superseded by the sheets of the *Geologischer Atlas der Schweiz* at a scale of 1:25,000. Both geological and tectonic maps are published:
Switzerland. Geologischen Kommission. *Carte Géologique Générale de la Suisse = Geologische Generalkarte der Schweiz = Carta Geologica Generale della Svizzera*. 1:200,000; 8 sheets. (Kummerly and Frey, 1956–65). Explanatory brochures are published for each sheet.
Spicher, A *Geologische Karte der Schweiz = Carte Géologique de la Suisse*. 1:500,000. (Schweizerische Geologische Kommission, 1980)
Spicher, A *Tektonische Karte der Schweiz = Carte Tectonique de la Suisse*. 1:500,000. (Schweizerische Geologische Kommission, 1972)

United Kingdom

The Geological Survey of Great Britain is another Geological Survey that has published prolifically over the years, so much so that it is difficult to know where to begin. The Survey issues several series:
1:25,000 (*Classical Areas of British geology*, 1969–), 1:63,360/1:50,000 (England and Wales in 353 sheets, 1:63,360 1907–75, 1:50,000 1972–; Scotland in 116 sheets, 1:63,360 1900–, 1:50,000 1972–; issued in "Solid" and "Drift", new series), and 1:250,000 (to total about 70 sheets; 1977–; replaces *Quarter-Inch series* 1:253,440, England and Wales 24 sheets and Scotland in 17 sheets, 1907–74). The former Institute of Geological Sciences (IGS) has issued overall maps and these are now the responsibility of the British Geological Survey (BGS):
Geological map of the United Kingdom. 3rd edn solid. 1:625,000; 2 sheets. (1979)
Quaternary map of the United Kingdom. 1:625,000; 2 sheets. (1977)

Dunning, F W *Tectonic Map of Great Britain and Northern Ireland.* 1:1,584,000. (Director General of the Ordnance Survey for the Institute of Geological Sciences, 1966)

Beyond the map series noted in the previous section, for England and Wales there are hydrogeological maps at 1:63,360 and 1:126,720, and an overall hydrogeological map:

Great Britain. Institute of Geological Sciences. *Hydrogeological Map of England and Wales.* 1:625,000. (IGS, 1977)

Scotland is covered in the map series listed in first section. Maps for the Outer Hebrides are part of a series of five papers by T J Jehu and R M Craig, published in the *Transactions of the Royal Society of Edinburgh,* 1924–34. Arran, Assynt, Glasgow, northern Skye, northern Shetland, and western Shetland are covered by special 1:63,360 sheets.

Wales is covered in the map series listed above and there is a useful guide by Bassett, D A *A Source-book of Geological, Geomorphological and Soil Maps for Wales and the Welsh Borders (1800–1966).* (National Museum of Wales, 1967)

Northern Ireland has one overall map

Northern Ireland. Geological Survey. *Geological Map of Northern Ireland.* Solid ed. 1:250,000. (Department of Commerce, 1977)

Yugoslavia

Geological map series at 1:50,000 and other scales have been issued; besides a tectonic (1:2,500,000, 1960) map, there is an overall geological map:

Yugoslavia. Savezni Geologoški Zavod. *Geološka Karta = Geological Map.* 1:500,000; 6 sheets. (1970) Text: *Tumac Geoloske Karte SFR Jugoslavije* (1971)

Union of Soviet Socialist Republics (USSR)

Principal map series are at scales of 1:200,000 and 1:1,000,000, with an obsolete 1:420,000 series covering many parts of Russia in Europe. Overall maps are published in a fairly large array, all by the Ministerstva Geologii SSSR as a general rule:

Geologicheskaia Karta SSSR. 1:2,500,000; 16 sheets. (1968)

Geologicheskaia Karta SSSR. 1:10,000,000 (Vsesoiuznoe Aerogeologicheskoe Nauchno-provizvodstvennoe ob' Edinenie Aerogeologiia, 1976)

Karta Magmaticheskikh Formatsii SSSR. 1:2,500,000; 17 sheets. (1971). Text.

Karta Chetvertichnykh Otlozhenii SSSR. 1:2,500,000; 16 sheets. (Vsesoiuznoe Aerogeologicheskoe Nauchno-provizvodstvennoe ob'Edinenie Aerogeologiia, 1976)

Gidrogeologicheskaia Karta SSSR. 1:2,500,000; 16 sheets. (1969) Text

Inzhenernoe-geologicheskaia Karta SSSR. 1:2,500,000; 16 sheets. (1972) Text

Tektonicheskaia Karta Fundamenta Territorii SSSR. 1:5,000,000; 4 sheets. (1974) Text.

Tektonicheskaia Karta SSSR. 1:2,500,000; 19 sheets (1966).

OCEANIA

The area covered by *Geological Map of the World, Australia and Oceania* (1:5,000,000, 13 sheets, beginning 1965) is bounded by latitudes 24°N and 48°S, and longitudes 108°W and 132°E, omitting part of the northwest and southeast.

Australia

The Geological Surveys of the various States have in the past mapped, and are presently carrying out mapping of, their respective States; the Bureau of Mineral Resources, Geology and Geophysics (BMR) has been active in producing overall maps such as *Earth Science Atlas of Australia* at a scale of 1:10m, a series at 1:5,000,000 in preparation in early 1983, a 1:2,500,000 map in 4 sheets (1976; *BMR Bulletin* **181**), and three major series, 1:250,000 (old series 1:253,440), which is about 75 percent complete, 1:100,000, and 1:50,000 (old series 1:63,360). Geological maps of different types have been published by other groups, such as the Geological Society of Australia, Tectonic Map Committee's, *Tectonic Map of Australia and New Guinea*, at a scale of 1:5,000,000 (1971). A metamorphic map of Australia at 1:5,000,000 was in preparation in early 1983.

All the States issue maps covering their own territories, with scales ranging from 1:506,880 to 1:2,534,440.

Fiji

Fiji. Geological Survey Department. *Provisional Geological Map of Fiji.* (1965; final edn in 1967)

French Polynesia

France. Bureau de Recherches Géologiques et Minières. *Territoire de la Polynesie Française, Carte Géologique.* 1:40,000; 6 sheets. (1965)

Guam

The US Geological Survey issued a 1:50,000 map in 1964, with text.

New Caledonia

The French Bureau de Recherches Géologiques et Minières has two series for this area, 1:50,000 (to total 40 sheets; 1956–), and 1:100,000 (in 10 sheets), with a 1:200,000 map issued in its *Mémoire* **113**, *Géologie de la Nouvelle-Calédonie*.

New Hebrides

The British Ministry of Overseas Development (formerly the Directorate of Overseas Surveys) has worked on a 1:100,000 series (to total 11 sheets; 1972–; *D.O.S. (Geol.)* **1181**), and has also issued a 1:1,000,000 map of the area, in 1975, as its *D.O.S. (Geol.)* **1196**.

New Zealand

The New Zealand Geological Survey has several series in hand, including 1:25,000 (1976–), 1:50,000 (1979–; from 1953 to 1979, 1:63,360), and 1:250,000 (28 sheets; 1960–69). A geological map in 2 sheets, one for North Island, one for South Island, both at 1:1,000,000, was published in 1972. The Survey also publishes maps in its *Miscellaneous series*.

Pacific Ocean

Heezen, Bruce C *Geological map of the Pacific Ocean.* 1:35,000,000. Scripps Institution of Oceanography (1975). In: *Initial reports of the Deep Sea Drilling Project*, vol. **30**.

Geologicheskaia Karta Tikhookeanskogo Podvizhnogo Poiasa i Tikhogo Okeana = Geological Map of the Pacific Mobile Belt and Pacific = Mapa Geológico del Cinturón Movil del Pacifico y del Oceano Pacífico. 1:10,000,000; 10 sheets. (Ministerstva Geologii SSSR).

Akademiia Nauk SSSR. Geologicheskii Institut. *Tektonischeskaia Karta Tikho-okeanskogo Segmenta Zemli = Tectonic Map of the Pacific Segment of the Earth*. 1:10,000,000; 6 sheets. (Glavnoe Upravlenie Geodezii i Kartografii, 1970)

Papua New Guinea

The Australian Bureau of Mineral Resources, Geology and Geophysics has issued sheets at 1:250,000 and 1:63,360. The national survey has issued a 1:1,000,000 map in 4 sheets (1972) and a 1:100,000 series (preliminary sheets; 1979–). BMR has also issued an overall map, at a smaller scale than the nationally issued map:
D'Addario, G W *Geology of Papua New Guinea*. 1:2,500,000. (1976)

Réunion see under **Africa.**

Solomon Islands

There is a 1:50,000 series issued by the Solomon Islands Geological Survey, and:
Great Britain. Directorate of Overseas Surveys. *Geological Map of the British Solomon Islands*. 2nd edn 1:1,000,000. (1969) (*D.O.S. (Geol.)* **1145A**) which accompanies: *British Solomon Islands Geological Record*, vol. **3**.

OCEANS

For Arctic Ocean, see **Arctic regions.**

For Pacific Ocean, see **Oceania.**

Pitman, Walter C *The Age of the Ocean Basins*. 1:40,000,000; 2 sheets. (Geological Society of America, 1974). (*Geological Society of America, Map and chart series* **MC-6**)

Atlantic Ocean

Emel'ianov, E M *Types of Bottom Sediments of the Atlantic Ocean; oceanological researches = Tipy Donnykh Osadkov Atlanticheskogo Okeana; okeanolog-icheskie issladovaniia*. (Mezhduvedomstvennyi Geofizicheskii Komitet pri Prezidiume Akademii Nauk SSSR, 1975). Includes 6 maps at 1:20,000,000.

Indian Ocean

Heezen, Bruce C *Geological Map of the Indian Ocean*. 1:29,000,000. (Unesco, 1978)
Pepper, J F *The Indian Ocean, the geology of its bordering lands and the configuration of its floor*. 1:13,655,000. (US Geological Survey, 1963) (*USGS Miscellaneous Investigations Series* **I-380**)

Remote sensing

Almost anyone who works with remote sensing has his own definition of the phrase; it is most broadly considered to be the products and techniques that obtain from a distance reliable information about the properties of surfaces and objects, in this case the surface of the earth. The vast majority of all remote sensing is based on the electromagnetic spectrum (EMS), and thus includes such sensors as photographic cameras, radar, lasers, infrared scanners, radiometers and imagers, spectrometers, micro-wave radiometers, television, and multi-spectral scanners (MSS); each is used for a certain portion of the EMS. The remote sensing techniques most often used in geology are aerial photography and multispectral scanning, the latter in the specific form of Landsat imagery.

Aerial photographs are a traditional product: geological maps have been prepared by photointerpretation (photogeology) for some years, see J A E Allum's *Photogeology and Regional Mapping* (Pergamon, 1966), V C Miller's *Photogeology* (McGraw-Hill, 1961), and the most current editions of the American Society of Photogrammetry's *Manual of Photogrammetry, Manual of Photographic Interpretation, Manual of Color Aerial Photography*, and *Manual of Remote Sensing*, and its periodical, *Photo-grammetric Engineering and Remote Sensing* (1934–; slight title changes over the years). Aerial photographs for geological applications are usually taken of snow-free ground between midmorning and midafternoon, when the sun is at a high angle and shadows have minimal effect, although photography taken of snow-covered areas, or at a low-sun angle, may be helpful in that topographic aspects may be enhanced.

Landsat imagery entered the geology world in 1972, when the US National Aeronautics and Space Administration launched what was then called ERTS-1 (Earth Resources Technology Satellite), later to become Landsat-1. Three Landsats have been launched since, with the current Landsat-4 sent up in 1982. The last, whose Thematic Mapper was to produce imagery of consider-able interest to the geologist, has had serious problems, such as power cable failures. Recently, Landsat has been turned over to a commercial firm, EOSAT, and there has been some question whether satellites will continue to be launched. Also, the French have sent up an imaging satellite, SPOT; the resulting images have higher resolution than those from Landsat imagery.

Initially Landsat applications to geology were mainly thought to be only that of reconnaissance, but this has evolved to a feeling that new important insights are possible when using such imagery (Siegal 1976; Geosat Committee, 1976; Australian Mineral Found-ation 1983; Bélanger 1983; Lillesand 1979; Siegal, 1980). Landsat imagery may be used for recognition of rock types, mapping of major geological units, revising geological maps, mapping igneous intrusions, mapping linear features (one of the most striking and successful of uses), and much more. An enormous amount has been written on remote sensing, and especially on Landsat; any citations given are the tip of a very large iceberg.

One benefit of Landsat — the vast amount of data — is also a major problem, which is increasingly dealt with by utilizing digital image processing, since all Landsat imagery is initially digital data (Castleman, 1979; Hord, 1982; Moik, 1980; Swain, 1978). Persons working with Landsat are well advised to peruse Nicholas Short's *The Landsat Tutorial Workbook: basics of satellite remote sensing* (US National Aeronautics and Space Administration 1982; *NASA Reference Publication* **1078**) and the US Geological Survey's *Landsat Data Users Handbook* (USGS, 1979). Information on a continuing basis is best provided by *EOSAT Landsat Data Users Notes* (1986–; previously called *Landsat Data Users Notes* and issued by US National Oceanic and Atmospheric Administration/US EROS Data Service, 1978–1986) and *Information Update for Users of NOAA/NESDIS Satellite Data* (US National Environmental Satellite, Data, and Information Service, 1981–). Questions may be directed to EROS Data Center, Sioux Falls SD 57198 USA or to EOSAT, 4300 Forbes Blvd., Lanham MD 20706 USA.

References

Aune, Quinton A (1960) The geologic map *Mineral Information Service* **13**(8), 1–8

Australian Mineral Foundation (1983). *Geological Interpretation of Aerial Photographs and Satellite Images, Parts 1 and 2*. Australian Mineral Foundation (*Workshop Course* **210/82**)

Barnes, John Wykeham (1981). *Basic Geological Mapping*. Milton Keynes: Open University Press

Bélanger, Robert (1983). Cartography: satellites to the rescue. *Association of Canadian Map Libraries Bulletin* **46**, 7–11

Bishop, Margaret S (1960). *Subsurface Mapping*. Wiley

Castleman, Kenneth R (1979). *Digital Image Processing*. Prentice-Hall

Geosat Committee. Ad Hoc Geological Committee on Remote Sensing from Space (1976?). *Geological Remote Sensing from Space*. The Committee

Hageman, B P (1968). The reliability of geological maps. *In*: C C Albritton (ed) *The Fabric of Geology*. Freeman

Harrison, J M (1963). Nature and significance of geological maps. *In*: C C Albritton (ed) *The Fabric of Geology*. Freeman

Hord, Richard (1982). *Digital Image Processing of Remotely Sensed Data*. Academic Press

Kay, M (1945). Paleogeographic and palinspastic maps. *Bulletin American Association of Petroleum Geolgists*. **29** (4), 426–50

Lillesand, Thomas M (1979). *Remote Sensing and Image Interpretation*. Wiley

Lowman, Paul D (1976). Geologic structure in California: three studies with Landsat-1 imagery. *California Geology*. **29**,(3), 75–81

Martin, E L (1973). Geological maps. *In: Use of Earth Sciences Literature*. Butterworths

Moik, Johannes G (1980). *Digital Processing of Remotely Sensed Images*. US National Aeronautical and Space Administration (*NASA SP*-**431**)

Robertson, T (1956). The presentation of geological information in maps. *The Advancement of Science*, **13**(50), 31–41

Siegal, Barry S (1976). Geologic mapping using Landsat data. *Photogrammetric Engineering and Remote Sensing*, **42**,(3), 325–37.

—— (1980). *Remote Sensing in Geology*. Wiley

Swain, Philip H (1978). *Remote Sensing; the quantitative approach*. McGraw-Hill

United States. Geological Survey. (1978). *Geologic Maps : portraits of the Earth*. USGS

CHAPTER EIGHT

Stratigraphy and regional geology

ANN LUM

Stratigraphy

Stratigraphy, or 'historical geology', is the study of the strata and the total environment of the earth through its succession. It is almost impossible to detach the study of the stratigraphy of a region from its regional geology, and hence the coupling of the two disciplines in this chapter.

Stratigraphy is also interlinked with other disciplines: geo-chronology, palaeoclimatology, and palaeogeography will be discussed in this chapter, palaeontology is covered in Chapter 9.

Publications covering all aspects of stratigraphy are published, or sponsored, by the International Geological Correlation Pro-gramme (IGCP), a subsidiary body of the International Union of Geological Sciences (IUGS). The projects operating within the programme, over a hundred to date, are listed and reviewed in its journal, *Geological Correlation*, and in the journal of the parent body, *Episodes*.

General works

Introductory textbooks are *The Nature of the Stratigraphic Record*, by D V Ager (2nd edn, Macmillan, 1981) and *Méthodes de la Stratigraphie et Géologie Historiques*, by J Boulin (Masson, 1977). The book series *Stratigraphie et Paleogéographie* includes one on *Principes et Méthodes*, by C Pomerol (Doin, 1980). Although considered an 'older' textbook, D T Donovan's *Stratigraphy: an introduction to principles* (Thomas Murby, 1966) is still highly regarded.

Other more specialized or advanced texts include *Facies Interpretation and the Stratigraphic Record*, by A Hallam (Freeman, 1981), *Dynamic Stratigraphy : an introduction to sedimentation and stratigraphy*, by R K Matthews (2nd edn, Prentice-Hall, 1984) and *Abriss der Historischen Geologie*, edited by K A Troger (Akademie Verlag, 1984). *Concepts and Methods of Biostratigraphy*, edited by E G Kauffman and J E Hazel (Dowden Hutchinson and Ross, 1977) is a volume of readings, and another series of papers are published in *Stratigraphy Quo Vadis?*, edited by E Seibold and J D Meulenkamp (*IUGS Publication*, **14**, 1984).

The journals covering this subject are:

Acta Stratigraphica Sinica (1966–), since 1980 renamed as the *Journal of Stratigraphy*
Geobios (1968–), and its *Memoires Special* (1977–)
Geological Correlation (1973–)
Lethaia (1968–)
Newsletter in Stratigraphy (1970–) which contains short, original papers
Notes et Contributions, Centre d'Etudes et de Recherches de Paleontologie et Biostratigraphie, Orsay (1971–)
Paleontologia Stratigraphica ed Evoluzione (1980–)
Professional Papers of Stratigraphy and Palaeontology, Beijing (1975–)
Rivista Italiana di Paleontologia e Stratigrafia (1895–)
Seminarios de Estratigrafia: serie monografias, Universidad Complutense de Madrid (1977–)
Studies in Paleontology and Stratigraphy, University of Louisville (1973–)
Voprosy Stratigrafii, Leningrad (1974–)

in addition, *Special Report of the Geological Society of London* contains only stratigraphic papers.

Stratigraphy is covered by the following bibliographies: *Bibliography and Index of Geology*, section 12: Stratigraphy, Historical Geology and Palaeoecology; *GeoAbstracts*, which began a separate series *Palaeontology and Stratigraphy* in 1986; *Referativnyi Zhurnal, Geologiya* section 6: Stratigrafiya Paleontologiya; and *Zentralblatt fur Géologie und Palaontologie, Teil I*, 4: Historische Géologie (see also Chapter 5).

Geochronology

Textbooks on this subject include *Geologic Time*, by D L Eicher (2nd edn, Prentice-Hall, 1976) and the important *A Geologic Time Scale*, by W B Harland *et al* (Cambridge University Press, 1982). Other publications include proceedings of a symposium published under the title *Contributions to the Geologic Time Scale*, edited by G V Cohee *et al* (American Association of Petroleum Geologists, *Studies in Geology*, **6**, 1978), *The Chronology of the Geological Record*, edited by N J Snelling (Geological Society of London, *Memoir* **10**, 1985) and *Stratigraphy: an interdisciplinary symposium* *report of a dialogue on the role of stratigraphy in geology and archaeology*, edited by S G H Daniels and S J Freeth (*Institute of African Studies, University of Ibadan, Occasional Publication* **19**, 1970).

Palaeoclimatology

Recent books include *The Coevolution of Climate and Life*, by S H Schneider and R Londer (Sierra Book Club, 1984), *Climate and Evolution*, by R Pearson (Academic Press, 1978), *Climates Throughout Geologic Time*, by L A Frakes (Elsevier, 1979), *Climate in Earth History*, by the United States National Research Council (National Academy Press, 1982) and *Milankovitvich and Climate: understanding the response to astronomical forcing*, edited by A Berger (Reidel, 1984), a 2-volume work resulting from a NATO Advanced Research Workshop.

Publications relating to the palaeoclimatology of the Quaternary era are *The Periglacial Environment*, by H M French (Longman, 1976), *The Ice Age : past and present*, by B S John (Collins, 1977), *The Winters of the World: earth under the ice ages*, edited by B S John (David and Charles, 1979), *The Last Great Ice Sheets*, edited by G H Denton and T J Hughes (Wiley 1981), *Reconstructing Quaternary Environments*, by J J Lowe and M J C Walker (Longman, 1984) and *Quaternary Paleoclimatology: methods of paleoclimatic reconstruction*, by R S Bradley (Allen and Unwin, 1985).

The journal for this discipline is *Palaeogeography, Palaeoclimatology, Palaeoecology* (1965–). Current titles in palaeoclimatology have been listed in this journal since 1973.

Relevant literature is indexed in *Palaeoclimatology: a bibliography with abstracts*, compiled by G W Reimherr (US National Technical Information Service, 1980), and in *Holocene Paleoclimates: an annotated bibliography*, compiled and edited by M Andrews (2 vols. *Institute of Arctic and Alpine Research, Occasional Paper* **41**, 1984).

Stratigraphic techniques

Relevant works recently published are *Quantitative Stratigraphy*, by F M Gradstein *et al* (Reidel, 1985), *Quantitative Stratigraphic Correlation*, edited by J M Cubitt and R A Reyment (Wiley, 1982), *Numerical Dating in Stratigraphy*, edited by G S Odin (Wiley Interscience, 1982) and *Sedimentation Models and Quantitative Stratigraphy*, by W Schwarzacher (*Developments in Sedimentology*, **19**, 1975).

Stratigraphic terminology and nomenclature

Guidelines for a code of practice can be found in *Code of Stratigraphic Nomenclature*, compiled by the American Commission on Stratigraphic Nomenclature (American Association of Petroleum Geologists, 1970). Attempts at instituting an international system have resulted in *International Stratigraphic Guide: a guide to stratigraphic classification, terminology and procedure*, edited by H D Hedberg for the International Subcommission on Stratigraphic Classification of the IUGS Commission on Stratigraphy (Wiley, 1976), which makes this book an essential reference work.

The International Subcommission on Stratigraphic Classification also publishes a *Circular* from time to time to update information.

Undoubtedly the major reference work in stratigraphy world wide is the series *Lexique Stratigraphique International* (1956–) published by the Centre National de la Recherche Scientifique. Roughly, each continent (or major area) is designated a volume number; each part, for a country or small group of countries, contains a dictionary of stratigraphic terms for the area, and lists bibliographies; volume, **8**, lists major stratigraphic terms. Individual volumes and parts will be listed under the continents in the sections on regional geology. They are, vol. **1**, Europe; vol. **2**, USSR; vol. **3**, Asia; vol. **4**, Africa; Vol. **5**, Latin America; vol. **6**, Oceania; vol. **7**, North America.

World stratigraphy and geologic systems

Coverage of the world and the whole geologic column is an ambitious task and is attempted in *Phanerozoic Geology of the World*, edited by M Moullade and A E M Nairn (Elsevier, 1978–), which provides a geographical treatment of each era. Envisaged in three sections, I, *The Palaeozoic*, II, *The Mesozoic*, and III, *The Cenozoic*, only two parts of *The Mesozoic*, A (1978) and B (1983), have been published to date.

The publications, books and journals, for each geologic system, are listed below.

Cenozoic

Ere Cenozoique: Tertiare et Quaternaire, by C Pomerol, from the series *Stratigraphie et Paleogéographie* (Doin, 1973), is the main reference work: this work is also available in translation as *The Cenozoic Era: Tertiary and Quaternary* (Ellis Horwood, 1982).

Relevant papers are published in the journals *Afzettingen Werkgroep voor Tertiare en Kwartaire Geologie* (1979–) and *East Asia Tertiary-Quaternary Newsletter* (1984–).

Quaternary

Books on the Quaternary include *Correlation of Quaternary Chronologies*, edited by W C Mahaney (Geo Books, 1984), *Quaternary Geology: a stratigraphic framework for multidisciplinary work*, by D Q Bowen (Pergamon, 1978) and *The Pleistocene: geology and life in the Quaternary Ice Age*, by T Nilsson (Reidel, 1983).

Proceedings of the meetings of the International Association for Quaternary Research are important publications (see also Chapter 17).

Current journals on the Quaternary are:

Boreas (1972–)
Bulletin Association Quebecoise pour l'Etude du Quaternaire (1975–)
CANQUA Newsletter (1986–)
Circular Quaternary Research Association (1879–)
Current Research in the Pleistocene (1984–)
Etude Quaternaire, Mémoire (1972–)
Folia Quaternaria (1960–)
Quartar (1938–)
Quartar Bibliothek (1952–)
Quartarpalaontologie (1975–)
Quaternary Australasia (1983–)
Quaternary Geology Bulletin, Geological Survey of Malaysia (1985–)
Quaternary Glaciations in the Northern Hemisphere (1974–)
Quaternary Newsletter (1970–)
Quaternary Research, New York (1970–)
Quaternary Research, Tokyo (1957)
Quaternary Science Reviews (1982–)
Quaternary of South America and Antarctic Peninsula (1983–)
Quaternary Studies in Poland (1979–)
Striae (1975–).

Tertiary

Publications on the Tertiary include *Earth's Pre-Pleistocene Glacial Record*, edited by M J Hambrey and W B Harland (Cambridge University Press, 1981) and *The Miocene Ocean*, edited by J P Kennett (*Geological Society of America, Memoir* **163**, 1985).

Current journals are *Tertiary Research* (1970–), which also contains references to recent publications, *Tertiary Research Special Paper* (1976–), and *Ter-qua Symposium Series* (1985).

Mesozoic

The main reference work published is *Ere Mesozoique*, by C Pomerol, from the series *Stratigraphie et Paleogéographie* (Doin, 1975).

Cretaceous, Jurassic, and Triassic

A recent publication for the first period is *Aspects of Mid-Cretaceous Regional Geology*, edited by R A Reyment and P Bengtson (Academic Press, 1981). The specialist journal is *Cretaceous Research* (1980–), which also contains *Bibliographia Cretacea*, compiled by J Schoebel (1980–), a list of current publications.

A long-established standard work for the Jurassic is *The Jurassic System*, by W H Arkell (Oliver and Boyd, 1956): more recent publications are *Jurassic Environments*, by A Hallam (Cambridge University Press, 1975) and the proceedings of the *International Symposium on Jurassic Stratigraphy, Erlangen, 1984* (3 vols. Geological Survey of Denmark, 1984).

New publications and review articles on the Triassic period appear in the journal *Albertiana* (1983–).

Palaeozoic

The major publication for this era is *The Lower Palaeozoic Rocks of the World*, edited by C H Holland (Wiley Interscience, 1971–1985). The four volumes in this series are 1, *Cambrian of the New World* (1971), 2, *Cambrian of the British Isles, Norden, Spitzbergen* (1974), 3, *Lower Palaeozoic of the Middle East, Eastern and Southern Africa, and Antarctica* (1981), and 4, *Lower Palaeozoic of the North-Western and West-Central Africa* (1985).

The journal *Biostratigraphie du Paleozoique* (1984–) is the main source of current information.

Carboniferous

An important source of information is the publications of the International Congress of Carboniferous Stratigraphy and Geology. A series, begun recently, is entitled *Carboniferous of the World*, edited by C Martinez Dias; Part I is on China, Korea, Japan and S E Asia (*IUGS Publication*, **16**, 1983).

The main bibliography, *Bibliography of Carboniferous Geology* parts 1–37, covered the literature from 1959 to 1979, and was compiled by the Geologisch Bureau voor het Mijngebied, Rijks Geologische Dienst, and the current journal is *Newsletter on*

Carboniferous Stratigraphy (1980–), which also contains lists of current publications.

Devonian

Recent publications are *The Devonian System*, by M R House *et al* (*Special Papers in Palaeontology*, **25**, 1979), *Aspects of a Stratigraphic System: the Devonian*, by D L Dineley (Wiley, 1984) and older publications include the *Proceedings of the International Symposium on the Devonian System*, edited by D H Oswald (2 vols, Alberta Society of Petroleum Geologists, 1967).

Ordovician

Recent works include *The Silurian-Ordovician Boundary*, edited by A Martinsson (E Schweizerbart'sche, 1977), *Aspects of the Ordovician System*, by D L Bruton (Universitetsforlag, Oslo, 1984) and *The Ordovician System*, by M G Bassett (University of Wales Press, 1976).

A number of important publications on the Ordovician are in the *Publication* series of the International Union of Geological Sciences, all with correlation charts and explanatory notes: **1**, *The Ordovician System in China*, by S F Sheng (1980); **2**, *The Ordovician System in the Near and Middle East*, by W T Dean (1980); **6**, *The Ordovician System in Australasia, New Zealand and Australia*, by B D Webby *et al* (1981); **8**, *The Ordovician System in Canada*, by C R Barnes *et al* (1981); **11**, *The Ordovician System in Southwestern Europe (France, Spain and Portugal)*, by W Hamman *et al* (1982); **12**, *The Ordovician System in the United States*, by R J Ross *et al* (1982); and also in the same series is **10**, *The Cambrian-Ordovician Boundary: sections, fossil distributions, and correlation*, edited by M G Bassett and W T Dean (1982).

Cambrian

Relevant publications are *The Cambrian System in Australia, Antarctica and New Zealand: correlation charts and explanatory notes*, edited by J H Shergold and A R Palmer (*IUGS Publication*, **19**, 1985), *The Cambrian System in the Near and Middle East*, by R Wolfart (*IUGS Publication*, **15**, 1983), and *The Precambrian — Cambrian Boundary*, edited by J W Cowie for the IGCP project no. 29 (*Precambrian Research*, **17**(2), special issue, 1982).

Precambrian

Recent publications on the Precambrian are *Geological Evolution of the Earth during the Precambrian*, by L J Salop, translated by V

P Grundina (Springer, 1983), *Precambrien Ere Paleozoique*, by C Pomerol, from the series *Stratigraphie et Paleogéographie* (Doin, 1977), *History of Concepts in Precambrian Geology*, edited by W O Kupsch and W A S Sarjeant (*Geological Association of Canada, Special Paper* **19**, 1979) and *Pre-Cambrian of the Southern Hemisphere*, edited by D R Hunter (*Developments in Precambrian Geology*, **2**, 1981).

On the evolution of early life forms *Chemical Evolution of the Early Precambrian*, edited by C Ponnamperuna (Academic Press, 1977), *Life in the Precambrian*, edited by M R Walter (*Precambrian Research*, **5**(2), 1977), *Micropalaeontology of the Precambrian and its importance for correlation*, edited by G Vidal (*Precambrian Research*, **15**(1), 1981) and *Earth's Earliest Biosphere: its origin and evolution*, edited by J W Schopf (Princeton University Press) are all relevant publications.

The key journal for this system is *Precambrian Research* (1974–).

Archaean

Informative publications are *Archaean Geochemistry*, edited by B F Windley and S M Naqvi (Elsevier, 1978) on the origin and evolution of the Archaean continental crust, and *The Archaean: the search for the beginning*, edited by G J H McCall (Dowden Hutchinson Ross, 1977).

Palaeogeography

Recent publications dealing with this subject include *Meere und Lander im Wechsel der Zeiten: die Palaogeographie als Grundlage fur die Biogeographie*, by E Thenius (Springer, 1977), *Vicariance Biogeography: a critique*, edited by G Nelson and D E Rosen (Columbia University Press, 1981), *Historical Biogeography, Plate Tectonics and the Changing Environment*, edited by J Gray and A J Boucot (Oregon State University Press, 1979), *Paleogeographic Provinces and Provinciality*, edited by C A Ross (*Society of Economic Palaeontologists and Mineralogists, Special Publication*, **21**, 1974) and a compilation of papers on *Paleobiogeography*, also edited by C A Ross, in *Benchmark Papers in Geology*, **31**, 1976.

The key journal for this discipline is *Palaeogeography, Palaeoclimatology, Palaeoecology* (1965–), which also lists current titles periodically.

Palaeogeographical regions are outlined in *Atlas of Palaeobiogeography*, edited by A Hallam (Elsevier, 1973), and mapped in *Phanerozoic Paleocontinental World Maps*, by A G Smith *et al* (Cambridge University Press, 1981), a revised and enlarged version of *Mesozoic and Cenozoic Palaeocontinental Maps*, by A Smith and G Briden (Cambridge University Press, 1977). Regions are outlined also in *Atlas of Continental Displacement 200 Million Years to the Present*, by H G Owen (Cambridge University Press, 1983).

Gondwana, Pangea, and Tethys

The published proceedings of the International Gondwana Symposium are key publications: the most recent are the third, *Gondwana Geology*, edited by K S W Campbell (Australia National University Press, 1975), the *Proceedings* of the fourth, in two volumes, edited by B Lasker and C S Raja Rao (Hindustan, 1979), and the fifth, *Gondwana Five*, edited by M M Crewswell and P Vella (Balkema, 1981). The results of a symposium entitled *Gondwana* are published in *Bulletin Société Géologique de France*, **23**(6), 1981.

The literature is covered by the *Bibliography on the Gondwana Geology, 1827–1955 (Gondwana Newsletter*, **3**, 61–114, 1978) and periodically updated in the same journal (*Gondwana Newsletter* 1969–).

The Position and Climatic Changes of Pangea and five Southeast Asian Plates during Permian and Triassic Times is a bibliography compiled by G O W Kremp (*Paleo Data Banks*, **7**, 1977).

Recent publications on Tethys are *Evolution of the Tethys*, edited by J Aubouin *et al* (*Tectonophysics*, **123**(1–4), 1986), *Paleooceanography of the Mesozoic Alpine Tethys*, by K J Hsu (*Geological Society of America, Special Paper* **170**, 1976), *The Tethys: her paleogeography and paleobiogeography from Paleozoic to Mesozoic*, edited by K Nakazawa and J M Dickens (Tokai University Press, 1985), *Tethys: the ancestral Mediterranean*, edited by P Sonnenfeld (Hutchinson Ross, Van Nostrand Reinhold, 1981) and *Neogene of the Mediterranean Tethys and Parathethys*, edited by F F Steininger *et al* (2 vols, Institute of Palaeontology, University of Vienna, 1985).

Regional geology

The exploitation of mineral and other natural resources, so important in the economic development of nations, led to the establishment of national and provincial geological surveys, and their official publications constitute the major geological literature of each area. Literature of regional geology is also in national and international journals, and in the publications of geological societies, national and provincial, and of special interest groups.

The publications of congresses, especially those of the International Geological Congress, held every four years, provide an important source of publications in regional geology, in the form of excursion/field trip guidebooks, abstracts, and proceedings. International organizations also publish on regional geology, for example the output from the International Union of Geological Sciences, already mentioned in this chapter, which are reviewed in the quarterly journal *Episodes*. Also there are publications on the International Geological Correlation Programme (IGCP) which, apart from regional geology, cover many aspects of stratigraphy, tectonics, etc.

The literature on regional geology is abstracted in the following publications: *Bibliography and Index of Geology*, section 13, Areal Geology; *Referativnyi Zhurnal, Geologiya*, section A, Regional'aya Geologiya; and *Zentralblatt fur Geologie und Palaontologie, Teil I*, section 3, Regionale Geologie.

WORLD-WIDE REGIONAL GEOLOGY

The scarcity of monographs attempting to cover regional geology worldwide is an indication of the difficulty of the task. *Beitrage zur Regionalen Geologie der Erde* (Gebruder Borntraeger, 1961–), and *Regional Geology Series* (Wiley, 1964–) are rare examples, and the publications in these series will be cited in their respective areas.

The *Encyclopedia of World Regional Geology*, edited by R W Fairbridge (Dowden Hutchinson Ross, 1975–) is in three parts, with articles on continents, regions, and countries written by specialists: they are Part 1 *Western Hemisphere (including Antarctica and Australia)*, Part 2 *Europe and Asia*, and Part 3 *Africa and the Middle East*.

Directories include *Directory of Geoscience Departments in Universities in Developing Countries*, 3rd edn, compiled by B K Tan and S C Kumar (AGID, 1983), and *Worldwide Directory of National Earth-Science Agencies and Related International Organizations*, compiled by W E Bergquist et al (*US Geological Survey, Circular 834*, 1981). The journal *Geotimes* annually publishes a revised list of geological societies in its August issue.

OCEANS, SEAS AND CONTINENTAL MARGINS

Books on oceans and continental margins include *The Geology of Continental Margins*, edited by C A Burk and C L Drake (Springer, 1974), *Géologie de Marge Continentale*, by G Boillot (Masson, 1979), and *Encyclopedia of Earth Sciences*, edited by R W Fairbridge (Dowden Hutchinson Ross, 1966) (See Appendix I, Chapter 4). Other publications include *Geological and Geophysical Investigations of Continental Margins*, by J Watkins et al (*American Association of Petroleum Geologists, Memoir* 29, 1979) *Problème Géomorphologique de Marge Continentale*, by J R Varney (*Annales de l'Institut Océanographique*, 56(supplement), 1980), *Géologie de Marges Continentales*, by R Blanchet and L Montadert (*Oceanologica Acta*, 4(supplement), 1981), and *Géologie des Océans*, by X le Pichon et al, in the same volume.

An important series of monographs is the *The Ocean Basins and Margins*, edited by A E Nairn and F G Stehli (Plenum, 1973–) with eight volumes published to date. The individual volumes are listed in the sections immediately following, of the various oceans and seas. Also important is the series *Géochimie Organique des Sédiments Marins Profonds: ORGON*, by the Comité de'Etudes Géochimiques Marins, and published by CNRS: the four published parts are **1**: *Mer de Norvège* (1977); **2**: *Atlantique* (1978); **3**: *Mauritanie, Sénégal, Iles du Cap-Vert* (1979); and **4**: *Golfe d'Aden, Mer d'Oman* (1981).

The work of the Deep Sea Drilling Project (DSDP) is described in *The Deep Sea Drilling Project: a decade of progress*, edited by J E Warme *et al* (*Society of Economic Paleontologists and Mineralogists, Special Publication* **32**, 1981). The project regularly publishes *Initial Report* (1969–), which contains substantial papers.

The current publication of the Joint Oceanographic Institutions for Deep Earth Sampling (JOIDES) is the *JOIDES Journal* (1975–). Other current journals include *Marine Geology* (1964–) and *Geo-Marine Letters* (1981–); relevant papers can also be found in a wide range of oceanographic journals.

Arctic Ocean

In addition to the literature found in the section on the Arctic, there is the publication *The Arctic Ocean*, (A E Nairn and F G Stehli, *The Ocean Basins and Margins*, vol. **5**, 1981) and *Marine Geology and Oceanography of the Arctic Seas*, edited by Y Herman (Springer, 1974).

Atlantic Ocean

The region is comprehensively covered by *The Geology of the Atlantic Ocean*, edited by K D Emery and E Uchupi (Springer, 1984). *Geology of the Middle Atlantic Islands*, by R Mitchell-Thome (*Beitrage zur Regionalen Geologie der Erde*, **12**, 1976) covers the Azores, Madeira, Canary and Cape Verde Islands, and there is a bibliography *Geologic Summary and Literature Guide to Mid-Atlantic Bight . . . (bibliography through to 1977)*, compiled by W J Rogers *et al* (*New York State Museum, Circular* **51**, 1982).

Publications on the important North Atlantic area include *The North Atlantic* (Nairn and Stehli, *The Ocean Basins and Margins*, vol. **2**, 1974), *North Atlantic Palaeooceanography*, edited by C P Summerhayes and N J Shackleton (*Geological Society of London, Special Publication* **21**, 1986), *Geologiya Atlanticheskogo Okeana*, by M V Klenova and V M Lavrov (Nauka, Moskow, 1975) and *Geology of the North Atlantic Borderlands*, edited by J M Kerr and A J Fergusson (*Canadian Society of Petroleum Geologists, Memoir* **70**, 1981).

For the various regions of the North Atlantic, there are the following publications *Geodynamics of Iceland and the North Atlantic Area*, edited by L Kristjansson (Reidel, 1974) which is the proceedings of a NATO Advanced Studies Institute and no. C 11 of that series; for the North-east Atlantic, publications include *Petroleum Geology of the Continental Shelf of North-west Europe*, the proceedings of a conference organized by the Institute of Petroleum, edited by L V Illing and G D Hobson (Heyden, 1981), *North-west European Shelf Seas: the sea bed and the sea in motion*, edited by F T Banner *et al* (2 vols, Elsevier, 1979–1980), and *Geology of the North-west European Continental Shelf, Vol.1 The West British Shelf*, by D Taylor and S M Mounteney (Graham and Trotman, 1975). For the North-west Atlantic, there are the publications *Geology of the Scotian Shelf*, by L H King and B MacLean (*Geological Survey of Canada, Paper* **74–31**, 1974), *Western North Atlantic Ocean*, by K O Emery and E Uchupi (*American Association of Petroleum Geologists, Memoir* **17**, 1972); in addition there are relevant publications in the series *Decade of North American Geology* described later in this chapter in the section on North America.

The geology of the South Atlantic area is covered in *The South Atlantic* (Nairn and Stehli, *The Ocean Basins and Margins*, vol. 1, 1973) and *South Atlantic Paleooceanography*, edited by K J Hsu and H J Weissert (Cambridge University Press, 1985). The *Geology of the South Atlantic Islands*, by R Mitchell-Thome (*Beitrage zur Regionalen Geologie der Erde*, 10, 1970) covers, amongst others, Ascension Island, St Helena, Tristan da Cunha and Falkland Islands. In addition, the Falkland Islands are covered in the publications of the British Antarctic Survey, *The Geology of the Falkland Islands*, by M E Greenway (*Scientific Report*, 76, 1972).

Baltic Sea

The geology of the Baltic Sea is covered by *Geologiya Baltiiskogo Morya*, edited by V Gudelis and E Emelyanov (Mokslas, 1976), and *The Quaternary of the Baltic*, edited by V Gudelis and L K Koniggson (University of Uppsala, Department of Quaternary Geology, 1979). The journal *Baltica* has been published since 1963.

Caribbean Sea

The geology is described in *The Gulf of Mexico and the Caribbean* (Nairn and Stehli, *The Ocean Basins and Margins*, vol. 3, 1975), and *Contributions to the Geological Oceanography of the Gulf of Mexico*, edited by V J Henry (Texas A & M University, 1972). Other relevant literature can also be found in the section on the Caribbean Region (see p.197–199)

Caspian Sea

The geology of the Caspian Sea is described in *Geologiya Kaspiiskogo Morva*, by A N Alikhamov (Elm, 1978).

Indian Ocean

Relevant publications include *The Indian Ocean* (Nairn and Stehli, *The Ocean Basins and Margins*, vol. 6, 1982), *Synthesis of Deep Sea Results in the Indian Ocean*, edited by C C Van der Borch (*Marine Geology*, 26(1–2), 1978), *Compte Rendu de l'Excursion de la Société Géologique de France dans l'Océan Indien, Juillet, 1978*, edited by G Leinhardt *et al* (*Bulletin Société Géologique de France*, 21(4): 415–511, 1979), *Indian Ocean Geology and Biostratigraphy*, edited by J R Heirtzler *et al* (American Geophysical Union, 1977), *The Evolution of the Indian Ocean since the Late Cretaceous*, by D McKenzie and J G Sclater (*Geophysical Journal*, 24, 437–528, 1971), and *Marine Geology and Oceanography of the Arabian Sea and Coastal Pakistan*, by B U Haq and J D Milliman (Van Nostrand Reinhold, 1985).

The islands of the Indian Ocean, Mauritius and Reunion, have their geology described in *The Geology and Mineral Resources of Mauritius*, by E S W Simpson (HMSO, 1951), *The Petrology of Mauritius*, by F Walker and L O Nicholson (HMSO, 1954), and *Etude Géologique de l'Ile de la Réunion*, by P Bussière (*Travaux, Bureau Géologique, Madagascar*, 84, 1958).

Mediterranean

The Mediterranean is covered in two volumes of Nairn and Stehli, *The Ocean Basins and Margins*: vol. 4A, *The Eastern Mediterranean* (1977), and 4B, *The Western Mediterranean* (1978); also there are *Geological Evolution of the Mediterranean Basin: R Selli Commemorative volume*, edited by D J Stanley and F

C Wezel (Springer, 1985), *The Mediterranean was a Desert*, by K J Hsu (Princeton University Press, 1984) and *Structural History of the Mediterranean Basins*, edited by B Biju-Duval and L Montadert (Technip, 1977).

For the various regions of the Mediterranean *Metallogeny and Plate Tectonics in the Northeastern Mediterranean*, edited by S Jankovic (Belgrade University, 1977), *Geology of the Southwestern Mediterranean Sea*, by D Neev *et al* (*Bulletin Israel Geological Survey*, **68**, 1976) and *La Mer Pelagiénne*, edited by P F Burollet, P Clairefond and W Winnock (*Géologie Mediterranéene*, **6**(1), 1979) cover most aspects.

Current important journals are *Géologie Mediterranéene* (1974–) and *Mediterranea* (1983–).

Literature is covered in the bibliography *Documentation of Earth-Scientific Literature for the Mediterranean Area*, compiled by H Closs, P F Burollet and H Glashoff (*Rapport et Procès-Verbaux des Réunions, Commission Internationale pour l'Exploration Scientifique de la Mer Mediterranée*, **22**(2b), 1974).

North Sea

Major publications are *The Nordic Seas*, edited by B C Hurdle (Springer, 1986), *The Geology of the North-west European Continental Shelf; vol. 2, the North Sea*, by R M Pegrum (Graham and Trotman, 1975), *Introduction to the Petroleum Geology of the North Sea*, edited by K W Glennie (Blackwell Scientific, 1984), *Quaternary Stratigraphy of the North Sea*, edited by I Aarseth and H P Sejrup (University of Bergen, 1984) and *Quaternary History of the North Sea*, edited by E Oele *et al* (Uppsala University, 1979).

Pacific Ocean

Books covering this ocean are *The Pacific Ocean* (Nairn and Stehli, *The Ocean Basins and Margins*, vol. **7A**, 1985 and **7B**, 1988. In addition there is a publication in Russian, *Geologiya Tikhookeanskogo Podvizhnogo Poyasa i Tikhogo Okeana* on the geology of the Pacific mobile belt and Pacific Ocean which is edited by L I Krasnyi and K M Khudolei (Nedra, Leningrad, 1978). Volume 1 of this work is on stratigraphy and palaeobiogeography, and volume 2 is on magmatism and tectonics.

Other major publications are *Tectonostratigraphic Terranes of the Circum-Pacific Region*, edited by D G Howell (Circum Pacific Council for Energy and Mineral Resources, 1985), *Accretion Tectonics in the Circum-Pacific Regions*, edited by M Hashimoto and S Uyeda (Reidel, 1983), *Circum-Pacific Energy and Mineral Resources*, by M T Halbouty *et al* (*American Association of Petroleum Geologists, Memoir* **25**, 1976), *Comparative Studies on the Geology of the Circum-Pacific Orogenic Belt in Japan and Chile*, edited by T Ishikawa and L Aguirre (Japan Society for the Promotion of Science, 1977), *Pacific Neogene Datum Planes: contribution to biostratigraphy and chronology*, edited by N Ikebe and R Tsuchi (University of Tokyo Press, 1984) and *Proceedings of the International Congress on Pacific Neogene Stratigraphy, 1st, Tokyo*, edited by T Saito and H Ujiie (Japan Society for the Promotion of Science, 1977).

On the different regions of the Pacific, there are *Geodynamics of the Eastern Pacific Region, Caribbean and Scotia Arcs*, edited by S J Ramon Cabre (*American Geophysical Union Geodynamics Series* **9**, 1983), *The Tectonic and Geologic Evolution of Southeast Asian Seas and Islands*, edited by D E Hayes (*American Geophysical Union, Geophysical Monographs* **27**, 1982), *Geology of the North Philippine Sea: geological results of the GDP Cruise of Japan*, edited by T Shiki (Tokai University Press, 1985), *Geodynamics in South-west Pacific*, proceedings of a symposia held in Noumea, 1976 (Technip, 1977), and other literature, indexed in

Bibliography of Geology and Geophysics of the South Pacific, compiled by L W Kroenke and E Bardsley (*United Nations Commission for the Coordination of Joint Prospecting for Mineral Resources in South Pacific Offshore Areas, Technical Bulletin* 1, 1975), of which there is a second edition, entitled *Bibliography . . . Southwestern Pacific*, edited by C Jouannic and R M Thompson (*Technical Bulletin*, 5, 1983).

The current journal is *Tikhookeanologiya Geologiya* (1982–) which is available also in English translation edition under the title *Geology of the Pacific Ocean*.

AFRICA

General works on the geology of Africa include *Géologie de Afrique*, by B Bessoles (*Mémoires BRGM*, 88 & 99, 1977–1980), *The Geochronology and Evolution of Africa*, by L Cahen *et al* (Oxford University Press, 1983), *Continental Terminal of Africa*, edited and compiled by C A Kogbe (*Department of Geology, Ahmadu Bello University, Occasional Publication* 7, 1978), the proceedings of the 2nd Working Conference of the International Geological Correlation Programme Project no.127, and *Evolution Géologique de l'Afrique*, edited by R Black (*Centre International pour la Formation et les Echanges Géologiques, Publications Occasionelles*, 4, 1985). Publications from conferences containing a wealth of papers include the proceedings from the Conference on African Geology, e.g. the Fifth, in Cairo, 1979 published in *Annals of the Geological Survey of Egypt*, 9 & 10, 1979–1980, and Colloque de Géologie Africaine, e.g. Tenth, in Montpellier, 1979, with papers published in *Revue de Géologie Dynamique et de Géographie Physique* 21(5), 1979.

The reference work for the stratigraphy of Africa is the *Lexique Stratigraphique International*, the parts for Africa being volume 4; Fasc. 12 is the part on *General Stratigraphy* and fasc. 13 on *Geochronology*. Other parts relating to countries in Africa are: fasc. 1a, Morocco (1956); 1b, Algeria (1956); 1c, Tunisia (1962); 2, Sahara, French & Portuguese West Africa (1956); 3, British West Africa, Sierra Leone, Nigeria (1956); 4a, Libya (1960); 4b, Egypt and Sudan (1966); 5a, British Somaliland (1956); 5b, French Somali (1956); 5c, Italian Somali (1956); 5d, Ethiopia (1956); 6, French Equatorial Africa, French Cameroons, Spanish Guinea and São Tomé (1956); 7a, Belgian Congo (1956); 7b, Angola (1956); 8a, Kenya (1963); 8c, Tanganyika 9a, Nyasaland (1956); 9b, Northern Rhodesia (1956); 9c, Southern Rhodesia (1956); 10a, Mozambique (1960); 10b, Union of South Africa, Bechuanaland, Swaziland and Basutoland (1960); 11, Madagascar (1956) and supplement (1960). No. 1 of the new series of the *Lexique* is *West Africa*, edited by J Fabre (Pergamon, 1983).

Works with a broad geographical scope include the recent publications, *The Archean of Equatorial Africa*, edited by L Cahen (Wiley, 1976), *Environmental History of East Africa: a study of the Quaternary*, by A C Hamilton (Academic Press, 1982), *Rift Valleys — Afro-Arabian*, edited by A M Quennell (Scientific and Academic Editions, 1982), *Geodynamic Evolution of the Afro-Arabian Rift System* (*Atti dei Covegni Lincei*, 47, 1980), *Geology of the Eastern Rift System of Africa*, by B H Baker (*Geological Society of America, Special Paper* 136, 1972), *Géologie de l'Afrique: le craton ouest africain*, by B Bessoles (*Mémoires BRGM*, 88, 1977), and *Geology and Mineral Resources of West Africa*, edited by J B Wright (Allen & Unwin, 1985) covering Eastern, Central and Western Africa.

Publications on Northern Africa include *The Precambrian in North Africa*, by H M E Schurmann (Brill, 1974), *Geology of the Northwest African Continental Margin*, edited by U von Rad (Springer, 1982), and *Geology and Oil Fields of Libya, Algeria and Tunisia*, being selected papers from *Bulletin American Association of Petroleum Geologists* 14 & 16 1976. On the Sahara and the Nile regions, there are the publications *The Sahara and the Nile: Quaternary environments and prehistoric occupation in Northern Africa*, edited by M A J Williams and H Faure (Balkema, 1980), and *Geological Evolution of the River Nile*, by R Said (Springer, 1981).

Mesozoic and Tertiary Geology of Southern Africa, by R V Dingle *et al* (Balkema, 1983), *Crustal Evolution of Southern Africa: 3.8 billion years of earth history*, by A J Tankard *et al* (Springer 1982), and *Southern Africa Prehistory and Paleoenvironments*, edited by R G Klein (Balkema, 1984) are recent publications on the geology of Southern Africa.

The literature of the geology of Africa is indexed annually in *Géologie Africaine* (1976–) and, in *Geology of Africa South of the Sahara* (1974–), abstracts of the literature received by the Geological Survey of South Africa are published.

Current journals on African geology are:
African Geology (1980)
African Geoscience Newsletter (1981–)
Annales Musée Royale de l'Afrique Centrale, série in 8vo. Sciences Géologiques (1961–)
Journal of African Earth Sciences (1983–)
Preliminary Report of African Studies, Association for African Studies, Nagoya University (1975–)
Rapport Annuel du Département de Géologie et de Minéralogie du Musée Royale de l'Afrique Centrale, Tervuren (1957–)
Recherches Géologiques en Afrique (1972–).

Algeria

Recent publications include *La Chaine Alpine d'Algérie Orientale et des Confins Algéro/Tunisiens*, by J M Villa (Thesis, Université de Paris VI, 1980) and *Introduction à la Géologie du Sahara Algérien et des Régions Voisines I: la couverture Phanerozoique*, by J Fabre (SNED, Algeria, 1976).

Useful also are the publications, although older, from the 19th International Congress of Geology, held in Algiers in 1952, which include Algeria in a series entitled *Monographies Régionales*.

The earlier literature is indexed in *Bibliographie de l'Algérie du Sud (Sahara) et des Régions Limitrophes*, by D Merabet (*Bulletin Service de la Carte Géologique de l'Algérie*, n.s. **37**, 1967). More recent publications are listed in *Publications du Service Géologique de l'Algérie, Nouvelle serie, Bulletin* (1953–).

Angola

Works on this country published since the first edition of this book are *Reconhecimento Cientifico de Angola: estudos de geología de palaeontología e de micrología*, by M Collignon *et al* (Academía des Ciencias, Lisboa, 1979), *Données Nouvelles sur la Stratigraphie, la Géochemie et la Géochronologie des Formations Precambriennes de la Partie Méridionale de Haut Plateau Angolais*, by P Bassot *et al* (*Bulletin BRGM*, sect.II,4, 1981) and *Etude Géologique et Cartographique des Formations Precambriennes du Haut Plateau Méridional d'Angola*, by M Pascal (Thesis, Université de Grenoble I, 1980).

The current serials published by the geological survey are *Boletim Servicos de Geologia e Minas, Angola* (1937–) and *Memoria Direccão Nacional de Geologia e Minas* (1971–).

The literature on the geology of Angola can be found in *Bibliographia Geologica de Angola*, by A T S Ferreira da Silva (*Memoria Servicos de Geologia e Minas*, **10**, 1971), and in an update for the years 1970–1972 published in the same series in 1973.

Benin (formerly Dahomey) and Togo

Etude Géologique et Structurale du Nord Ouest Dahomey, du Nord Togo et du Sud-Est de Haute-Volta, by P Affaton (*Travaux Laboratoire de Science de la Terre, St. Jerome*, série B, **10**, 1975) and *Prospection Minière Commune au Benin et au Togo, entre les Parallèlles 9 et 10 N*, by J L Lasserre and P Matheus (BRGM, 1980) are some of the more recent publications covering the two countries; older, but still relevant, publications are *Contribution à l'Etude du Bassin Sédimentaire Côtier du Dahomey et du Togo*, by M Slansky (*Memoir BRGM*, **11**, 1962), *Le Precambrien du Togo et du Nord-Ouest Dahomey*, by P Aicard (*Bulletin Direction Federal Mines et Géologie, Dakar*, **3**, 1957) and *Le Precambrien du Dahomey*, by R Pougnet (*Bulletin Direction Federal Mines et Géologie, Dakar*, **22**, 1957).

Botswana (formerly Bechuanaland)

The economic importance of the region is reflected in the literature, as in the *Mineral Resources Reports* series: *Resources inventory of Botswana; metallic minerals, mineral fuels and diamonds*, by J W Baldock (no **4**, 1977) and *Resource Inventory of Botswana: industrial rocks and minerals*, by N W D Massey (no **3**, 1973) are two such publications. Other literature relevant to the region includes *The Proceedings of a Seminar Pertaining to the Limpopo Mobile Belt*, by I F Ermanovics *et al* (*Geological Survey Department, Bulletin* **12**, 1977).

Serial publications of the Geological Survey are: *Bulletin, Geological Survey Department, Botswana* (1965–), *District Memoir, Geological Survey of Botswana* (1973–), *Mineral Resources Report* (1973–) and *Report of the Geological Survey Department, Botswana* (1953–).

To locate relevant publications there is the *Annotated Bibliography and Index of the Geology of Botswana, 1971–1976*, by G McEwen (Geological Survey of Botswana, 1983) whilst an earlier bibliography, of the same title, covering the years 1967–1970, by J D Bennett, was published in 1971 and the literature to 1966, compiled by C A Laughton, in 1967.

Burundi

In the absence of recent literature from Burundi, the reference work on the region is *Etude Géologique du SE du Burundi*, by A Waleffe (*Annales de Musée Royale Afrique Centrale*, **48**, 1965).

Cameroon

Geological publications for the country have an economic bias: *Plan de Recherches Minérales*, by B Bessoles (2 vols. République Federal Cameroun, 1968), *Index Mineraux et Ressources Minerales du Cameroun: synthèse simplifiée des connaissances sur la géologie du Cameroun*, by L Laplaine and B Bessoles (*Bulletin Direction Mines et Géologie Territoire de Cameroun*, **5**, 1969) and *Géologie et Géochimie de Sources Thermominérales du Cameroun*, by A Le Maréchal (*Travaux et Documents ORSTOM*, **59**, 1976). The older literature of Cameroon is covered by *Bibliographie Géologique du Cameroun*, compiled by A Denaeyer and F Blondel (*Bulletin Société Camerounaise*, **6**, 1944).

Serial publications of the Geological Survey are *Bulletin de la Direction des Mines et de la Géologie Territoire du Cameroun*, (1954–) and *Rapport Annuel du Service Géologique, Territoire du Cameroun* (1954–).

Central African Republic

Publications on the geology of the country are *Géologie et Resources Minérales de la République Centrafricaine. Etat des Connaissances à fin 1963*, by J L Mestraud and B Bessoles (*Memoires BRGM*, **60**, 1982) and *Les Gisement de Diament de la Republique Centrafricaine dans leur Contexte Géologique de l'Afrique Centrale; cas particulier de M'Zako*, by A Latou (Thesis, Université de Paris VI, 1982).

Chad

The geological publications include two field conference publications from the Petroleum Exploration Society of Libya: *Excursion to Chad* (1961) and *South-Central Libya and Northern Chad: a guidebook to the geology and prehistory*, edited by J J Williams (1966). In addition there are *Géochimie du Bassin du Lac Chad*, by J Y Gac (*Travaux et Documents ORSTOM*, **123**, 1980) and *Les Formations Sédimentaires Tertiares et Quaternaires de Cuvette Tchadienne et les sols qui en Derivent*, by J Pias (*Mémoires ORSTOM*, **43**, 1970).

Congo see Zaire

Congo (Brazzaville)

The geology is described in the explanatory text accompanying the geological map of the country, *Notice Explicative de la Carte Géologique de la République du Congo Brazzaville au 1:500,000*, by P Dadet (*Mémoires BRGM*, **70**, 1969).

Dahomey see Benin

Djibouti

Publications covering this country's geology are *Géologie de la Région Sud-Ouest du TFAI (Région Lac Abhe — Lac Asal)*, by J Demange and L Stieltjes (*Bulletin BRGM*, sect. IV,**2**, 1975), *Afar between Continental and Ocean Rifting; Proceedings, International Symposium Afar Region and Related Rift Problems, 1974*, edited by A Pilger and A Rosler (E. Schweizerbart'sche, 1975–1976), *L'Afar*, a special issue in *Revue de Géographie Physique et Géologie Dynamique* **15**(4), 1973 and *Geology of Central and Southern Afar (Ethiopia and Djibouti Republic)*, edited by J Varet (2 vols, CNRS, 1978).

Egypt

Recent, and older, works by R Said include *Review of Egyptian Geology*, by Said and E M El-Shazly (*Egyptian Reviews of Science*, **1**, 1957), *The Geology of Egypt* (Elsevier, 1962), *Studies on Some Mineral Deposits of Egypt* (Geological Survey of Egypt, 1970), *The Basement Complex of Egypt* (*Annals Geological Survey of Egypt*, **2**, 1972), *Mineral Deposits of Egypt* (Geological Survey of Egypt, 1975) and *The Geological Evolution of the River Nile* (Springer, 1981).

Current journals are *Annals of the Geological Survey of Egypt* (1971–), *Egyptian Journal of Geology* (1957–) and *Papers, Geological Survey of Egypt* (1955–).

There is no shortage of bibliographies, and full coverage is given by the following: *The Geology of Egypt: an annotated bibliography*, by F El-Baz (Brill, 1984), *Bibliography of Scientific and Technical Literature relating to Egypt up to the end of 1939*, by E H Koldani (Survey and Mines Department, Egypt, 1941), *Bibliography of Geology and Related Sciences concerning Egypt for the period 1960–1973*, compiled by R Said *et al* (Geological Survey of Egypt, 1976), *Geologic Literature on Egypt, 1933–1978*, compiled by C R Glenn and J M Denman (*US Geological Survey, Open-File Report*, **80–930**, 1980) and *Bibliography of Geology and Related Sciences concerning the Western Desert, Egypt (1732–1984)*, compiled by A A B Salman (Egyptian Geological Consulting Office, 1984).

Equatorial Guinea

Geological information on the country can be obtained from *Plan Mineral: programme quinquennal de prospection des minerais de la République Révolutionnaire de Guinée* (BRGM, 1980) and from the older bibliography *Geologia y Geografia Fisica de la Guinea Continental Española*, by J D de Lizaur y Roldan (Direccion General de Marruencoe y Colonias, Madrid, 1945).

Ethiopia

Books covering this country include *The Geology of Ethiopia*, by P A Mohr (University College, Addis Ababa, 1962) reprinted with a new foreword in 1970, *A Field Guide to Ethiopian Minerals, Rocks and Fossils*, by B Morton (Addis Ababa University Press, 1978), and *Mineral Occurrences of Ethiopia*, by D A Jelenc (Ministry of Mines, Addis Ababa, 1966). The current journal is *Bulletin, Geological Survey of Ethiopia* (1975–). In addition, for literature on Afar, see under the section on Djibouti.

Gabon

The geology of Gabon is described in the following works: in the explanatory notes to *Carte Géologique de la République Gabonaise au 1:1,000,000 (Mémoires BRGM, 72*, 1970) and *Eléments de Géologie Gabonais* (Service de la Recherche Appliquée aux Enseignements, 1979).

Gambia

The *Report on a Rapid Geological Survey of Gambia*, by W G Cooper (*Bulletin of the Geological Survey of the Gold Coast, 3*, 1925) is the standard work on this country, in the absence of anything more recent.

Ghana

Relevant publications include *Symposium on the Occasion of the Fiftieth Anniversary of the Ghana Geological Survey, 1970 (Ghana Geological Survey Bulletin, 38*, 1970) and *Contributions to the Geology of Ghana, vol 1*, edited by G O Kesse and E Jones (*Report of the Geological Survey of Ghana, 75–8*, 1977).

The serial publications of the Geological Survey of Ghana are the *Bulletin* (1925–), and *Report* (1913–).

Guinea

Géologie de la Guinée Française, by R Furon (*Publications du Bureau d'Etudes Géologiques et Minières Coloniales, 19*, 1943) is the only published work available for this area.

Guinea Bissau

The geology of this country is described in *Geologia da Guine Portuguesa*, by J E Texeira (*Curso de Geologia do Ultramar 1* : 55–103, 1968) published by the Junta de Investigacoes do Ultramar, Lisbon.

Ivory Coast

The most recent publications include *Géologie et Minéralisations du Sud-Ouest de la Côte d'Ivoire*, by A Papon *et al* (*Mémoires BRGM, 80*, 1973) and *Esquisse Structurale de la Côte d'Ivoire*, by B Tagini (Société d'Etat pour le Developpement Minier de la Côte d'Ivoire (SODEMI), Abidjan, 1971).

Bulletin de la Direction de Mines et de Géologie de Côte d'Ivoire (1961–) and *Rapport Annuel de la Direction des Mines et de la Géologie de Côte d'Ivoire* (1966–) are publications currently published by the national geological survey. The geological literature is extensively covered in *Bibliographie de la Géologie et de la Recherche Minière en Côte d'Ivoire (1855–1983)*, compiled by H Madon (2 vols, SODEMI, Abidjan, 1985) and in *Bibliographie de la Côte d'Ivoire, vol 3: Sciences Physique et de la Terre*, by G Janvier (3 vols, Université d'Abidjan, 1972–1975).

Kenya

Sources of information for this country include *The Geology and Mineral Resources of Kenya*, by W Pulfrey, 2nd edition revised by J Walsh (*Bulletin Geological Survey of Kenya*, **9**, 1969), *Geology of the Olduvai Gorge*, by R L Hay (University of California Press, 1976) and *Minerals of Kenya*, by C G B DuBois, revised by J Walsh (*Bulletin Geological Survey of Kenya*, **11**, 1970).

The current publications of the Geological Survey of Kenya are the *Bulletin* (1954–), *Memoirs* (1953–) and *Report* (1949–).

The literature is listed in *Bibliography of the Geology of Kenya, 1859–1968*, compiled by N P Dosaj and J Walsh (*Bulletin Geological Survey of Kenya*, **10**, 1970) and in *A Bibliography on Kenya: Geosciences*, by I O Nyambok (Scandinavian Institute of African Sciences, 1977).

Lesotho see under South Africa

Liberia

Publications covering this country of Africa include *Stratigraphy and Structure of Basins on the Coast of Liberia*, by R W White (*Liberia Geological Survey, Special Paper* **3**, 1972) *Geologic Reconnaissance in Western Liberia*, by R W White and G W Leo (*Liberian Geological Survey, Special Paper* **1**, 1969) and *Geophysical Surveys of Liberia and Tectonic and Geologic Interpretations*, by J C Behrendt and C S Wotorson (*US Geological Survey, Professional Paper*, **810**, 1974).

Current journals publishing papers on Liberian geology are *Bulletin, Geological, Mining and Metallurgical Society of Liberia* (1966–), *Bulletin, Liberian Geological Survey* (1967–), *Memorandum Report, Liberian Geological Survey* (1970–) and *Special Paper, Liberian Geological Survey* (1969–).

Libya

The major work on the country is *The Geology of Libya*, edited by M J Salem and M T Busrewil, which is the publication of the proceedings of the 2nd Symposium on the Geology of Libya, held in Tripoli, 1978 (3 vols, Academic Press, 1981). Of interest also is *Geology and Mineral Resources of Libya*, edited by G H Goudarzi (*US Geological Survey, Professional Paper* **660**, 1970), whilst the stratigraphy of Libya is described in *Stratigraphic Lexicon of Libya*, by S Banerjee (*Department Geology and Mining Bulletin*, **13**, 1980) and *Stratigraphic Nomenclature of the Northwestern Offshore of Libya*, by O S Hammuda *et al* (Earth Sciences Society of Libya, 1985). The work *Bibliography of the Geology of Libya*, compiled by J Azzouz Ettalhi and D Krokovics (*Industrial Research Centre, Geological and Mining Department, Bulletin* **11**, 1978) is a documentation of the country's geological literature.

Madagascar

Publications include *Géologie de Madagascar: les terrains sédimentaires*, by H Besaire (*Annales Géologiques de Madagascar*, **35**, 1972).

Current publications of the geological survey are *Annales Géologiques de Madagascar* (1931–) and *Documentation du Bureau Géologique, Service Géologique de Madagascar* (1956–).

The literature is covered in *Bibliographie Géologique de Madagascar (1940–1973)*, compiled by H Besaire (*Documentation Bureau de Géologie, Republic of Malagasy*, **189**, 1974).

Malawi (formerly Nyasaland)

The Geology and Mineral Resources of Malawi, by G S Carter and J D Bennett (rev. edn, *Bulletin Geological Survey of Malawi*, **6**, 1973) is the main publication describing the geology of the country.

Current publications of the Geological Survey Department of Malawi are its *Bulletin* (1965–), *Memoirs* (1958–) and *Report* (1923–).

Mali

Geological details of the country can be found in *Plan Mineral de la République de Mali*, edited by J P Bassot (Direction National Géologie et Mines République Mali, 1978).

Mauritania

Publications include a special issue on Mauritania in *Bulletin Société Géologique de France* 7 series, vol **11**, 1969 and others are *Plan Minéral de la République Islamique de Mauritanie (Rapport BRGM*, 1975), *Les Mauritanides, Afrique Occidentale. Essai de Synthèse*, by J Sougy and J P Lecorche (*Geological Survey of Canada Paper*, **78–13**, 1977), *Le Precambrien Supérieur et le Paléozoique de l'Afar de Mauritanie*, by R Trompette (*Travaux Laboratoire Science de la Terre, St. Jerome*, **B 7**, 1973), *Etude Géologique de la Chaine des Mauritanide*, by J C Chiron (*Mémoires BRGM*, **84**, 1974) and *Les Grandes Unités des Mauritanides, aux confins du Sénégal et de la Mauritanie: l'évolution structurale de la chaine du Precambrien Superieur au Devonien*, by A Le Page (2 vols, Thesis, Université de Marseilles, 1983).

Morocco

Publications include *Eléments de Géologie Marocaine*, by A Michard (*Notes et Mémoires Service Géologique de Maroc*, **252**, 1976), *Histoire Géologique du Maroc*, by E K Saadi (Mohamed V Culturelle et Universitaire, Fez, 1983), *Géologie de Gîtes Minéraux Marocains*, by J Acard *et al* (*Notes et Mémoires Service Géologique de Maroc*, **87**, 1952) which is updated in *Notes et Mémoires*, **276**, 1980, and a special issue entitled *Géologie Marocaine* in *Sciences Géologiques Bulletin*, **38**(2), 1985.

Current journal publications are *Mines et Géologie, Bulletin Trimestriel de la Direction des Mines et de Géologie* (1958–) and *Notes et Memoires du Service Géologique de Maroc* (1927–).

Literature is indexed in *Bibliographie Analytique de Sciences de la Terre, Maroc et Régions Limitrophes (depuis le recherches géologiques à 1964)*, compiled by P Morin in 2 volumes (*Notes et Memoires, Service Géologique de Maroc*, **182**, 1965). Updates of this bibliography can be found in later issues of the same series: *1965–1969* in no.**212**, 1970; *1970–1976* in no.**270**, 1979.

Mozambique

The publications for the geology of this country are not recent. They are *Noticia Explicativa do Esboco Geologico de Mocambique 1:2,000,000*, by A J de Freitas

(*Boletim, Servicos de Industria, Minas e Geologia*, **23**, 1957), and *A Geologia e o Desenvolvimento Economico e Social de Mocambique*, by F Goncalves and J Caseiro (Junta de Investigacoes do Ultramar, Lisboa, 1959).

Boletim, Servicos de Industria, Minas e Geologia Mocambique (1937–) is currently published by the state survey.

Namibia

Major publications include *Intracontinental Fold Belts: case studies in the Variscan belt of Europe and the Damara Belt in Namibia*, edited by H Martin and F W Eder (Springer, 1983) and *The Precambrian of South West Africa and Namaqualand*, by H Martin (University of Capetown, Precambrian Research Unit, 1965).

The geological survey publishes *Communications* (1985–) and *Memoirs, Geological Survey, South West Africa* (1934–).

The literature is well covered in *A Bibliography of Geological and Allied Subjects, South West Africa*, by H Martin (*Bulletin, Precambrian Research Unit, University of Capetown*, **1**, 1965), *Bibliography of South West Africa: geography and related fields*, compiled by R F Logan (South West Africa Scientific Society, 1969) in the series *Scientific Research in South West Africa, 8th series*, and in *Bibliography of South-West African-Namibian Earth Science*, compiled by P J Hugo *et al* (Geological Survey of South West Africa, 1983).

Niger

Publications on the geology of Niger are *Essai de Description des Formations Géologiques de la République du Niger*, by J Greigert and R Pougnet (*Mémoires BRGM*, **48**, 1967), *Géologie et Paléontologie du Gisement de Gaddufaouna (Aptian du Niger)*, by P Taquet (CNRS, 1976) in the series *Cahiers de Paléontologie*.

Nigeria

Publications on Nigeria include *Geology of Nigeria*, edited by C A Kogbe (Elizabethan Publishing, 1976), *Nigeria: its petroleum, resources and potential*, edited by A Whiteman (2 vols, Graham and Trotman, 1982) and *The Geological Sciences in Service of Nigeria*, by P G Cooray (University of Ife Press, 1973).

Journals currently published on Nigerian geology are *Bulletin, Department of Geology, Ahmadu Bello University* (1978–), *Occasional Paper, Department of Geology, Ahmadu Bello University* (1972–), *Occasional Paper, Geological Survey of Nigeria* (1921–) and *Report, Geological Survey of Nigeria* (1930–).

The earlier geological literature was indexed in *Nigeria*, compiled by N W Posnett *et al* (3 vols, *Overseas Development Administration, Land Resources Bibliography*, **2**, 1971).

Portuguese Guinea see Guinea Bissau

Rhodesia see Zimbabwe

Rwanda

Minéralisation et Inventaire des Minéraux du Rwanda, by B Baudin (*Etudes Rwandaise*, **1**, 1979) provides an outline of the geology of the country.

The state survey publishes the journal *Bulletin du Service Géologique, Ruhengeri* (1964–).

Senegal

Les Grandes Unités des Mauritanides, aux confins du Sénégal et de la Mauritanie:

l'évolution structurale de la chaine du Precambrien Superieur au Devonien, by A Le Page (2 vols, Thesis, Université de Marseilles, 1983).

Current journals are *Annuaire de l'Association Sénégalaise pour l'Etude du Quaternaire de l'Ouest Africain* (1967–) and *Bulletin de Liaison, Association Sénégalaise pour l'Etude du Quaternaire de l'Ouest Africain* (1966–).

Sierra Leone

Recent publications are *The Geology and Minerals of Sierra Leone*, by S W Morel (University of Sierra Leone, 1976) and *The Geology and Mineral Resources of Northern Sierra Leone*, by A MacFarlane *et al* (*Institute of Geological Sciences, Overseas Memoir*, **7**, 1981).

The Geological Survey of Sierra Leone currently publishes the *Bulletin* (1958–) and *Report* (1918–).

Somalia

Geology of Somalia, by A S Osman *et al* (Ministry of Mineral and Water Resources, 1976) provides a brief description, but still a major reference is *The Geology and Palaeontology of British Somaliland*, edited by W A MacFadyen (2 vols, Government Press, 1933–1935).

The Geological Survey of the Somali Republic publishes the *Report* (1956–).

South Africa

Major publications on South Africa are *The Geology of the Republic of South Africa*, by D R Van Eeden (*Geological Society of South Africa Special Publications*, **18**, 1972), *An Introduction to the Historical Geology of South Africa*, by J F Truswell (Purnell, 1970) and *The Geological Evolution of South Africa*, by J F Truswell (Purnell, 1977).

Current journals on South Africa are:

Annale van die Universiteit van Stellenbosch, serie Al: Geologie (1975)
Annals of the Geological Survey of South Africa (1962–)
Bulletin, Chamber of Mines, Precambrian Research Unit, University of Capetown (1965–)
Bulletin, Geological Survey of South Africa (1934–)
Geobulletin, Geological Society of South Africa (1963–)
Handbook, Geological Survey of South Africa (1959–)
Memoirs, Geological Survey of South Africa (1905–)
Report, Chamber of Mines, Precambrian Research Unit, University of Capetown (1963–)
Special Publications, Geological Society of South Africa (1970–)
Transactions and Proceedings Geological Society of South Africa (1934–).

For coverage of the literature, apart from the *Bibliography and Subject Index of South African Geology* (1957–), there is the *Bibliography of South African Geology, 1936–1956: Subject Index*, by S H Haughton (1973) and *Author Index*, by P T Wilson (1972), all published by the Survey.

South West Africa see Namibia

Spanish Guinea see Equatorial Guinea

Spanish Sahara see Western Sahara

Sudan

Publications of interest include *The Geology of the Sudan Republic*, by A J Whiteman (Clarendon Press, 1971) and *Outline of the Geology and Mineral*

Deposits of the Democratic Republic of the Sudan and Adjacent Areas, by J Vail (*Overseas Geology and Mineral Resources*, **49**, 1978). More recent publications of interest are *Late Proterozoic Stratigraphy of North-East Africa and Arabia*, by F A Ibrahim (Faculty of Earth Sciences, King Abdul Aziz University, Jeddah, 1982) and *Géologie de l'Afrique: la chaine panafrique "zone mobile d'Afrique Central et zone mobile soudanaise"*, by P Bessoles and R Trompette (*Mémoires BRGM*, **92**, 1980).

Bulletin, Geological and Mineral Resources Department, Sudan (1911–) is the journal of the state survey.

Swaziland

The Geology of Swaziland, by D R Hunter (Geological Survey and Department of Mines, 1961), an older publication, has never been updated.

Bulletin of the Geological Survey and Mines Department, Swaziland (1961–) and *Report of the Geological Survey and Mines Department, Swaziland* (1942–) are two journals publishing information on the geology of this country.

Tanzania (formerly Tanganyika)

The comprehensive publication is *The Summary of the Geology of Tanzania*, by A M Quennel *et al* (*Memoir Tanganyika Geological Survey*, 1956–1970). Part I covers introduction and stratigraphy, part II, a geological map, part IV, economic geology, and part V, the structure and geotectonics of the Precambrian.

The Geological Survey of Tanzania currently publishes the *Bulletin* (1927–).

Literature is indexed in *A Bibliography of the Mineral Resources of Tanzania*, compiled by O Nilsen (Scandinavian Institute of African Studies, Uppsala, 1980) and there is the earlier *Bibliography of the Geology and Mineral Resources of Tanzania to December 1967* (Bureau of Resources Assessment and Land Use Planning, Dar-es-Salaam, 1969).

Togo see under Benin

Tunisia

Publications on Tunisia are *Geology of Tunisia and Adjacent Parts of Algeria and Libya*, by W F Bishop (*Bulletin American Association of Petroleum Geologists*, **59**, 413–450, 1975) and the earlier but useful *Guidebook to the Geology and History of Tunisia*, edited by L Martin (9th Annual Field Conference of the Petroleum Exploration Society of Libya, Guidebook, 1967).

Notes du Service Géologique Tunisie (1951–) is the current journal publication of the state survey.

Bibliographie Analytique des Sciences de la Terre, Tunisie et Régions Limitrophes, by P Morin (*Notes du Service Géologique Tunisie*, **33**, 1973) covers the geological literature of the country.

Uganda

Publications on Uganda are *Mineral Resources of Uganda*, by J W Barnes (*Bulletin Geological Survey of Uganda*, **4**, 1961) and *The Physiographic Development of Uganda*, by A M J Swardt and A F Treadwell (*Overseas Geology and Mineral Resources*, **10**, 1969).

The Geological Survey and Mines Department of Uganda publishes the *Bulletin* (1933–), *Memoirs* (1925–), and *Report* (1959–).

Upper Volta

Publications on Upper Volta are *Le Potentiel Minier de la République de Haut Volta*, by G Hottin *et al* (Direction de Mines et de Géologie, République de Haute-

Volta, 1975) and *La Géologie et les Resources Minières de la Haute Volta Méridionale*, by Y Sagatsky (*Bulletin, Direction des Mines et de Géologie*, **13**, 1954).

Western Sahara

The geology is described in *Lithostratigraphy of the Northern Spanish Sahara*, by L K Ratschiller (*Memorie del Museo Tridentino di Scienze Naturali*, **18**(1), 1970).

Zaire

The geology of Zaire is covered by the book *Etude Géologique du Haut-Zaire; genèse et évolution d'un segment lithosphere Archéen*, by J Lavreau (Musée Royale de l'Afrique Centrale, 1982) and the earlier *Géologie du Congo Belge*, by L Cahen (H Vaillant-Carmanne, 1954).

The journal currently published is *Annales Faculté de Sciences Université National du Zaire, Géologie et Géographie* (1975–).

Zambia (formerly Northern Rhodesia)

An Outline of the Geology of Western Zambia, by N J Money (*Record, Zambia Geological Survey* **12**, 103–123, 1972) and *The Geology of the North-eastern Region of Zambia*, by H Schandelmeier et al (University of Zambia, 1976) are the most recent literature on Zambia, in addition to the special issue in *Géologie en Mijnbouw*, **51**(3), 1972.

Earlier publications include *A Summary of the Provisional Geological Features of Northern Rhodesia*, by T D Guernsey and J A Bancroft (*Colonial Geology and Mineral Resources*, **1**(2), 1950) and *The Geology and Mineral Resources of Northern Rhodesia*, by W H Reeve (*Bulletin of the Geological Survey of Northern Rhodesia*, **3**, 1963).

The Geological Survey Department of Zambia publishes the *Economic Report* (1963–), *Records* (1956–) and *Report* (1963–).

The *Annotated Bibliography and Index of the Geology of Zambia* (Geological Survey of Zambia, 1965–1977) covers the geological literature for the period 1931–1973.

Zimbabwe (formerly Rhodesia)

Recent publications are *An Outline of the Geology of Rhodesia*, by J C Stagman et al (*Bulletin Geological Survey of Rhodesia*, **80**, 1978) and *The Palaeontology of Rhodesia* (*Bulletin Geological Survey of Rhodesia*, **70**, 1973).

The main publications of the Zimbabwe Geological Survey are *Annals* (1975–), *Bulletin* (1920–) and *Short Report* (1919–). Others include *Special Publications, Geological Society of Zimbabwe* (1984–).

Geological literature is indexed in *Rhodesian Geology: a bibliography and brief index to 1968*, compiled by C C Smith and H E van der Heyde (*Occasional Paper, National Museum of Rhodesia*, **4** (31B), 1971).

THE AMERICAS

LATIN AMERICA, CENTRAL AMERICA

Latin America is described in the *Lexique Stratigraphique International*, volume **5**: fasc, 2a, Central America (1960); 2b, Antilles (except Cuba) (1966); 2c, Cuba (1959); 3a, Venezuela (1956) of which there was a second edition in 1978; 4a, Colombia I, Precambrian-Tertiary (1968); 4b, Colombia II, Tertiary to Quaternary (1974); 5a, Ecuador (1956), with a second edition in 1977; 5b, Peru (1956); 7, Chile

(1957); 9a, Uruguay (1958); 9b, Paraguay (1958); 9c, Falkland Islands (1964); 10b, Guiana (1962), are the fascicules published to date.

The major work for Central America is *Die Geologie der Mittelamerikas*, by R Weyl (*Beitrage zur Regionalen Geologie der Erde*, **1**, 1961) and, by the same author, *Geology of Central America*, (*Beitrage zur Regionalen Geologie der Erde*, **15**, 1980). Other important publications are the *Stratigraphic Atlas of North and Central America*, edited by T D Cook and A W Bally (Princeton University Press, 1975), *Estructura Geologica*, *Historica Tectonica y Morfologica de America Central*, by G Dengo (2nd rev. edn, Central Regional de Ayuda Technica, 1973), and *Metallogenesis in Latin America*, edited by J L L Moreno *et al* (*IUGS Publication*, **5**, 1981).

Papers from conferences are a source of information on the region; titled *Geowissenschaftliches Latinamerika-Kolloquium*, the 5th was published in *Munster Forschung zur Geologie und Palaontologie*, 44–45, 1978, the 6th (1980) in Heft 51 of the same journal, the 7th in *Zentralblatt fur Geologie und Palaontologie, Teil I* Parts 3/4, 1981, and the 8th in Parts 3/4 1983 of the above journal; there are also publications from the Congreso Latinamericano de Geologia.

Reference works include *Directory of Geological Services of Latin America and the Caribbean, 1981*, edited by V R Ricaldi (Ministry of Energy and Mines of Venezuela, 1982), *Sources of Geological Information for the Caribbean and Latin America*, by the Geological Information Group of the Geological Society of London, 1982, and *Bibliografia Geologica y Paleontologia de Centralamerica y el Caraibe*, by L D Gomez (Museo Nacional de Costa Rica, 1975).

Mexico

The third edition of *Geologia de Mexico*, by E Lopez Ramos (2 vols, D F Enero, 1983) covers the country comprehensively; other publications are *Geology of Northwestern Mexico and Southern Arizona*, edited by L Ortlieb and J Roldan (Instituto de Geologia, UNAM, 1981) and *Libro-guia de la Excursion Mexico-Oaxaca*, edited by L R Segura and R Rodriguez-Torres (Sociedad Geologia Mexicana, 1970).

Journals currently published on Mexican geology are:
Anales del Instituto Geologico de Mexico (1917–)
Boletin de la Asociacion Mexicana de Geologs Petroleros (1959–)
Boletin, Instituto de Geologia de la Universidad Nacional Autonoma de Mexico (1895–)
Boletin de la Sociedad Geologica Mexicana (1905–)
Revista, Instituto de Geologia Universidad Nacional Autonoma de Mexico (1977–).

Belize (formerly British Honduras)

Publications describing the geology of Belize are *Geology of Southern British Honduras with notes on adjacent areas*, by C G Dixon (Government of Belize, 1955), *Geology of Northern British Honduras*, by G Flores (*Bulletin American Association of Petroleum Geologists*, **2**, 1952) and the more recent work *The Geology of the Maya Mountains, Belize*, by J H Bateson and I H S Hall (*Institute of Geological Sciences, Overseas Memoir*, **3**, 1977).

Costa Rica

Magmatism and Crustal Evolution in Costa Rica, by H Pichler and R Weyl (*Geologische Rundschau*, **64**(2), 1975) and *Bibliografia de la Geologia de Costa Rica*, compiled by G Dengo (*Publicación Costa Rica Universidad, Serie Ciencias Naturales*, **3**, 1962) cover the geology of this area.

El Salvador

The geology is described in the older work, *Reconnaissance Geology and Vertebrate Palaeontology of El Salvador, Central America*, by R A Stirton and W K Gealey (*Bulletin Geological Society of America*, **60**, 1949) and the literature of the country is indexed in *A Bibliography of the Geology Relating to El Salvador, Central America, 1576–1973*, compiled by A P Humphreys (privately published, 1973). The state survey currently publishes *Anales del Servicio Geologico Nacional de El Salvador* (1955–).

Guatemala

The book for the geology of the country is *Guatemala, where plates collide: a reconnaissance guide to Guatemalan geology*, by F Nagle *et al* (Miami Geological Society, 1977) and the literature is listed in the bibliography *Bibliografia Geologica de Guatemala, America Central* (Edicion, 1966).

Honduras

The following works, *Mesozoic Stratigraphy of Honduras*, by R A Mills *et al* (*Bulletin American Association of Petroleum Geologists*, **51**(9), 1967) and *Volcanic History of Honduras*, by H Williams and A R MacBirney (*University of California Publications in Geology*, **85**, 1969) summarize the geology of Honduras.

Nicaragua

Volcanic History of Nicaragua, A R MacBirney and H Williams (*University of California Publications in Geology*, **55**, 1965) is the only work which gives a geological description for Nicaragua.

Panama (including Canal Zone)

Publications for the region are *Geology and Paleontology of the Canal Zone and adjoining parts of Panama*, by W P Woodring (*US Geological Survey, Professional Papers*, **306 D-E**, 1970–1973) and *A Geological Reconnaissance of Panama*, by R A Terry (*Occasional Paper California Academy of Sciences*, **23**, 1956).

CARIBBEAN REGION

Publications on the Caribbean Region include *Geologia Structurale de la Region des Caraïbes (Mexique, Amerique Central, Antilles, Cordillère Caraïbe)*, by J Butterlin (Masson, 1977), *Geology, Geophysics and Resources of the Caribbean*, edited by J D Weaver (Intergovernmental Oceanographic Commission, 1977), *Outline for a Geophysical and Geological Study of the Caribbean Sea* (Woods Hole Oceanographic Institute, 1973), *Contributions to the Geology and Paleobiology of the Caribbean and Adjacent Seas*, edited by P Jung (*Verhandlungen Naturforschende Gesellschaft, Basel*, **83**(1), 1974), *Geodynamique de Caraibes*, by A Mascle (Technip, 1985) and *West Indies Island Arcs*, edited by P H Mattson (Dowden Hutchinson Ross, 1977). In addition, reference can be made to the transactions of the Caribbean Geological Conference.

Antilles

The major work on the area is *Geologie de Antillen*, by R Weyl (*Beitrage zur Regionalen Geologie der Erde*, **4**, 1966); also useful is the publication *Paleogeography and Geological History of the Greater Antilles*, by K M Khudoley and A A Meyerhoff (*Geological Society of America, Memoir* **129**, 1971).

Bahamas

Relevant papers can be found in *Proceedings of the 1st Symposium on the Geology of the Bahamas, 1982* (College Centre of the Finger Lakes Bahamian Field Center, San Salvador, 1983).

Cuba

Comprehensive texts on the country have not been published since *Geologia de Cuba*, by G Furrazola-Bermudez *et al* (2 vols, Consejo Nacional de Universidad, Havana, 1964).

Current journals are:
Actas Instituto Geologia Academia Ciencias (1972–)
Ciencias de la Tierra y del Espacio, Academia de Ciencias de Cuba (1977–)
Jornada Cientifica de Instituto de Geologia y Paleontologia, La Habana (1982–)
Publicaciones Especial, Departamento Cientifico de Geologia, La Habana (1964–)
Publicaciones Especial, Instituto de Geologia y Paleontologia (1974–)
Publicaciones Especial, Ministerio de Mineria, La Habana (1971–).

Dominican Republic

Papers published on the country include *Geology of the Central Dominican Republic*, by C O Bowen (*Geological Society of America, Memoir* **98**:11–84, 1966) and *Hispaniola: tectonic focal point of the Northern Caribbean; three geological studies in the Dominican Republic*, compiled and edited by B Lidz and F Nagle (Miami Geological Society, 1979).

Guadeloupe and Martinique

A comprehensive geology can be found in *Martinique Guadeloupe*, by D Westercamp and H Tarzieff (Masson, 1980) in the series *Guides Géologiques Régionaux*.

Haiti

The country is described in *Géologie Générale et Régionale de la République d'Haiti*, by J Butterlin (*Travaux et Mémoires, Institute des Hautes Etudes de l'Amerique Latine*, **6**, 1960) and, more recently, in *Survey of the Geology of Haiti: guide to field excursions in Haiti*, by F Maurrasse (Miami Geological Society, 1982).

Jamaica

Recent publications on the geology of Jamaica are *Minerals and Rocks of Jamaica: a guide to the identification, location, occurrence and geological history*, by R D Porter *et al* (Jamaica Publishing House, 1982) and *Field Guide to Selected Jamaican Geological Localities and Mines*, edited by R M Wright (Geological Survey Department, 1974). *Bibliography of Jamaican Geology*, edited by M Kinghorn (GeoAbstracts Ltd, 1977) provides an index of the literature on the area.

Journal of the Geological Society of Jamaica (1964–) is currently published.

Martinique see under **Guadeloupe**

Puerto Rico

Major geological publications are *The Geography of Puerto Rico*, by R Pico (Aldine Publishing Co, 1974) and the earlier *A Survey of the Geology of Puerto Rico*, by R C Mitchell (*Technical Paper, Puerto Rico Agricultural Experiment Station*, **13**, 1954). The literature is documented in *Bibliography and Index of the*

Geology of Puerto Rico and Vicinity, 1866–1968, by M Hooker (Geological Society of Puerto Rico, 1969).

Trinidad and Tobago

The old publication, still referred to, is *The General and Economic Geology of Trinidad, B.W.I*, by H H Suter (2nd edn, HMSO, 1960).

SOUTH AMERICA

Publications on the geology of the continent are *Metallogenetische Provinzen in Sudamerika*, by H Putzer (E. Schweizerbart'sche, 1976), *Lateinamerika*, by W Zeil *et al* (*Geologische Rundschau*, **59**(3), 1970), an issue devoted, almost exclusively, to South America, *Geology and Tectonics of Northern South America*, by G Belizzia *et al* (*Geodynamics Commission Bulletin*, **3**, 1981) and the older work, *Handbook of South American Geology*, edited by W F Jenks (*Geological Society of America, Memoir* **65**, 1956). There is also the series *Quaternary of South America and Antarctic Peninsula* (1983–).

Andes

The major work on the mountain range is *The Andes: a geological review*, by W Zeil (*Beitrage zur Regionalen Geologie der Erde*, **13**, 1979); in addition there is *Cordillère des Andes*, edited by J Aubouin (*Revue de Géographie Physique et de Géologie Dynamique*, 2nd series, **15**(1–2), 1973). The series *Report of Andean Studies* (1978–) is published by the Association of Andean Studies, Shizuoka University in Japan.

Argentina

Major publications on Argentina are *Geologia Regional Argentina*, by A F Leanza (Academia Nacional de Ciencias, Cordoba, 1972), the results of the first Simposio de Geologia Regional Argentina, 1969, which also includes a bibliography, *Lexico Estratigrafico de la Republica Argentina* (1976–) in the series *Publicacion Especial, Servicio Geologico Nacional, Argentina*, *Geologia Regional Argentina: segundo simposio de Geologia Regional Argentina, 1976* (2 vols, Academia Nacional de Ciencias, Cordoba, 1979–1980), and *Geologia del Noroeste Argentina*, by F G Acenilaza and A J Toselli (Universidad Nacional de Tucuman, Facultad de Ciencias, 1981).

There are a number of current geological journals:
Acta Geologica Lilloana (1956–)
Actas Congresso Geologico Argentino (1965–)
Anales Subsecretaria de Minero Argentina (1947–)
Anuario Asociacion Geologica Argentina (1983–)
Boletin de la Asociacion Geologica de Cordoba (1970–)
Boletin Comite Argentino para el Programa Internacional de Correlacion Geologica (1982–)
Boletin Informativo, Direccion Nacional de Geologia y Mineria (1957–)
Boletin Servicio Geologico Nacional Argentina (1913–)
Comunicaciones de Museo de Mineralogia y Geologica, Universidad Nacional de Cordoba (1946–)
Contribuciones Cientificas Facultad de Ciencias Exactas, Universidad de Buenos Aires, seria Geologia (1950)
Notas Instituto de Fisiografia y Geologia, Rosario (1963–)
Publicaciones, Centro de Investigaciones en Recursos Geologicos, Buenos Aires (1984–)

Publicaciones, Servicio Nacional Minero Geologico, Argentina (1924–)
Resumes Congreso Argentino de Paleontologia y Biostratigrafica (1974–)
Revista de la Asociacion Geologica Argentina (1946–)
Revista de la Facultad de Ciencias Exactas, Fisicas y Naturales, Universidad Nacional de Cordoba, Serie Ciencias Geologicas (1970–)
Revista Museo Argentino Ciencias Naturales, Bernardino Rivadavia Instituto Nacional de Investigacion de las Ciencias Naturales, Ciencias Geologicas (1957–)
Revista Museo de la Plata, Seccion Geologia (1936–)
Revista Servicio Nacional Minero Geologico, Republica Argentina (1965–).

Bolivia

A comprehensive book on the country is the older publication *Geologia de Bolivia*, by F Ahfeld and L Bransia (Instituto Boliviano del Petroleo, 1960); reference can be made also to the later publication *Sinopsis Estratigrafica de Bolivia, 1. Parte de Paleozoico*, by G L A Rodrigo and A Castanos (Academia Nacional Ciencias, La Paz, 1978).

The literature is indexed in *Bibliographie Géologique de la Bolivie*, compiled by R Naif and J Munoz-Reyes (Servicio Geologico Bolivia, 1978) and the state survey publishes the journal *Boletin del Servicio Geologico de Bolivia* (1961–).

Brazil

The comprehensive work on Brazil is *Geologie von Brasilien*, by K Beurlen (*Beitrage zur Regionalen Geologie der Erde*, **9**, 1970). Useful reviews are *The Geology of Brazil*, edited by J M Mabesoone *et al* (*Earth Science Reviews*, **17**, 1981) and *Geologia do Brasil*, by J C Mendes do S Petri (*Enciclopedia Brasileira, Biblioteca Universitaria, Geociencias, Geologia*, **9**, 1971). Other publications of interest are *Minerals of Brazil*, by R R Franco *et al* (3 vols, Elsevier, 1975) and *Geologia do Brasil (Fanerozoico)*, by S Petri and V J Fulfaro (Quieroz, 1983). An important source of information are the papers published from the *Congresso Brasileiro de Geologia*.

Current geological journals from Brazil are:
Acta Geologica Leopoldensia (1976–)
Boletim, Departamento de Geologia, Universidade Federal do Rio Grande do Norte (1981–)
Boletim Instituto de Geociencias Universidade de Sao Paulo (1972–), and its *Publicacao Especial* (1984–)
Boletim Instituto de Geociencias, Universidade Federal de Rio de Janeiro (1967–)
Boletim Instituto Geografico e Geologico do Estado de Sao Paulo (1939–)
Boletim Instituto Geologico, Sao Paulo (1976–)
Boletim do Museu Nacional do Rio de Janeiro, Nova Serie, Geologia (1943–)
Boletim Museu Paranaense Emilio Goeldi, Nova Serie Geologica (1957–)
Boletim Paranaense de Geociencias (1959–)
Cadernos Ciencias da Terra (1969–)
Iheringia, Museu Rio-Grandense de Ciencias Naturais, Serie Geologia (1967–)
Pesquisas, Instituto de Geociencias, Universidade Federal do Rio Grande do Sul, Porto Alegre (1978–)
Publicacoes Nucleo de Sao Paulo Sociedade Brasileira de Geologia (1979–)
Revista Brasileira de Geociencias (1971–)
Revista do Instituto Geologico, São Paulo (1980–).

The literature is documented in *Bibliografia e Indice da Geologia do Brasil, 1641–1940*, compiled by D Iglesias and M de L Meneghezzi (*Boletim Divisao de Geologia e Mineralogia*, **204**, 1959); this bibliography is kept up-to-date, e.g. for the year *1966–1967*, compiled by D Iglesias and M de L Meneghezzi (*Boletim Ministerio das Minas e Energia*, **256**, 1976), and for *1968–1970*, compiled by M da G

Taveres Price (1978); in addition, there is the useful *Bibliography and Subject Index of Brazilian Geology (languages other than Portuguese)*, compiled by M N Womack and edited by A Button (University of Witwatersrand, 1976).

Chile

Geologie von Chile, by W Zeil (*Beitrage zur Regionalen Geologie der Erde*, **3**, 1964) covers the geology of the country comprehensively. Papers from the national geological congress, *Actas Congreso Geologico Chileno* (1976–) provide useful information also.

Current journals on geology are *Boletin Servicio Nacional de Geologia y Mineria, Chile* (1958–), *Comunicaciones, Departmento de Geologia, Universidad de Chile* (1960–), *Publicaciones, Instituto de Geologia, Universidad de Chile* (1956–) and *Revista Geologica de Chile* (1974–).

Bibliografia Geologica de Chile (1927–1953), compiled by J Munoz Cristi and J K Korot (*Publicacion Instituto de Geologia Chile Universidad*, **5**, 1955) and *Bibliografia Geologica del Norte de Chile*, in 2 parts, by D Srytrova (*Repertorio Bibliografico, Centro de Documentacion, Universidad del Norte*, **5** and **7**, 1975–1976) index the geological literature of the country.

Colombia

The main work covering the country is *Zur Geologie von Kolumbien, Sudamerika*, by F B Stibane (*Geotektonische Forschungen*, **30**, 1968); other important publications are *Precambrian to the Middle Cretaceous Stratigraphy of Colombia*, by H Burgl, translated from the Spanish by C G Allen and N R Rowlinson (C G Allen, Bogota, 1973), *Permian to Recent Stratigraphic Paleontology of Colombia*, by H Burgl (C G Allen, Bogota, 1974) and *Géologie des Andes Colombiennes*, by L Radelli (*Travaux et Mémoires, Laboratoire de Géologie de la Faculté des Sciences, Université de Grenoble*, 1967).

Current journals are *Boletin de Geologia, Facultad de Petroleos, Universidad Industrial de Santander* (1958–), *Boletin Geologico, Instituto Geologico Nacional, Colombia* (1954–), *Geologia Colombiana* (1962–), *Geologia Norandina* (1984–) and *Memorias Colombiano de Geologia* (1976–).

The earth sciences literature is indexed in *Bibliografia de la Biblioteca del Instituto Geofisico de los Andes Colombianos Sobre Geologia y Geofisica de Colombia*, compiled by J E Ramirez (2nd edn, *Boletin Instituto Geofisico de los Andes Colombianos; serie C, Geologia*, **6**, 1957) with a supplement in *Publicacion* of the same institution, no,**18**, 1973.

Ecuador

The geology of the country is comprehensively covered in *Geologie von Ecuador*, by W Sauer (*Beitrage zur Regionalen Geologie der Erde*, **11**, 1971) and for a brief description there is *Outline of the Geology of Ecuador*, by B Kennerly (*Overseas Geology and Mineral Resources*, **55**, 1980). Other major publications are *Notes on the Geology of Ecuador with special reference to the Western Cordillera*, by W Kohler and G van der Kaaden (*Geologische Jahrbuch*, Reihe B, **35**, 1979) and *The Glaciation of the Ecuadorian Andes*, by S Hastenrath (Balkema, 1980). The literature is indexed in *An Annotated Bibliography of Ecuadorian Geology*, by C R Bristow (*Overseas Geology and Mineral Resources*, **58**, 1981).

The Guianas and Surinam

Papers covering the area of French Guiana, Guyana and Surinam can be found in the proceedings of the *Inter-Guiana Geological Conference* (1961–) but there are also specific regional works.

The geology of French Guiana is described in two publications by B Choubert: *Géologie et Pétrographie de Guyane Française* (ORSTOM, 1949), and *Le Precambrien des Guyanes (Mémoires BRGM*, 81, 1974).

Two publications provide information on Guyana: *The Geology of Southern Guiana*, by J P Berrange (*Institute of Geological Sciences Overseas Memoirs*, 4, 1977) and *The Geology of British Guyana*, by G M Stockley (*Bulletin, British Guyana Geological Survey*, 25, 1955). Further information is listed in *Bibliography of the Geology and Mining of British Guyanas*, by C G Dixon and H K George (*Bulletin, British Guyana Geological Survey*, 32, 1964).

The national geological survey publishes *Bulletin, Geological Survey of Guyana* (1933–), *Record, Geological Survey of Guyana* (1962–) and *Report, Geological Survey of Guyana* (1947–).

The geological survey of Surinam publishes a monographic series, *Contribution to the Geology of Surinam* (1969–), within the journal *Mededelingen Geologisch Mijnbouwkundige Dienst van Suriname*. No.8 (1984) of this series is a recent comprehensive study of the area, *Geology of Surinam*, by D R de Vletter. The current journal *Mededelingen Geologisch Mijnbouwkundige Dienst van Suriname* was first published in 1948.

A bibliography of the country is provided by *De Geologische Literatur over of van belang voor Nederlandsch-Guyana (Surinam) en de Nederlandsche Westindische Eilandern (Antillen, 1934–1950)*, compiled by J F Steenhuis (Mouton, 1951).

Paraguay

Geologie von Paraguay, by H Putzer (*Beitrage zur Regionalen Geologie der Erde*, 2, 1962) is still the only major publication on this country.

Peru

Publications on Peru include *The Geology of the Western Cordillera of North Peru*, by E J Cobbing et al (*Institute of Geological Sciences, Overseas Memoirs*, 5, 1981), *Géologie des Andes Peruviénnes*, by G Laubacher (*Travaux et Documents ORSTOM*, 95, 1978), *Etude Géologique des Andes du Peru Central*, by F Medard (*Mémoires ORSTOM*, 86, 1978) and the publications from the *Congreso Peruano de Geologia*.

Boletin de la Sociedad Geologica del Peru (1925–) is the current journal.

The literature is indexed in *Bibliografia Geologica del Peru*, by L Castro Bastos (the author, 1960).

Surinam see **The Guianas**

Uruguay

Major publications are *Geologia del Uruguay*, by J Bossi (2nd edn, *Coleccion Ciencias Universidad de la Republica, Montevideo*, 2, 1969) and *Recursos Minerales del Uruguay*, by J Bossi (Ed. Daniel Aljanti, 1978).

The geological journal for the country is *Boletin Instituto Geologico del Uruguay* (1927–).

Coverage of the literature is given in *Bibliografia Sobre Geologia y Paleontologia del Uruguay*, by A Mones (Museo Nacional de Historia Natural, 1981) with a supplement in 1982.

Venezuela

Publications for the country include *Synthèse Paleogéographique et Pètrolière du Venezuela Occidental*, by E Zambrano et al (Technip, 1972) and *Géologie de la*

Chaine Caraibe au Méridien de Caracas (Venezuela), by C Beck (*Publication
Société Géologique du Nord*, **14**, 1984).
Current journals are:
Boletin Asociacion Venezolana de Geologia, Mineria y Petroleo (1958–)
Boletin de Geologia, Direccion de Geologia, Publicacion Especial, Venezuela
(1956–)
Boletin de Geologia, Ministerio de Minas e Hidrocarburos, Venezuela (1951–)
Boletin de la Sociedad Venezolana de Geologos (1965–)
Publicaciones Especiales Asociacion Venezolana de Geologia (1968–).
The geological bibliography is *Internal Publications on the Geology of Venezuela,
1958 to mid-1965*, by R M Stainforth (*Bulletin, American Association of Petroleum
Geologists*, **49**, 2289–2294, 1965).

NORTH AMERICA

The major series of publications on North America is *The Decade of North America*
published by the Geological Society of America. Publication began in 1982 with the
Centennial Special Volume **1**: *Geologists and Ideas: a history of North American
Geology*. In all, 19 volumes are envisaged for *Geology of North America (US &
Mexico)*; 9 for *Geology of North America (Canada)*; 6 for *Centennial Field Guides*;
4 for *Centennial Special Volumes*; 1 volume and 24 packets of *Continent-Ocean
Transects*, and 1 volume and 7 four-sheet maps for *Continent-scale Geologic Maps
of North America*.
Major publications, covering the whole continent are *Geological Evolution of
North America*, edited by C W Stearn *et al* (3rd edn, Wiley, 1979), *The Evolution of
North America*, revised edition, by P B King (Princeton University Press, 1977) and
Historical Geology of North America, by N S Petersen *et al* (W C Brown, 1980).
The relevant volume of *Lexique Stratigraphique International* for North America
is **7**. Fasc. 1 (1967), the only one published in this volume, is in 3 parts, all on the
United States. Stratigraphic works are *Correlation of North American Silurian
Rocks*, edited by W B N Berry and A J Boucot (*Geological Society of America,
Special Paper*, **102**, 1970), *Stratigraphic Atlas of North and Central America*, edited
by T D Cook and A W Bally (Princeton University Press, 1975), *The Mesozoic of
Middle North America*, edited by D F Stott and D J Glass (*Canadian Society of
Petroleum Geologists, Memoir* **9**, 1984), *Quaternary Stratigraphy of North America*,
edited by W C Mahaney (Dowden, Hutchinson, Ross, 1976) and *North America
and the Great Ice Age*, by C L Matsch (McGraw-Hill, 1976). Others of interest are
C W Chesterman's *The Audubon Society Field Guide to North American Rocks and
Minerals* (A A Knopf, 1978) and *Gemstones of North America*, by J Sinkankas (2
vols, Van Nostrand Reinhold, 1957-1976).
Of economic interest are the publications *North American Oil and Gas Fields*,
edited by J Braustein (*American Association of Petroleum Geologists, Memoir*, **24**,
1976), *Papers on Mineral Deposits of Western North America*, edited by J D Ridge,
(*Nevada Bureau of Mines and Geology, Report*, **33**, 1979) and *Tectonics and
Mineral Resources of Southwest North America*, by L A Woodward (*New Mexico
Geological Society, Special Publication*, **6**, 1976).
The literature, 1923–1970, is indexed and published annually in *Bibliography of
North American Geology* which is now incorporated in the *Bibliography and Index
of Geology* (see Chapter 5). The wealth of regional literature, contained in field
trip guidebooks, is indexed in *Geologic Field Trip Guidebooks of North America: a
union list incorporating monographic titles*, compiled and edited by the Geoscience
Information Society (Phil Wilson Publishing Co., 1973), with a later edition, *Union
List of Geologic Field Trip Guidebooks of North America* published by the
American Geological Institute in 1978.

Canada

Useful publications for Canada are *Natural Landscapes of Canada: a study in regional earth science*, by J B Bird (2nd edn, Wiley, 1980), *Field Trip Guidebook*, edited by A L Currie and W O MacKasey (Geological Association of Canada, 1978), *Canadian Mineral Occurrence Index (CANMINDEX) of the Geological Survey of Canada*, edited by D D Picklyk *et al* (*Geological Survey of Canada, Paper* **78–8**, 1978), *Canada's Continental Margins and Offshore Petroleum Exploration*, by C J Yorath *et al* (*Canadian Society of Petrologists and Geologists, Memoirs*, **4**, 1975) *Offshore Geology of Eastern Canada*, edited by B R Pelletier (*Geological Survey of Canada, Paper* **74–30**, 1974), the *Lexicon of Canadian Stratigraphy* (Canadian Society of Petroleum Geologists, 1981–) of which two volumes are published to date: **1**, *Arctic Archipelago*, and **2**, *Yukon Territory, The Ordovician System in Canada*, by C R Barnes *et al* (*IUGS Publication*, **8**, 1981) and *Proterozoic Basins of Canada*, edited by F H A Campbell (*Geological Survey of Canada, Paper* **81–10**, 1981).

Current journal publications on Canadian geology are:
Bulletin of Canadian Petroleum Geology (1963–)
Bulletin, Geological Survey of Canada (1913–)
Canadian Journal of Earth Sciences (1964–)
Geolog (1972–)
Geos (1972–)
Geoscience Canada (1972–)
Memoir, Geological Survey of Canada (1910–)
Miscellaneous Report, Geological Survey of Canada (1962–)
Memoir, Canadian Society of Petroleum Geologists (1973–)
Papers, Geological Survey of Canada (1939–)
Special Papers, Geological Association of Canada (1956–).

Geological literature from the provinces of Canada, is omitted from this publication. The literature of the provinces is extensively covered in *Geologic Reference Sources*, by D C Ward *et al* (2nd edn, Scarecrow Press, 1981).

Greenland

A recent publication on the geology of Greenland is *Palaeontology and Stratigraphy of Greenland: short contributions*, compiled by J S Peel (*Rapport Gronlands Geologiske Undersogelse*, **101**, 1980).

Currently published journals are *Greenland Geoscience* (1979–), formerly part of the series *Meddelelser om Gronland, Bulletin Gronlands Geologiske Undersogelse* (1976–) and *Rapport Gronlands Geologiske Undersogelse* (1964–).

United States

Not many publications attempt to cover the whole country in a single volume. Publications dealing with various aspects of United States geology include *Fieldguide to Landforms in the United States*, by J A Shiner (Collier MacMillan, 1972), *The Ordovician System of the United States*, edited by R J Ross *et al* (*IUGS Publication*, **12**, 1982), *The Mississippian and Pennsylvanian (Carboniferous) Systems in the United States*, (*US Geological Survey, Professional Papers*, **1110 A-L**, 1979), *The Caledonides in the USA*, edited by J W Skehan and P H Osberg (Western Observatory, 1979) and *Geological Evolution of the United States Atlantic Margin*, edited by C W Poag (2 vols, Van Nostrand Reinhold, 1985).

The stratigraphy of the United States is covered by vol.7, fasc. 1 of the *Lexique Stratigraphique International* (1967). Names are regularly published and updated in *Lexicon of Geological Names of the United States* in the *US Geological Survey Bulletin*, e.g. *1936–1960* in *Bulletin* **1200**, 1966; *1961–1970* in no.**1350**, and *1968–1975* in no.**1520**, 1981.

Publications on the economic geology of the United States include *United States Mineral Resources*, edited by D A Brobst and W P Pratt (*US Geological Survey, Professional Paper* **820**, 1973), *Mines and Minerals of the Great American Rift (Colorado-New Mexico)*, by R W Holmes and M B Kennedy (Van Nostrand Reinhold, 1982) and *Petroleum Geology of the United States*, by M J Landes (Wiley, 1970).

The main, current, serial publications of the United States Geological Survey are *Bulletin* (1983–) and *Professional Paper* (1902–). Those of the Geological Society of America are *Abstracts with Programs* (1969–), *Bulletin* (1980–), *Geology, Boulder* (1973–), *Memoir* (1934–), *Memorials* (1973–), and *Special Paper* (1934–).

Other periodicals include:
American Association of Petroleum Geologists, Studies in Geology (1975–)
American Association of Petroleum Geologists, Bulletin (1918–)
American Journal of Science (1818–)
Bulletin Southern Geological Society (1972–)
Fieldiana Geology (1910–)
Journal of Geology (1893–)
Memoir, American Association of Petroleum Geologists (1962–)
Northeastern Geology (1980–)
Publication Southern Geological Society (1981–)
Report American Geological Institute (1981–)
Smithsonian Contribution to Earth Science (1969–).

Publications on the various states are omitted; state publications are extensively covered in *Geologic Reference Sources* (see previous page).

POLAR REGIONS

Antarctic

The major publications on Antarctic geology are *Geology of the Antarctic Peninsula*, by G E Grikorov (Amerind, 1978), which is a translation of *Geologiya Antarkticheskogo Poluostrova* (Nauka, 1973), *Antarctic Geoscience*, edited by C Craddock, the proceedings from the Symposium on Antarctic Geology and Geophysics, held at Madison in 1977 (University of Wisconsin Press, 1982), *Proceedings of the Second Symposium on Antarctic Geoscience, Tokyo, 1980*, edited by T Nagata (*Memoirs, National Institute of Polar Research (Tokyo), Special Issue*, **21**, 1982), *International Symposium on Antarctic Earth Sciences, Adelaide, 1982*, edited by R L Oliver *et al* (Cambridge University Press, 1983) and *Antarctic Glacial History and World Palaeoenvironments*, edited by E M Van Zindern Bakker (Balkema, 1978).

The current journals of Antarctic science, including the sub-Antarctic islands are the publishing vehicles for the national research teams working in the regions:
Antarctic Journal of the United States (1966–)
Antarctic Meteorite Newsletter (1978–)
Antarctic Record (New Zealand) (1978–)
Antarctic Record, Reports of the Japanese Antarctic Research Expedition (1957–)
Antarctic Research Series, Washington (1964–)
Antardia, Buenos Aires (1971–)
Boletin de Difusion, Instituto Antartico Chileno (1965–)
Bulletin, British Antarctic Survey (1963–)
CNFRA, Publications du Comité National Français de Recherches Antarctiques (1962–)
Memoirs, National Institute of Polar Research (Tokyo), Series C: Earth Sciences (1975–)
Quaternary of South America and Antarctic Peninsula (1983–)

Report. British Antarctic Survey (1975–)
Scientific Reports, British Antarctic Survey (1962–).
Literature, including the geological, is indexed in *Antarctic Bibliography* (1965–).

Arctic see also **Greenland, Iceland and Spitzbergen**

Publications on the Arctic include *Marine Geology and Oceanography of the Arctic Seas*, edited by Y Herman (Springer, 1974), *The Ocean Basins and Margins, vol.* **5**: *The Arctic Ocean*, edited by A E M Nairn *et al* (Plenum, 1981), *Arctic Geology*, edited by M G Pitcher (*American Association of Petroleum Geologists, Memoir* **19**, 1973), which is the proceedings of the Second International Symposium on Arctic Geology, held in 1971. The proceedings of the First Symposium, in 1960, are published in *Geology of the Arctic*, edited by G O Raasch (2 vols, University of Toronto Press, 1960) and there are also *The Arctic Basin*, by J E Sater (rev. edn, Arctic Institute of North America, 1969), *Arctic Pleistocene History and the Development of Submarine Permafrost*, by M Vigdorchik (Westview Press, 1980) and *The Alaskan Shelf: hydrographic, sedimentary and geochemical environment*, by G D Sharma (Springer, 1979).
Journals currently published are:
Arbok Norsk Polarinstitut (1960–)
Arctic (1948–)
Arctic and Alpine Research (1969–)
Arctic Bulletin (1973–)
Polar News (1960–)
Polar Notes (1959–)
Polar Record (1931–)
Polarhandbook (1964–)
Skrifter Norsk Polarinstitut (1948–)
Tektonika Artikii (1975–)
Trudy Nauchno-Issledovatel'skogo Instituta Geologii Arktikii (1958–).
Geological literature was included in the compilation *Arctic Bibliography* (1953–1975) and in *Publications of the Arctic Islands by the Geological Survey of Canada*, by R L Christie (*Canada Geological Survey, Papers* **71–10, 73–11, 76–28**, 1971–1977).

ASIA

The size and extensiveness of the continent prohibits the coverage of its geology in a single comprehensive volume. There is no shortage of literature on its constituent regions, e.g. Southeast Asia, the Middle East.
The countries are covered by volume **3** of the *Lexique Stratigraphique International*: fasc.1, China (1964–1971); 2a, Korea (1956); 2b, Manchuria (1956); 3a, Japan (1956); 3b, Ryukyu (1957); 4, Taiwan (1957); 5, Philippines (1957); 6a, Indochina (1956); 6b, Malaya (1956); 6c, Thailand (1956); 6d, Burma (1956); 7a, Indonesia (1956); 7b, British Borneo (1956), with a second edition in 1961; 7c, Malay Archipelago (1956); 8a, India, Pakistan, Nepal, Bhutan (1957); 8b, Burma (1957); 8c, Ceylon (1957); 9, Turkey, (1960); 9a, Afghanistan (1961); 9b, Iran (1972); 10a, Iraq (1959); 10b, Saudi Arabia, Aden Protectorate and Dhufar, Qatar Peninsula (1968–1975); 10c(1) Lebanon, Syria, Jordan (1963), with a revised edition on Jordan in 1976; 10c(2), Israel (1960).
The Economic and Social Commission for Asia and the Pacific (ESCAP), a subsidiary body of the United Nations, published the *ESCAP Atlas of Stratigraphy* in 3 volumes (1978–1982); in this work the countries covered are Burma, Malaysia, Thailand, Indonesia and Philippines in vol.1, Japan in vol.2, and Bangladesh, Fiji, India, Indonesia and Nepal in vol.3.

Far East & Southeast Asia

A range of papers on the region can be found in the series *Geology and Paleontology of South East Asia*, complete in 24 volumes (University of Tokyo Press, 1964–1983) and, from the same publisher, *Geology and Mineral Resources of the Far East* in 3 volumes (1967–1971). Major publications are the *Proceedings of the Regional Conference on the Geology of South East Asia*, edited by B K Tan (*Geological Society of Malaysia, Bulletin* **6**, 1973), *The Tectonic and Geologic Evolution of Southeast Asian Seas and Islands (Geophysical Monographs*, **27**, 1982), *The Evolution of the East Asian Environment*, edited by R O Whyte (*Occasional Papers and Monographs, Centre for Asian Studies, University of Hong Kong*, **59**, 1984), and *Carboniferous of the World I: China, Korea, Japan and S E Asia*, edited by C Martinez Diaz (*IUGS Publication*, **16**, 1983).

Current journals are *Modern Quaternary Research in Southeast Asia* (1975–) and *Journal of Southeast Asian Earth Sciences* (1986–).

For reference, there is *Sources of Geological Information for S E Asia*, by G B Wolfenden *et al* (*Geological Society for London, Miscellaneous Paper* **9**, 1978).

Himalayas

Some of the many publications on the region are, *Geology of the Himalayas*, edited by A E T Atkinson (Cosmo, 1980), *Himalayas*, 2 volumes in the series *Colloques Internationaux*, **268**, 1977, *Himalayan Shears*, by P S Saklani (Himalayan Books, 1983), *Contemporary Geoscientific Research in Himalaya*, edited by A S Sinha (2 vols, Bishen Singh Mahrendra Pal Singh, 1981–1983), *Geology of Vindhyanchal*, edited by K S Valdiya *et al* (Hindustan, 1982), *Stratigraphy and Correlations of the Lesser Himalayan Formations*, edited by K S Valdiya and S B Bhatia (Hindustan, 1980) and *Geology of the Kumaun Lesser Himalaya*, by K S Valdiya (Wadia Institute, 1980).

There are also the monographic publications in the series *Current Trends in Geology*, edited by P S Saklani (Today and Tomorrow's, 1978–). The series includes *Tectonic Geology of the Himalaya* (**1**, 1978), *Structural Geology of the Himalaya* (**2**, 1980), and *Himalaya Thrusts and Associated Rocks* (**9**, 1986). The series *Contributions to Himalayan Geology*, edited by V J Gupta (Hindustan, 1981–) includes vol.**1**, *Upper Paleozoic of the Himalayas*, (1981), vol.**2**, *Stratigraphy and Structure of Kashmir and Ladakh Himalaya* (1983) and vol.**3**, *Geology of Western Himalaya* (1986), which have been published to date.

Important publications are the papers which appear in *Himalayan Geology*; this publication presents the proceedings of annual seminars organized by the Wadia Institute of Himalayan Geology (1971–).

Papers published are indexed in *Bibliography on Himalayan Geology*, compiled by S K Kapoor *et al* (*Geological Survey of India, Miscellaneous Publication*, **28**, 1976) and *Bibliography on Himalayan Geology (1970–1975)*, by G N Tripathi (Institute of Petroleum Geology, 1976).

MIDDLE EAST

Publications on the Middle East region are *The Environmental History of the Near and Middle East Since the Last Ice Age*, edited by W C Brice (Academic Press, 1978), *Persian Gulf*, edited by B H Purser (Springer 1973) and *Geology of the Arabian Peninsula* (*US Geological Survey, Professional Paper*, **560 A-I**, 1966–1975).

The literature of the region is documented in the following bibliographies: *Bibliography of Levant Geology, including Cyprus, Hatay, Israel, Jordania, Lebanon, Sinai and Syria*, compiled by M A Avnimelech (2 vols, Israel Program for Scientific Translations, 1965–1969) — vol.**1** covers the period from the 16th century to 1963, and vol.2, 1963–1968; *Geological Bibliography of the Arabian Gulf*, by F A Sharief (King Saud University Press, 1982), and *The Dead Sea and its Surroundings: bibliography and geological research*, by V Arad *et al* (*Israel Geological Survey, Special Publication*, **3**, 1984). Current literature is indexed in the monthly *Current Bibliography of Middle East Geology* (1976–) published by the Israel Geological Survey.

Aden

Publications on the area are *The Stratigraphy and Structure of the Eastern Aden Protectorate*, by Z R Beydoun (*Overseas Geology and Mineral Resources Supplement Series*, **5** 1964), *Geology of the Arabian Peninsula: Eastern Aden Protectorate and Part of Dhufar*, by Z R Beydoun (*US Geological Survey, Professional Paper*, **560-H**, 1966) and *Geology of the Arabian Peninsula: Aden Protectorate*, by J E G W Greenwood and D Bleackley (*US Geological Survey, Professional Paper*, **560-C**, 1967).

Afghanistan

Major publications include the comprehensive *Geologie von Afghanistan*, by R Wolfart and H Wittekindt (*Beitrage zur Regionalen der Erde*, **14**, 1980) and for the regions of the country, *Beitrage zur Geologie von Zentral und Sud Afghanistan*, by G Andritzky *et al* (*Beihefte zum Geologischen Jahrbuch*, **96**, 1971), *Zur Geologie von Sudost Afghanistan*, by O Gans (*Beihefte zum Geologischen Jahrbuch*, **84**, 1970) and *On the Geological Development of Central and South Afghanistan*, by D Weippert *et al* (*Afghan Geological and Mineral Survey, Bulletin*, **4**, 1970).

The current publication of the Afghan Geological and Mineral Survey is its *Bulletin* (1964–).

Documentation of the literature is covered by *Bibliographie zum Geologie Afghanistans*, compiled by H Kastner (*Beihefte zum Geologischen Jahrbuch*, **114**, 1971), *Bibliography on the Geology of Afghanistan and Immediately Adjacent Areas (as at the end of 1970)* (*Afghan Geological and Mineral Survey, Bulletin*, **6**, 1971) and *Progress in the Geology of Afghanistan, 1972–1978*, by M F Buchroithner (*Zentralblatt fur Geologie und Palaontologie*, Teil I, (5/6), 328–376, 1979).

Bahrain

Publications on this country are *Geology, Geomorphology and Pedology of Bahrain*, edited by J C Doornkamp *et al* (GeoAbstracts Ltd, 1980) and *Geology of the Arabian Peninsula: Bahrain* by R P Willis (*US Geological Survey, Professional Paper*, **560-E**, 1967).

Bangladesh

Earlier publications for this country are included in the regional geological works for British India and for Pakistan.

Bhutan

Apart from literature on the Himalayas, specifically for this country is the *Geology of the Bhutan Himalaya*, by A Gansser (*Denkschriften der Schweizerischen Naturforschenden Gesellschaft*, **96**, 1983).

Brunei see Malaysia

Burma

The most recent and comprehensive work is the *Geology of Burma*, by F Bender (*Beitrage zur Regionalen Geologie der Erde*, **16**, 1983).

Literature for the country, 16th century to 1978, has been indexed in *Earth Sciences Bibliography of Burma, Yunnan and Andaman Islands*, compiled by P J Goosens (*Proceedings, Regional Conference on the Geology and Mineral Resources of Southeast Asia*, **3**, 493–536, 1978).

Cambodia see under **Vietnam**

Ceylon see **Sri Lanka**

China

The most recent publication *The Geology of China*, by Z Y Yang *et al* (Clarendon Press, 1986) overcomes the language problem usually found in the majority of the literature on Chinese geology. Other English language publications include *Quaternary Geology and Environment of China*, edited by T S Liu, for the Quaternary Research Association of China (Springer, 1985), *Micropalaeontology of China*, by P X Wang *et al* (Springer, 1985), *Geological and Ecological Studies of Qinghai-Xizang (Tibet) Plateau; proceedings of the symposium on Qinghai-Xizang (Tibet), Beijing, 1980* (2 vols, Science Press, Gordon & Breach, 1981) and *Proceedings of the International Symposium on the Continental Shelf, with special reference to the East China Sea* (2 vols, China Ocean Press, Springer, 1984).

The majority of publications, however, are still in Chinese, although English titles, contents pages, and abstracts are now generally part of the publication. Amongst the major publications from China are the *Palaeontological Atlas* of all parts of China, projected in approximately 45 volumes (Geological Publishing House), *Stratigraphy of China (Zhongguo Dizhi)* (Geological Publishing House, 1979–) in 14 volumes, with an English edition anticipated, *Stratigraphic Correlation in China* by the All-China Stratigraphic Committee (4 parts, Science Press, 1979) and their *Stratigraphic Correlation Chart in China with Explanatory Text* (1982).

Regional geological literature includes *The Stratigraphy and Palaeontology in W.Sichuan and E.Xizang* by the Sichuan Bureau of Geology and the Nanjing Institute of Geology and Palaeontology (Sichuan Peoples' Press, 1982), *Tertiary Paleontology of North Continental Shelf of South China Sea* by South Sea Branch of the Petroleum Corporation, Peoples' Republic of China (Guangdong Science and Technology Press, 1981), *Stratigraphy of Xizang*, by B G Zhang *et al* (Science Press, 1984) and *Palaeontology of Xizang (Series of the Scientific Expedition to the Qinghai-Xizang Plateau* (5 vols, Science Press, 1973-1976).

Since 1980 there has been a sharp increase in the number of journals published:
Acta Geologica Sinica (1973–)
Acta Geophysica Sinica (1952–)
Acta Petrologica Mineralogica et Analytica (1981–)
Acta Sedimentologica Sinica (1983–)
Bulletin of the Institute of Geology, Chinese Academy of Geological Sciences (1980–)
Bulletin of the Institute of Geomechanics, Chinese Academy of Geological Sciences (1980–)
Bulletin of the Institute of Mineral Deposits, Chinese Academy of Geological Sciences (1980–)
Bulletin of the Nanjing Institute of Geology and Mineral Resources (1980–)
Bulletin Nanjing Institute of Geology and Palaeontology (1980–)
Bulletin of the Shengyang Institute of Geology and Mineral Resources (1980–)
Bulletin of the Tianjian Institute of Geology and Mineral Resources (1980–)
Bulletin of the Xi'an Institute of Geology and Mineral Resources (1980–)
Bulletin of the Yichang Institute of Geology and Mineral Resources (1980–)
Contributions to the Geology of the Qinghai-Xizang Plateau (1983–)
Earth Science, Wuhan College of Geology (1981–)

Gansu Geology (1983–)
Geochimica (1974–)
Geological Review (1943–)
Journal of the Changchun Geological Institute (1977–)
Journal of Nanjing University, Natural Science B; Journal of Stratigraphy (1966–)
Memoirs, Nanjing Institute of Geology and Palaeontology (1958–)
Mineral Deposits (1982–)
Professional Papers of Stratigraphy and Palaeontology (1975–)
Scientia Geologica Sinica (1963–)
Seismology and Geology (1979–).

Current literature is listed in *Abstracts of Chinese Geological Literature* (1984–) compiled by The National Geological Library, Beijing, and in *China Science and Technology Abstracts, Series 2: Chemistry, Earth Science, Energy Science*. Recent literature is discussed in *Annotation of Principal Geologic Literature published from 1970 to 1982 in China*, by K Y Lee (*US Geological Survey, Open File Report*, **84–107**, 1984).

Cyprus

An Outline of the Geology and Geomorphology of Cyprus, by T M Pantazis (Cyprus Geographical Association, 1971) gives a brief description of the island and a bibliography of its geology: *A Revised Bibliography of Cyprus Geology*, by the same author was published as *Cyprus Geological Survey, Bulletin*, **2**,57–81, 1969.

The Geological Survey Department currently publishes the serials, *Bulletin* (1963–), *Memoir* (1959–), and *Report* (1955–).

Hong Kong

Publications for the colony are *The Geology of Hong Kong* (Government Printer, 1952) and *The Economic Geology of Hong Kong* (Hong Kong University Press, 1964), both by S G Davis, both rather old.

India

Some recent books are revised editions of older works: *Geology of India*, by D N Wadia (4th edn, McGraw-Hill, 1975) and *Geology of India and Burma*, by M S Krishnan (6th edn, CBS, Delhi, 1982); *Geology of the Indian Subcontinent*, by J Calder and J D Herbert (Cosmo, 1981), is a reprint of the 1922 edition.

Recent publications include *Fundamentals of Historical Geology and Stratigraphy of India*, by R Kumar (Wiley Eastern, 1986), *Geomorphology of India*, by S Ray (KLM, 1978), *Geological History of the Platform Areas of India* (Geological Mining and Metallurgical Society, 1978) and *Structure and Tectonics of Precambrian Rocks of India*, edited by S Sinha-Roy (*Recent Research in Geology*, **10**, 1983). There is also the 4-volume work, *Indian Stratigraphy*, edited by V J Gupta (Hindustan, 1973–1977), a series which is respectively numbered: **1**, *Indian Palaeozoic Stratigraphy* (1973); **2**, *Indian Mesozoic Stratigraphy* (1975); **3**, *Indian Cenozoic Stratigraphy* (1976); **4**, *Indian Precambrian Stratigraphy* (1977).

Current journal publications of the Geological Survey of India are: *Bulletin, Series A: Economic Geology* and *Series B: Engineering Geology and Groundwater* (1950–), *Memoirs* (1856–), *Miscellaneous Publications* (1968–) and *Records* (1868–).

Other journals for Indian geology are:
Bulletin of the Geological Mining and Metallurgical Society of India (1937–)
Geoscience Journal, Lucknow (1980–)
Geoviews (1975–)

Indian Journal of Earth Sciences (1974–)
Journal of the Geological Society of India (1959–)
Memoirs Geological Society of India (1963–)
Publications of the Centre of Advanced Study in Geology, Panjab University (1966–)
Quarterly Journal of the Geological and Mining and Metallurgical Society of India (1926–)
Recent Research in Geology (1973–).

For bibliographies, *Indian Geoscience Abstracts*, starts with the literature from 1974 (Geological Survey of India, 1981–) whilst *Indian Geological Index* begins from 1971 (Geological Survey of India, 1975–). Specialist bibliographies are *Bibliography on the Palaeontology and Stratigraphy of India 1947–1980*, by S K Kapoor (*Geological Survey of India, Miscellaneous Publication* **52**, 1981) and *Micropaleontologia Indica: an annotated bibliography of microfossils from India*, compiled by S S Gowda (Bangalore Central College, 1971). The *Bibliography of Indian Geology*, is published by the Geological Survey of India in the *Miscellaneous Publications* series; so far, only various sections of part 4, *Palaeontological Index*, have been published.

Indonesia

Major publications on Indonesia are *Tectonics of the Indonesian Region*, by W Hamilton (*US Geological Survey, Professional Paper* **1078**, 1979), *The Geology of Indonesia*, by B W Van Bemmelen (2nd edn, Nijhoff, 1970) and *The Geology and Tectonics of Eastern Indonesia*, edited by A J Barber and S Wiryosujono (*Geological Research and Development Centre, Bandung, Special Publication* **2**, 1981).

Current journals are:
Buku Tahunan, Pertambangan Indonesia (1960–)
Bulletin Geological Survey of Indonesia (1964–)
Contributions from the Department of Geology, Institute of Technology, Bandung (1959–)
Publikasi Chusus Direktorat Geologi Republik Indonesia (1963–)
Publikasi Teknik Djawatan Geologi which are in various series, *Seri Geofiska* (1970–), *Seri Geologi Ekonomi* (1960–), *Seri Geologi Umum* (1964–) and *Seri Paleontologi* (1960–)
Special Publications, Geological Research and Development Centre, Bandung (1978–)
Special Publications Geological Survey of Indonesia (1970).

A bibliography for the area is the *Geology of Indonesia: publications 1958–1975 and a few earlier publications*, by H D Tija (*Publikasi Chusus Direktorat Geologi Republik Indonesia*, **5**, 1977).

Iran

Stratigraphic Lexicon of Iran, part 1: Central North and East Iran, by J Stoecklin (*Geological Survey of Iran, Report* **18**, 1971) and *Stratigraphic Correlation of Turkey, Iran, Pakistan*, edited by S M I Shah and A M Quennel (2 vols, Overseas Development Agency, 1980) are rare publications on the geology of this country.

The journal publications of the state survey are: *Report, Geological Survey of Iran* (1964–) and *Special Publication, Geological Survey of Iran* (1969–).

The literature is indexed in *Bibliography of Geology of Iran*, by N C Rosen (*Geological Survey of Iran, Special Publication*, **2**, 1969) and in *Bibliographie der Geologischen Literatur des Iran bis 1978*, by M D Durkoopt *et al*, 2nd edn (*Bochumer Geologische und Geotechnische Arbeiten*, **2**, 1982).

Iraq

Publications include *Regional Geology of Iraq: stratigraphy and palaeogeography*, edited by I I M Kassah and S Z Jassim (State Organization for Minerals, 1980), *Geology of the Arabian Peninsula: Southwestern Iraq*, by K M Al Naqib (*US Geological Survey, Professional Paper*, **560-G**, 1967) and *Tertiary Microfacies of Iraq*, by A J Al Hashimi and A M Reda (Director General for Geological Survey and Mineral Investigation, 1983).

Journal of the Geological Society of Iraq (1968–) is the only journal on Iraqi geology.

Literature is documented in *Bibliography on the Geology of Iraq*, by S A Al Sinawi and A B Naqash (*Bulletin, Baghdad University, College of Science*, **17**(2): 517–569, 1976); part two of this bibliography is published in the *Iraqi Journal of Science* **21**(4), 1980. *A Guide to the Literature on Iraq's Natural Resources (containing Studies and Reports from 1833 to 1968)* (UNESCO, 1969) is also useful for reference.

Israel

Monographs on Israel geology include *Regional Stratigraphy of Israel: a guide for geological mapping*, by Y Bartov *et al* (Geological Survey of Israel, 1981), *The Quaternary of Israel*, by A Horowitz (Academic Press, 1979), *Studies in Stratigraphy of Israel: the Triassic*, by L Picard and A Flexer (Israel Institute of Petroleum, 1974) and *Géomorphologie d'Israel*, by D Nir (*Mémoires et Documents, Centre National de la Recherches Scientifiques*, **16**, 1975).

The Geological Survey of Israel currently publishes the *Bulletin* (1950–), *Current Research* (formerly *Summary of Activities*) (1949–) and *Special Publication* (1980–). Other journals are *Israel Journal of Earth Sciences* (1963–) and *Proceedings of the Annual Meeting, Israel Geological Society* (1973–).

An important bibliography is *Geological Research in Sinai: a bibliography*, by V Arad and Y Bartov (*Geological Survey of Israel, Special Publication*, **2**, 1981).

Japan

Books on the geology of Japan are *Outline of the Geology of Japan*, edited by T Yoshida (3rd edn, Geological Survey of Japan, 1975), *Geology and Mineral Resources of Japan, vol 1: Geology*, edited by K Tanaka and A T Nozawa (3rd edn, Geological Survey of Japan, 1977) and the older *The Geology of Japan*, edited by F Takai *et al* (University of California Press, 1963). *The Landforms of Japan*, by T Yoshikawa *et al* (University of Tokyo Press, 1981) and *The Neogene of Japan: its biostratigraphy and chronology*, edited by R Tsuchi (IGCP-114 National Working Group of Japan, 1981) are two more recent publications.

Currently the Geological Survey of Japan publishes the *Bulletin* (1886–), *Report* (1922–), and *Special Report* (1966–). There are many other current journals:
Bulletin of the National Science Museum, Series C (Geology) (1975–)
Chigaku Kenkyu (Geoscience Magazine) (1926–)
Chikyu Kagaku (Earth Science) (1949–)
Contributions, Department of Geology and Mineralogy, Niigata University (1966–)
Contributions from the Institute of Geology and Paleontology, Tohoku University (1924–)
Geological Report of the Hiroshima University (1951–)
Geological Report, Institute of Earth Science, Waseda University (1966–)
Geoscience Report, Shizuoka University (1975–)
Journal of Earth Sciences, Nagoya University (1953–)
Journal of the Faculty of Science, Hokkaido University, Series 4: Geology and Mineralogy (1930–) and *Series 7: Geophysics* (1973–)

Journal of the Faculty of Science, University of Tokyo, Section 2: Geology, Mineralogy, Geography, Geophysics (1925–)
Journal of the Geological Society of Japan (1897–)
Journal of Geosciences, Osaka City University (1950–)
Journal of the Japanese Association of Mineralogy Petrologists and Economic Geologists (1966–)
Journal of Science of the Hiroshima University, Series C: Geology and Mineralogy (1951–)
Kumamoto Journal of Science, Series B: Biology and Geology (1952–)
Memoirs of the Faculty of Science, Kochi University, Series E: Geology (1980–)
Memoirs of the Faculty of Science, Kyoto University, Series D: Geology and Mineralogy (1967–)
Memoirs of the Faculty of Science, Kyushu University, Series D: Geology (1940–)
Memoirs, Geological Society of Japan (1976–)
Report of the Institute of Geoscience, University of Tsukuba (1974)
Science Report of the Faculty of Geology, Kyushu University, Geology (1950–)
Science Reports of the Institute of Geoscience, University of Tsukuba, Section B: Geological Sciences (1980–)
Science Reports of the Tohoku University, Second Series: Geology (1912–) and *Third Series: Petrology Mineralogy and Economic Geology* (1921–)
Science Reports of the Yokohama National University, Section 2: Biological and Geological Sciences (1952–).
Literature is indexed in *Bibliography for the Geology of Japan, 1873–1955* (Chijin Shokan, 1956) and in *Bibliography of Palaeontology of Japan*, edited by K Kanmera and H Ujiie of which *1941–1950, 1951–1960 and 1961–1975* have been published in *Special Papers, Palaeontological Society of Japan*, nos 1 (1951), 9 (1962) and 22 (1978) respectively.

Jordan

Publications on this country are *Geologie von Jordanien*, by F Bender (*Beitrage zur Regionalen Geologie der Erde*, **7**, 1968), with an English translation *Geology of Jordan* published in 1974; and also by Bender, *Geology of the Arabian Peninsula: Jordan* (*US Geological Survey, Professional Paper*, **560-I**, 1975) and, more recently, *Geology of Jordan*, edited by A M Abed and H M Khaled (Jordanian Geological Association, 1983), which is the proceedings of the 1st Jordanian Geological Conference, held in 1982 have been published.
For retrospective reference, there is the series *Bibliography of Geoscientific Literature of the Dead Sea* (*Jordan Natural Resources Authority, Library Bibliography Series*, **1–6**, 1969–1972).

Korea

Recent books on the country are *The Geology of South Korea*, by A J Reedman and S H Um (Geological and Mining Institute of Korea, 1975), *Marine Geology of Korean Seas*, by Sung Kwun Chough (Reidel, 1983) and *Synthetic Research for Geology of Korea* (Korean Institute of Energy and Resources (KIER), 1983).
Current journals are *Bulletin, Geological Survey of Korea* (1957–), *Journal of the Geological Society of Korea* (1965–), *Report, Korea Institute of Energy and Resources* (1981–) and, from the same institution, *KIER Bulletin* (1978–) and *KIER Miscellaneous Reports* (1979–).

Kuwait

Geology of the Arabian Peninsula: Kuwait, by D I Milton (*US Geological Survey, Professional Paper*, **560-F**, 1967) is the only publication which covers the geology of the area.

Laos see under **Vietnam**

Lebanon

Geologie von Syrien und dem Libanon, by R Wolfart (*Beitrage zur Regionalen Geologie der Erde*, **6**, 1967) is the reference book for this country, and its literature is comprehensively covered in *The Geology of Lebanon: bibliography 1772–1984*, compiled by A Arad and A Ehrlich (Israel Geological Survey, 1985).

Malaysia

Books on Malaysia include *Geology of the Malay Peninsula: West Malaysia and Singapore*, edited by D J Corbett and C S Hutchinson (Wiley, 1973) and *Proceedings of a Workshop on Stratigraphic Correlation of Thailand and Malaysia* (Geological Society of Malaysia and Geological Society of Thailand, 1983).

Current journals are:
District Memoirs, Geological Survey of Malaysia (1937–)
Economic Bulletin, Geological Survey of West Malaysia (1958–)
Geological Papers, Geological Survey of Malaysia (1972–)
Professional Paper, Geological Survey of Malaysia (1962–)
Bulletin, Geological Society of Malaysia (1968–)
Warta Geologi (1975–).

The older literature is indexed in *Bibliography and Index of Geology of West Malaysia*, by D J Corbett (*Bulletin of the Geological Society of Malaysia*, **2**, 1968).

Mongolia (Outer)

The work for the region is *Geologiya Mongol'skoi Narodnoi Republik; the geology of the Peoples' Republic of Mongolia*, edited by N A Marinov (3 vols, Nedra, Moscow, 1973–1977).

Nepal

Books on the country are *Geology of Nepal (West of Nepal, Himalaya)*, by J M Remy (CNRS, 1975), *Geology of Nepal*, by C K Sharman (2nd edn, Education Enterprises, Katmandu, 1977) and *Etudes sur le Quaternaire de l'Himalaya: la haute vallée de la Buri Gandaki, Népal*, by M Fort (CNRS, 1979) in the series *Cahiers Népâlais*. In the same series is the publication, *Bibliographie du Népal*, compiled by I Cloitre Trincano; volume **3**, tome 3, is entitled *Géologie de l'Himalaya Central* (CNRS, 1984).

The current journal published is the *Journal, Nepal Geological Society* (1981–).

Oman

Geology of the Oman Mountains, by K W Glennie *et al* (*Verhandelingen, Nederlandsch Geologisch Mijnbouwkundig Genootschaap*, **31**, 1974) and *Geology and Mineral Resources of the Trucial Oman States*, by J E G W Greenwood (Institute of Geological Sciences, 1968) are two publications covering the geology of the area.

The literature is reviewed in *Bibliography of the Geologic Literature of the Sultanate of Oman, part I: 1824–1985*, by H Weier and G Stringer (*Zentralblatt fur Geologie und Palaontologie*, Teil **1**, (7/8), 779–798, 1986).

Pakistan

Publications on the geology of Pakistan are *Stratigraphy of Pakistan*, edited by S M I Shah (*Geological Survey of Pakistan, Memoir*, **12**, 1977) and *A Guide to the Stratigraphy of Pakistan*, by M W A Iqbal and S M I Shah (*Geological Survey of*

Pakistan, Records, **53**, 1980). *Stratigraphic Correlation of Turkey, Iran, Pakistan*, edited by S M I Shah and A M Quennel (2 vols, Overseas Development Agency, 1980) and *Geodyamics of Pakistan*, by A Farah and K A de Jong (Geological Survey of Pakistan, 1979) are two more specialized works.

Journals currently published are:
Contributions to the Geology of Pakistan (1980–)
Geological Bulletin of the Punjab University (1961–)
Geological Bulletin, University of Peshawar (1966–)
Geonews (1967–)
Memoirs of the Geological Survey of Pakistan (1956–)
Records of the Geological Survey of Pakistan (1948–).

The literature is indexed in the following bibliographies: *Preliminary Bibliography and Index of the Geology of Pakistan (Geological Survey of Pakistan, Records*, **12**, 1965) and *Bibliography of the Geology of Southern Pakistan*, by F Hussein and Z Ahmed (*Geological Survey of Pakistan, Records*, **54**, 1980).

Philippines

Current journals are *Report of the Bureau of Mines, Philippines* (1952–) and *Journal of the Geological Society of the Philippines* (1947–).

Documentation of the geological literature of Philippines is in the following publications: *Bibliography on Philippine Geology, Mining and Mineral Resource*, compiled by J Teves (Philippine Bureau of Mines, 1953) and, a continuation volume, *1953–1965*, compiled by B T Aquino and L G Santos (Philippine Bureau of Mines, 1971); there is also *Bibliography of Philippine Palaeontology and Stratigraphy, 1861–1957*, compiled by B A Daleon (*Philippine Geologist*, **12**, 16–32, 1967).

Qatar

The geology is outlined in the short publication *Geology of Qatar*, commissioned by Qatar Department of Petroleum Affairs (Schlumberger, 1981).

Saudi Arabia

Publications on the country are *Evolution and Mineralization of the Arabian-Nubian Sheild*, edited by A M S Al Shanti (4 vols, Pergamon, 1979–1980), *Geodynamic Evolution of the Afro-Arabian Rift System (Atti dei Convegni Lincei*, **47**, 1980), *Geology of the Arabian Peninsula: sedimentary geology of Saudi Arabia*, by R W Powers *et al* (*US Geological Survey, Professional Paper*, **560-D**, 1966) and *Quaternary Period in Saudi Arabia*, edited by S S Al-Sayani and J G Zotl (2 vols, Springer, 1978).

Annotated Bibliography, (Pre 1970) (Mineral Resources Bulletin, **19**, 1980), *1970–1975 (Mineral Resources Bulletin*, **20**, 1977) and *1976–1980 (Mineral Resources Bulletin*, **27**, 1981) is a useful series in which the geological literature is indexed. The Ministry of Petroleum and Mineral Resources, Saudi Arabia, publishes the *Mineral Resources Bulletin*.

Singapore

Geology of the Republic of Singapore (Public Works Department, 1976) is the most recent publication of the area.

Sri Lanka

The comprehensive works and bibliographies of the geology of Sri Lanka are not recent. They are *An Introduction to the Geology of Ceylon*, by P G Cooray (*Spolia*

Zeylanica, **31**(1), 1967) and, even older, *Bibliography of Ceylon Geology*, by D N Wadia (Department of Mineralogy, Colombo, 1943).

The only geological journals are *Professional Paper, Geological Survey Department, Sri Lanka* (1943–) and the *Administration Report of the Director of the Geological Survey* (1961–).

Syria

Geologie von Syrien und dem Libanon, by R Wolfart (*Beitrage zur Regionalen Geologie der Erde*, **6**, 1967) is still the comprehensive work for the country, and current papers are published in *Syrian Geological Magazine* (1978–).

Geological literature is indexed in *The Geology of Syria: bibliography*, compiled by V Arad (Israel Geological Survey, 1985).

Taiwan

The geology is outlined in *An Introduction to the Geology of Taiwan*, by C S Ho (Ministry of Economic Affairs, 1975).

The current geological journals are:
Acta Geologica Taiwanica (1947–)
Bulletin of the Geological Survey of Taiwan (1947–)
Memoirs, Geological Society of China, Taipei (1962–)
Petroleum Geology of Taiwan (1962–)
Proceedings of the Geological Society of China, Taipei (1957–)
Ti-chih (1978–).

Thailand

Publications for Thailand are *Proceedings of the Conference on the Geology of Thailand*, edited by R B Stokes *et al* (*Special Publications, Department of Geology, Chiang Mai University*, **1**, 1971), *Proceedings of a Workshop on Stratigraphic Correlation of Thailand and Malaysia* (Geological Society of Malaysia and Geological Society of Thailand, 1983), *Outline of the Geology and the Mineral Potential of Thailand*, by L Hahn *et al* (*Geologisches Jahrbuch*, Reihe B, **59**, 1986) and *Geology and Mineral Deposits of Thailand*, by D R Shawe (*US Geological Survey, Open File Report*, **84–403**, 1984).

The journals for Thailand are *Journal of the Geological Society of Thailand* (1975–) and *Memoirs, Geological Survey of Thailand* (1966–).

Geological literature is indexed in *Bibliography of the Geology and Mineral Resources of Thailand, 1951–1977 (Proceedings of the 3rd Regional Conference on Geology and Mineral Resources of Southeast Asia*, 805–843, 1978) and the earlier literature is to be found in *Geology of Thailand, 1916–1962* and *Palaeontology of Thailand, 1916–1962*, by T Kobayashi (*Geology and Palaeontology of South-east Asia*, **1**, 1–15, 17–29, 1964).

Turkey

Books on Turkish geology are *Geology of Turkey*, by R Brinkman (Enke, 1976), *Geology and History of Turkey*, edited by A S Campbell (Petroleum Exploration Society of Libya, 1971), *An Introduction to the Petroleum Geology of Southeast Turkey*, compiled by L Erdogan and A Akgul (Chamber of Turkish Geological Engineers, 1983) and *Stratigraphic Correlation of Turkey, Iran, Pakistan*, edited by S M I Shah and A M Quennel (2 vols, Overseas Development Agency, 1980).

Current journals are:
Aegean Earth Sciences (1981–)
Bulletin of the Mineral Research and Exploration Institute of Turkey (1936–)
Jeomorfoloji Dergisi (1969–)

Maden Tetkik ve Arama Enstitusu Yayinlaridan (1937–)
Turkiye Cumhuriyetinde Jeolojik Gorumler (1936–)
Turkiye Jeoloji Kurumu Bulteni (1947–)
Turkye Jeomorfologlar Dernegi Yayini (1973–)
Yerbilimleri Enstitusu Yayin Organi Hacettepe Universitesi (1976–).

United Arab Emirates see Saudi Arabia

Vietnam, Cambodia, Laos

The geology of these countries is outlined in *Geology of Laos, Cambodia, South Vietnam and the eastern part of Thailand*, by D R Workman (*Overseas Geology and Mineral Resources*, **50**, 1977), and covered by the bibliography *Bibliography of the Geology and Mineral Resources of Democratic Kampuchea, the Lao Peoples' Democratic Republic and the Socialist Republic of Viet-Nam 1951–1975 (with supplement 1976–1977)*, compiled by the Committee for Coordination of Investigation of the Lower Mekong Basin, 1977.

Yemen

A brief description of this country is in *Geology of the Arabian Peninsula: Yemen*, by F Geukens (*US Geological Survey, Professional Paper*, **560-B**, 1966).

EUROPE

The recent books for Europe are *The Geology of Europe*, by D V Ager (McGraw-Hill, 1980), *Europe from Crust to Core*, edited by D V Ager and M Brooks (Wiley, 1977), *The Structure of Western Europe*, by J G C Anderson (Pergamon, 1987), *Géologie de l'Europe*, by J Cogne and M Slansky (*Mémoires, BRGM*, **108**, 1980), and by the same authors in the same series, *Evolution Géologique de l'Europe*, (**107**, 1980), *Geology and Landscape in Britain and Europe*, by D John *et al* (Oxford University Press, 1983) and *Mineral Deposits of Europe* published by The Institute of Mining and Metallurgy and the Mineralogical Society from 1978 onwards.

Géologie des Pays Européens: Geology of the European Countries is a compilation of the field trip guidebooks issued for the 26th International Geological Congress in Paris, 1980, in cooperation with the Comité National Français de Géologie, which covers the geology of most of Europe in four volumes: *France, Belgique, Luxembourg; Espagne, Grèce, Italie, Portugal, Yougoslavie; Denmark, Finland, Iceland, Norway, Sweden*; and *Austria, Federal Republic of Germany, Ireland, the Netherlands, Switzerland, United Kingdom* (Dunod and Graham & Trotman).

The series *Sammlung Geologische Fuhrer* (1897–) provides useful guides to individual areas of Europe, whilst a periodical for Europe is *Terra Cognita* (1981–) published by the European Union of Geosciences.

For stratigraphic nomenclature, the relevant parts of the *Lexique Stratigraphique International* are in volume **1**: fasc. 1a, Greenland; 1b, Iceland; 1c, Faroe Islands; 1d, Svalbard; 2a, Norway; 2b, Finland; 2c, Sweden; 2d, Denmark; 3a, England, Wales and Scotland; 3b, Ireland; 4a, France, Belgium, Netherlands, Luxembourg; 5, Germany, 6a, Poland; 6b, Czechoslovakia; 7a-c, Switzerland; 8, Austria; 9, Hungary; 10a, Spain; 10b, Portugal; 11, Italy; 12a, Yugoslavia; 13a, Rumania; 13b, Bulgaria.

Reference works include *Sources of Information on the Geology of the Continental EEC Countries (Geological Society of London, Miscellaneous Paper*, **12**, 1980) and *Annotated Bibliographies of Mineral Deposits of Europe*, compiled by J D Ridge (Pergamon, 1984).

Albania

Major publications are not very recent; they are *Géologie de l'Albanie* (Albanian Institute of Geology, 1970), and *Regard sur la Géologie de l'Albanie et sa place dans la géologie des Dinarides*, by J Aubouin and I Ndojaj (*Bulletin, Société Géologique de France*, ser 7, **6**(5), 593–625, 1964).

The periodical publications are *Buletini Shkencave Gjeologijke* (1983–) and *Permbledhje Studimesh* (1965–) of the Institutu Studimeve dhe Projectimeve Gjeologo-Minerale, Tirane.

For the documentation of Albanian geology, there is *Bibliografie Botimeve Shgipe per Gjelgjine Minierat dhe Nafter 1944–1969*, compiled by V Meko and A Papa (*Permbledhje Studimesh*, **2**(15), 129–156, 1970), which supplements *Bibliografi Geologico e Geografico-fisica della Regionale Albanese*, by M Magnani (2nd edn, Reale Ufficio Geologico d'Italia, 1941).

Alps

Books and major articles on the Alps include *Géologie des Chaines Alpine Issues de la Tethys*, by J Aubouin *et al* (*Mémoires BRGM*, **115**, 1980), *Kleine Geologie der Ostalpen*, by H Bogel and K Schmidt (Ott, 1976), *Alpine Structural Elements: Carpathian-Balkan-Caucasus-Pamir Orogene Zone*, edited by M Mahel' (Veda, 1982) and the important work *Geological Atlas of Alpine Europe and Adjoining Alpine Areas*, edited by M Lemoine (Elsevier, 1978). Sections on Alpine geology can also be found in regional geological works of those countries in which the mountain rarge occurs.

The specialist journal for this area is *Géologie Alpine*, formerly *Travaux du Laboratoire de Géologie de la Faculté de Science de l'Université de Grenoble* (1985–).

Apennines

The geology of the region can be found in *Atti del Convegno sul Tema Moderne Vendute sulla Geologia dell'Apennino* (Accademia Nazionale de Lincei, 1973) and in *Development of the Northern Apennines Geosyncline* (*Sedimentary Geology*, **4**, 1970).

Austria

Recent books published are *Der Geologische Aufbau Osterreiches*, edited by R Oberhauser, (Springer, 1980) and *Abriss der Geologie von Osterreich*, by V del Nego (Geologische Bundesanstalt, 1977). An older, but substantial reference book, is *Geologie von Osterreich*, by F X Schaffer (2nd edn, Deuticke, 1951).

The current journal publications are the *Abhandlungen* (1852–), *Arbeitstagung* (1977–), and the *Jahrbuch* (1850–) of the Geologische Bundesanstalt; *Mitteilungen der Osterreichischen Geologischen Gesellschaft* (1908–) and the *Geologisch-Palaontologisch Mitteilungen Innsbruck* (1971–). Pre-1918 issues of the above journals also contain literature on the constituent countries of the Austro-Hungarian Empire.

The geological literature of Austria was documented in *Geologische Literatur in Osterreich 1945–1950* (*Verhandlungen der Geologischen Bundesanstalt, Sonderband* **B**, 1951), and this was regularly updated annually in issues of the *Verhandlungen* until just before the journal's cessation in 1982. Recent literature is indexed in *Bibliographie Geowissenschaftliche Literatur uber Osterreich fur die Jahre 1979–1983*, published in 1985 by the Geologische Bundesanstalt.

Azores see **Portugal**

Balearic Islands see **Spain**

Balkans

Journals covering the Balkans as a whole are the *Geoloski Anali Balkanskoga Poluostrva (Annales Géologiques de la Peninsule Balkaniques)* (1889–) and *Geologica Balcanica* (1949–).

Belgium

A recent book on the country is *Belgique*, by F Robaszynski and C Dupuis from the series *Guides Géologiques Régionaux* (Masson, 1983). Older works include *Géologie de la Belgique: une introduction*, by A Lombard (Naturalistes Belges, 1958) and *Prodrome d'une Description Géologique de la Belge*, edited by P Fourmarier (Société Géologique de Belge, 1954).

Journals currently published are:
Aardkundige Mededelingen, Leuven (1980–)
Annales de la Société Géologique de Belge (1874–)
Annales de la Société Géologique de Belge du Nord (1870–)
Bulletin Cercle de Géologique de Belgique (1974–)
Bulletin de l'Institut Royal des Sciences de Belgique, Sciences de la Terre (1972–)
Bulletin de la Société Belge de Géologie (1887–)
Bulletin Société Royale Belge d'Etudes Géologiques et Archéologiques (1957–)
Mémoires de l'Institut Géologique de l'Universite de Louvain (1913–)
Mémoires de la Société Géologique de Belge (1898–).

The literature of Belgian geology is indexed in *Abstracts of Belgian Geology and Physical Geology* (1967–).

Bulgaria

The only comprehensive work is *Geolgiya na Bulgariya*, by E Boncev (2 parts, Nauki i Izkustvo, 1955–1960). A useful, more recent article is *Geology of Bulgaria: a review*, by R M Foose and F Manheim (*Bulletin American Association of Petroleum Geologists*, **59**(2), 303–335, 1975).

The major journal for the country is *Izvestiya Geologicheskiya Institut Bulgarska Akademiya* (1951–), published in several series; since 1974, each series is published under its own title, namely *Geokhimiya, Mineralogiya i Petrologiya, Geotektonika, Tektonofizika i Geodinamika, Inzhenerna Geologiya i Khidrogeologiya, Paleontologiya, Stratografiya i Litologiya* and *Rudoobrazuvatelni Protsesi i Mineralni*. Other journals include *Godishnik na Sofijskiya Universitet Geologo-Geografski Fakultet* (1965–), and *Spisanie na Bulgarskoto Geologichesko Druzhestvo* (1927–).

There is, however, no shortage of publications, which are indexed in *Abstracts of Bulgarian Scientific Literature: Geosciences* published quarterly since 1957. Publications prior to this date are contained in the *Bibliography of Geological Publications in Bulgaria, 1828–1964*, by C Spasov *et al* (Bulgarska Akademiya na Naukite, 1978).

Czechoslovakia

Regional Geology of Czechoslavkia, edited by M Mahel and T Buday (Ustredni Ustav Geologicky, 1968) is a comprehensive work on the country in two volumes: **1**, *The Bohemian Massif* and **2**, *The West Carpathians*. More recent publications are *Czechoslovak Geology and Global Tectonics*, edited by M Mahel and P Reichwalder (Veda, 1979) and *Geological history of the territory of the Czech Socialist Republic*, by M Suk *et al* (Czechoslovak Academy of Sciences, 1984).

The journal publications for Czechoslovakia are:
Acta Geologica et Geographica Universitatis Comeniae, Geologica (1958–)
Acta Universitatis Carolinae, Geologica (1954–)
Casopis pro Mineralogii a Geologii (1956–)

Folia Musei Rerum Naturali Bohemiae Occidentalis, Geologia (1972–)
Geologicke Prace, Zpravy (1954–)
Geologicke Sbornik, Monografica Seria (1958–)
Geologicky Pruzkum (1959–)
Geologicky Sbornik (1950–)
Kninovna Ustredniho Ustav Geologickeho (1949–)
Nauki o Zemle, Seria Geologica (1965–)
Rozpravy Ustredniho Ustav Geologickeho (1950–)
Sbornik Geologickych Ved, Rada Geologie (1963–)
Rada Zk of the same series, (1964–), is now known as *Zapadne Karpaty* and has
been subdivided into several series: *Seria Mineralogia, Petrografia, Geochemia,
Metalogeneza* (1977–), *Seria Hydrogeologia a Inzinierska Geologia* (1974–), *Seria
Geologia* (1976–), and *Seria Paleontologia* (1975–)
Vestnik Ustredniho Ustavu Geologickeho (1951–).

The bibliography for Czech geological and mineralogical works is the
Mineralogicko-Geologicka Bibliographie CSR, published annually since 1955 by
the state survey, Ustredni Ustav Geologicky. Earlier literature is indexed under the
same title: *1897–1918* (published 1969) and *1919–1927* (published 1970).

Denmark

A recent publication is *The Geology of Denmark*, volume **1**: *The Development of
Denmark since the last Glacial* (Dansk Geologisk Undersogelse, 1973).

Despite the lack of large monographic works on Denmark, there is no shortage
of papers from the journal publications. The current state geological publications
include *Arbog Danmarks Geologiske Undersogelse* (1972–), *Danmarks Geologiske
Undersogelse Series, A* (1976–), *B* (1977–) and *C* (1984–) and *Rapport Danmarks
Geologiske Undersogelse* (1968–). Societies' publications are *Arsskrift Dansk
Geologisk Forening* (1969–) and *Bulletin of the Geological Society of Denmark*
(1894–).

Finland

Books on the geology of Finland are few and not so recent: *Suomen Geologia*,
compiled by K Rankama (Kirjaytyma, 1964) and *Pre-Quaternary Rocks in Finland*,
by A Simonen (*Bulletin de la Commission Géologique de la Finlande*, **191**, 1960).

The state geological survey publication is *Bulletin of the Geological Survey of
Finland* (1895–). The Geological Society of Finland, publishes the *Bulletin of the
Geological Society of Finland* (1968–) and *Geologi* (1949–). Another journal is
*Suomen Tiedeakatemian Toimituksia, Sarja III: Geologica-Geographica (Annales
Academiae Scientiarum Fennicae)* (1942–).

Literature is indexed in *Geologische Bibliographie Finnlands 1555–1933*,
compiled by A Laitakari (*Bulletin de la Commission de la Finlande*, **108** & **231**,
1943–1968), and *Geological Bibliography of Finland 1934–1970: an author list*, by
the same compiler (*Bulletin Geological Survey of Finland*, **270**, 1975); also there is
Guide to the Publications of the Geological Survey of Finland 1879–1960, by M
Okko and M Hannikainen (Geologinen Tutkimuslaitos, 1960).

France

The books covering the geology countrywide are *Géologie de la France*, by
J Debelmas *et al* (2 vols, Doin, 1974) and *Geology of France*, edited by C Pomerol
et al from the series *Guides Géologiques Régionaux* (Masson, 1980), and *Terroirs et
Vins de France: itinéraires oenologiques et géologiques*, edited by C Pomerol *et al*
(2nd edn, TOTAL & BRGM, 1983).

Two major series covering all regions of the country are *Géologie Regionale de la*

France (Hermann, 1942–) and *Guides Géologiques Régionaux* (Masson, 1969–): the latter also includes neighbouring areas, e.g. Belgium and Alpine regions, and former French territories overseas e.g. Martinique and Guadeloupe.

The major publisher of French geological literature is the Bureau de Recherches Géologiques et Minières (BRGM). Its journal publications are *Annales du Services d'Information Géologique du Bureau de Recherches Géologiques, Géophysiques et Minières* (1953–), *Bulletin du Bureau de Recherches Géologiques et Minières* (1961–) with its present component series: serie 2, section 2: *Géologie de la France* (1980–) and section 4: *Géochronique* (1980–); *Documents, Bureau de Recherche Géologiques et Minières* (1978–) and *Mémoires du Bureau de Recherches Géologiques et Minières* (1960–).

The journal publications of the Société Géologique de France are *Bulletin de la Société de Géologique de France* (1830–), *Mémoires de la Société Géologique de France* (1833–) and its *Hors-Series* (1962–).

Other current journal publications of France, both national and regional, are:

Annales Scientifiques de l'Université de Besançon, Géologie (1950–)

Annales de la Société Géologique du Nord (1870–)

Bulletin Association Géologique Auboise (1978–) and its series *Numéro Spécial* (1978–)

Bulletin de Centre de Recherche Exploration-Production Elf-Aquitaine, Pau (1967–) and its *Mémoire* (1979–)

Bulletin d'Information de Géologues du Bassin de Paris (1964–) and its *Hors-Séries* (1980–)

Bulletin de l'Institut de Géologie du Bassin d'Aquitaine (1966–)

Bulletin de Liaison Société Géologique de Normandie et des Amis du Musée du Harve (1978–)

Bulletin de la Société Géologique et Minéralogique de Bretagne (1920–)

Cahiers Géologiques (1950–)

Cahiers de l'Office de la Recherche Scientifique et Technique Outre-Mer, série Géologie (1969–)

Documents de Laboratoire de Géologie, Lyon (1962–) and its *Hors Séries* (1973–)

Documents et Travaux de l'Institut Géologique Albert de Lapparent (1980–)

Géobios (1968–) and its *Hors Mémoires Spéciales* (1977–)

Mémoires Géologiques de l'Université de Dijon (1973–)

Mémoires de l'Institut du Bassin Aquitaine (1971–)

Lithos (1980–)

Mémoires du Musée National d'Histoire Naturelle, série C: Sciences de la Terre (1950–)

Mémoires de la Société Géologique et Minéralogique de France (1924–)

Reunion Annuelle des Sciences de la Terre (1973–)

Revue de Géologie Dynamique et de Géographie Physique (1958–)

Revue de l'Institut du Pétrole (1946–)

Sciences Géologiques, Bulletin (1972–) and *Mémoire* (1971–)

Sciences de la Terre (1948–) and its *Informatiques Géologiques* (1973–) and *Mémoires* (1962–)

Travaux du Département de Géologie de l'Université de Picardie (1984–)

Travaux Laboratoire de Géologie, Ecole Normale Superieure, Paris (1967–)

Travaux du Laboratoire de Géologie Historique et de Paléontologie, Universite de Provence (1938–).

Germany

Books on Germany include *Geolgie von Deutschland und einigen Randgebeiten*, by G Knetsch (Schweizerbart'sche, 1963) and more recently, but more specialized, *Nordwestdeutschlands im Tertiar*, by H Tobien (*Beitrage zur Regionalen Geologie der Erde*, **18**, 1986).

The current journal publications for Germany are:

Abhandlungen des Geologischen Landesamtes in Baden-Wurttenburg (1953–)

Abhandlungen des Staatlichen Museums fur Mineralogie und Geologie zu Dresden, formerly *Jahrbuch* (1954–)

Arbeiten aus dem Institut fur Geologie und Palaontologie, Universitat Stuttgart (1947–)

Aufschluss (1950–)

Beihefte zur Zeitschrift Geologie (1952–)

Berliner Geowissenschaftliche Abhandlungen (1977–)

Clausthaler Geologische Abhandlungen (1975–) and its *Sonderband* (1975–)

Clausthaler Tektonische Hefte (1965–)

Erlanger Geologische Abhandlungen (1952–)

Forschungsarbeiten und Publikationen Geologisch-Palaontologisches Institut und Museum der Christian-Albrechts- Universitat, Kiel (1981–)

Forschungsbericht Geologisch-Palaontologisches Institut Universitat Frankfurt-am-Main (1976–)

Fortschritte in der Geologie von Rheinland (1958–)

Frankfurter Geowissenschaftliche Arbeiten, Serie A: Geologie- Palaontologie (1982–)

Freiberger Forschungshefte, Reihe C (1951–)

Fundgrube (1965–)

Geologica Bavarica (1949–)

Geologica et Palaeontologica (1967), and its *Sonderband* (1972–)

Geologie und Palaontologie in Westfalen (1983–)

Geologische Abhandlungen Hessen (1950–)

Geologische Blatter Nordost Bayern und Angrenzende Gebiete (1951–)

Geologische Rundschau (1910–)

Geologisches Jahrbuch (1943–), now in six series: *Reihe A: Allgemeine und Regionale Geologie Deutschland* (1972–), *Reihe B: Regionale Geologie Ausland* (1972–), *Reihe C: Hydrogeologie Ingenieurgeologie* (1972–), *Reihe D: Mineralogie Petrographie Geochimie Lagerstattenkunde* (1972–), *Reihe E: Geophysik* (1973–) and *Reihe F: Bodenkunde* (1973–)

Geologisches Jahrbuch Hessen (1950–)

Gottinger Arbeiten zur Geologie und Palaontologie (1969–)

Hallesches Jahrbuch fur Geowissenschaften (1949–)

Kleine Senckenberg-Reihe (1971–)

Jahrbericht und Mitteilungen des Oberrheinischen Geologischen Veriens, Neue Folge (1911–)

Jahrbericht des Niedersachischen Geologischen Vereins (1911–)

Jahresheft des Geologischen Landesamtes in Baden-Wurttemburg (1955–)

Mainzer Geowissenschaftliche Mitteilungen (1972–)

Meyniana (1952–)

Mitteilungen Alfred-Wegener-Stiftung (1981–)

Mitteilungen der Bayerischen Staatsammlung fur Palaontologie und Historische Geologie (1961–)

Mitteilungen aus dem Geologisch-Palaontologisch Institut dem Universitat Hamburg (1935–)

Mitteilungen aus dem Geologischen Institut der Universitat Hannover (1968–)

Munchener Geowissenschaftliche Abhandlungen (1984–)

Munstersche Forschungen zur Geologie und Palaontologie (1965–)

Nachrichten Deutsche Geologische Gesellschaft (1970–)

Neues Jahrbuch fur Geologie und Palaontologie (1833–), now in series *Abhandlungen* (1943–) and *Monatshefte* (1950–)

Oberrheinisches Geologische Abhandlungen (1929–)

Senckenbergiana Lethaea (1954–)

Stuttgarter Beitrage zur Naturkunde, Serie B: Geologie und Palaontologie (1972–)
Zeitschrift fur Angewandte Geologie (1955–)
Zeitschrift der Deutschen Geologischen Gesellschaft (1849–)
Zeitschrift fur Geologische Wissenschaften (1973–)
Zentralblatt fur Geologie und Palaontologie, Stuttgart (1950–)
Zitteliana (1969–).

Bibliographies include *Zur Geschichte der Geologie, Geophysik, Mineralogie und Palaontologie: Bibliographie und Reperatorium fur Deutsche Demokratische Republik*, by P Schmidt (*Veroffentlichungen der Bibliothek der Bergakademie Freiberg*, **40**, 1970) with an update for the literature of 1970–1976 (**71**, 1978); bibliographies for individual regions of Germany can also be found in the journal *Abhandlungen Zentrales Geologisches Institut*, and contains, notably, *Bibliographie der Geologischen Wissenschaften fur die Deutsche Demokratische Republik*, edited by H Lange and C Schroter (**16**, 1967–).

Greece

The most recent and comprehensive work on Greece is *Geologie von Greichenland*, by V Jacobshagen (*Beitrage zur Regionalen Geologie der Erde*, **19**, 1986).

The geological journals of Greece are *Bulletin of the Geological Society of Greece* (1953–), *Eidikai Melatai epi tes Geologica tes Ellados [The Geology of Greece]* (1951–), *Geologikai kai Geofizikai Meletai (Geological and Geophysical)* (1951–); and *Annales Géologiques des Pays Helléniques* (1947–).

Geological literature is documented in *Geological and Physicogeographic Bibliography of Greece* (Later title *Geoscience and Natural Science: Bibliography of Greece*) by D Haralambous. Volume **1** covers the period 1500–1959 (Institute for Geology and Subsurface Research, 1961), volume **2**, 1960–1973 and volume **3**, 1974–1979 (National Institute of Geological and Mining Research, 1975 & 1980).

Hungary

A geology of the country can be found in *Geologie von Ungarn*, by L Trunko (*Beitrage zur Regionalen Geologie der Erde*, **8**, 1969).

The current journals are:
Acta Geologica, Academiae Scientiarum Hungaricae (1952–)
Annales Universitatis Scientiarum Budapestinensis de Rolando Eotvos Nominate, Sectio Geologica (1957–)
Evi Jelentese a Magyar Allamai Foldtani Intezet (1925–)
Evkonyve a Magyar Allami Foldtani Intezet (1871–)
Foldtani Kozlony (1858–)
Geologica Hungarica, Series Geologica (1952–).

The literature of Hungary is indexed annually in *A Magyarorszagm Megjelent Foldtani Irodatm Szakbibliografija (Bibliography of Geological Literature published in Hungary)* (Magyar Allami Foldtani Intezet, 1965–). Under varying titles, listings have been published in *Foldtani Kozlony* since 1900.

Iceland

Recent major publications are *Iceland : a guidebook for a geology study tour*, by J W Perkins (University College Cardiff, 1983), *Geology of Iceland*, by K Saemundsson *et al* (Special Issue of *Jokull*, **29**, 1979) and *Iceland: evolution, active tectonics and structure*, by W Jacoby *et al* (*Journal of Geophysics*, **47**(1–3), 1980).

Ireland see also **United Kingdom for literature on British Isles, and United Kingdom: Northern Ireland**

Books on Ireland include *A Geology of Ireland*, by C H Holland (Scottish Academic Press, 1981), *Quaternary History of Ireland*, by K S Edwards and W P Warren (Academic Press, 1985) and *Geology and Scenery in Ireland*, by J B Whittow (Penguin Books, 1974).

For the coasts and seas around Ireland, there is the recent publication *Geology of Offshore Ireland and West Britain*, by D Naylor and P Shannon (Graham and Trotman, 1982) and *The Geology of the South Irish Sea*, by M R Dobson *et al* (*Institute of Geological Sciences, Report* **73/11**, 1973).

The Geological Survey of Ireland publishes the following journals:
Bulletin (1970–)
Guide Series (1976)
Memoirs (1858–)
Report Series (1975–)
Special Papers (1971–).
The *Irish Journal of Earth Sciences* (1978–) (called the *Journal of Earth Sciences Royal Dublin Society* until recently) is another current journal.

Italy

The major work on the area is *Geology of Italy*, edited by C H Squyres (2 vols, Earth Sciences Society of Libya, 1975). Other books include *Geologia dell'Italia*, edited by A Desio *et al* (Union Tipografica — Editrice Torinese, 1973), and *L'Italia Geologica: storia degli ultimi 230 milioni di anni*, by G Flores and M Pieri (Longanesi, 1981).

The current journal publications of Italy are:
Atti dell'Istituto di Geologia della Universita di Genova (1963–)
Atti dell'Istituto Geologico della Università di Pavia (1943–)
Bollettino del Servizio Geologica d'Italia (1845–)
Bollettino della Società Geologica Italiana (1882–)
Geologia Applicata e Idrogeologica (1966–)
Geologica Romana (1962–)
Giornale de Geologia (1926–)
Memorie e Note dell'Istituto di Geologia Applicata dell'Università degli Studi, Napoli (1947–)
Memorie di Scienze Geologiche (1912–)
Memorie della Società Geologica Italiana (1833–)
Publicazioni dell'Istituto di Geologia e Paleontologia dell'Università di Cagliari (1966–)
Quaderni dell'Istituto di Geologia dell'Universita di Genova (1980–)
Rediconti della Società Geologica Italiana (1978–)
Rivista Italiana di Paleontologia e Stratigrafia (1895–), and its *Memoria* (1934–)
Studi Geologici Camerti (1971–)
Studi Trentini di Scienze Naturali, Acta Geologica (1977–).
The literature is documented in *Bibliografia Geologica d'Italia* in 15 volumes, each covering a region of the country (Consiglio Nazionale de le Ricerche, 1956–1964). Pre-1930 literature is listed in *Giornale di Geologia*, **6**(2), 1931.

Luxembourg

Publications are the rather old *Beitrage zur Geologie von Luxembourg*, by M Lucius (2nd edn, *Publications du Service Géologique de Luxembourg*, **2**, 1955) and *Outline of the Geology of the Grand-Duche de Luxembourg*, by R C Mitchell-Thome (*Acta Geologica Academiae Scientiarum Hungaricae*, **20**, 259–286, 1976).

The journal currently published is that of the Service Géologique de Luxembourg, its *Bulletin* (1937–).

Malta

The only comprehensive work is the *Geology of the Maltese Islands*, by H P J Hyde (Lux Press, 1955), the other works available being *Field Guide to the Mid-Tertiary Carbonate Facies of the Maltese Islands*, by D W J Bosence *et al* (Palaeontological Association, 1981), and *An Outline of Maltese Geology and Guide to the Geology Hall of the National Museum of Natural History, Mdina*, by G Zammit-Maempel (1977).

Netherlands

The most complete work is still the older *Geologie van Nederland*, by F J Faber (4 vols, J Noorduijn, 1947–1965). More recent, but a less extensive publication, is *The Geology of the Netherlands*, by C J van Staalduinen *et al* (*Mededelingen Geologische Dienst*, **31–4**, 1979).

Geological journals currently published are:
Geologica Ultraiectina (1957–) and its *Special Publications* (1980–)
Geologie en Mijnbouw (1939–)
GUA Papers of Geology (1979–)
Gea (1971–)
Gronboor en Hamer (1960–)
Jaarverslag van der Rijks Geologischen Dienst (1963–)
Mededelingen van de Nederlandse Geologisch Vereining (1962–)
Mededelingen van's Rijks Geologischen Dienst (1941–)
Mededelingen van de Werkgroep voor Tertiare en Kwartaire Geologie (1964–)
Scripta Geologica (1971–)
Staringia (1971–)
Verhandelingen van het Koninklijk Nederlandsch Geologisch Mijnbouwkundig Genootschap, Geologische Serie (1912–).

Norway

Recent publications include *The Geology of Norway*, by C Oftedahl (*Norges Geologiske Undersogelse*, **356**, 1980), *Geological Field Study Guide to Southern Norway*, by R E Bevins (University College Cardiff, 1984) and *Field Excursion Guide* to the Fourth International Symposium on the Ordovician System in Oslo, edited by D L Bruton and S H Williams (*Paleontological Contributions, Universitet i Oslo*, **279**, 1982).

Journals published on Norwegian geology are *Norges Geologiske Undersogelse* (1981–), and *Norsk Geologisk Tidsskrift* (1905–).

Poland

The work *Geology of Poland*, edited by S Sokolowski (Instytut Geologiczny, 1970–1977), is a translation of *Budowa Geologiczna Polski* (1968); the three volumes cover **1**: Stratigraphy, **2**: Catalogue of Fossils, and **3**: Tectonics.

The current journal publications are:
Acta Geologica Polonica (1950–)
Acta Universitatis Wratislaviensis, Prace Geologiczno-Mineralogiczne (1962–)
Annales Societatis Geologorum Poloniae (1923–)
Annales Universitatis Mariae Curie-Sklodowska, Serie B: Geographa, Geologia, Mineralogia et Petrographia (1946–)
Biuletyn Geologiczny, Universytet Warsawski (1961–)
Biuletyn Instytutu Geologicznego (1938–)

Biuletyn Peryglajalny (1954–)
Bulletin de l'Académie Polonaise des Sciences, Série Sciences de la Terre (1953–)
Geologia Sudetica (1964–)
Kwartalnik Geologiczny (1957–)
Prace Geologiczne, Polska Akademia Nauk, Komisja Nauk Geologicznych (1961–)
Prace Instytut Geologiczny (1921–)
Przeglad Geologiczny (1953–)
Studia Geologica Polonica (1958–)
Zeszyty Naukowe, Akademii Gorniczo-Hutniczei, Geologia (1975–).

Portugal see also **Spain**

Books on Portugal include the recent publications, *Introducção a la Geologia de Portugal*, by C Teixeira and F Goncalves (Instituto Nacional Investigacão Cientifica, 1980) and *Introduction à la Géologie Générale du Portugal*, by A Ribeiro *et al* (Servicos Geologicos de Portugal, 1979).

The current journal publications are:
Boletim do Museu e Laboratorio Mineralogico e Geologico da Universidade di Lisboa (1931–)
Boletim da Sociedade Geologica de Portugal, (1941–)
Ciencias da Terra, Universidade Nova de Lisboa (1976–)
Comunicacões dos Servicos Geologicos de Portugal (1883–)
Garcia de Orta, Seria de Geologia (1973–)
Memorias y Noticias, Publicacões do Museu Mineralogico e Geologico da Universidade de Coimbra (1921–).

The older, geological literature of Portugal is covered by *Geologia de Portugal: ensaio bibliografico*, by L de Menezes Correa Acciaiuoli (2 vols, Direcção General de Minas e Servicos Geologicos de Portugal, 1957).

Pyrenees

Recent literature on the Pyrenees is *The Geological Evolution of the Pyrenees*, edited by E Banda and S M Wickham (*Tectonophysics*, **129**(4), 1986) and *The Geology of the Central Pyrenees*, by H J Zwart (*Leidse Geologische Mededelingen*, **50**, (1), 1979).

Romania

Geology of Romania, by B C Burchfiel and M Bleahu (*Geological Society of America, Special Paper* **158**, 1976) and *Geologia Romaniei*, by V Mutihac and L Ionesi (Technica, 1973) are recent outlines of Romanian geology. An older account can be found in *Geologia Republicii Populare Romine*, by N Oncescu (Technica, 1960).

Current journals for this country are:
Analele Universitatii Bucuresti, Seria Stiintele Naturii, Geologie-Geographie (1964–)
Anuar Institutului de Geologie si Geofizica (1907–)
Dari de Seama ale Sedintelor (1911–)
Mémoires Institut de Géologie et Géophysique, Bucharest (1956–)
Revue Roumaine de Géologie Géophysique et Géographie (1957–)
Studii si Cercetari de Geologie Geofizika Geografie, Seria Geologie (1956–)
Studii Technice si Economicae, Institutu Geologic, Bucuresti, Seria A-J (1952–).

Bibliografia Geologic a Republicii Socialiste Romania was originally published in 1926 (Institutul Geologic si Geofizica), and supplements have been published since in 1929, 1939, 1962, 1975, and 1981.

Scandinavia

Books and major articles on Scandinavia include *The Scandinavian Caledonides*, by T Strand and O Kulling (Wiley, 1972), *Geochemical Prospecting in Fennoscandia*, edited by A Kvalheim (Wiley Interscience, 1967) and *North Atlantic and Fennoscandia*, edited by R Meissner (*Geojournal*, **3**, (3), 1979).

Spain

A recent publication *Geologia de España: libro jubliar J M Rios*, edited by J A Comba (Instituto Geologico y Minero de España, 1983), covers the geology of the country in 3 volumes: **1** & **2** are entitled *Geologia de España* and **3** is on special themes.

Publications on the stratigraphy of the Iberian Peninsula are *Contributions to the Carboniferous Geology and Palaeontology of the Iberian Peninsula*, edited by M J Lemos de Sousa (Universidade de Porto, Faculdade de Ciencias Mineralogia y Geologia, 1983) and *Cretacio de la Peninsula Iberica*, by the Grupo Español de Mesozoico (*Cuadernos de Geologia Iberica*, **8**, 1982).

The current journals for the geology of Spain are:
Acta Geologica Hispanica (1966–)
Boletin Geologico y Minero (1967–)
Boletin Servicio Geologico, Ministerio de Obras Publicas (1954–)
Brevoria Geologica Asturica (1957–)
Cuadernos de Geologia Iberica (1970–)
Estudios Geologicos, Instituto de Geologia, Madrid (1945–)
Memorias del Instituto Geologico y Minero (1911–)
Publicaciones de Geologia, Universidad de Barcelona (1970–)
Studia Geologica Salmanticensia (1971–), with special volumes (1980–)
Temas Geologic Mineros (1978–)
Trabajos de Geologia, Facultad de Ciencias, Universidad de Oviedo (1967–)
Trabajos del Museo y Laboratorio de Geologica, Barcelona (1892–)
Trabajos sobre Neogeno-Cuaternario (1974–).

New literature for Spain and Portugal is listed with every issue of *Boletin Geologico y Minero*; previously the *Bibliografia Geologica Espanola* was issued in *Acta Geologica Hispanica* annually.

Spitzbergen

The geology is outlined in the following books and papers: *Geology of Spitzbergen* (1970) and *Stratigraphy of Spitzbergen* (1977), both translation editions edited by V N Sokolov (British Library Lending Division), and also by Sokolov with others, *The Main Features of Tectonic Structure of Spitzbergen* (*Geological Magazine*, **105**, 1968); *The Geological Development of Svalbard during the Precambrian, Lower Palaeozoic and Devonian*, edited by T S Winsnes (*Skrifter Norsk Polarinstitut*, **167**, 1979) and *Contribution to the Geology of North Western Spitzbergen*, by A Hjelle and Y Ohta (*Skrifer Norsk Polarinstitut*, **158**, 1974) are more specialized works.

Sweden

The comprehensive text on Swedish geology is the 3rd edition of *Sveriges Geologi*, by N H Magnusson *et al* (Svenska Boksforslag, 1963). Major, more recent, articles include *The Quaternary of Sweden*, by H Agrell (*Avhandlingar och Uppsatser, Sveriges Geologiska Undersokning*, serie C, **630**, 1979) and *The Precambrian of Sweden*, by T Lundqvist (*Avhandlingar och Uppsatser, Sveriges Geologiska Undersokning*, serie C, **768**, 1979).

Current journals of Swedish geology are:
Avhandlingar och Uppsatser, Sveriges Geologiska Undersokning (1868–)

Bulletin of the Geological Institution of the University of Uppsala (1892–)
Geologiska Foreningens i Stockholm Forhandlingar (1872–)
Lund Publications in Geology (1980–)
Stockholm Contributions in Geology (1957–).

Literature is compiled in *Swedish Geological Literature, 1958–1963*, edited by W Larsson (*Avhandlingar och Uppsatser, Sveriges Geologiska Undersokning*, serie C, **630**, 1976): earlier compilations are in the journal *Geologiska Forenings i Stockholm Forhandlingar*.

Switzerland

A recent book published is the *Geology of Switzerland: a guidebook*, edited by R Trumpy for the Schweizerischen Geologischen Commission (Wepf, 1980): in 2 volumes, the first outlines the geology of Switzerland, and the second, contains details of geological excursions. Other publications include *Geologische Wanderungen in der Schweiz*, by H Hierli (Ott, 1974) and *Der Schweiz Jura und seine Fossilien; geographie, geologie und palaontologie der Nordostschweiz*, by K Karsh and E Mutwiler (Frankh'sche, 1981).

Journals currently published are:
Beitrage zur Geologie der Schweiz, Geotechnische Serie (1899–), and *Kleinere Mitteilungen* (1931–)
Beitrage zur Geologie der Schweiz (1982–)
Bulletin d'Information de la Section des Sciences de la Terre de l'Universitè de Genève (1969–)
Bulletin des Laboratoires de Géologie, Géographie Physique, Minéralogie et du Musée Géologique de l'Université de Lausanne (1901–)
Bulletin der Vereinigung Schweizerischen Petroleum-Geologen und-Ingenieure (1935–)
Eclogae Geologicae Helveticae (1888–)
Schweizerische Mineralogische und Petrographische Mitteilungen (1921–)
Schweizerische Palaontologische Mitteilungen (1874–).

United Kingdom

There is no shortage of recent publications on the geology of the British Isles: *The Structure of the British Isles*, by J G C Anderson and T R Owen (2nd edn, Pergamon, 1980), and also by Anderson, *Field Geology in the British Isles: a study in crustal evolution* (Allen and Unwin, 1979); *Plate Tectonics and the Evolution of the British Isles*, edited by G C Brown *et al* (*Journal of the Geological Society of London*, **139**, 1982), *The Variscan Fold Belt in the British Isles*, edited by P Hancock (Adam Hilger, 1983), *Geomorphology of the British Isles*, by D K C Jones (Methuen, 1981), *The British Isles Through Geological Time: a northward drift*, by J P B Lovell (Allen & Unwin, 1977), *Geological Evolution of the British Isles*, by T R Owen (Pergamon, 1976) and *Igneous Rocks of the British Isles*, by D J Sutherland (Wiley, 1982), to list but a few.

On British palaeontology (see also the next chapter) there are *Fossils and the Geological History of the British Isles*, by D N Knight (Methuen, 1973), a set of three volumes *British Paleozoic Fossils, British Mesozoic Fossils*, and *British Caenozoic Fossils* (British Museum (Natural History) 1965–1983) and the dated, but useful, *Dictionary of British Fossiliferous Localities* (Palaeontographical Society, 1954); the specialist publications in the series *Palaontographical Society Monographs* (1847–) are definitive works, and there is also *Field Guide to Fossils* (1983–) published by the Palaeontological Association.

On stratigraphy, publications include *A Dynamic Stratigraphy of the British Isles: a study in crustal evolution*, by R Anderton *et al* (Allen & Unwin, 1979), *The Quaternary in Britain* edited by J Neale and J Flenley (Pergamon, 1981), *The*

Stratigraphy of the British Isles, by D R Rayner (Cambridge University Press, 1981), *The Ice Age in Britain*, by B W Sparks and R G West (Methuen, 1981) and *Pleistocene Geology and Biology: with especial reference to the British Isles*, by R G West (Longman, 1977).

Referring also to the section on the North Sea, publications on the surrounding seas and basins are *The Geology of the Sea Around the British Isles*, by A J Smith (Pergamon, 1981) and *Atlas of Offshore Sedimentary Basins in England and Wales: post Carboniferous tectonics and stratigraphy*, edited by A Whittaker (Blackwell, 1985).

The regional treatment of the geology of the United Kingdom can be found in several series. There are the volumes published in the series *British Regional Geology* (1935–) by the British Geological Survey listed in the next section, and the series of *Guides* (1958–), of the Geologists' Association: both these series are regularly updated with new editions. Field guides, e.g. of the Quaternary Research Association and Palaeontological Association, are useful, as is the *Geology Explained* . . . series of books published by David and Charles (which includes Lake District, South Wales and the Forest of Dean, etc).

The current publications of the British Geological Survey (until recently the Institute of Geological Sciences, and prior to that the Geological Survey of Great Britain) are their *Reports* (1969–), *Mineral Assessment Reports* (1975–), and the previously mentioned *British Regional Geology* (1935–), *Economic Memoirs, Coalfield Memoirs, District Memoirs* and the *Sheet Memoirs* which describe the geological maps issued by the Survey.

Publications of the Geological Society of London are the *Newsletter* (1952–), *Journal* which replaced the *Quarterly Journal* in 1972, *Memoirs* (1958–), *Miscellaneous Papers* (1974–), *Special Publications* (1964–), and *Special Reports* (1971–) which are stratigraphic reports.

Some journals for British geology are:

Amateur Geologist (1966–)
British Geologist (1975–)
Bulletin of the British Museum (Natural History): Geology (1949–)
Circular, Geologists' Association (1875–)
Geological Curator (1979–)
Geological Magazine (1864–)
Geology Today (1985–)
Proceedings of the Geologists' Association (1859–).

The new literature is indexed regularly in *British Geological Literature* (1964–) and in the major abstracting journals mentioned in Chapter 5.

England & Wales

Recent books include *Geology of England and Wales*, edited by P McL Duff and A J Smith (Academic Press, 1982), the monographic series *The Geomorphology of the British Isles*, published by Methuen, which consists of *Southeast and Southern England*, by D J C Jones (1981), *Northern England*, by C A M King (1976) and *Eastern and Central England*, by A Straw and K M Clayton (1979). Covering regions within England and Wales are *Geological Highlights of the West Country*, by W A MacFadyen (Butterworths, 1970), *The Geology of North East England*, edited by D A Robson (Natural History of Northumbria, 1980), *The Peak District*, by M Simpson (Unwin, 1982) and *The Lake District*, by R V Davis (Unwin, 1982). The areas covered by *British Regional Geology* are Northern England, the Pennines, Eastern England, North Wales, South Wales, the Welsh Borderlands, Central England, East Anglia, Bristol and Gloucester District, London and Thames Valley, South-west England, the Hampshire Basin, and the Wealden District.

Current, mostly regional, journals are:
Black Country Geologist (1981–)
Geological Journal (1964–)
Journal of the Arthur Holmes Society (1972–)
Journal of the Harker Geological Society (1962–)
Mercian Geologist (1964–)
Proceedings Cumberland Geological Society (1962–)
Proceedings of the North-East Lancashire Group of the Geologists' Association
    (1964–)
Proceedings of the Reading Geological Society (1975–)
Proceedings of the Shropshire Geological Society (1980–)
Proceedings of the Ussher Society (1962–)
Proceedings of the Westmorland Geological Society (1974–)
Proceedings of the Yorkshire Geological Society (1905–)
Transactions of the Leeds Geological Society (1883–)
Transactions of the Royal Geological Society of Cornwall (1818–), known briefly as
    the *Journal of Earth Sciences.*

Scotland

Books on this country are *The Geology of Scotland*, edited by G Y Craig (2nd edn,
Scottish Academic Press, 1983), *Scotland's Environment During the Last 30,000
years*, by R J Price (Scottish Academic Press, 1983) and *Geology and Scenery in
Scotland*, by J B Whittow (Penguin Books, 1983).

   British Regional Geology (see above) includes the following areas: the Northern
Highlands, Grampian Region, the Midland Valley of Scotland, the South of
Scotland, Orkney and Shetland, and the Tertiary volcanic districts. *Sheet Memoirs,
Economic Memoirs*, etc., for Scotland are published by the British Geological
Survey, as they do for England and Wales.

   The journals for Scottish geology are *Edinburgh Geologist* (1977–), *Proceedings
of the Geological Society of Glasgow* (1966–), *Scottish Journal of Geology* (1965–),
and *Transactions of the Edinburgh Geological Society* (1972–).

Northern Ireland see also **Ireland**

The book for the province is *Regional Geology of Northern Ireland*, by H E Wilson
(Northern Ireland Geological Survey, 1972).

   The Geological Survey of Northern Ireland currently publishes the series of *Sheet
Memoirs, Special Memoirs, Geophysical Papers, Geomagnetic Bulletins*, and
Seismological Bulletins, for the province.

Yugoslavia

There are no recent books published on Yugoslavia. There is, however, a number
of journal publications:
Geologija Razprave in Porcilia (1953–)
Geoloski Anali Balkanskogo Poluostrva (1889–)
Geoloski Glasnik, Titograd (1956–)
Geoloski Glasnik, Sarajevo (1955–)
Geoloski Vjesnik, Zagreb (1947–)
*Glasnik Prirodnjaukog Muzeja u Beogradu, Serija a Mineralogija, Geologija,
    Paleontologija* (1958–)
Trudovi Geoloski Zavod Sotsialistichka Republika Makedonija (1947–)
Vesnik Geologija, Beograd (1932–)
Zapisnici Srpskog Geologskog Drustva (1897–).
   The geological literature of Yugoslavia is indexed in *Geoloska Bibliografija*

Jugoslavije published periodically as an annexe of *Geoloski Anali Balkanskogo Poluostrva*; dates covered are, to 1944, 1944–1958, 1959–1962, 1963–1967, and 1968–1970.

Union of Soviet Socialist Republics (USSR)

There is an abundance of geological literature on the Soviet Union, not only in Russian, but in the various languages of the Soviet States. However, few major journals and books are available in translation.

The main works include *Geology of the USSR*, by D V Nalivkin, translated and edited by N Rast and T S Westoll (University of Toronto Press, 1973), and the recently published *Geology of the USSR, first part: old cratons and Paleozoic fold belts*, by V E Khain (*Beitrage zur Regionalen Geologie der Erde*, **17**, 1985).

The publications of the 27th International Geological Congress, held in Moscow in 1984, include excursion guidebooks to all parts of the Soviet Union; volume **1** of the *Report* is entitled *Geology of the USSR*, and many papers of relevance are found in the 23 *Proceedings* volumes.

Although not available in translation, these works are worth the mention: *Geokhronologiya SSSR*, edited by N I Polevoi (3 vols, Nedra, Leningrad, 1973–1974), *Stratigrafiya SSSR*, edited by D V Nalivkin, in 14 volumes which began in 1963 (Nauka, Moscow) and *Paleografiya SSSR*, edited by A P Vinogradov (4 vols, Nedra, Leningrad, 1974–1975).

Stratigraphic terms are named in *Lexique Stratigraphique International*, volume **2** (CNRS, 1958–1959) in 2 volumes, with an additional index volume. The dictionary for geological terms is *Stratigraficheskii Slovar' SSSR*, by I E Zannia and B K Likharev (Nedra, Leningrad, 1975).

Journals currently published in the Soviet Union are:
Byulletin' Komissii po Ichucheniyu Chervertichnogo Perioda (1936–)
Byulletin' Moskovskogo Obshchestva Ispytalei Prirody, Otdel Geologicheskii (1922–)
Doklady Akademiya Nauk SSSR, Seriya Geologiya (1922–) (available in translation as *Doklady of the Academy of Sciences of the USSR, Earth Sciences* (1959–)
Geofizicheskie Metody Ravedki v Arktike (1966–)
Geofizicheskii Zhurnal (1961–)
Geologicheskii Sbornik, L'vov (1956–)
Geologiya i Geofizika (1969–)
Geologiya i Geografia (1979–)
Geologiya Sibiri i Dal'nego Vostoka (1974–)
Geologiya Uzberezhya i dna Chornogo ta Azov'skogo Moriv u Mezhak URSR (1970–)
Geoloogiline Komik (1961–)
Geoloogilised Markmed (1961–)
Geomorfologiya (1969–)
Issledovanie Zemli iz Kosmosa (1980–)
Itogi Nauki i Tekhniki, Seriya Geologiya (1972–), and *Seriya Geokhimiya Mineralogiya i Petrografiya* (1975–)
Izvestiya Akademii Nauk Armyanskoi SSR, Nauk o Zemle (1958–)
Izvestiya Akademii Nauk Azerbaidzhanskoi SSR, Seriya Nauk o Zemle (1966–)
Izvestiya Akademii Nauk SSSR, Seriya Geologicheskaya (1936–)
Izvestiya Vysshikh Uchebnykh Zavedenii, Geologiya i Razvedka (1967–)
Izvestiya Zapadno-Sibirskogo Otdeleniya Geologicheskogo Komiteta (1921–)
Ocherki po Istorii Geologicheskikh Znanii (1953–)
Paleontologiya i Stratigrafiya Pribaltiki i Belorussi (1961–)
Trudy Geologicheskogo Instituta, Akademiya Nauk SSSR (1932–)
Trudy Instituta Geologii, Akademiya Nauk Tadzhikskoi SSR (1958–)

Trudy Instituta Geologii i Geofiziki, Sibirskoe Otdelenie (1960–)
Trudy Instituta Geologii i Geokhimii, Akademii Nauk SSR, Ural'skii Nauchnyi Tsentr (1972–)
Trudy Litovskogo Nauchno-Issledovatel'skogo Geologorazvedochnogo Instituta (1965–)
Trudy Mezhvedomstvennyi Stratigraficheskii Komitet SSSR (1969–)
Trudy Sovrmestnaya Sovetsko-Mongol'skaya Nauchno-Issledovatel'skaya Geologicheskaya Ekpeditsiya (1970–)
Trudy Upraveleniya Geologii Sovieta Ministrov Tadzhikskoi SSR, Paleontologiya i Stratigrafia (1965–)
Trudy Vsesoyuznogo Neftyanogo Nauchno-Issledovatel'skogo Geologo-Razvedochnogo Instituta (VNIGRI) (1950–)
Trudy Vsesoyuznogo Nauchno-Issledovatel'skii Geologorazvedochnyi Institut (VNIGNI) (1951–)
Uzbekskii Geologicheskii Zhurnal (1965–)
Vestnik Moskovskogo Universiteta, Seriya 4, Geologiya (1962–)
Visnyk Kyyivskoho Universytetu, Seriya Geologiya (1959–)
Voprosy Chetvertichnoi Geologii (1962–)
Voprosy Paleontologii i Stratigrafii Azerbaizhana (1976–)
Voprosy Stratigrafii (1974–)
Zhizn Zemli (1961–).

Apart from the coverage of Soviet literature in *Referativnyi Zhurnal, Geologiya*, the literature is indexed annually in *Geologicheskay Literature SSSR* (Tsentral'naya Geologicheskaya Biblioteka, 1934–). The bibliographic series *Geologicheskay Izuchenost' SSSR* (1966–) covers the literature retrospectively, and treats each region separately.

OCEANIA

Volume 6 of the *Lexique Stratigraphique International* is concerned with Oceania and is the major reference work for the region. The parts completed to date are 2: Polynesia and Micronesia, 1956; 3a: New Guinea, 1966; 4: New Zealand, 1959; 5: Australia; 5aQueensland, 5b: New South Wales, 5c: Victoria; 5d: Tasmania; 5e: South Australia; 5f: Western Australia; 5g: Northern Territory; 5h: Australia, Generalities, 1956–1975.

Books covering specific aspects of the region include *Volcanism in Australasia*, edited by R W Johnson (Elsevier, 1976) and *Correlation of the Silurian rocks of Australia, New Zealand and New Guinea*, by J A Talent *et al* (*Geological Society of America, Special Paper,* **150**, 1975). Publications on the area can also be found in the section on the Pacific Ocean.

Australia

Important works published recently on Australia are *Phanerozoic Earth History of Australia*, edited by J J Veevers (Clarendon, 1984), *Guide to the Geology of Australia*, by W D Palfreyman (*Bureau of Mineral Resources, Geology and Geophysics, Bulletin,* **181**, 1984) and *Ancient Australia: the story of its past geography and life*, by C Laserson, revised by R O Brunnschweiler (Angus & Robertson, 1984).

Current geological journal publications for Australia are mainly from the Bureau of Mineral Resources, Geology and Geophysics:
BMR (formerly the *Annual Report*), (1971–)
BMR Journal of Australian Geology and Geophysics (1976–)
Bulletin, Bureau of Mineral Resources, Geology and Geophysics (1932–)
Report, Bureau of Mineral Resources, Geology and Geophysics (1947–).

The journal publications of the Geological Society of Australia are:
Alcheringa (1975–)
Australian Journal of Earth Sciences (formerly the *Journal*) (1953–)
Special Publications of the Geological Society of Australia (1968–).
Major publications of the various regions of Australia, and the journals of their respective geological surveys and institutions are listed below.

New South Wales

Volcanism in Eastern Australia: with case histories from New South Wales, edited by F L Sutherland *et al* (*Publication, Geological Society of Australia, New South Wales Division*, **1**, 1985) and *Geology of the Sydney Basin: a review*, by S J Mayne *et al* (*Bureau of Mineral Resources, Geology and Geophysics, Bulletin*, **149**, 1974) are the main papers for this state.
Journals are:
Annual Report, Department of Mineral Resources and Development, New South Wales (1875–)
Bulletin of the Geological Survey of New South Wales (1922–)
Memoirs of the Geological Survey of New South Wales, Geology Series (1887–)
Records of the Geological Survey of New South Wales (1889–).

Northern Territory

The Geology of the Northern Territory, by G W D'Addario *et al* (*Bureau of Mineral Resources, Geology and Geophysics, Bulletin*, **180**, 1976) is the publication of the geological survey of the Northern Territory.

Queensland

The Geology and Geophysics of Northeastern Australia, edited by R A Henderson and P J Stephenson (Geological Society of Australia, Queensland Division, 1980) and *Queensland Field Geology Guide*, by N C Stevens (Geological Society of Australia, Queensland Division, 1984) are the most important works.
There are three journals: *Publication Geological Survey of Queensland* (1902–), *Papers University of Queensland, Department of Geology, New Series* (1937–) and *Queensland Government Mining Journal* (1902–).

South Australia

Southern Aspect: an introductory view of South Australian geology, by A R Alderman (South Australian Museum, 1973) and *A Guide to the Geology and Mineral Resources of South Australia*, by N H Ludbrook (D J Woolman, 1980), are the two main works.
Journals from the state survey are: *Bulletin, Geological Survey of South Australia* (1912–), *Quarterly Geological Notes, Geological Survey of South Australia* (1970–), *Report of Investigations, Geological Survey of South Australia* (1954–), and *Annual Report, Director General Mines and Energy, South Australia* (1977–).
Older literature of the region is indexed in *Bibliography of South Australian Geology*, compiled by E N Teesdale-Smith (Geological Survey of South Australia, 1959).

Tasmania

Geology and Mineral Resources of Tasmania, by I B Jennings and E Williams (*Tasmania Geological Survey, Bulletin*, **50**, 1967) and the comparatively recent *The Tasman Geosyncline: a symposium in honour of Dorothy Hill*, edited by A D Denhead *et al* (Geological Society of Australia, 1974) are the main sources of geological information.

Journals are *Bulletin, Geological Survey of Tasmania* (1907–) and *Report of the Director of Mines, Tasmania* (1940–).

Victoria

Regional Guide to Victorian Geology, edited by J MacAndrew and M A H Marsden (2nd edn, University of Melbourne, School of Geology, 1973) and *Geology of Victoria*, edited by J G Douglas and J A Ferguson (*Geological Society of Australia, Special Publication*, 5, 1976) are the main works.

Journals are *Bulletin, Geological Survey of Victoria* (1903–), *Memoirs, Geological Survey of Victoria* (1903–) and *Mining Geology and Energy Journal of Australia* (1937–).

Western Australia

Geology of Western Australia (*Geological Survey of Western Australia, Memoirs*, 2, 1975) is the chief publication.

Journals are: *Bulletin, Geological Survey of Western Australia* (1898–); *Memoir, Geological Survey of Western Australia* (1975–); *Report, Geological Survey of Western Australia* (1969–); *Annual Report of the Geological Survey of Western Australia* (1969–), all published by the state's Geological Survey.

Fiji

Geological literature on Fiji is indexed in *Bibliography of the Geology of Fiji*, compiled by R F Duberal and P Rodda (2 vols, Fiji Department of Geological Surveys, 1969–1980) with a supplementary vol.2 in *Report, Fiji Mineral Resources Department*, 8, 1982, and vol.3 by D Greenbaum, in the same series, 1983.

The publications of the Geological Survey, the Mineral Resources Division, Fiji, are the *Bulletin* (1973–), *Memoir* (1964–), *Report* (1972–) and *Annual Report* (1980–).

New Caledonia

Literature on the region includes an outline, *The Geology of New Caledonia*, by A R Lillie and R N Brothers (*New Zealand Journal of Geology and Geophysics*, 13, 145–183, 1970) and *Géologie de la Nouvelle-Calédonie*, by J P Paris (*Mémoires BRGM*, 113, 1981).

New Hebrides

The most recent papers include *Tectonic Framework of the New Hebrides Island Arc*, by D E Karig and J Mammerickx (*Marine Geology*, 12, 187–205, 1972), *Geological Evolution of the New Hebrides Island Arc*, by A H G Mitchell and A J Warren (*Journal Geological Society of London*, 127, 501–529, 1971) and *Development of the New Hebrides Archipelago*, by D I J Mallick (*Philosophical Transactions Royal Society of London*, (B) 272, 227–285, 1975).

Survey reports currently published are in the *Annual Report of the Geological Survey of New Hebrides* (1959–).

New Zealand

A variety of publications has appeared on the geology of this country including *The Geological History of New Zealand and its Life*, by C A Fleming (Oxford University Press, 1979), *Legend in the Rocks: an outline of New Zealand geology*, by M Gage (Whitcoulls, 1980), *The Geological Structure of New Zealand*, by J T Kingma (Wiley, 1974), *Strata and Structure in New Zealand*, by A R Lillie (Tohunga Press, Auckland, 1980), *Illustration of New Zealand Fossils: a New*

Zealand Geological Survey guidebook, compiled by G Speden and W Keys (*New Zealand Department of Scientific and Industrial Research Information Series*, **150**, 1981), *New Zealand Adrift: the theory of continental drift in a New Zealand setting*, by G R and D L Stevens (Reed Ltd, 1980), *The Geology of New Zealand*, edited by R P Suggate *et al* (2 vols, New Zealand Geological Survey, 1978), *Field Guide to New Zealand Geology: an introduction to the rocks, minerals and fossils*, by J Thornton (Reed Methuen, 1985) and *Economic Geology of New Zealand*, by G J Williams (*Australian Institute of Mining and Metallurgy*, **4**, 1974).

Current journals are *Bulletin, New Zealand Geological Survey* (1906–), *Memoirs, New Zealand Geological Survey* (1928–), *New Zealand Journal of Geology and Geophysics* (1958–) and *Newsletter, Geological Society of New Zealand* (1959–).

The literature is indexed in *A Bibliography of New Zealand Geology, 1951–1969*, compiled by G Warren (*New Zealand Geological Survey, Bulletin*, **93**, 1977). Coverage of earlier periods, as well as a subject index and an update of the above, is published in the same journal, under the title *Index to the Bibliography of New Zealand Geology*, compiled by D L Jenkins: periods covered are, to 1950 in *Bulletin*, **65A**, 1976; 1951–1969 in *Bulletin*, **93A**, 1982.

Papua New Guinea

An outline of the geology of the area is in *A Geological Synthesis of Papua and New Guinea*, by D B Dow (*Bureau of Mineral Resources, Geology and Geophysics* (Australia) *Bulletin*, **201**, 1977), while papers relevant to the area may be found in the *Memoirs* series of the Geological Survey of Papua New Guinea (1972–).

Literature of the region is comprehensively indexed in the following compilations: *Earth Sciences Abstracts, Papua New Guinea to 1970* by W Manser (*Bureau of Mineral Resources, Geology and Geophysics, Bulletin*, **143**, 1974); *1972–1973* by W Manser and N M Reynolds (*Geological Survey of Papua New Guinea, Memoir*, **4**, c. 1975) and *1977–1979*, by W Manser (*Geological Survey of Papua New Guinea, Memoir* **8**, 1983).

Solomon Islands

Papers on the islands can be found in the volumes constituting the *British Solomon Islands Geological Record* (1957–1969).

The journals currently published are *Bulletin, Solomon Island Geological Survey* and the *Memoir* (1958–) series of the Geological Survey.

CHAPTER NINE

Palaeontology

ANN LUM

Palaeontology is the science of fossils, the preserved remains of plants and animals, as well as the record of their activity — trace fossils, fossil footprints and trails. Palaeontology is the study of life in the past, a study which is linked inextricably with the rocks which contain the fossils, and hence with sedimentology, stratigraphy, and other branches of the geological sciences. In turn, fossils provide crustal evidence of the history of life on the earth, the evolutionary processes it has undergone, and the climates of the past.

The interests of palaeontology and archaeology overlap: the palaeontologist looks at remains up to several thousand million years old, whilst the archaeologist limits his attention to the most recent 5000 years or so. The study of fossil man and his ancestors is thus considered the province of the palaeontologist.

The observation of fossils has been noted throughout written history: the ancient Greek philospher Xenophanes (560–480 BC) recognized fossils as remains of living organisms, myths of dragons in China, and giants in Europe, were perhaps attempts at a rational explanation of fossil mammal remains. Fossils, both vertebrate and invertebrate, were used for decorative and symbolic purposes in antiquity (K P Oakley, *Occasional Papers in Technology, Pitt Rivers Museum*, **12–13**, 1975–1985). However, the serious study of palaeontology developed late compared with some of the exact sciences.

The Renaissance brought little advance to palaeontology, although 'figured stones' were often included in the antiquarian's cabinet of curiosities. The nature of fossils referred to by Georg Bauer (1495–1555), better known as Agricola, in his work *De*

Natura Fossilium, Lib. *X*, 1546, and by Conrad Gesner (1516–1565) in his *De Rerum Fossilium*, 1565, is somewhat broader than the modern definition of organic remains; they include any distinctive object found in the earth.

The work of John Ray (1627–1705), Georges Buffon (1707–1788) and Carl Linnaeus (1707–1778) outlining the classification of organisms based on the species concept, the order of living organisms, and the establishment of a common nomenclature, was an essential step in the development of the science. Serious study began in the late eighteenth century, parallel to the other developments in the science of geology, with the detailed contributions of Jean Baptiste Lamarck (1744–1829) on vertebrates, Adolphe Brongniart (1801–1879) on fossil plants, and Christian Gotfried Ehrenberg (1795–1896) on microfossils.

The literature of palaeontology centres on the collection and identification of the fossil, to place it in its geographic and stratigraphic range, and to understand its environment and mode of life. Interest in the literature may come from the amateur collector, the student, the museum curator, the taxonomist, the academic or industrial research. Also the literature of descriptive palaeontology does not date; there is a constant need to refer to the original descriptions and a regular demand for retrospective bibliographies, however short. The first edition of this chapter, by Sarjeant and Harvey, dealt in great detail with the various bibliographies in the subject: those works continue to be useful for reference and the authors' comments are still valid, so reference to the first edition will be useful. This edition will not attempt to duplicate this material though some of the major references will be included with the new material.

The trends in publishing since the last edition, have changed emphasis. Textbooks are less description orientated, concentrating more on a biological approach: publications from conferences and proceedings are now commonplace, presenting papers on current research and state-of-the-art reviews.

Although there are quite a number of dictionaries in the earth sciences (see Chapter 4), very few are devoted solely to palaeontology, and of these none is in the English language. The few exceptions are *Päläontologisches Wörterbuch*, by U Lehmann (2nd edn, Enke, 1977) and *Paleontologicheskii Slovar'*, by G A Bezhosova and F A Zhuraleva (Nauka, 1965). V G Telberg's *Russian-English Dictionary of Paleontological Terms* (Nauka, 1968) is useful, and is the only bilingual dictionary in the field. One therefore has to rely on the geological and biological dictionaries.

Useful for quick reference are R Steel and A P Harvey's *The Encyclopaedia of Prehistoric Life* (Mitchell Beazley, 1979) and

The Enclyopedia of Paleontology, edited by R W Fairbridge and D Jablonski (*Encylopedia of Earth Sciences*, vol.8) (Dowden Hutchinson & Ross, 1979).

Directories are an essential reference requirement, whether they are of persons currently active in the field, or of organizations or societies. The *Directory of Palaeontologists of the World* is published under the aegis of the International Palaeontological Association. The fifth edition is at preparatory stage for publication in 1989. Lists of members of societies are a useful back-up to this, e.g. the *Directory of the All-Union Paleontological Society of the USSR* (Nauka, 1984), but societies generally issue more informal lists in their newsletters.

Palaeontology is actively studied in most earth science and natural history institutions and academic departments. Information can be obtained from *A Directory of Societies in Earth Science*, a listing published annually in *Geotimes*, in their August issue, or *Worldwide Directory of National Earth-Science Agencies and Related International Organisations* (US Geological Survey, Circular **834**, 1981). The *Directory of Natural History and Related Societies in Britain and Ireland*, compiled by A Meenan (British Museum (Natural History), 1983) contains many entries for societies with a palaeontological interest.

There is also a need to locate and refer to palaeontological collections; C D Sherborn's *Where is the _ _ _ _ Collection?* (Cambridge University Press, 1940) has been updated by R J Cleevely's *World Palaeontological Collections* (Mansell & British Museum (Natural History), 1983); this work lists the collections by the collectors, with an index of institutions by country. Also useful as a source of information is T Sharpe's *Geology in Museums: a bibliography and index* (National Museum of Wales Geological Series, **6**, 1983).

The reference work for the history and biography of palaeontology can be found in W A S Sarjeant's *Geologists and History of Geologists* (5 vols, MacMillan, 1980). The older *Palaeontologi Catalogus Bio-bibliographicus*, compiled by K Lambrecht and A Quenstedt (*Fossilium Catalogus I, Animalia*, **72**, 1935) is still useful. The *Annual Bibliography of the History of Natural History 1982–* (British Museum (Natural History), 1985–) includes palaeontology within its scope, and can be accessed from the alphabetical author list and through the subject and biographical index.

Taxonomy

Modern classification and nomenclature of all living organisms, began with the publication of the 10th edition of *Systema Naturae* by Carl Linnaeus in 1758, in which he adopted the binomial system for plant and animal species. The rules which govern present day botanical and zoological nomenclature have evolved through international agreement, resolving the differing practices which had developed in different places; the *International Code of Botanical Nomenclature* (rev. edn, 1983) and the *International Code of Zoological Nomenclature, adopted by the XX General Assembly of the International Union of Biological Sciences* (International Trust for Zoological Nomenclature and University of California Press, 1985) are the main reference works. These rules are applied equally to fossil and recent plants and animals, a requirement which was recognized in the latter half of the nineteenth century, and discussed at the International Congress of Geology in Paris, 1878, and Bologna, 1881. However, the palaeontologist does have special needs: zoological names based on the activity of fossils (trace fossils, tracks) have only just received recognition in the latest edition of the Code.

The codes are complex documents. Works of less complexity on taxonomy include: the journal for botanical nomenclature, *Taxon* (1951–) and for zoological nomenclature, *Systematic Zoology* (1952–). The official organ of the International Trust for Zoological Nomenclature is the *Bulletin of Zoological Nomenclature* (1943–), which publishes the proposed changes in names and the decisions reached.

The massive task of listing all genera and species of animals, recent and fossil, was undertaken by Charles Davies Sherborn (1861–1942) in *Index Animalium* (Cambridge University Press and British Museum (Natural History), 1902–1933); this work covered all published names up to 1850, together with the reference to the first description. The *Nomenclator Zoologicus*, compiled by S A Neave (Zoological Society of London, 1939–1940) lists the genera and subgenera from 1758 (Linnaeus, 10th edition) to 1935; there are further updates, 1936-1945 (1950) and 1956–1965 (1966), after which they were continued, on an annual basis, as Section 20 of the *Zoological Record* (see p.269).

Palaeontological texts

Palaeontological techniques

The literature of collection, preparation and curation of fossils is essential to those for whom palaeontology is more than a theoretical subject. For students embarking on their practical work, S P Tunnicliff's *I am Beginning my Research: what shall I do with my geological specimens?* (National Environment Research Council, 1983) provides useful guidelines.

General instructions can be found in *Fossils, Minerals and Rocks: collection and preservation*, by R Croucher and A R Woolley (British Museum (Natural History) and Cambridge University Press, 1982). Most works on collection are intended for the amateur market, such as *Fossils and Fossil Collecting*, by R Hamilton (Hamlyn, 1975), *Finding Fossils*, by R Hamilton and A N Insole (Penguin Books, 1977), *Collecting Fossils*, by A Major (Bartholomew, 1974), and *Hunting for Fossils*, by M Murray (MacMillan, 1974). Most are biased towards the countries in which they are marketed, with the inclusion of locality information, e.g. R L Casanova and R P Ratkevich's *An Illustrated Guide to Fossil Collecting* (3rd rev. edn, Naturegraph, 1981), J R MacDonald's *The Fossil Collector's Handbook: a paleontology field guide* (Prentice Hall, 1983) and R P MacFall and J Wollin's *Fossils for Amateurs: a handbook for collectors* (2nd edn, Van Nostrand Reinhold, 1983) all have an American bias, whilst G Krumbiegel & H Walther's *Fossilien: Sammlungen, Präparieren, Bestimmen Ausweten* (Enke, 1977) has a European emphasis. However, this may not necessarily detract from their value as general instruction manuals, and additional local information can be obtained from geological guides. Most of these publications contain elementary classification and identification guides.

For more advanced work, there is A E Rixon's *Fossil Animal Remains: their preparation and conservation* (Athlone Press, 1976) but *Handbook of paleontological techniques*, edited by B Kummel and D M Raup (W H Freeman, 1965) is still the major reference work. It deals with techniques for collecting and preparation (mechanical, chemical, radiation, casting, moulding and illustration) for each major group of fossils. It also contains a substantial bibliography for identification purposes and further study. The specialized techniques required for palaeopalynology are given in I Doher's *Palynomorph Preparation Procedures currently used in Paleontology and Stratigraphy Laboratories (US Geological Survey, Circular* **830**, 1980).

Current discussion of preparation and curation can be found in the journals *Geological Curator* (1974–) and *Der Präparator* (1955–). The new publication covering this topic, *Guidelines for Curation of Geological Materials*, edited by C H C Brunton *et al* (*Geological Society of London, Miscellaneous Papers*, **17**, 1985), specially produced in a looseleaf form with binder, includes codes of practice, acquisition, documentation, preservation, occupational hazards, uses of collections (information retrieval, scientific research, exhibitions), with a useful appendix listing adhesives, suppliers, and an outline of the Geological Site Documentation Scheme.

Popular and elementary textbooks

Almost everyone has seen a fossil, whether in a museum, in the countryside, or on the sea shore. Many have succumbed to the magic appeal of dinosaurs and other prehistoric monsters, or to the curiosity to understand our own origins. There is a proliferation of popular books, not only for children, but also for interested adults, which can serve as useful introductions to the subject. One hesitates to mention any of them as they have a tendency to go out-of-print very quickly, although some may be available for loan from some libraries. Those currently in print are available from those booksellers with a good natural history selection.

Amongst popular books, C L and M A Fenton's *The Fossil Book* (Doubleday, 1958) and J Augusta and Z Burian's *Prehistoric Animals* (Spring Books, 1958) have managed to withstand the challenge from younger rivals; the latter is particularly useful for the illustrations it contains. Of the recent ones, L B Halstead's *Hunting the Past: fossils, rocks, tracks and trails, the search for the origin of life* (Hamish Hamilton, 1982) takes the evolutionary point of view. Some, like *A Pictorial Guide to Fossils*, by G R Case (Van Nostrand Rheinhold, 1982), are elementary identification manuals. Others, like R Fortey's *Fossils: the key to the past* (Heinemann & British Museum (Natural History), 1982), deal with the various aspects of palaeontology, together with brief descriptions of the different groups of fossils.

A pocket size book, *The Observer's Book of Fossils*, by R M Black (Frederick Warne, 1977) is useful for the amateur on a country walk. For the young, or newcomer to palaeontology, D Lambert's *The Cambridge Guide to Prehistoric Life* (Cambridge University Press, 1985) offers a concise key to fossils of all groups, with brief chapters on the fossil record and on fossil hunting — techniques, methods, as well as a very brief history.

Intermediate and advanced textbooks

General palaeontology texts tend to fall into two categories: those with emphasis on the palaeobiology, and those with a systematic approach, dealing with all, or a range of, fossil groups.

The Natural History of Fossils, by C Paul (Weidenfeld & Nicholson, 1980) is an example of the former; its contents include preservation history, palaeoecology and palaeogeography, faunal succession and evolution, origin of life and of life on land. Another is G G Simpson's *Why and How: some problems and methods in historical biology* (Pergamon, 1980), which deals with the fossil record, morphology, palaeoecology, systematic taxonomy, biometrics and biogeography. New textbooks include S M Stanley's *Earth and Life through Time* (Freeman, 1985).

Some of the introductory texts for students, also in this category, are updated editions and translations of well-tested and readable textbooks. They include *Principles of Paleontology*, by D M Raup and S M Stanley (2nd edn, W H Freeman, 1978), *Introduction to Paleobiology: general paleontology*, by B Ziegler (Ellis Horwood, 1983), a translation from his book in German, *Allgemeine Paläontologie* (E. Schweizerbart'sche, 1972) and *Fossils and Life of the Past*, by E Thenius (English Universities Press, 1973), again a translation.

Introductory texts, with a bias towards a systematic approach, include R M Black's *The Elements of Palaeontology* (rev. edn, Cambridge University Press, 1973); A M Davies' *An Introduction to Paleontology* (3rd edn, Allen & Unwin, 1976), despite its title, concentrates on invertebrate groups. Fewer textbooks of this kind are being published, reflecting an emphasis on the former approach, and placing reliance, perhaps, on the more specialized systematic texts for information.

Several descriptive systematic works are monumental works which do not date; J Pivateau's *Traite de Paleontologie* (7 vols: Vol.1–3, Invertebres; vol.4–7, Vertebres, Masson, 1952–1969), and Yu A Orlov's *Osnovy Paleontologii* (15 vols, Akademiya Nauk SSSR, 1959–1963), of which some parts are available in English translation. These multi-volume works are produced by specialists on each fossil group of which the volumes are constituted. By far the most comprehensive and most important index to published generic and specific names is the *Fossilium Catalogus* (Junk, 1913–), published in two series, *I: Animalia* and *II: Plantae*; the parts are periodically added to and revised.

Compared to the many identification keys published for the life sciences, there are very few which cover the whole spectrum of palaeontology; *Determination Pratique des Fossiles*, by A Chauvin A Chauvin and A Calleux, (2nd edn, Masson, 1977) is one example.

From a recognition that the majority of geologists do not have access to good libraries, an industry has developed around the publication, in microfiche format, of the classic works in palaeontological literature. The older literature is essential to those with any taxomonic content in their work, for identification and first descriptions. *The Paleontological Microfiche Library*, edited by

A Hallam (Oxford Microform Publications, 1985) gives a choice of several collections; Palaeozoic corals, Crinoids, Trilobites, Mesozoic bivalves, Mesozoic brachiopods, and Mesozoic ammonites, the keyworks in each group selected by a specialist and accompanied by a historical commentary.

Palaeoecology and palaeoclimatology

Palaeoecology is the study of fossils and their environments, based on changes in the fossil record. The subject is inevitably linked with changes in the environment: the climate and geography of the geological past. Thus the specialist journal in this subject is *Palaeogeography, Palaeoclimatology, Palaeoecology* (1965–).

Palaeoecology

Introductory books to palaeoecology include *Ancient Sedimentary Environments and the Habitats of Living Organisms*, by J-C Gall, a translation from the French (Springer Verlag, 1983), *Ecology and Earth History*, by R N Fiennes (Croom Helm, 1976) and L F Laporte's *Ancient Environments* (2nd edn, Prentice Hall, 1979). Selected readings from the journal *Scientific American*, on *Paleontology and Paleoenvironments*, edited by B J Skinner (W H Freeman, 1981) provides useful reading for the student. *Life History of a Fossil: an introduction to taphonomy and paleoecology*, by P Shipman (Harvard University Press, 1981) is another student text.

Paleoecology, Concepts and Applications, edited by J R Dodd and R J Stanton (Wiley Interscience, 1981) is an important textbook, replacing D V Ager's *Principles of Palaeoecology* (McGraw-Hill, 1963), which, together with another older work, *Introduction to Quantitative Palaeoecology*, by R A Reyment (Elsevier, 1971) is still useful for techniques and methodology. *The Ecology of Fossils: an illustrated guide*, edited by W S KcKerrow (Duckworth, 1978) describes a number of communities through the fossil record, although the rocks illustrated are European. *Structure and Classification of Paleocommunities*, edited by R W Scott and R R West (Dowden Hutchinson & Ross, 1976), the papers from a symposium of the Paleontological Society, discusses mostly communities from soft sediments: another book in this area is *Communities of the Past*, by J Gray, A J Boucot and W B N Berry (Dowden Hutchinson Ross, 1981). Books on palaeoecology, but with a marine emphasis, include J W Valentine's *Evolutionary Paleoecology of the Marine Biosphere* (Prentice Hall, 1973), A J Boucot's *Principles of Benthic Marine Paleoecology* (Academic Press, 1981), and *Biotic Interactions in Recent and Fossil Benthic*

Communities, edited by J S Tevesz and P L McCall (Plenum, 1983).

Specialist reading for the vertebrate palaeoecologist includes *Fossils in the Making: vertebrate taphonomy and paleoecology*, by A K Behrensmeyer and A P Hill (University of Chicago Press, 1980), *Quaternary Palaeoecology*, by H J B and H H Birks (Edward Arnold, 1980), and *Palaeoecology of Beringia*, edited by D M Hopkins (Academic Press, 1982).

Palaeoclimatology

Books on the subject include L A Frakes' *Climates Through Geologic Time* (Elsevier, 1979) and *Climates of the Past*, by M Schwarzbach (American Meteorological Society Press, 1977). A recent publication, for advanced reading, is *Milanovich and Climate: understanding the response to astronomical forcing*, edited by A Berger *et al* (Reidel, 1984).

Palaeogeography

Palaeogeography is dealt with in detail elsewhere in this book (see chapter 8). Some works especially relating to fossils are *Wandering Lands and Animals*, by E H Colbert (Dutton, 1973) for introductory reading, and for advanced reading *Paleobiogeography*, edited by C A Ross (Dowden Hutchinson Ross, 1976).

Palaeogeochemistry

Publications presenting studies on the chemical processes of fossilization, include *Biogeochemistry of Amino Acids*, by P E Hare, T C Hoering and K King (Wiley, 1980), and *Biomineralization and Biological Metal Accumulation: biological and geological perspectives*, edited by P Westbroek and E W De Jong (Reidel, 1983). The former is the proceedings of a conference of that name, and the latter, a collection of fifty papers from the Fourth International Symposium on Biomineralization held in 1982, which includes a discussion on the global geochemical significance of biomineralization.

Fossil animal studies within this topic are *The Biochemistry of Animal Fossils*, by R W G Wyckoff (Scientechnica, 1972) and *Paleobiogeokhimiya Morskikh Bespozvonochnych* [Palaeobiogeochemistry of marine invertebrates], by E V Krasnov (*Trudy Instituta Geologii i Geofiziki, Akademiya Nauk SSSR*, **379**, 1980).

Evolution and the fossil record

The fossil evidence is an essential element in the study of evolution. Evolution studies encompass a much wider field, the literature of which is a study on its own: within the context of palaeontology, the aspects of the fossil record itself, phylogenetics, and the origin of life, extinctions, and living fossils are relevant.

The fossil record and stratigraphy

For elementary reading on the history of the evolution of plants and animals, there is R Cowan's *History of Life* (McGraw-Hill, 1976). A general textbook on the subject is *Evolution of the Earth*, by R H Dott *et al* (McGraw-Hill, 1981). *Fossil Record: a symposium with documentation*, edited by W B Harland *et al* (Geological Society of London, 1967), provides a number of papers on evolutionary aspects of various faunas and floras in the first section, followed by a documentation on the fossil record in three parts: Plantae, Invertebrata, and Vertebrata.

Evolution and the Fossil Record, edited by L F Laporte (W H Freeman, 1978), provides a selection of readings from the *Scientific American*; another selection is published in *Patterns of Evolution as Illustrated by the Fossil Record*, edited by A Hallam (*Developments in Palaeontology and Stratigraphy*, **5**, 1977). Physical and chemical changes which influence trends in evolution, such as cycles and sudden changes in the earth's environment, biological radiation and extinctions, can be found in *Patterns of Change in Earth Evolution*, edited by H D Holland and A F Trendall (*Dahlem Workshop Report*, **5**, Springer Verlag, 1984).

Phylogenetics

The most important work in phylogenetics is Willi Hennig's *Phylogenetic Systematics*, translated, with revisions (University of Illinois Press, 1966).

A work with specific reference to palaeontology can be found in a collection of papers from a symposium "Phylogenetic Models" held at the *Second American Paleontological Convention, Phylogenetic Analysis and Paleontology*, edited by J Cracraft and N Eldredge (Columbia University Press, 1979).

Origin of life

The biogeochemical origin of life has attracted much popular interest in recent years, with a number of publications aimed at that market. For serious study, *Cosmochemistry and the Origin of Life*, edited by C Ponnamperuma (*NATO Advanced Study Institute Series*, Reidel, 1983) provides basic information from an expert panel of contributors. An important work in this area is *Earth's Earliest Biosphere: its origin and evolution*, edited by J W

Schopf (Princeton University Press, 1983). An earlier collection of papers can be found in *Geochemistry and the Origin of Life*, edited by K A Knenvolden (*Benchmark Papers in Geology*, **14**, 1974).

Extinctions

Extinctions have received widespread attention in recent years. General reviews of the subject, the intensity, rate and causes, and evidence in the fossil record, are given in the work edited by M H N Nitecki, *Extinctions* (University of Chicago Press, 1984), also in *Dynamics of Extinction*, edited by D K Elliot (Wiley, 1986), and *Evolution and Extinction Rate Controls*, edited by A J Boucot (*Developments in Paleontology and Stratigraphy*, **1**, 1975).

Extinction as a result of sudden change is expounded in *Catastrophe and Earth History: the new uniformitarianism*, edited by W A Berggren and J A Van Couvering (Princeton University Press, 1984). Extraterrestrial causes of extinctions are expounded in *Geological Implications of Large Asteroids and Comets on the Earth*, edited by L Y Silver and P H Schultz (*Geological Society of America, Special Papers*, **190**, 1982). Examining the possible causes for the end of the age of reptiles, *Cretaceous-Tertiary Extinctions and Possible Terrestrial and Extraterrestrial Causes*, represents the proceedings of a workshop held by the K-TEC Group in 1976 (*Syllogeus*, **12**, 1977). M Allaby and J Lovelock also look at the same period in *The Great Extinction* (Secker & Warburg, 1983).

The extinction of some mammals in the Pleistocene is examined in *Quaternary Extinctions: a prehistoric revolution*, by P S Martin and R G Klein (University of Arizona Press, 1984).

Living fossils

"Living fossils" is a concept which dates from Darwin. A collection of papers, covering a range of views on their recognition, can be found in N Eldredge and S M Stanley's work entitled *Living Fossils* (Springer Verlag, 1984).

Precambrian fossils

This field of study, until fairly recently, has been included under micropalaeontology. As a study on its own, it reflects interests in the origin of life and the earliest life forms. Among the books published are M F Glaessner's *The Dawn of Animal Life: a biohistorical study* (Cambridge University Press, 1984) and *Earth's Earliest Biosphere, its Origin and Evolution*, which represents the collaborative results of a project, organized and edited by J W Schopf (Princeton University Press, 1984). Glaessner provides a substantial bibliography to his book, and in addition, *Bibliography of Precambrian Paleontology and Paleobiology*, compiled by S M Awramik and S Barghoorne, can be located in *Leaflets, Harvard University Botanical Museum*, **26**, p.65–175, 1978. The journal *Precambrian Research* (1974–) is a relevant source of information.

Micropalaeontology

Micropalaeontology is the study of microfossils, and, of necessity, includes a cross-section of taxonomically unrelated groups. The study of microfossils dates back to the discovery of the microscope, but, as a separate discipline, it only came into being in the first quarter of this century as the result of the growing petroleum industry, and the application of microfossils, especially Foraminifera, as a correlation tool.

An introductory textbook for students on micropalaeontology is M D Brasier's *Microfossils* (Allen & Unwin, 1980). For intermediate studies, there is G Bignot's *Les Microfossiles: les differents groupes, exploitation, paléobiologique et géologique* (Bordas, 1982). For further studies, a collection of papers, which includes major ones on planktonic foraminifera, and edited by F T Banner and A R Lord, *Aspects of Micropalaeontology* (Allen & Unwin, 1982), is useful at research level.

The economic importance of micropalaeontology inevitably led to intensive study and extensive publication. *The Bibliography and Index of Micropaleontology*, published monthly by the American Museum of Natural History with the American Geological Institute, has been listing the literature of micropalaeontology since 1972, arranged by microfossil group, with an annual author and subject index.

The essential journals are:

Micropaleontology (1955–), with its associated *Special Publications* (1975–)
Journal of Micropalaeontology (1982–)
Revue de Micropaléontologie (1958–)
Cahiers de Micropaléontologie (1974–)
Utrecht Micropalaeontological Bulletin (1969–) and its *Special Bulletin* (1974–).

Journals with a regional speciality are *Revista Espanola di Micropaleontologia* (1969–), *Voprosy Mikropaleontologii* (1956–), *British Micropalaeontologist* (1976–), and *Acta Micropaleontologica Sinica* (1984–). The publications of specialist institutions are *Travaux Laboratoire Micropaléontologie* (1972–) and *Tübingen Mikropaläntologischen Mitteilungen* (1983–), whilst a very specialized journal is *Microfossiles Organiques du Paléozoique* (1967–).

The European Micropalaeontological Colloquium and the African Colloquium of Micropalaeontology are held regularly, and the publications from these conferences are useful sources of regional information.

Marine micropalaeontology

The Deep Sea Drilling Project has generated an abundance of material and interest in marine micropalaeontology. Their *Initial Reports* contain detailed papers on groups of microfossils: a review of the Project is *The Deep Sea Drilling Project: a decade of progress*, edited by J E Warme *et al* (*Society of Economic Paleontologists and Mineralogists, Special Publications*, **32**, 1981).

One of the first textbooks on the subject, and indeed for micropalaeontology as a whole, was *Introduction to Marine Micropaleontology*, edited by B U Haq and A Boersma (Elsevier, 1978). An intermediate text is *The Micropalaeontology of the Oceans*, edited by B M Funnel and W R Riedel (Cambridge University Press, 1971). An advanced text is A T S Ramsey's *Ocean Micropalaeontology* (2 vols, Academic Press, 1977).

The comprehensive reference volume on the study of planktonic microfossils is *Planktonic Stratigraphy*, edited by H M Bolli *et al* (Cambridge University Press, 1985). It covers all the important taxa in plant and animal groups. The papers presented at an earlier conference, the 2nd International Conference on Planktonic Microfossils, Rome, 1970 were published as *Proceedings* (Edizioni Tecnoscienza, 1971) and are still in use. The interpretation of marine microfossils, most of which are planktonic, is dependent on the knowledge of the biogeography of their modern counterparts. A major contribution to the understanding of this can be found in *Zoogeography and Diversity in Plankton*, edited by S van der Spoel and A C Pierrot-Butts (Edward Arnold, 1979).

Reviews of research can be found in *Microfossils from Recent and Fossil Shelf Seas*, edited by J W Neale and M D Brasier (Ellis Horwood, for the Micropalaeontological Society, 1980).

The journal for the subject is *Marine Micropalaeontology*, published by Elsevier (1976–).

Conodonts

The major text on conodonts can be found in *Treatise of Invertebrate Paleontology, part W, Miscellanea (Supplement 2): Conodonta* (University of Kansas and the Geological Society of America, 1981). A relevant collection of papers, *Conodont Paleozoology*, is edited by F H T Rhodes (*Geological Society of America, Special Papers*, **141**, 1973).

Catalogue of Conodonts, volume I, edited by W Ziegler (E. Schweizerbart'sche, 1974), is projected in 5 volumes, in loose-leaf format. It aims to cover all stratigraphically important conodonts, each section undertaken by a specialist. Each genus and species is detailed with references to original descriptions, type specimens, synonyms and geographical distribution, and illustrated, usually with a photograph. F J Collier's *Catalog of Type Specimens of Invertebrate Fossils: Conodonta (Smithsonian Contributions to Paleobiology*, **9**, 1971) contains an alphabetical listing and index by strata.

The ecology of conodont-bearing animals, their distribution and mode of life, were discussed at a symposium in 1975; a volume of papers from that meeting, *Conodont Paleoecology*, edited by C R Barnes (*Geological Association of Canada, Special Volume*, **15**, 1976) is useful to the specialist. *Taxonomy, Ecology and Identity of Conodonts* is the title of the publication of the Third European Conodont Symposium (ECOS III) in *Fossils and Strata*, **15**, 1982.

The Fourth European Conodont Symposium (ECOS IV) produced *A Stratigraphic Index of Conodonts*, edited by A C Higgins and R L Austin (Ellis Horwood for the British Micropalaeontological Society, 1985). Each strata is described by an expert giving details of localities, collections, and bibliography, and the publication is an essential reference work.

The main published bibliographies of conodont literature are a little outdated. S R Ash's *Bibiliography and Index of Conodonts* covers the period 1949–58 (in *Micropaleontology* **7**, 213–244, 1961), and 1959–63 (in *Brigham Young University Geology Studies* **10**, 3–50, 1963). S P Ellison's *Annotated Bibliography and Index of Conodonts* covers 1962–67 (published in *Publication, Texas Bureau of Economic Geology* **6210**, 128p., 1962)

Calcareous microfossils

The *Catalogue of Calcareous Nannofossils*, edited by A Farinacci, in 12 vols, to date (Edizioni Tecnoscienzia 1969–) is an important reference work. *A Stratigraphic Index of Calcareous Nannofossils*, edited by A R Lord (Ellis Horwood, for the British Micropalaeontological Society, 1982) is also a reference work for the reseacher, and contains a comprehensive bibliography.

A current update of the subject, *Bibliography and Taxa of Calcareous Nannoplankton*, can be found in the *Newsletter of the International Nannoplankton Association* (1979–).

Foraminifera

The relevant part of the *Treatise of Invertebrate Paleontology* for the group is part *C, Protista 2: Sarcodina*, chiefly "Thecamoebians" and Foraminiferida (University of Kansas, 1964). A general text is J R Haynes' *Foraminifera* (MacMillan, 1981). It covers all aspects, but with emphasis on the stratigraphic application of the group, and includes a substantial bibliography. The *Manual of Planktonic Foraminifers*, by J A Postuma (Elsevier, 1971) covers the species important for stratigraphic correlation. E Boltovskoy and R Wright's *Recent Foraminfera* (Junk, 1978) is a comprehensive textbook covering all aspects of their biology.

The periodical publications for the subject are both from the Cushman Foundation for Foraminiferal Research. They are the *Journal of Foraminiferal Research* (1971–) and the *Special Publications* of the Foundation (1952–). The *Journal* regularly publishes lists entitled 'Recent Literature on Foraminifera', a useful source of current information.

The major bibliographies and indexes of Foraminifera are published by the American Museum of Natural History. They are B F Ellis' *Catalogue of Index Foraminfera* (3 vols, 1965–1967) which covers the larger Foraminifera, the *Catalogue of Index Smaller Foraminifera* (3 vols, 1968–1969), B F Ellis and A R Messina's *Catalogue of Foraminifera* (1940–) and *Catalogue of Planktonic Foraminifera*, edited by T Saito *et al* (1976–). Others include H E Thalmann's *Bibliography and Index to New Genera, Species and Varieties of Foraminfera* (published annually in the *Journal of Paleontology*, vol. **7–33** for the years 1933 to 1959), and *An Index to the Genera and Species of the Foraminfera, 1890–1950* (Stanford University, 1960).

A compilation of the taxa of Foraminifera, each briefly described, illustrated, and placed in their stratigraphic context, with bibliography, is the *Stratigraphic Atlas of Fossil Foraminifera*, edited by D G Jenkins and J W Murray (Ellis Horwood, for the British Micropalaeontological Society, 1981).

For reseachers, a collection of reviews can be found in a three-volume work, *Foraminifera*, edited by R H Hedley and C G Adams (Academic Press, 1974–1978), which includes systematic and subject indexes and some biographical material. Papers, also for the researcher, from the *Second International Symposium on Benthic Foraminfera, 1983; Benthos '83*, is edited by H J Oertli (Elf Aquitaine, 1984).

Ostracods

H V W Howe's *Ostracod Taxonomy* (2nd edn, Louisiana State University Press, 1962) provides a basic text. A more recent work is F F Helmdach's *Leitfaden zur Bestimmung Fossilier und Rezenter Ostrakoden* (De Gruyter, 1977), which provides a systematic treatment, an illustrated key index with information on type, synonym, strata and distribution, but is poorly illustrated. Identification is facilitated by the use of the publication *Stereoscan Atlas of Ostracod Shells* (1973–), originally published by the University of Leicester, now by the British Micropalaeontological Society, and issued bi-annually.

The part of the *Treatise of Invertebrate Paleontology* dealing with Ostracoda is *Q, Arthropoda 3: Crustacea Ostracoda* (University of Kansas, 1961). A supplement to this is in preparation. *Fossil and Recent Ostracoda*, edited by R M Bate *et al* (Ellis Horwood, for the British Micropalaeontological Society, 1982) is for the specialist, with sections on structure, experimentation and techniques, systematic reviews, distribution and ecology, concepts and history.

The major compilation on Ostracoda is another of the American Museum of Natural History's publications — the *Catalog of Ostracoda*, by B F Ellis and A R Messina (1952–). Other bibliographies are H N Coryell's *Bibliographic Index and Classification of Mesozoic Ostracoda* (2 vols, University of Dayton Press, 1963), and E K Kempf's *Index and Bibliography of Nonmarine Ostracoda* (Universtat Koln, 1980).

Publications from meetings include the *Sixth International Symposium on Ostracoda*, which has a wider coverage than its title suggests, *Aspects of Ecology and Zoogeography of Recent and Fossil Ostracoda*, edited by H Loffler and D Danielopal (Junk, 1977). *The Biology and Paleobiology of Ostracoda; a symposium held at Delaware in 1972*, edited by F M Swain *et al*, is published in the *Bulletins of American Paleontology*, **65** (282), 1975, and contains a variety of papers ranging from morphological and physiological studies, to microscopy, classification and nomenclature. The *Colloquium on the Paleoecology of Ostracods*, edited by H J Oertli (Société Nationale des Pétroles d'Aquitaine, 1971), contains papers on morphology and stratigraphy.

Current information can be obtained in a specialist newsletter, the *Ostracodologist* (1963–).

Flagellates

The proliferation of descriptions of dinoflagellates in the past two to three decades have made up-to-date catalogues and indexes an essential working tool. A Eisenack's *Katalog der Fossilien Dinoflagellaten, Hystrichosphaeren und verwandten Mikrofossilien* (2 vols, E. Schweizerbart'sche, 1964–1975), and the publications of J K Lentin and G L Wiliams in the *Report Series of the Bedford Institute of Oceanography*, BI-R-77–8, 1977), are useful publications.

A Systematic Guide to Fossil Organic-walled Dinoflagellate

Genera, by D Artzner *et al* (*Life Sciences Miscellaneous Publications, Royal Ontario Museum*, 1979) provides an illustration of each genera to facilitate the use of catalogues; its limitations are that it excludes calcareous and siliceous cysts.

An illustrated book on the subject is J D Dodge's *Atlas of Dinoflagellates* (Farrand Press, 1985). General textbooks include *Fossil and Living Dinoflagellates*, by W A S Sarjeant (Academic Press, 1974). *A Glossary of Terminology Applied to Dinoflagellate Amphiesmae and Cysts and Acritarchs* (in *American Association of Stratigraphic Palynologists Foundation Series*, **2a**, 1978) is a useful aid to study.

Algae

A good introduction and practical guide to fossil algae can be found in J L Wray's *Calcareous Algae (Developments in Palaeontology and Stratigraphy*, **4**, 1977), which deals with their origin, occurrence, stratigraphy and ecology.

The proceedings of the first International Symposium of Fossil Algae in 1975 resulted in a collection of papers useful for further study in the subject: *Fossil Algae*, edited by E Flügel (Springer Verlag, 1977). The publication from the second symposium, in 1979, is narrower in scope: *Phanerozoic Stromatolites*, edited by C Monty (Springer Verlag, 1981).

Stromatolites are presumed algal deposits found in abundance in Precambrian strata, although they have no trace of filaments, and there are problems in their classification and nomenclature. *Stromatolites*, edited by M R Walter (*Developments in Sedimentology*, **20**, 1976) is a comprehensive work on the subject. The publication from *The Workshop on Stromatolites: characteristics and utility, Udaipur, 1978*, is in the *Miscellaneous Publication of the Geological Survey of India*, **44**, 1980. Current literature on stromatolites is listed periodically in the journal *Precambrian Research* (1974–).

The Palaeobiology of Plant Protists, by H Tappan (W H Freeman, 1980) is a compilation of knowledge of fossil and living pro- and eucaryotic algae. Its contents include procaryotes (bacteria and blue-green algae), rhodophyta, acritarchs, dinoflagellates, silicoflagellates, diatoms, calacareous nannoplankton, and green algae, to name some of the groups the book covers. The comprehensiveness of the volume makes it a basic work for the subject, and the full glossary and extensive bibliography makes it especially valuable.

A source of current information can be found in *Algal Newsletter* (1977–), *Stromatolite Newsletter* (1972–) and *Acritarch Newsletter* (1984–).

Palynology (see also Palaeobotany)

The literature of palaeopalynology is also found within the wider spectrum of palaeobotany. Of those works exclusive to palynology, a notable bibliography is G O W Kemp's listings of palynological literature, from early Precambrian through to the Quaternary, in several parts, and with supplements, which can be found in the periodical publication *Palaeo Data Banks* (1977–1983): this information is now available on a commercial database called Palynodata.

Textbooks on palynology include *An Illustrated Guide to Pollen Analysis*, by P D Moore and J A Webb (Hodder & Stoughton, 1978) and K Faegri & J Iverson's *Textbook of Pollen Analysis* (3rd edn, Hafner, 1975). For papers on palynology assembled for the use of students, *Palynology*, edted by M D Muir and W A S Sarjeant (*Benchmark Papers in Geology*, **46 & 47**, 1977) devotes the first part to spores and pollen, and the second to dinoflagellates, acritarchs and other fossils. A major collection of papers for the researcher is found in *Interpreting Pollen Structure and Function*, edited by I K Ferguson and J Muller (*Review of Palaeobotany and Palynology*, **35**(1), 1981).

For the student of higher plants, works on their palynology are *Fossil Pollen Records of Extant Angiosperms*, by J Muller in *Botanical Review*, **47**(1), 1981, and *A Key to the Genera of Fossil Angiosperm Pollen*, by J Jansonius in *Review of Palaeobotany and Palynology*, **26**, 1978.

The *Genera File of Fossil Spores and Pollen* is in the form of three thousand index cards (in two boxes), compiled by J Jansonius and L V Hills (Department of Geology, University of Calgary, 1976), arranged in alphabetical order, and covering genera from Silurian to the Miocene. Each generic name is accompanied by its bibliographic citation, geological age, nomenclature problems, and usually, a line drawing of the holotype.

The journals of palynology are *Grana* (1965–), *Journal of Palynology* (1965–), *Palynology* (1977–), *Pollen et Spores* (1959–), *Review of Paleobotany and Palynology* (1967–). Important too, are the publications of the American Association of Stratigraphic Palynologists, in the *Contribution Series*.

An annual compilation of publications in palynology can be found in *Bibliographie: Palynologie* (1974/5–) as a supplement to the journal *Pollen et Spores*.

Useful for reference is G O W Kemp's *Morphologic Encyclopedia of Palynology: an international collection of definitions and illustrations of spores and pollen* (University of Arizona Press, 1986) and, for Quaternary palynology, B Huntley and H J B Birks'

An Atlas of Past and Present Pollen Maps for Europe; 0–13,000 years ago (Cambridge University Press, 1983).

Palaeobotany

For the non-specialist, *The Evolution of Plants and Flowers*, by B Thomas (Peter Lowe, 1981) explains plant morphology and evolution with colour illustrations and line drawings. An intermediate textbook is *Paleobotany and the Evolution of Plants*, by W N Stewart (Cambridge University Press, 1983), which concentrates on the fossil plant record and its evolution, and an advanced textbook is *Paleobotany: an introduction to fossil plant biology*, by T N Taylor (McGraw-Hill, 1981). This book has a very wide brief: classification, correlation, morphology and a full systematic treatment of the plant kingdom from Precambrian non-vascular plants to angiosperms: it is supplemented by a useful bibliography and a taxonomic index. A collection of papers useful at this level, edited by D L Dilcher & T N Taylor, is *Biostratigraphy of Fossil Plants: successional and paleoecological analysis* (Dowden, Hutchinson Ross, 1980).

An evolutionary emphasis is noticeable in *Evolution — Naturgeschichte Höherer Planzen*, edited by R Daber (Akademie Verlag, 1980), a volume of contributions from specialists, useful for advanced studies, and with abstracts in English, Russian and French. Another publication on the same theme is *Geological Factors and the Evolution of Plants*, edited by B H Tiffney (Yale University Press, 1985).

The journals specializing in palaeobotany are:

Acta Palaeobotanica (1960–)
Argumenta Palaeobotanica (1966–)
Boletin Asociacion Latinamericana de Paleobotanica y Palinologica (1973–)
Bonner Paläobotanische Mitteilungen (1974–)
Geophytology: a journal of palaeobotany and allied sciences (1971–)
Birbal Sahni Institute of Palaeobotany publications include the *Monographs* (1964–), *Memorial Lectures* (1973–), and *Special Publications* (1974–)
Palaebotanist (1952–)
Palaeontolographica, Abteilung B: Paläophytologie (1933–)
Paleobotanica Latinamericana (1979–)
Paleobotanika (1956–)
Review of Paleobotany and Palynology (1967–).

The major reference work for all groups of fossil plants is the *Fossilium Catalogus, pars Plantae* (s'Gravenshaage, 1914–). This multivolume work is occasionally published, and devotes each part to a group. Also important as a reference text is *Traité de Paléobotanique*, edited by E Boureau (Masson, 1964–); nine volumes have been published so far.

The major bibliographic tool for palaeobotany is the *Biblio-*

graphy and Index to Palaeobotany and Palynology, 1971–1975, edited by H Tralau and B Lundblad (Swedish Natural Science Research Council, 1983). In two volumes, the first is a bibliography and the second an index: this work and its companion volume, the *Bibliography and Index for 1950–1970* (published 1974), provides the most comprehensive compilation of the literature. In addition, E Boureau's *Rapport sur la Paléobotanique dans de Monde 1956–1974*, is published in *Regnum Vegetabile*, volumes **7**, **11**, **19**, **24**, **35**, **42**, **57**, **78**, and **89**.

The *Index of Figured Plant Megafossils, 1971–1975*, compiled by M Boersma and L M Broekmeyer, is in four volumes: Triassic, Jurassic, Permian, and Carboniferous (Laboratory of Palaeobotany and Palynology, University of Utrecht, 1979–1982). It lists species alphabetically, together with their bibliographic citation, strata and locality: they are also separately arranged into their major plant groups, geochronological age and country of origin.

For the higher plants G E Dolph has compiled the *Bibliography of Angiosperm Paleobotany* since 1978, and published it as *Miscellaneous Publications of The International Organization of Palaeobotany*, 1979–.

Regional bibliographies of palaeobotany include *Bibliography of American Paleobotany for 1962/63*, published by the Paleobotanical Section of the Botanical Society of America (1965–) and *Bibliografia Paleobotanica y Palinologica Latinamericana*, published periodically in *Boletin Asociacion Latinamericana de Paleobotanica y Palinologica* for the New World. For Europe there is *Rapport sur la Paléobotanique et la Paléopalynologie: France, Belgique, Suisse* for the years 1972/1976– (Laboratoire de Paleobotanique, Université Pierre et Marie Curie, 1977–). For Australasia there is the *Bibliography of Australasian Palynology and Palaeobotany from 1977/78–* (Palynological and Palaeobotanical Association of Australasia, 1979–), and for China the *Bibliography of Chinese Palaeobotany to 1980–* (Academia Sinica, Institute of Geology and Palaeontology, Nanjing, 1981–).

Palaeoecology of plants

An important book in this subject is V A Krasilov's *Palaeoecology of Terrestrial Plants* (translated from the Russian by the Israel Program for Scientific Translation, Wiley, 1975). The book is divided into four sections: an introduction to the principles, methods of burial, transport and fossilization, morphological restoration from plant parts, and the reconstruction of vegetation and environments; it also contains a substantial bibliography.

Keyworks in various palaeobotanical groups include; for the

cryptogamic plants, *Paleozoic Origins of the Cycads*, by S Mamay (*US Geological Survey, Professional Paper*, **934**, 1976) and *Comparative Anatomy of Vegetative Organs of the Pteridophytes* (2nd edn, Bornträger, 1972); for gymnosperms, *Patterns in Gymnosperm Evolution*, edited by T N Taylor (*Review of Palaeobotany and Palynology*, **21** (1), 1976).

Of the books which deal with the origin and evolution of angiosperms, some have a morphological approach, e.g. K R Sporne's *The Morphology of Angiosperms* (Hutchinson, 1974); *Flowering Plants, Evolution Above Species Level*, by G L Stebbins (Arnold, 1974), takes a genetical approach, and N F Hughes' *Palaeobiology of Angiosperms Origins* (Cambridge University Press, 1976) puts the palaeobotanical evidence into geological and taxonomic perspective. A number of papers covering the various aspects can be found in *Origin and Early Evolution of Angiosperms*, edited by C B Beck (Columbia University Press, 1976).

Invertebrate palaeontology

The single, most important, compilation is the *Treatise of Invertebrate Paleontology*, edited by R C Moore, published by the Geological Society of America and the University of Kansas. This publication, with each part authored by a specialist, has been issued and revised irregularly since 1953. *Part A*, the *Introduction*, edited by R A Robinson and C Teichert (1979) covers the entire fossil record. The rest, numbered alphabetically in the systematic sequence, are dealt with in their respective groups.

A Manual of Invertebrate Paleontology, by J C Dyer and F R Schram (Stipes Publishing Co., Illinois, 1983) advertises itself as the "beginners' guide to the identification of fossils". Other introductory textbooks on invertebrates include P Tasch's *Palaeobiology of the Invertebrates: data retrieval from the fossil record* (2nd edn, Wiley, 1979), with chapters on habits, morphology, and a wide treatment of phyla, and *Fossil Invertebrates*, by U Lehmann and G Hillmer (Cambridge University Press, 1983).

For the more advanced student, E N K Clarkson's *Invertebrate Palaeontology and Evolution* (Allen & Unwin, 1979) deals with the principles of palaeontology, evolution and the invertebrate phyla, with a list of references, and there is also B Ziegler's *Specielle Paläontologie: Protisten, Spongien und Coelenteraten, Mollusken* (Schweizerbart'sche, 1983). *Invertebrate Fossils*, by R C Moore *et al* (McGraw-Hill, 1952) although old, is still a major textbook.

A recent publication, useful for both the amateur and the student is the *Atlas of Invertebrate Macrofossils*, edited by J W Murray (Longman, 1985).

A classic work on invertebrates is A M Davies' *Tertiary Faunas*; the emphasis of the book is shown by its subtitle, 'a textbook for the oilfield palaeontologist and students of geology'. The second edition, revised by F E Eames and R J G Savage, published in two volumes: *The Composition of Tertiary Faunas*, and *The Sequence of Tertiary Faunas* (Allen & Unwin, 1971–1975), is a useful reference work for the student.

On the evolutionary perspective, *The Origin of Major Invertebrate Groups*, edited by M R House (Academic Press, 1979) (*Systematics Association Special Volume*, **12**) contains a useful collection of papers covering the range from Precambrian fossils to the origin of chordates.

Protozoa see *Micropalaeontology*

Archaeocyantha, Porifera

The part of the *Treatise* dealing with fossil sponges is *E, Archaeocyantha, Porifera* (1954); the volume on Archaeocyantha was revised and enlarged in a second edition in 1972. For an introductory text, *Living and Fossil Sponges*, by W D Hartman *et al* (*Sedimenta*, **8**, 1980) can be consulted.

Coelenterata

Coelenterata is covered in the *Treatise of Invertebrate Paleontology, part F* (1956) with a supplement on the Rugosa and Tabulata (1981).

Recent Advances in the Paleobiology and Geology of the Cnidaria, edited by W A Oliver *et al* (*Palaeontographica Americana*, **54**, 1984), is the proceedings from the Fourth International Symposium, Washington, D.C., 1983, and contains a large number of papers useful for reference.

The Rugose Coral Genera, by G Cotton (privately published by the author, 1984) is a replacement of the volume of the same name (Elsevier, 1973) and its supplements of 1974, 1977 and 1980, by integrating all the information in these volumes. It is an essential reference tool, listing all genera and species and their references, especially in combination with *The Rugose Coral Species*, by the same author (also privately published, 1983).

Fossil Cnidaria, with the parallel title of *Cnidaires Fossiles* (Centre National de la Recherche Scientifique, 1972–), provides up-to-date information on the subject as well as a current bibliography in each issue.

Bryozoa

For reference texts in bryozoology, *Advances in Bryozoology*, edited by G P Larwood and M B Abbott (Academic Press, 1979) (*Systematics Association Special Volume* **13**), contains a number of review articles covering a wide range of ideas and methods. The part for Bryozoa in the *Treatise of Invertebrate Paleontology* is *G* (published 1953) and revised, in part, in 1983.

Combining zoology and palaeontology in the same volume are *Recent and Fossil Bryozoa*, edited by G P Larwood and C Nielsen (Olsen and Olsen, 1981), and *Living and Fossil Bryozoa: recent advances in research*, edited by G P Larwood (Academic Press, 1973).

Brachiopoda

An index for brachiopods is *Living and Fossil Brachiopod Genera 1775–1979; lists and bibliography*, compiled by R A Doescher (*Smithsonian Contributions to Paleobiology*, **42**, 1981).

The major landmark in the treatment of the study of the group is M J S Rudwick's *Living and Fossil Brachiopods* (Hutchinson University Library, 1970). It is an interpretive study of the phylum as functioning and evolutionary systems. It is an easy text to read, without having to resort to the more comprehensive part in the *Treatise of Invertebrate Paleontology, Part H*. (2 vols, 1965).

British Brachiopods, by C H C Brunton and G B Curry (Linnean Society and Academic Press, 1979) (*Synopses of the British Fauna*, new series, **17**) covers both the recent and fossil specimens.

Mollusca

The coverage for the molluscan groups in the *Treatise of Invertebrate Paleontology* is quite comprehensive: *part I* covers general features and Gastropoda (published 1960); *part K* deals with Cephalopoda (1960); *part L* with Ammonoidea (1957); and *part N* with Bivalvia (1971); *part J*, for Gastropoda, and *M*, for Coleoidea, and a revision of *part L* are in preparation.

An important reference work for the Bivalvia is the *Genera of the Bivalvia: a systematic and bibliographic catalogue* (revised and updated) by H E Vokes (Paleontological Research Institution, 1980). This catalogue is a revision of his earlier catalogue in *Bulletins of American Paleontology*, **51**(232), 1967: each entry consists of a generic name, author, date and reference, synonyms and homonyms. *Evolutionary Systematics of Bivalve Molluscs* in

Philosphical Transactions of the Royal Society of London, B, **284**, (1001), 1978 contains useful papers.

The ammonites are an attractive group, related to the nautilus, with much popular appeal. For an introduction to the subject, *The Ammonites: their life and their world*, by U Lehmann (Cambridge University Press, 1981, translated from the German edition of 1976) examines the fossil against the recent cephalopod groups and their biology. For advanced reading, *The Ammonoidea: the evolution, classification, mode of life and geological usefulness of a major fossil group*, edited by M R House and J R Senior (Academic Press, 1981) (*Systematic Association Special Volume*, **18**) is part of the proceedings of an international symposium held in York in 1979. The current use of ammonites as guide fossils in stratigraphic studies makes this reference work essential for study.

The periodical publications for Mollusca cover both recent and fossil groups; they include *Journal of Molluscan Studies* (1976–), formerly the *Proceedings of the Malacological Society, London* (1893–1975), and the *Journal of Conchology* (1874–). *Cephalopod Newsletter* (1977–) is an informal publication for current awareness, and includes in its distribution a *Directory of World Cephalopod Workers* (University of Durham, 1977).

Arthropoda

The parts of the *Treatise of Invertebrate Paleontology* dealing with Arthropoda are: *O; Arthropoda General Features*, (1960); *P; Chelicerata, Pycnogonida, Palaeoisopus*, (1955); *R; volume 1–2, Crustacea exclusive of Ostracoda, Myriapoda, Hexapoda*, (1969); volume 3 of this part, Hexapoda, is still in preparation.

Trilobita

A good illustrative book (unfortunately a limited edition) is Levi-Setti's *Trilobites: a photographic atlas* (University of Chicago Press, 1975). Apart from the attractive appearance, it gives a brief summary on trilobite morphology.

Some of the more important papers for trilobites come from the serial, *Fossils and Strata*, published by the Universitets Forlaget in Oslo : *Organization, Life and Systematics of Trilobites*, by J Bergström in no.**2**, 1973, and *Evolution and Morphology of Trilobita, Trilobitoidea and Merostomata*, edited by A Martinsson in no.**4**, 1975. The University also publishes *Trilobite News* (1971–).

Another publication, this one by the Palaeontological Association, is *Trilobites and Other Early Arthropods: papers in honour of*

Professor H B Whittington, edited by D E G Briggs and P D Lane (*Special Papers in Palaeontology*, **30**, 1983). A volume devoted to trilobites in the *Treatise on Invertebrate Paleontology* is planned but not published.

Crustacea

A collection of papers, entitled *Crustacean Phylogeny*, edited by F Schram (Balkema, 1983) provides stimulating reading for the student and research worker. Also good reading are the *Papers from the Conference on the Biology and Evolution of Crustacea, Sydney, 1980*, edited by J K Lowry (*Australian Museum Memoir*, **18**, 1983).

Insecta

W Hennig's *Insect Phylogeny* (English translation, Wiley, 1981), is considered a milestone in the study of insect evolution, and it also includes localities of Palaeozoic and Mesozoic insects. Other publications for this enormous group can be found in *Entomology: a guide to information services*, by P Gilbert and C J Hamilton (Mansell, 1983).

Echinodermata

The sections in the *Treatise of Invertebrate Paleontology* for this group are *S: Echinodermata, General Features, Homalozoa, Crinozoa exclusive of Crinoidea*, (1968), *T: Crinoidea* (3 vols, 1978) and *U: Asterozoans, Echinozoans* (1966).

An intermediate text, *Echinoderms: notes for a short course*, edited by T W Broadhead and J A Waters (*University of Tennessee Studies in Geology*, **3**, 1980), provides a number of review papers. A textbook with an interpretive approach to echinoderms is the recently published *Echinoid Palaeobiology*, by A Smith (Allen & Unwin, 1984), the first of a series of monographic works in *Special Topics in Palaeontology*.

The study of fossil echinoderms is closely related to the living. Important in literature are the proceedings of three International Conferences: one is edited by D Zavodnik *et al*, and published in *Thalassia Jugoslavica* **12** (1), 1978 and there are *Echinoderms: past and present*, edited by M Jangoux (Balkema, 1980) and *Echinoderms*, edited by J M Lawrence (Balkema, 1982).

Echinoderm workers and students can subscribe to *Echinoderm Newsletter* (1968–), published by the United States National Museum, which runs a regular column of recent publications.

The bibliographies for echinoderms are *Index of Living and Fossil Echinoids, 1924–1970*, edited by P Kier and M H Lawson

(*Smithsonian Contributions to Palaeobiology*, **34**, 1977), N E Weisbrod's *Bibliography of Cenozoic Echinoidea including some Mesozoic and Palaeozoic Titles (Bulletins of American Paleontology*, **59**(263), 1971), and G D Webster's *Bibliography and Index of Paleozoic Crinoids, 1942–1968 (Geological Society of America Memoir*, **137**, 1973).

Graptolithina

Part V in the *Treatise of Invertebrate Palaeontology*, first published in 1955, with a second revised and enlarged edition in 1970, is the most authoritative work on Graptolithina. Readings on the group can be found in a memorial volume in honour of D M B Bulman; 20 papers, edited by R B Rickards, D E Jackson and C P Hughes, is entitled *Graptolite Studies*, in *Special Papers in Palaeontology*, **13**, 1974).

Miscellanea

Conodonts (see under 'Marine Micropalaeontology', p.248)

Worms

Worms are "treated" as part of a wider group of *Miscellanea* in *part W* of the *Treatise of Invertebrate Paleontology* (published 1962). A revised supplement is in preparation.

Trace fossils

Recent years have seen an intensification of publication on trace fossils, largely because of their interpretive value. *The Study of Trace Fossils: a synthesis of principles, problems and procedures in ichnology*, edited by R W Frey (Springer Verlag, 1975) contains original research as well as review papers on the history of ichnology, classification, preservation, significance in palaeontology, palaeoecology, stratigraphy and sedimentology. However, the major reference work is the *Treatise of Invertebrate Paleontology W: Miscellanea, supplement 1, Trace Fossils and Problematica* (revised and enlarged, 1975).

For the student, a useful collection of papers is *Terrestrial Trace Fossils*, edited by W A S Sarjeant (*Benchmark Papers in Geology*, **76**, 1983). Two volumes, entitled *Trace Fossils*, edited by T P Crimes and J C Harper, in *Geological Journal Special Issues* **3** (1970) and **9** (1976) contain papers of general interest. For the specialist, there are *Biogenic Structures: their use in interpreting depositional environments*, edited by H A Curran (*Society of Stratigraphic Paleontologists and Mineralogists Special Publica-*

tion, **35**, 1985), *Ichnology: the use of trace fossils in sedimentology and stratigraphy*, by A A Ekdale *et al* (Society of Economic Paleontologists and Mineralogists, 1984), and *Recent Advances in Paleoecology and Ichnology*, edited by J D Howard *et al* (American Geological Institute, 1971).

Ichnology, the study of trace fossils, has its own periodical, *Ichnology Newsletter* (1979–), which contains regular updates of recent literature.

Vertebrate palaeontology

Introductory textbooks include B J Stahl's *Vertebrate History: problems in evolution* (McGraw-Hill, 1974), which centres on the concepts surrounding vertebrate palaeontology, and E Colbert's *Evolution of Vertebrates* (3rd edn, revised, Wiley, 1980), which covers the requirements for the popular and student market.

At the intermediate level is *Précis de Paléontologie des Vertebrates*, by J Piveteau, J P Lehman and C Deschaseaux (Masson, 1978) and *Vertebrate Life*, by W N Farland *et al* (Collier MacMillan, 1979).

An important work for both advanced students and researchers is *Problems in Vertebrate Evolution*, edited by S M Andrews *et al* from the Linnean Society Symposium Series, and the problems treated in this volume are the current controversies in morphology and evolution. A two-volume work, *Basic Structure and Evolution of Vertebrates*, by E Jarvik (Academic Press, 1980–1981), is a summary of the author's life's work, reflecting vertebrate history and morphology. Other books on vertebrate evolution include S Lovtrup's *The Phylogeny of Vertebrates* (Wiley, 1977) which takes a cladistic approach. Other important works in vertebrate phylogeny for the researcher are *Problèmes Actuels de Paléontologie: evolution des vertébrés* (2 vols, *Colloques Internationaux du Centre National de la Recherche Scientifique*, **218**, 1975), and *Phylogenie et Paléobiogeographie: livre jubilaire en l'honneur de Robert Hoffstetter* published in *Geobios, Memoirs Speciales*, **6**, 1982.

A book on land adaptation, *The Terrestrial Environment and the Origin of Land Vertebrates*, edited by A L Panchen (Academic Press, 1980) (*Systematics Association Special Volume*, **15**), is the proceedings of a symposium on the ecology, systematics and functional morphology of early land vertebrates and their ancestors. With special reference to respiration, *The Evolution of Air Breathing in Vertebrates*, by D J Randall *et al* (Cambridge University Press, 1981) provides a useful book for functional morphologists. With reference to the limb a useful volume is *The*

Development of the Vertebrate Limb: an approach through experiment, genetics and evolution, by J R Hinchcliffe and D R Johnson (Oxford University Press, 1980).

Major regional studies include the four-part work, *Fossil Vertebrates of Africa*, under the general editorship of L S B Leakey, and later together with R J G Savage and S C Corydon (Academic Press, 1969–1976), and *The Fossil Vertebrate Record of Australasia*, by P V Rich and E M Thompson (Monash University Press, 1982).

The journals for vertebrate palaeontology are *Palaeovertebrata* (1967–), *Journal of Vertebrate Palaeontology* (1981–), *Contributions to Vertebrate Evolution* (1977–), and *Mesozoic Vertebrate Life* (1980–). Regional publications are *Vertebrata Palasiatica* (1957–), *Memoirs, Institute of Vertebrate Palaeontology and Palaeoanthropology* (1957–) from China, and *Alaskan Journal of Pre-Pleistocene Vertebrate Paleontology* (1980–). Current awareness publications include *News Bulletin, Society of Vertebrate Palaeontology* (1941–).

A documentation of publications in the subject can be found in the *Bibliography of Vertebrate Palaeontology*, published by Geosystems Ltd., from the database GeoArchive, from 1973 to 1978. It was previously published by the Society of Vertebrate Palaeontology, under this, and former titles, from 1945 to 1971. The bibliography comprises an author catalogue, subject index, list of serials searched, geographic index, stratigraphic index, and a list of new taxa.

A more formal, but irregular publication, compiled by the Society of Vertebrate Palaeontology in cooperation with the Museum of Paleontology, University of California, Berkeley, and published by the American Geological Institute, is the *Bibliography of Fossil Vertebrates*. It is a continuation of the bibliography, 1969–1972, by J T Gregory *et al*, in *Geological Society of America, Memoirs*, **141**, 1973. The volumes published to date are 1973–1977, 1978, 1979 and 1980. The layout of the bibliography is similar to the *Bibliography of Vertebrate Palaeontology*.

Palaeoichthyology

A specialist series of monographs, edited by H P Schultze, *Handbook of Palaeoichthyology* (Gustav Fischer, 1978–) is projected in ten volumes and is being published erratically. Each volume is written by a specialist, and covers general morphology and systematics. The volumes published to date are *Placodermi* (**2**, 1978), *Chondrichytes I: Paleozoic Elasmobranchii* (**3A**, 1981), *Acanthodii* (**5**, 1979), and *Otolithi piscium* (*10*, 1985).

The second edition of *Palaeozoic Fishes*, by J A Moy-Thomas, substantially revised by R S Miles (Chapman & Hall, 1971), is arranged systematically, each chapter dealing with the classification, time range, morphology, biology, and evolution of the fishes, and is suitable as an intermediate text. For the beginner there is *An Age of Fishes: the development of the most successful vertebrate*, by J Spoczynska (Charles Scribner, 1976).

Review papers in palaeoichthyology can be found in *Interrelationships of Fishes*, edited by P H Greenwood et al (Academic Press, 1973), each group being treated by an expert.

A journal for fossil fish researchers is *Palaeoichthyology* (1983–).

Herpetology

The major reference work for fossil amphibians and reptiles is *Handbuch der Paläoherpetologie: Encyclopedia of Palaeoherpetology*, edited by D Kühn (Gustav Fischer, 1969–), which is projected in 19 volumes.

Elementary texts include W E Swinton's *Fossil Amphibians and Reptiles* (British Museum (Natural History), 1973), while A S Romer's *Osteology of the Reptiles* (University of Chicago Press, 1976, a reprint of the 1956 edition) is still useful. *Morphology and Biology of Reptiles*, edited by A Bellairs and C B Cox (Academic Press, 1976), in the Linnean Society Symposium Series, deals with both fossil and recent species in papers covering a wide range of topics on reptile biology.

Dinosaurs

Dinosaurs, of all reptiles, and indeed of all the fossil groups, seem to excite the most popular interest. Of the many elementary dinosaur books with which the market is inevitably flooded, a few examples are L B Halstead's *The Evolution and Ecology of the Dinosaurs* (Peter Low, 1976), which contains good illustrations as well as discussion on interrelationships between each group of dinosaurs, and their reconstruction from the fossil evidence, A Charig's *A New Look At Dinosaurs* (revised edition, Heinemann and British Museum (Natural History), 1983) which starts dinosaur studies from basic principles, *The World of Dinosaurs*, by M Tweedie (Weidenfeld & Nicholson, 1977) offers a well-balanced account, and L B and J Halstead's *Dinosaurs* (Blandford Press, 1981) provides ample illustrations. Last in the series is D Norman's *The Illustrated Encyclopedia of Dinosaurs* (Salamander Press, 1985) which gives a systematic review of each group, including anatomical details, and chapters on their origins and extinction.

Another category of popular dinosaur publication is the dictionary *The Illustrated Dinosaur Dictionary*, compiled by H R Battler (Lothrup, Lee and Shepard, 1983) and D F Glut's *The New Dinosaur Dictionary* (Citadel Press, 1982) which are for the advanced amateur market. In the style of a dictionary is M Benton's *Pocket Book of Dinosaurs* (Kingfisher Books, 1984) which is suitable for the younger reader.

The evolution of dinosaurs, in particular the controversial subject of warm-blooded dinosaurs, was the topic for an American Association for the Advancement of Science Symposium in 1978. The proceedings, edited by R D K Thomas and E C Olson, *A Cold Look At the Warm-Blooded Dinosaurs* (Westview Press, 1980), provides a comprehensive review of the subject, including bibliographies. An earlier work on the subject is A J Desmond's *The Hot-Blooded Dinosaurs: a revolution in palaeontology* (Dial, 1976).

The extinction of dinosaurs is another controversial topic, with many contributed theories. The gradual warming of the earth, at the end of the Cretaceous, giving rise to cataracts in the dinosaurs' eyes, is the theme of *The Last Dinosaurs: a new look at the extinction of dinosaurs*, by L R Croft (Elmwood Books, 1982), one of many publications; J L Cloudsley Thompson's *Why Dinosaurs Became Extinct* (Meadowfield Press, 1978) is another.

Fossil birds

The Age of Birds, by A Feduccia (Harvard University Press, 1980) on the history of birds through their dinosaur ancestry, the evolution of extant groups, including flightless birds, is suitable for the popular audience as well as the student market. Also on the elementary level is W E Swinton's *Fossil Birds* (3rd edn, British Museum (Natural History), 1975).

Fossil mammals

A book for the junior and elementary undergraduate market is *The Evolution of the Mammals*, by L B Halstead (Eurobook Limited, 1978): for the intermediate level there is B Kurten's *The Age of Mammals* (Columbia University Press, 1972) and a book for all interested readers is A J Sutcliffe's *On the Track of Ice Age Mammals* (British Museum (Natural History), 1985), which is also well illustrated.

The study of the reptilian ancestry is an integral part in the evolution of mammals. An introductory text is J C McLoughlin's *Synapsida: a new look into the origin of mammals* (Viking Press,

1980). Two books, for the undergraduate and research market, treat the evolution of mammals with differing theories: T S Kemp's *Mammal-like Reptiles and the Origin of Mammals* (Academic Press, 1982) concentrates on the ancestry through the cynodonts; *The Evolution of Mammalian Characters*, by D M Kermack and K A Kermack (Croom Helm, 1984) deals with the convergent evolution in mammal-like reptile groups.

An indispensable reference work is D E Savage and D E Russell's *Mammalian Paleofaunas of the World* (Addison-Wesley, 1983) which is an accumulation of data on the fossil record of mammals, particularly that of the Tertiary period. It includes a comprehensive listing of species for each period and provides major bibliographic citations. The Mesozoic is covered in detail by J A Lillegraven, Z Kielan-Jaworowska and W A Clemens' *Mesozoic Mammals: the first two-thirds of mammalian history* (University of California Press, 1979), and the Pleistocene by B Kurten's *Pleistocene Mammals of Europe* (Weidenfeld and Nicholson, 1968) and B Kurten and E Anderson's *Pleistocene Mammals of North America* (Columbia University Press, 1980).

Cetacea

The transition from terrestrial to marine environment of the Cetacea (whales, dolphins and porpoises) is a fascinating study. G A Mchedlidze's *General Features of the Paleobiological Evolution of Cetacea* (English translation, Balkema, 1985), covers a group from the USSR.

Primates

A comprehensive book on primate palaeontology is *Evolutionary History of Primates*, by F S Szalay and E Delson (Academic Press, 1980) which avoids the usual bias to human evolution.

Palaeoanthropology

Apart from the literature on fossil primatology, books on the evolution of early humans from primate origins, for the researcher, include *New Interpretations of Ape and Human Ancestry*, edited by R S Ciochon and R S Corruccini (Plenum, 1983), published in the series *Advances in Primatology: primate brain evolution*, edited by E Armstrong and D Falk (Plenum, 1982). Books for the student include *The Human Primate*, by R E Passingham (W H Freeman, 1982), and *Primate Evolution and Human Origins*, by R L Ciochon and J G Fleagle (Benjamin/Cumming Press, 1985).

Fossil hominids

Textbooks on palaeoanthropology include *The Fossil Evidence for Human Evolution*, by W E Le Gros Clark, revised and enlarged by B G Campbell (3rd edn, University of Chicago Press, 1978), *The Study of Human Evolution*, by R B Eckhardt (McGraw-Hill, 1979), and *Fossil Evidence; the human evolutionary journey*, by F E Poirier (2nd edn, C V Mosby, 1977). Students may find *Hominid Fossils: an illustrated key*, by T W Phenice and N J Sauer (Wm C Brown, 1977), and *Atlas of Radiographs of Early Man*, by M F Skinner and G H Sperber (A R Liss, 1982), useful.

For reading at the elementary level, or for general interest, *Origin*, by R E Leakey and R Lewin (MacDonald & Jane's, 1977), *The Making of Mankind*, also by R E Leakey (Michael Joseph, 1981), and *Missing Links: the hunt for the earliest man*, by J Reader (Collins, 1981) are suitable introductions.

The last couple of decades have seen the discovery of more fossil hominids, especially in Africa. An introduction to this can be found in M G Leakey and R E Leakey's edited work, *The Fossil Hominids and an Introduction to their Context* (Clarendon Press, 1978) based on the Koobi Fora Research Project, in which the specimens from the area are catalogued. Introducing the subject through one of the most significant finds is *Lucy: the beginning of humankind*, by D C Johanson and M A Edey (Granada, 1981).

Literature on the geological setting of fossil man includes *Hominid Sites: their geologic settings*, edited by G Rapp and C F Fondra (Westview Press, 1981), and, with special reference to the East African Rift Valley, *Geological Background to Fossil Man*, edited by W W Bishop (Geological Society of London, 1977).

Most of the journals for anthropology cover a wide range of subjects, physical and biological anthropology, archaeology, social anthropology and ethnography, as well as palaeoanthropology, usually without any clear distinctions. One of the most important journals for palaeoanthropologists is the *American Journal of Physical Anthropology* (1981–) and journals which are mostly on palaeoanthropology are *Acta Anthropologica Sinica* (1982–) and *Journal of Human Evolution* (1972–).

Others of wider coverage with palaeoanthropological content are:

American Anthropologist (1960–)
Annual Review of Anthropology (1972–)
Anthropologia Hungarica (1957–)
Anthropological Papers of the American Museum of Natural History (1907–)
Anthropological Papers, Museum of Anthropology, University of Michigan (1949–)
Anthropological Records, University of California (1937–)

L'Anthropologie (1890–)
Anthropologie, Brno (1923–)
Anthropos, Athens (1974–)
Anthropos, Brno (1958–)
Antiquity (1927–)
Anthropologischer Anzeiger (193?–)
Antropologia y Paleoecologia Humana (1979–)
Antropologischeskii Sbornik (1956–)
Archives de l'Institut de Paléontologie Humaine (1927–)
Biometrie Humaine (1983–)
Current Anthropology (1960–)
Eastern Anthropologist (1948–)
Fieldiana: Anthropology (1845–)
L'Homme (1961–)
Homo (1950–)
Man (1901–)
Memoirs of the Institute of Vertebrate Palaeontology and Palaeoanthropology (1957–)
Voprosy Antropologiya (1957–).

Apart from general bibliographies on anthropology, the literature on fossil hominids is documented in *A Bibliography of Fossil Man*, compiled in 4 parts by G E Fay (Southern State College, Arkansas for part 1, *and* Museum of Anthropology, Colorado State College for the rest, 1959–1969). It covers the literature for 1845 to 1968, including that for the Old World hominids.

Current bibliographies include *Abstracts in Anthropology* (1970–), *Abstracts in German Anthropology* (1980–) and *Anthropological Index* (1968–).

Hominid specimens are documented and described in *Catalogue of Fossil Hominids*, edited by K P Oakley, and in the later parts collaborating with B G Campbell and T I Molleson (British Museum (Natural History), 1967–1975). Part 1 covers Africa, 2 Europe, and 3 Americas, Asia and Australasia. *Guide to Fossil Man: a handbook of human palaeontology*, by M H Day (4th rev. edn, Cassel, 1986) is a hominid catalogue for students and researchers. Both the above are important reference works as is *The Origins of Modern Humans: a world survey of the evidence*, edited by F H Smith and F Spencer (A R Liss, 1984), which is a substantial review of all aspects of palaeoanthropology, with a bibliography.

Much of the useful literature for researchers arises from conferences and symposia. From the Taung symposia, landmarks and current research are presented in *Hominid Evolution: past, present and future*, edited by P V Tobias (A R Liss, 1985); *Aspects of Human Evolution*, edited by C B Stringer (Taylor and Francis, 1981) is volume **21** of the Symposia series of the Society for the Study of Human Evolution; *The Emergence of Man: a discussion* in *Philosophical Transactions, Royal Society of London, series B*,

292, (1057), 1981, was from a meeting of the Royal Society and the British Academy; *Current Argument on Early Man*, edited by L K Konigsson (Pergamon, 1980) is the proceedings of a Nobel Symposium organized by the Royal Swedish Academy of Sciences.

Abstracting and indexing services

The literature of palaeontology straddles the publications of both earth and life sciences. This is reflected in the kind of bibliographies that palaeontologists have to search, and the information services they have to use.

The most important progress in recent years is the extensive mechanization of abstracting services, not only as a method of producing bibliographies, but in making available online these databases (see also Chapter 6). Some of these databases have their retrospective files online, making them more valuable, especially for the palaeontologists, who often have the need to refer to early literature.

As described in Chapter 6, the *Bibliography and Index of Geology* is a product of the database GeoRef: the relevant sections for palaeontology in this wide-ranging geological bibliography are 08–11 (General palaeontology, palaeobotany, invertebrate and vertebrate palaeontology). Similarly, *Geotitles* is produced from GeoArchive, but the usefulness to the palaentologist lies in the systematic palaeontological terms laid out in its thesaurus for online searching.

General scientific databases which publish geological sections as a separate product are Pascal Thema (formerly *Bulletin Signalétique*) and *Referativnyi Zhurnal*. In the former service, section 227 is Paléontologie: although the palaeontological papers are set out together in classified order, the geographical, subject and author indexes are combined for the entire volume. The palaeontological section in *Referativnyi Zhurnal* **08**. *Geologiya* is "Г", with citations arranged by the Universal Decimal Classification system; it is not available online, but is an unparalleled source of palaeontological literature from the Soviet Union, even though the bibliography is by no means exclusive to that region.

The most important life science database for palaeontology is the *Zoological Record*, a bibliography which dates back to 1865. Papers on fossil and extinct species are cited together and the recent volumes, in 20 parts, are published annually; the parts are — 1. Comprehensive zoology; 2. Protozoa; 3. Porifera; 4. Coelenterata; 5. Echinodermata; 6. Vermes; 7. Brachiopoda; 8. Bryozoa; 9. Mollusca; 10. Crustacea; 11. Trilobita; 12. Arachnida; 13. Insecta; 14. Protochordata; 15. Pisces; 16. Amphibia; 17. Reptilia; 18. Aves; 19. Mammalia; 20. List of new genera and subgenera (an annual update of S A Neave's *Nomenclator Zoologicus*). The author arrangement is supplemented by subject and geographic indexes. BIOSIS took over the production of *Zoological Record* from the Zoological Society of London in 1982, and the bibliography from 1973 is available online through DIALOG.

Zentralblatt für Geologie und Paläontologie is a bibliography of a different kind; each issue consists of a literature review on a selected theme, listings and detailed abstracts of recent book and journal publications arranged in subject order, with emphasis on substantial abstracts on each, rather than a comprehensive compilation. Four issues (7 parts) are published annually, with an author index as a separate issue.

A new abstracting service in palaeontology, launched in 1986, is the *Geographical Abstracts: palaeontology and stratigraphy*, an expansion of the services already offered by Geo Abstracts, and available online in Geobase. The arrangement is primarily systematic, with age and geographic subdivisions, and issued six times a year with an annual subject and author index.

Book reviews and lists of publications of palaeontological institutions provide informal current awareness service in the literature: these appear in journals and newsletters like *Geotimes, Geochronique*, the *Circular of the Palaeontological Association*, just to name but a few.

APPENDIX: Currently Published Palaeontological Journals

Acta Palaeontologica Polonica (1956–)
Acta Palaeontologica Sinica (1953–)
Alcheringa (1975–)
Ameghiniana (1957–)
Annales de Palaeontologie (1906–) Series divided between v.50–67 (1964–1981) into *Vertébrés and Invertébrés*; recombined from v.68– (1982–)
Archaeopteryx (1983–)
Beiträge zur Paläontologie Osterreich (1976–)
Biostratigraphie du Paleozoïque (1984–)
Bolletino della Societá Paleontologia Italiana (1960–)
Bulletin of American Paleontology (1895–)
Bulletin of the British Museum (Natural History), Geology Series (1947–)
Bulletin Mizunami Fossil Museum (1974–)
Bulletin Nanjing Institute of Geology and Palaeontology (1980–) ·
Bulletin of the Southern California Paleontological Society (1968–)

Col-Pa, Coloquios de la Catedra de Paleontologia, Facultad de Ciencias, Universidad de Madrid (1964–)
Communicaciones del Museo Argentino de Ciencias Naturales 'Bernardino Rivadavia' e Instituto Nacional de Investigacion de la Ciencias Naturales, Paleontologia (1966–)
Communicaciones Paleontologicas del Museo de Historia Natural de Montevideo (1970–)
Communicaciones Paleontologicas del Museo Municipal Real de San Carlos, Colonia, Uruguay (1977–)
Contributions Institute of Geology and Palaeontology, Tohoku University (1924–)
Contributions from the Museum of Palaeontology, University of Michigan, Ann Arbor (1928–)
Cranium (1984–)
Developments in Palaeontology and Stratigraphy (1975–)
Documenta Naturae (1981–) First part issued as *Documenta* in 1976
Facies (1979–)
Field Guidebook Series, Society of Economic Paleontologists and Mineralogists (1978–)
Fieldiana: Geology (1945–) *Memoirs* (1947–)
Folia Musei Rerum Naturalium Bohemiae Occidentalis, Geologia (1972–)
Forschungsarbeiten und Publikationen, Geologisch-Paläontologisch Institut und Museum der Christian-Albrechts- Universität Kiel (1981–)
Forschungsbericht Geologisch-Paläontologisches Institut Universität Frankfurt-am-Main (1976–)
Fossil Nytt (1962–)
Fossilien (1984–)
Fossils (Japan) (1967–)
Fossils Magazine (1976–)
Fossils Quarterly (1982–)
Fossils and Strata (1972–)
Fragmenta Mineralogica et Palaeontologica (1969–)
Geobios (1968–) Mémoire Speciale (1977–)
Geologica et Palaeontologica (1967–) *Sonderband* (1972–)
Geologica Hungarica, Series Palaentologia (1929–)
Geological Curator (1979–)
Geologie und Paläontologie im Westfalen (1983–)
Geologisch-Paläontologische Mitteilungen Innsbruck (1971–)
Geology and Palaeontology of Southeast Asia (1964–)
Huashi (19??–)
Iheringia, Serie Geologica (1967–)
Illustrated Catalogue of the Type Specimens of the Palaeontological Museum of the University of Uppsala (1973–)
Jornada Cientifica Instituto Geologia Paleontologia Cuba (9, 1982–)
Journal of the Palaeontological Society of India (1956–)
Journal of Paleontology (1927–)
Kleine Senckenberg Reihe (7, 1977–)
Leitfossilien (1975–)
Lethaia (1968–)
Memoir, The Paleontological Society (1968–)
Memoirs of the Association of Australian Palaeontologists (1983–)
Memoirs Bernard Price Institute for Palaeontological Research (1977–)
Memoirs Geological Survey of India (Palaeontologica Indica) (1861–)
Memoirs of the Geological Survey of New South Wales, Palaeontology (1888–)
Memoirs Nanjing Institute of Geology and Palaeontology (1958–)

Memorias del Museo Paleontologico de la Universidad Zaragoza (1985–)
Minéraux et Fossiles (1975–)
Mitteilungen aus dem Geologisch-Palaontologischen Institut der Universität Hamburg (1969–)
Monde et Minéraux, Minéralogie, Paléontologie, Gemmologie (1957–)
Monographs of the Palaeontological Society of India (1958–)
Mosasaur (1983–)
Munstersche Forschungen zur Geologie und Paläontologie (1965–)
Neues Jahrbuch für Geologie und Paläontologie, Abhandlungen (1950–) *Monatshefte* (1950–)
Nomen Nudum (1971–)
Original Report Okamura Fossil Laboratory (1973–)
Öslenytani Vitak (Discussiones Palaeontologicae) (1973–)
Palaeoecology of Africa, and of Surrounding Islands and Antarctica (1966–)
Palaeontographica (1851–) *A: Paläozoologie, Stratigraphie* (1933–) *B: Paläophytologie* (1933–)
Palaeontographia Italica (1895–)
Palaeontographica Americana (1916–)
Palaeontographica Canadiana (1983–)
Palaeontographical Society Monographs (1847–)
Palaeontologia Africana (1953–)
Palaeontologia Cathayana (1983–)
Palaeontologia Jugoslavica (1958–)
Palaeontologia Polonica (1929–)
Palaeontologica Sinica Series A–D (1922–)
Palaeontology (1957–)
Palaios (1986–)
Paläontologische Zeitschrift (1913–)
Paleo-Quebec (1974–)
Paleobiologie Continentale (1970–)
Paleobiology (1974–)
Paleobios, Berkeley (1967–)
Paleontologia y Evolucion (1965–)
Paleontologia Mexicana (1954–)
Paleontological Bulletin, New Zealand Geological Survey (1913–)
Paleontological Journal (1967–) cover to cover translation of *Paleontologicheskii Zhurnal*
Paleontological Monographs, Society of Economic Paleontologists and Mineralogists (1978–)
Paleontologicheskii Sbornik (1961–)
Paleontologicheskii Zhurnal (1957–)
Paleontologiya, Stratigrafiya i Litologiya (1975–)
Professional Papers in Stratigraphy and Palaeontology (1975–)
Publicaties van de Belgische Vereniging voor Paleontologie (1978–)
Publication, Geological Research and Development Centre, Indonesia, Paleontology Series (1981–)
Quaderni Museo Comunale di Paleontologia (1983–)
Quatärpaläontologie (1975–)
Revue de Paléobiologie (1982–)
Revista del Museo Argentino de Ciencias Naturales 'Bernardino Rivadavia' e Instituto Nacional de Investigacion de la Ciencias Naturales, Paleontologia (1963–)
Revista del Museo de la Plata, Nuevo Serie, Seccion Paleontologia (1956–)
Rivista Italiana Paleontologia Stratigrafia (1935–) *Memoria* (1952–?)
Sbornik Geologickych Ved, Rada P: Paleontologie (1963–)

Schweizerische Paläontologische Abhandlungen (1874–)
Senkenbergiana Lethaea (1854–)
Smithsonian Contributions to Paleobiology (1969–)
Special Papers, Palaeontological Society of Japan (1951–)
Special Papers in Palaeontology (1967–)
Special Publication, Paleontological Society, University of Tennessee (1984–)
Special Publications, Geological Curators Group (1976–)
Special Publications, Palaeontological Society of India (1982–)
Special Publications, Society of Economic Paleontologists and Mineralogists (1957–)
Special Publications, Southern California Paleontological Society (1980–)
Special Topics in Palaeontology (1984–)
Special Volume, Palaeontological Institution of the University of Uppsala (1973–)
Transactions and Proceedings of the Palaeontological Society of Japan (1951–)
Travaux du Laboratoire de Paléontologie, Faculté de Sciences, Université de Paris Orsay (1971–)
Travaux du Laboratoire de Géologie Historique et de Paléontologie, Université de Provence, Marseilles (1972–)
Trudy Paleontologicheskogo Instituta, Akademiya Nauk SSSR (1932–)
Trudy Sessii Vsesoyuznogo Paleontologicheskogo Obshchestva (1957–)
Trudy Sovremestnaya Sovetsko Mongolskaya Paleontologicheskaya Expeditsiya (1974–)
Trudy Vsesoyuzoe Paleontogicheskoe Obshchestvo (1957–)
Tulane Studies in Geology and Paleontology (1962–)
Uchenye Zapiski Nauchno-issledovatel'skii Institut Geologii Artiki, Paleontologiya i Biostratigrafiya (1960–)
University of Kansas Paleontological Contributions, Articles (1947–) *Papers* (1965–) *Monograph* (1982–)
Veröffentlichungen Naturkunde Museum Bielenfeld (1981–)
Voprosy Paleontologii (1950–)
Zapadne Karparty, Seria Paleontologia (1975–)
Zentralblatt für Geologie und Paläontologie (1963–)
Zitteliana (1969–).

CHAPTER TEN

Mineralogy and crystallography

ANTHONY HALL

In the early days of each subject, mineralogy and crystallography developed together, and much of the basic knowledge of crystallography was amassed by mineralogists working on well-formed mineral specimens. With the crystallographic properties of the natural minerals now known in great detail, the majority of present-day crystallographers are working on substances unrelated to mineralogy, ranging over the whole field of organic and inorganic chemistry. Thus while most mineralogists have an interest in crystallography, the reverse is not necessarily true. The majority of mineralogists view themselves as specialists within the larger field of earth science, whereas the majority of professional crystallographers owe their broader allegiance to either physics or chemistry.

The main areas of interest to mineralogists are the chemical composition of minerals, their crystalline structure, their physical properties, and their occurrence in nature. Important areas of specialization are in clay mineralogy, gemmology, and applied mineralogy (including ores and ceramic materials). The major concern of crystallographers is to determine the atomic structure of crystalline substances ("crystal structure analysis") but the study of physical properties is also important, and the techniques of crystallography are being applied increasingly to the study of non-crystalline materials such as glasses, liquids and amorphous substances.

Organizations

The main organizations concerned with the encouragement of mineralogy as a subject are the national mineralogical societies, which publish journals and arrange scientific meetings. The principal ones are:

Mineralogical Society of Great Britain and Ireland
Mineralogical Society of America
Société Française de Minéralogie et de Cristallographie
Deutsche Mineralogische Gesellschaft
All-Union Mineralogical Society (USSR)
Mineralogical Association of Canada
Societa Italiana di Mineralogia e Petrologia
Schweizerische Mineralogischen und Petrographischen Gesellschaft
Mineralogical Society of Japan

Smaller societies or groups exist in other countries, but because of the relatively small number of professional mineralogists the smaller groups do not generally publish their own journals.

The International Mineralogical Association (IMA) exists to co-ordinate the work of the national societies and to organize international meetings, but it has no individual membership, and although it sponsors occasional symposia it does not publish a journal: it does, however, produce the *World Directory of Mineralogists* (3rd edn, 1985). The Group of European Mineralogists (GEM) co-ordinates the work of the European mineralogical societies and publishes a directory of institutions.

Crystallography has developed as a separate subject more recently than mineralogy, and the subject has no regional or topographical aspects, so its organization has tended to be more international in character. The most important international organization is the International Union of Crystallography, and there are smaller groups affiliated to national physical and chemical societies. There is a well-established American Crystallographic Association, and the British Crystallographic Association was inaugurated in 1982.

Several of the specialized areas of mineralogy have their own organizations, namely those concerned with ceramics (British Ceramic Society, American Ceramic Society), clay mineralogy (Clay Minerals Society), and gemmology (Gemmological Association of Great Britain, Gemological Institute of America). The ceramic field is also served by industrial research organizations, such as the British Ceramic Research Association, although their work is not primarily mineralogical. There are several societies catering for the interests of amateur mineralogists (i.e. mineral collectors); some are rather ephemeral in character, but a few are of interest to professional mineralogists.

There are mineral collections in most national museums, such as the British Museum (Natural History), and in some local and university museums.

Textbooks and reference works

By far the most important reference book in mineralogy is *Rock Forming Minerals*, by W A Deer, R A Howie and J Zussman (Longman, 1962–3), which was originally published in five volumes. A second edition is in course of production and the first three volumes have appeared; in 1978 (**2A** — single-chain silicates), 1982 (**1A** — orthosilicates) and 1986 (**1B** — disilicates and ring silicates). A shorter, one-volume version, by the same authors, has been published under the title *Introduction to the Rock Forming Minerals* (Longman, 1966) as a textbook for undergraduates.

Dana's System of Mineralogy, an exhaustive compendium of mineralogical information first published at the end of the last century, has been partly revised over the years but has now been superseded by Deer, Howie and Zussman, as far as the rock-forming minerals are concerned. The three volumes of the 7th edition, by C Palache, H Berman and C Frondel, remain an important source of information for certain mineral groups, i.e. Vol. **1** (Wiley, 1944) on the elements, sulphides, sulphosalts and oxides, Vol. **2** (1951) on the other non-silicates, and Vol. **3** on silica minerals (1962).

Two comprehensive reference works which deal with all the known minerals, but give only the basic details, are the *Encyclopedia of Minerals*, by W L Roberts *et al* (Van Nostrand Reinhold, 1974), and the *Encyclopedia of Mineralogy*, edited by K Frye (Hutchinson Ross, 1981). The former work contains more about the individual minerals and includes useful references to key papers, while the latter contains articles on various aspects of mineralogy.

In addition to the works mentioned above, the following monographs and reference books deal with specific groups of minerals:

Feldspar Minerals, by J V Smith (2 vols, Springer-Verlag, 1974)
Feldspar Mineralogy, edited by P H Ribbe (Mineralogical Society of America, 1975)
Orthosilicates, edited by P H Ribbe (*Reviews in Mineralogy*, **5**, 1980)
Amphiboles: crystal chemistry, phase relations and occurrence, by W G Ernst (Springer-Verlag, 1968)
Micas, edited by S W Bailey (*Reviews in Mineralogy*, **13**, 1984).
Natural Zeolites, by G Gottardi and E Galli (Springer-Verlag, 1985)
Zeolites: occurence, properties, use, edited by L B Sand and F A Mumpton (Pergamon Press, 1978)

Emerald and Other Beryls, by J Sinkankas (Chilton Book Co., 1981)
Oxide Minerals, edited by D Rumble (Mineralogical Society of America, 1976)
Mineral Chemistry of Metal Sulfides, by D J Vaughan and J R Craig (Cambridge University Press, 1978)
Sulphide Minerals: crystal chemistry, parageneses and systematics, by I Kostov & J M Stefanova (E. Schweizerbart'sche Verlagsbuchhandlung, 1982)
Platinum-group Elements: mineralogy, geology, recovery, edited by L J Cabri (*Canadian Institute of Mining and Metallurgy, Special Volume* **23**, 1981)
Carbonates: mineralogy and chemistry, edited by R J Reeder (*Reviews in mineralogy, number* **11**, 1983)
The Mineralogy of the Diamond, by Yu L Orlov (Wiley-Interscience, 1977)
Apatite: its crystal chemistry, mineralogy, utilization, and geologic and biologic occurrences, by D McConnell (Springer-Verlag, 1973)
Phosphate Minerals, edited by J O Nriagu and P B Moore (Springer-Verlag, 1984)

The clay minerals have an extensive literature all of their own. Clay mineralogy is a very specialized field of study, of great importance to soil scientists and sedimentary petrologists. Special techniques are needed to identify and study clay minerals and there are numerous books devoted to this subject (see also Chapter 19). The standard textbook is *Clay Mineralogy*, by R E Grim (McGraw-Hill, 1968). More recent works are *The Chemistry of Clay Minerals*, by C E Weaver and L D Pollard (Elsevier, 1973), and *Clays and Clay Minerals in Natural and Synthetic Systems*, by B Velde (Elsevier, 1977). Texts dealing with specific methods of clay mineral identification are:

Phyllosilicates and Clay Minerals: a laboratory handbook for their X-ray diffraction analysis, by J Thorez (G. Lelotte, 1975)
The Electron-optical Investigation of Clays, edited by J A Gard (Mineralogical Society, 1971)
Advanced Techniques for Clay Mineral Analysis, edited by J J Fripiat (Elsevier, 1981)
Electron Micrographs of Clay Minerals, by T Sudo, S Shimoda, H Yotsumoto and S Aita (Elsevier, 1980)
Atlas of Infrared Spectroscopy of Clay Minerals and their Admixtures, by H W van der Marel and H Beutelspacher (Elsevier, 1976)
The Differential Thermal Investigation of Clays, edited by R C Mackenzie (Mineralogical Society, 1957)
Crystal Structures of Clay Minerals and their X-ray Identification, edited by G W Brindley and G Brown (Mineralogical Society, revised reprint, 1983).

There are many textbooks of mineralogy for students. The leading British one, which deals with the principles of mineralogy as well as with details of the individual groups of minerals, is *Mineralogy for Students*, by M H Battey, (2nd edn, Longman, 1981), and the leading American one is *Manual of Mineralogy*, by C Klein and C S Hurlbut (20th edn, Wiley, 1985). *Introduction to the Rock Forming Minerals*, by Deer, Howie and Zussman, referred to above, is also very widely used as a student textbook. It contains descriptions of the rock-forming minerals but does not deal with the principles of mineralogy; it has been translated into several languages. A textbook which places more emphasis on minerals, other than the abundant rock-forming ones, is *Mineralogy: concepts, descriptions, determinations*, by L G Berry, *et al* (2nd edn, W H Freeman, 1983). Another work which is out-of-date but still comprehensive and valuable, is *Dana's Textbook of Mineralogy*, by W E Ford (4th edn, Wiley, 1932).

Morphological crystallography was many years ago one of the most important branches of mineralogy and crystallography and although it has now declined in importance it remains an essential part of undergraduate mineralogy courses. Apart from its treatment in the general mineralogy textbooks there are three widely-used student textbooks specifically devoted to this subject: *An Outline of Crystal Morphology*, by A C Bishop (Hutchinson, 1967), *An Introduction to Crystallography*, by F C Phillips (4th edn, Oliver & Boyd, 1971), and *Laboratory Manual of Crystallography for Students of Mineralogy and Geology*, by G Tunnell and J Murdoch (Wm C Brown, 1957).

An important branch of mineralogy that is very well served by textbooks and reference books is optical mineralogy, dealing with the principles of crystal optics and their application to the identification of minerals under the microscope. Some of the more comprehensive or better-illustrated books on optical mineralogy and optical techniques are:

Crystals and the Polarising Microscope by N H Hartshorne and A Stuart (4th edn, Arnold, 1970)

Introduction to the methods of optical crystallography, by F D Bloss (Holt, Rinehart & Winston, 1961)

Practical optical crystallography, by N H Hartshorne and A Stuart (2nd edn, Arnold, 1969)

Optical crystallography, by E E Wahlstrom (5th edn, Wiley, 1979)

The microscopic study of crystalline materials, such as rocks and minerals, requires a special type of microscope, the *polarizing* or *petrological* microscope, and as well as being briefly described in the books listed above it is dealt with in more detail in a handbook published by one of the leading microscope manufacturers: *The Polarising Microscope*, by A F Hallimond (3rd edn, Vickers Instruments, York, 1970).

Tables of optical data intended for use in determinative mineralogy are given in *Identification Tables for Minerals in Thin Section*, by E P Saggerson (Longman, 1975), *Optical Mineralogy: the non-opaque minerals*, by W R Phillips and D T Griffen (W H Freeman, 1981), and in *Optical Determination of Rock-forming Minerals. Part I: Determinative tables*, by H U Bambauer, F Taborszky and H D Trochim (E Schweizerbart'sche, 1979). The last work is the first English translation of the fourth edition of a classic German reference book originally compiled by W E Tröger, and it contains many useful diagrams. The French work *Caractères Optiques des Minéraux Transparents: tables de détermination*, by J Girault (Masson, 1980) is more complete than other compilations in containing optical data on all the known transparent minerals.

For student use, several textbooks have been written which combine the principles of optical crystallography with tabulations of optical data. *Elements of Optical Mineralogy*, by A N Winchell (3 vols, 4th edn, Wiley, 1951) is the classic of this type and is very comprehensive in its coverage, but unfortunately does not use the standard modern symbols for refractive indices. *Optical Mineralogy*, by P F Kerr (4th edn, McGraw-Hill, 1977) is more elementary and is widely used, but it is badly in need of revision and correction. *Optical Mineralogy*, by D Shelley (2nd edn, Elsevier, 1985) is the standard modern textbook. The beautifully illustrated *Atlas of Rock-forming Minerals in Thin Section*, by W S MacKenzie and C Guilford (Longman, 1980) contains coloured photomicrographs of the common minerals and is widely used by students. One of the few elementary textbooks to deal with the optical properties of opaque, as well as transparent, minerals is *A Practical Introduction to Optical Mineralogy*, by C D Gribble and A J Hall (Allen & Unwin, 1985).

The specialized techniques of studying orientated crystals under the microscope are described in *The Universal Stage*, by R C Emmons (*Geological Society of America, Memoir*, **8**, 1943) and *The Spindle Stage: principles and practice*, by F D Bloss (Cambridge University Press, 1981), both of which contain a more extensive discussion of advanced optical methods than their specific titles might suggest.

Different techniques are used for the optical examination of the opaque minerals, which include most of the ore minerals. There are several textbooks in this field, including *Ore Microscopy*, by E N Cameron (Wiley, 1961), and *Ore Microscopy and Petrography*, by J R Craig and D J Vaughan (Wiley, 1981). Two important reference works are *Tables for Microscopic Identification of Ore Minerals*, by W Uytenbogaardt and E A J Burke (2nd edn, Elsevier, 1971), and *The Ore Minerals and Their Intergrowths*, by P Ramdohr (2 vols, 2nd edn, Pergamon Press, 1981).

An essential data source for ore microscopy is the *Quantitative Data File for Ore Minerals*, edited by A J Criddle and C J Stanley (British Museum (Natural History), 1986). This is the second edition of a compilation of optical data on the opaque minerals produced under the auspices of the International Mineralogical

Association's Commission on Ore Microscopy. It is in the form of a card index with entries for 420 minerals, and gives details of reflectance at various wavelengths, micro-hardness and chemical composition.

The standard textbooks of gemmology are *Gem Testing*, by B W Anderson (9th edn, Butterworths, 1980), *Gemstones*, by F G H Smith (14th edn, Chapman and Hall, 1972), *Gemology*, by C S Hurlbut and G S Switzer (Wiley, 1979), *Gems: their source, descriptions and identification*, by R Webster (2nd edn, Butterworths, 1970), and *Handbook of Gem Identification*, by R T Liddicoat (10th edn, Gemological Institute of America, 1975). A valuable compendium of gemmological information is *Gemstones of North America*, by J Sinkankas (2 vols, Van Nostrand, 1959 & 1976). The identification of synthetic, as opposed to natural, gemstones is discussed in *Identifying Man-made Gems*, by M O'Donoghue (NAG Press, 1983).

There are many introductory and general textbooks on crystallography, varying quite a lot in emphasis according to whether they were written by physicists, chemists, or mineralogists. A modern advanced treatment is *Geometrical and Structural Crystallography*, by J V Smith (Wiley, 1982): this is relatively theoretical in tone. *Chemical Crystallography: an introduction to optical and X-ray methods*, by C W Bunn (2nd edn, Clarendon Press, 1961), is a more elementary and readable account of the methods used to study crystals, although it is a little out-of-date. A book which particularly meets the needs of geology students is *Crystallography: an introduction for earth science and other solid state students*, by E J W Whittaker (Pergamon Press, 1981); this combines the morphological and structural aspects of crystallography in a way that most of the chemical and physical crystallography textbooks do not. The relationship between crystallography and crystal chemistry is emphasized in *Crystallography and Crystal Chemistry: an introduction*, by F D Bloss (Holt, Rinehart and Winston, 1971), and specific crystal structures are described in *Structural Inorganic Chemistry*, by A F Wells (3rd edn, Oxford University Press, 1962). A more practical emphasis is taken in *Crystallography and its Applications*, by L S Dent Glasser (Van Nostrand Reinhold, 1977).

The individual crystal structures of actual minerals are illustrated in *Crystal Structures of Minerals*, by W L Bragg and G F Claringbull (Bell, 1965), and reviewed in *Crystal Chemical Classification of Minerals*, by A S Povarennykh (2 vols, Plenum Press, 1972).

The physical properties of minerals are described in *Physics of Minerals and Inorganic Materials: an introduction*, by A S

Marfunin (Springer-Verlag, 1979), *The Physics and Chemistry of Minerals and Rocks*, edited by R G J Strens (Wiley-Interscience, 1976), *Mineralogical Applications of the Crystal Field Theory*, by R G Burns (Cambridge University Press, 1970), and *Structure-property Relations*, by R E Newnham (Springer-Verlag, 1975). Both chemical and physical aspects are considered in *Principles of Mineral Behaviour*, by A Putnis and J D C McConnell (Blackwell, 1980).

Physical techniques that can be used to study minerals are described in the literature of physics and materials science, but some specific mineralogical applications are listed below:

Electron Microscopy in Mineralogy, edited by H R Wenk *et al* (Springer-Verlag, 1976)

Physicochemical Methods of Mineral Analysis, edited by A W Nicol (Plenum Press, 1975)

Differential Thermal Analysis. Application and results in mineralogy, by W Smykatz-Kloss (Springer-Verlag, 1974)

Infrared and Raman Spectroscopy of Lunar and Terrestrial Minerals, edited by C Karr (Academic Press, 1975)

Infrared Spectra of Minerals and Related Inorganic Compounds, by J A Gadsden (Butterworths, 1975)

The Infrared Spectra of Minerals, edited by J D Farmer (Mineralogical Society, 1974)

Physical Methods in Determinative Mineralogy, edited by J Zussman (2nd edn, Academic Press, 1977)

Spectroscopy, Luminescence and Radiation Centres in Minerals, by A S Marfunin (Springer-Verlag, 1979)

X-ray diffraction is the most important technique used in the study of crystalline materials. There are many good textbooks on the subject, for example *Elements of X-ray Diffraction*, by B D Cullity (Addison-Wesley, 1956), and *Elements of X-ray Crystallography*, by A J C Wilson (Addison-Wesley, 1970). One particular book which is written with the geological or mineralogical reader in mind is *X-ray Diffraction Methods*, by E W Nuffield (Wiley, 1966).

The most important application of X-ray diffraction is in mineral identification, and the particular technique used for this purpose is called the X-ray powder method. It is described in *The Powder Method in X-ray Crystallography*, by L V Azaroff and M J Buerger (McGraw-Hill, 1958), and *Interpretation of X-ray Powder Diffraction Patterns*, by H Lipson and H Steeple (Macmillan, 1970). An essential source of data for mineral identification is the *X-Ray Powder Diffraction File*. This file contains listings of the X-ray powder patterns of all organic and inorganic compounds, including minerals, and the mineralogical entries contain much additional information. The *X-Ray Powder Diffraction File* was initially compiled by the American Society for Testing and Materials (and is hence colloquially known as the 'ASTM Index') on behalf of a consortium of physical and mineralogical societies. It was first issued on index cards (38 sets published up to 1989), and is now also available on microfiche, as computer tapes or discs, and partly in book form. The mineralogical data are also issued separately in book form and are periodically updated. The *X-Ray Powder Diffraction File* is currently being published by the Joint Committee on Powder Diffraction Standards and is available from the International Centre for Diffraction Data in Swarthmore, Pennsylvania; it is not normally held by libraries, but most well-equipped mineralogical laboratories have a set. X-ray diffraction data on two of the major groups of minerals have also been brought together in *Calculated X-ray Powder Patterns for Silicate Minerals*, by I Y Borg and D K Smith (*Geological Society of America, Memoir*, **122**, 1969), and *X-ray Powder Data for Ore Minerals*, by L G Berry and R M Thompson (*Geological Society of America, Memoir*, **85**, 1962), in addition to the more specific works on clay minerals referred to in an earlier paragraph.

Single crystal X-ray methods are mainly used for crystal structure analysis, i.e. the determination of the detailed atomic structure of a crystalline substance, and are of more interest to the specialist crystallographer than to the general mineralogist. The classic work in this field is *X-ray Crystallography* by M J Buerger (Wiley, 1942, reprinted 1980). More modern works include *X-ray Crystal Structure Determination*, by G H Stout and L H Jensen (Macmillan, 1968), *Crystal Structure Analysis: a primer*, by J P Glusker and K N Trueblood (Oxford University Press, 1972), and *X-ray Crystallography: an introduction to the theory and practice of single-crystal structure analysis*, by G H W Milburn (Butterworths, 1973). Up-to-date accounts of specialized X-ray applications appear in the conference volumes entitled *Advances in X-ray Analysis*, which are published annually by Plenum Press.

Periodicals

The main primary sources for mineralogy are the journals which are specifically devoted to mineralogy, or to mineralogy and petrology. The former are mostly published by the national mineralogical societies, although a few are maintained by commercial publishers. A list of mineralogical journals is given below, starting with the most prestigious.

American Mineralogist (1916–)
Canadian Mineralogist (1957–)
Mineralogical Magazine (UK) (1876–)

Bulletin de Minéralogie (France; formerly entitled *Bulletin de la Société Française de Minéralogie et Cristallographie* (1878–)
Neues Jahrbuch für Mineralogie (Germany — published in two series, *Monatshefte* and *Abhandlungen*, (1950–); formerly entitled *Neues Jahrbuch für Mineralogie, Geologie und Palaeontologie* (1833–58)
Fortschritte der Minèralogie (Germany) (1911–)
Contributions to Mineralogy and Petrology (1966–); formerly *Beitrage zur Mineralogie und Petrographie* (1957–65) and *Heidelberger Beitrage zur Mineralogie und Petrographie* (1947–57)
Mineralogy and Petrology (Austria) (1987–); formerly *Tschermaks Mineralogische und Petrographische Mitteilungen* (1948–86) and *Mineralogische und Petrographische Mitteilungen* (1878–1943)
Schweizerische Mineralogische und Petrographische Mitteilungen (Switzerland) (1921–)
Mineralogical Journal (Japan) (1953–)
Journal of the Mineralogical Society of Japan (1952–)
Journal of the Japanese Association of Mineralogists, Petrologists and Economic Geologists (1929–)
Periodico di Mineralogia (Italy) (1930–)
Rendiconti della Societa Italiana di Mineralogia e Petrologia (Italy) (1941–)
Mineralogica et Petrographica Acta (Italy) (1954–)
Zapiski Vsesoyuznogo Mineralogischeskogo Obshchestva (USSR) (1948–)
Mineralogicheskii Sbornik (USSR) (1947–)
Mineralogicheskii Zhurnal (Kiev, USSR) (1979–)
Trudy Mineralogicheskogo Muzeya, Akademiya Nauk (USSR) (1949–)
Mineralogia Polonica (Poland) (1970–)
Archiwum Mineralogiczne (Poland) (1926–)
Indian Mineralogist (1960–)

There have been recent moves to co-ordinate the publishing activities of the major national mineralogical societies and groups in western Europe. As a first step, they have adopted a common format for their journals, and for several years have produced a joint index in addition to the separate indexes of the individual journals. This has been distributed by the Société Française de Minéralogie et de Cristallographie to the subscribers to all the national journals under the title of *"European Journal of Mineralogy (Index)"*, which has caused much confusion to librarians because there is as yet no journal of that title.

Two very important, commercially published, journals which combine mineralogy with petrology are *Contributions to Mineralogy and Petrology* and *Journal of Petrology*. Mineralogical contributions also appear in many geological publications of a more general nature, including academic journals and the reports of national Geological Surveys. Papers dealing with certain specialized aspects of mineralogy, particularly on crystal structures and the physical properties of minerals, are scattered throughout the literature of inorganic chemistry and materials science. This is especially true of those minerals that have technically significant properties, such as the magnetic minerals, or those that belong to structural groups with many synthetic varieties, such as the zeolites or garnets.

A special role is played by the International Mineralogical Association, which not only sponsors regular international meetings of which the proceedings are published by national societies, but also has a committee on new mineral names which vets reports of new minerals submitted to all the leading mineralogical journals. The work of this committee ensures that new mineral names are not bestowed on previously known or inadequately characterized material.

Much of the primary literature in crystallography is not relevant to earth science, consisting of descriptions of substances that do not occur geologically. Crystal structure analyses of minerals are frequently published in the mineralogical journals mentioned above, and also in journals specifically devoted to crystallography. The most important of these is *Acta Crystallographica* (1948–), which is now published in three series: section A devoted to theoretical crystallography, section B to crystal structure, and section C to the structures of specific compounds. Other leading journals are the *Journal of Applied Crystallography* (1968–), *Zeitschrift für Kristallographie* (1921–45; 1954–), *Journal of Crystallographic and Spectroscopic Research* (1982–) formerly *Journal of Crystal and Molecular Structure* (1971–81), *Journal of Crystal Growth* (1967–), and *Kristallografiya* (1956–). Crystallographic descriptions of minerals also appear occasionally in the various journals of inorganic chemistry. One journal specifically devoted to the crystallographic study of minerals is *Physics and Chemistry of Minerals* (1977–).

The most important specialized field of mineralogy to have its own journals is clay mineralogy, which supports *Clays and Clay Minerals* (1953–), and *Clay Minerals* (formerly *Clay Minerals Bulletin*) (1947–), the latter being sponsored jointly by seven European groups of clay mineralogists. Some information on clay mineralogy is also to be found in the literature of soil science (see Chapter 19). Original articles on topographical mineralogy and gemmology appear in *Journal of Gemmology* (1947–), *Gemmologist* (1931–), *Rocks and Minerals* (1926–), *Mineralogical Record* (1970–), *Gems and Gemology* (1934–) and *Zeitschrift der Deutschen Gemmologischen Gesellschaft* (1964–). Information of mineralogical interest is occasionally published in the specialized literature of ceramics, such as the *Journal of the American Ceramic Society* (1918–), *Transactions of the British Ceramic Society* (1939–) formerly *Transactions of the Ceramic Society* (1916–38)), and *Bericht der Deutschen Keramischen Gesellschaft* (1920–).

Abstracts and bibliographies

In general, mineralogy is a relatively small branch of geology, and it falls within the scope of the more generalized geological bibliographies and information retrieval services (see Chapters 5 and 6).

The main abstracting service specific to mineralogy is *Mineralogical Abstracts* (1920–), published jointly by the Mineralogical Societies of Great Britain and America. *Mineralogical Abstracts* is issued in four issues a year and each volume contains approximately 4000 abstracts, divided into sections on Mineral Data, Petrology, Geochemistry, etc. Each volume is thoroughly indexed, so that *Mineralogical Abstracts* is the most rapid and effective method of gaining access to the recent mineralogical literature. The abstracts are larger than those of most other abstracting journals, but the coverage of relevant publications is incomplete, especially for papers not written in English. *Mineralogical Abstracts* is now available online.

The other important abstracting service which caters for mineralogists is *Chemical Abstracts* (1907–), which has a more extensive coverage of the non-English language literature but has less informative abstracts, and indexes which are difficult to use. *Chemical Abstracts* is linked to a system of computerized information retrieval.

There are now several thousand known minerals and several individuals and publishers have undertaken the task of keeping track of the growing total. Regular summaries of new mineral descriptions are published in the *American Mineralogist* and *Bulletin de Minéralogie*. Every few years the *Mineralogical Magazine* publishes lists of new mineral names with notes of synonyms and discredited species, the most recent list being in *Mineralogical Magazine*, **50**, 741–761 (1986); the chemical formula and basic crystallographic and optical data are given for each of the valid new species. *A Manual of New Mineral Names 1892–1978*, edited by P G Embrey and J P Fuller (British Museum (Natural History) and Oxford University Press, 1980) is a collected edition of the first 30 of these lists.

Comprehensive lists of the known minerals are given in: *An Index of Mineral Species and Varieties, Arranged Chemically*, by M H Hey (2nd edn, British Museum (Natural History), 1955) supplemented by the *Appendix to the Second Edition of Index of Mineral Species and Varieties, Arranged Chemically*, by M H Hey (British Museum (Natural History), 1963); in the *Encyclopedia of Minerals*, by W L Roberts *et al* (Van Nostrand Reinhold, 1974); in *Mineralogische Tabellen*, by H Strunz (6th edn, Akademische

Verlag, Leipzig, 1977); and in *Glossary of Mineral Species 1983*, by M Fleischer (*Mineralogical Record*, Tucson, Arizona, 1983) supplemented by *Additions and Corrections to the Glossary of Mineral Species 1983*, by M Fleischer (*Mineralogical Record*, **15**, 51–54, 1984 and **16**, 155–158, 1985).

CHAPTER ELEVEN

Geochemistry

JOHN N WALSH

The literature of geochemistry has shown a dramatic and sustained growth since the first edition of this book was published. Geochemistry is now clearly identified as one of the major branches of the earth sciences. The implications of progress made in this field extend into other areas of geology, and indeed other disciplines. There are several reasons for the growth of geochemistry during the last 15 years. The most important of these has been a large increase in the available data base for the subject. Improvements in analytical instrumentation have dramatically increased the range and the levels of the elements (and their isotopes) that can be determined: these improvements have contributed substantially to our understanding of geological processes, and have also led to the increased use of geochemistry in the detection of economically viable mineral deposits. Sustained demand for metals has produced an increase in geochemical exploration activity. Other areas of geochemistry have shown equally dramatic expansion — isotope geochemistry, organic geochemistry and geochemistry applied to the interpretation of petrogenetic processes. Low temperature geochemistry (sedimentary and aqueous geochemistry) shows clear signs of a rapid growth during the next few years.

This expansion in activity has brought with it an ineluctable increase in the literature of geochemistry. So great has been the increase in geochemical literature that it is difficult to keep pace even in specialized branches of the subject. It is necessary to subdivide geochemistry in an attempt to cover the literature, and thus this chapter contains two major sections. In the first section 'general' geochemistry is discussed, listing textbooks, etc., that attempt to provide a broad overview of the subject, including some of the classic texts of geochemistry. The second part of the chapter covers the literature of specific subjects in geochemistry: these topics are chosen as widely as possible, but inevitably some aspects of the subdivisions used must remain arbitrary and unsatisfactory because subdivisions within such a subject as geochemistry are inherently artificial.

Textbooks, journals and abstracting services are the major sources of information for geochemists. The many textbooks now published in the various aspects of geochemistry are covered in some detail in the following sections. Textbooks remain the most useful general sources of information. The journals are invaluable for information on specific aspects of the subject and several journals now publish review papers — for example *Earth Science Reviews* (Elsevier, 1966–); titles relevant to geochemists are listed later in this chapter.

General literature of geochemistry, and the early textbooks

Many of the early books published in geochemistry have become "classic" texts in the field, and indeed have served as models for some of the more recent textbooks. It should also be noted that in the early days of the development of geochemistry several chemistry textbooks included much interesting and useful geochemical information. *The Physical Chemistry of Igneous Rock Formation*, published by the Faraday Society (1925) is just one example of an early source of useful information. However, the most important early work in geochemistry was undoubtedly F W Clarke's *The Data of Geochemistry*, published in 1924 as *US Geological Survey Bulletin* **770**. The first edition of this work had been published in 1908 as *US Geological Survey Bulletin* **330**, and later editions were published in 1911, 1916, 1920, and 1924 as *Bulletins* **491**, **616**, **695**, and **770** respectively. These publications, especially the 1924 edition, contain a fund of information that is useful even today.

Two other early texts should also be noted; *Geochemistry*, by K Rankama and Sahama, published by the University of Chicago Press in 1950, and V I Goldschmidt's *Geochemistry*, published by Oxford University Press in 1954: despite their age these books are still frequently cited as reference sources. Rankama and Sahama's book gives a general account of the subject together with an account of the detailed geochemistry of the elements. Goldschmidt's classic book was published posthumously, Goldschmidt having died in 1947 leaving the bulk of the manuscript completed: some parts of the book (finally published in 1954) were contributed by other authors, and the whole text was carefully edited by Dr A Muir. The book drew together much of the existing data of geochemistry, including data on the individual elements or groups of elements, and was seminal to the development of much of modern geochemistry. The lasting impact of Goldschmidt's book has been considerable: it brought together a vast body of detailed information on the distribution of the elements in the geological environment, and, more fundamentally, it developed the concept that the distribution of the elements was not a haphazard event. It demonstrated that there existed basic "rules" or "laws" controlling element distribution patterns, and much of the rationale of present-day geochemical research derives from ideas suggested by Goldschmidt.

The 1960s saw an increase in the output of geochemical publications. One major reference work was a new edition of *The Data of Geochemistry* as *US Geological Survey, Professional Paper* **440**. This sixth edition is published as individual chapters and edited by M Fleischer. Those published to date include:

Chapter B-1 *Cosmochemistry. Part 1. Meteorites*, by B Mason (1979)
      D *Composition of the Earth's Crust*, by R L Parker (1967)
      F *Chemical Composition of Subsurface Waters*, by J D Hem and G A Waring (1963)
      G *Chemical Composition of Rivers and Lakes*, by D A Livingstone (1963)
      K *Volcanic Emanations*, by D E White and G A Waring (1963)
      L *Phase Equilibrium Relations of the Common Rock-forming Oxides Except Water*, by G W Morey (1964)
      N *Chemistry of Igneous Rocks. Part 1. The chemistry of the peralkaline oversaturated obsidians*, by R Macdonald and D K Bailey (1973)
      S *Chemical Composition of Sandstones — excluding carbonate and volcanic sands*, by F J Pettijohn (1963)
      T *Nondetrital Siliceous Sediments*, by E R Cressman (1962)
     W *Chemistry of the Iron-rich Sedimentary Rocks*, by H L James (1966)
      Y *Marine Evaporites*, by F H Stewart (1963)
    JJ *Composition of Fluid Inclusions*, by E Roedder (1972)
    KK *Compilation of Stable Isotope Fractionation Factors of Geochemical Interest*, by I Friedman and J R O'Neil (1977).

A somewhat comparable undertaking has been the publication of the *Handbook of Geochemistry*, edited by K H Wedepohl, published by Springer Verlag, from 1969–1978, as two separate volumes. Volume 1 contains a series of keynote articles on selected topics in the field of general geochemistry. Volume 2 is a systematic documentation of the geochemistry of the elements: this volume details aspects of the behaviour of the elements (crystal chemistry, isotopic abundances, concentration in common rock types and meteorites, and chemical behaviour in igneous, metamorphic and sedimentary rocks, etc.). Another, less well-known work dating from the 1960s is the *Discovery of the Elements*, by M E Weeks and H M Leicester, published by Chemical Education Publishing Co., Easton, Pa., in 1968: this contains a fascinating account of the discovery of the elements and much information relevant to the early history and development of geochemistry.

Several general textbooks of geochemistry have been published: these have varied in scope and in the approach adopted, but most have attempted to present an "overview" of geochemistry. B Mason's *Principles of Geochemistry* was first published in 1952 by Wiley: a second edition was published in 1958, a third in 1966, and the fourth, considerably revised, by B Mason and C B Moore, was published in 1982. The original book developed from a university undergraduate lecture course and adopted a broad approach to geochemistry, including information on crystal chemistry, phase equilibria, and other useful background information. For many thousands of undergraduate students the book has been their standard text on geochemistry and the approach adopted is certainly readable and comprehensible: the extent of its popularity is reflected in the number of editions and reprintings. The fourth edition has undergone a considerable re-write of some of the sections, and includes the introduction of a chapter on 'Isotope Geochemistry'. The continued success of the book is due to its attempt to present a general, and somewhat popular, view of geochemistry, rather than any attempt to cover specific aspects of the subject in great depth.

Another well used textbook is *Introduction to Geochemistry*, by K B Krauskopf (McGraw-Hill, 1979). Here the approach adopted is different from Mason's book: more emphasis is placed on basic chemical principles and their application to geochemical examples, and this background information is valuable. In contrast to many geochemistry textbooks there is more emphasis placed on sedimentary and low temperature geochemistry; igneous geochemistry is less well served.

The introductory text *Geochemistry*, by A H Brownlow (Prentice-Hall, 1979) will be of value to those approaching the subject with a more limited background knowledge and will probably provide a good introduction to the subject for those less concerned with developments in detailed aspects of the subject. Another relatively elementary textbook by K H Wedepohl, *Geochemistry*, was published by Holt, Rinehart and Winston, in 1971: this is a clearly presented introduction to some of the basic aspects of geochemistry.

The Geochemistry of Solids, by W S Fyfe (McGraw-Hill, 1964) was an attempt to introduce some of the modern concepts of chemical bonding into geochemical theory. The book does not cover the field of geochemistry in as broad a manner as textbooks such as Mason's *Principles of Geochemistry*. It does, however, present a more advanced approach to the theory of trace element partitioning into mineral structures. C J Allegre and G Michard's *Introduction to Geochemistry* (Reidel, 1974) is a brief but valuable review of selected aspects of geochemistry. Although lacking a broad coverage of the subject, the selected aspects of geochemistry are explained with considerable finesse. The book will be valuable to those seeking information in those areas of geochemistry that are covered by the book.

The recent publication of *Inorganic Geochemistry*, by P Henderson (Pergamon, 1982) has brought to the subject a modern textbook of considerable stature. This book provides a first class text with an up-to-date and advanced coverage of many areas of geochemistry; it is at present probably the most useful of the books on general geochemistry. It should not, however, be seen as an all-embracing book and serious students of geochemistry will need to refer to other books in specific areas of the subject: also Henderson's book is probably not the publication most suited to those seeking an introductory book on geochemistry. The extent of the expansion of geochemistry within the last 20 years has been such that no one textbook can adequately cover the subject, consequently, there are inevitably areas omitted from *Inorganic Geochemistry*; nevertheless much of "mainstream" geochemistry is described in a lucid and comprehensive manner.

No account of the "general" literature would be complete without reference to the many symposium volumes and compilations of papers which have been published from time to time but it is impossible to cover these satisfactorily. Some of these collections attempt to present an overview of recent advances in significant parts of geochemistry: many are collections of papers on specific aspects of geochemistry with only a limited common theme. It is perhaps appropriate to select just two of the most useful of these compilation volumes for special mention: they are probably the most useful, and any account of the literature of geochemistry should make reference to them. *Researches in Geochemistry*, edited by P H Abelson (2 vols, Wiley, 1959 and 1967) contains a series of keynote papers on a broad spectrum of geochemical topics; these papers were state-of-the-art summaries when published. They are now somewhat dated but will nevertheless provide useful information. A more extensive compilation of papers has been published in *Origin and Distribution of the*

Elements, edited by L H Ahrens (Pergamon, 1969) as a volume in the series on the *Physics and Chemistry of the Earth*; several volumes of this on-going series have now been published. The series is reasonably comprehensive and covers a considerable proportion of the subject of geochemistry. Many of the articles published are up-to-date and comprehensive summaries of the topics covered, and good literature coverage has been provided in many of the articles. The continuing nature of the series has enabled many of the areas of geochemistry that have become important in recent years to be discussed and summarized.

Cosmic and lunar geochemistry

Progress in extra-terrestrial geochemistry has advanced substantially in recent decades, as space travel and manned flights to the Moon have made available lunar material for detailed analytical work. This has had a real impact on our knowledge and understanding of the mode of formation of the planets of our solar system. The most recent work is available in the specialized journals. Many of the geochemical periodicals have published articles on cosmic and lunar geochemistry but *Geochimica et Cosmochimica Acta (GCA)* (Pergamon) has placed particular emphasis on this aspect of geochemistry. *GCA* published in 1979 a comprehensive collection of research papers, as part of a special 'Moon' volume.

A comprehensive account of element synthesis and elemental abundances in solar and stellar material was given in 1961 by L H Aller, *The Abundances of the Elements* (Interscience). *The Solar System* (1953–1963) edited by G P Kuiper and published by University of Chicago Press is a major reference work, with contributions by many authors. More recent work on the solar system is covered in books such as *The Origin of the Solar System* (Wiley, 1978) edited by S F Dermott and *The Solar System* (Prentice-Hall, 1979) by J A Wood. A detailed account of the suggested mechanisms of nucleosynthesis and element synthesis is given by D D Clayton in *Principles of Stellar Evolution and Evolution* (McGraw-Hill, 1968).

A fertile field of geochemical research has been studies on the composition and origin of meteorites: prior to the lunar landings, meteorites provided our only direct evidence of the composition of extra-terrestrial material. *The Nature and Origin of Meteorites*, by D W Sears (Oxford University Press and Adam Hilger, 1978) provides an overview of many aspects of these extra-terrestrial bodies, including their geochemistry. An earlier work, *Meteorites*, by B Mason (Wiley, 1962) provided background information, including a detailed classification in which meteorite geochemistry plays a significant part. A valuable source of information, specifically on the geochemistry of meteorites, is Mason's contribution to the re-issued *Data of Geochemistry. Geochemistry. Part 1. Meteorites (USGS Geological Survey Professional Paper **440-B-1**, 1979).* Two more specialized books on particular types of meteorites are *Tektites and Their Origin,* by J A O'Keefe (Elsevier, 1976), and *Carbonaceous Meteorites,* by B Nagy (Elsevier, 1975): both these books contain much relevant information.

The geochemistry of other planets in the solar system has received much attention in recent years, with an understandable emphasis on lunar geochemistry. The review by S R Taylor *Lunar Science: a post-Apollo view* (Pergamon, 1975) represents a compilation of the results from the Apollo missions. The availability of lunar material for modern high quality analytical work has had a profound effect on our knowledge of the origin and evolution of the Moon, and has also brought substantial advances in understanding the origin of the other planets of the solar system. These advances can be seen in other important books, e.g. *Origin of the Earth and Moon,* by A E Ringwood (Springer-Verlag, 1979). Further useful information on the geochemistry of the planets of the solar system is given in the review *Mineralogy of the Planets: a voyage in space and time,* by J V Smith (Mineralogical Society, 1979). Two papers that give useful data on estimations of planetary compositions and planetary evolution were published in 1977: A Anders 'Chemical composition of the Moon, Earth and eucrite parent body', *Philosophical Transactions Royal Society London, A* **285**, 23–40, and A Anders and T Owen 'Mars and Earth: origin and abundance of volatiles', *Science,* **198**, 452–465.

The bulk composition of the Earth, and of the major divisions of the Earth (core, mantle, and crust) are discussed in many of the standard textbooks of geochemistry, see earlier pages of this chapter. Henderson's *Inorganic Geochemistry* (Pergamon, 1982) and Mason and Moore's *Principles of Geochemistry* (Wiley, 1982) both contain excellent accounts of this aspect of geochemistry. Several other publications provide additional information: *The Nature of the Solid Earth,* edited by E C Robinson (McGraw-Hill, 1972) is one such publication, presenting an overview of the geochemistry (and geophysics) of the Earth's interior; similarly, *Structure of the Earth,* by S P Clark (Prentice Hall, 1971) gives a brief yet comprehensive account, including geochemical evidence; M W McElhinney's *The Earth: its origin, structure and evolution* (Academic Press, 1975) is another book likely to be useful to geochemists, although it covers other geological fields; and A E

Ringwood's *Composition and Petrology of the Earth's Mantle* (McGraw-Hill, 1975) includes much geochemical information on our (somewhat limited) knowledge of the Earth's mantle and its possible composition.

The relative accessibility of the Earth's crust in comparison with the core and mantle has inevitably led to a less speculative approach to its geochemistry. A classic early paper by F W Clarke and H S Washington (*US Geological Survey Professional Paper* **127**, 1924) contains an extended summary of available analytical data. This was the first scientific attempt to derive a "bulk chemical composition" for the Earth's crust, and there have been many subsequent attempts: the literature for these is well summarized in the standard geochemical textbooks. A paper by S R Taylor that deserves special mention was published in 1964 in *Geochimica et Cosmochimica Acta*, **28**, 1273–1285 entitled 'Abundance of chemical elements in the continental crust: a new table'. This contains a most valuable compilation of data, especially on minor and trace elements.

High temperature geochemistry (igneous and metamorphic geochemistry)

Many of the earlier textbooks placed much emphasis on igneous geochemistry, and especially on the role of trace elements in igneous geochemistry. There is no doubt that this has been a fertile field of research for geochemistry and has contributed much to the development of the subject. It remains an area where much new data is being produced but other fields of geochemistry, notably aqueous and sedimentary geochemistry, are now of growing importance. Many of the standard textbooks in geochemistry provide valuable sources of information in igneous and metamorphic geochemistry; this applies to the earlier textbooks and to the more recent general books. There are also several other books, and especially articles in periodicals, which provide essential information and data summaries.

A paper that was of great importance in studies of element distribution patterns in magmatic fractionation was published in *Geochimica et Cosmochimica Acta*, (**1**, 129–208, 1951) written by L R Wager and R L Mitchell, 'The distribution of trace elements during strong fractionation of basic magma — a further study of the Skaergaard intrusion, E. Greenland'. This provided extensive analytical data on the behaviour of many trace elements through a strongly differentiated intrusive body. Developments beyond the basic concepts of trace element distribution that had been propounded by Goldschmidt were suggested in a paper by A E

Ringwood published in 1955, 'The principles governing trace element distribution during magmatic crystallisation' (*Geochimica et Cosmochimica Acta*, **7**, 189–202 and 242–254), emphasizing the possible importance of electronegativity, in addition to ionic size, in controlling trace element substitution. This theme was also evaluated by L H Ahrens in a paper in the series *Physics and Chemisty of the Earth*, **5**, 1–54, on 'The significance of the chemical bond for controlling the geochemical distribution of the elements'. The distribution of the transition metals in igneous processes has been extensively covered in the geochemical literature: the book by R G Burns, *Mineralogical Applications of Crystal Field Theory* (Cambridge University Press, 1970) provides a summary of the application of crystal field theory to the geochemistry of the transition metals, and the emphasis in this study is very much on the distribution of these elements in igneous rocks.

It is also worth considering that many textbooks on "igneous petrology" have contained a significant geochemical contribution (see also Chapter 12). Thus books such as *Igneous Petrology*, by I S E Carmichael, F J Turner and J Verhoogen (McGraw-Hill, 1974) have excellent summaries of much of igneous geochemistry. *Basalts*, by H H Hess and A Poldervaart (Wiley, 1967) and *The Evolution of the Igneous Rocks*, edited by H S Yoder (Princeton University Press, 1979) contribute to the geochemical data base in this area also.

Several books, and papers, have provided basic data for the geochemical interpretation of high temperature processes. Thus *Layered Igneous Rocks*, by L R Wager and G M Brown (Oliver & Boyd, 1968), provided comprehensive information on classic intrusive sequences, notably the Skaergaard intrusion. An early study by S R Nockolds and R Allen 'The geochemistry of some igneous rock series', published as three papers in *Geochimica et Cosmochimica Acta* (**4**, 105–142, **5**, 245–285 and **9**, 34–77, 1953–56) is also worthy of mention.

Geochimica et Cosmochimica Acta also published two important papers on the interpretation of trace element behaviour in igneous rocks: P W Gast 'Trace element fractionation and the origin of tholeiitic and alkaline magma types' in volume **32**, 1057–1086, 1982, and J Hertogen and R Gijbels, 'Calculation of trace element fractionation during partial melting' in volume **40**, 313–322, 1976. The paper by S R Taylor, 'The application of trace element data to problems in petrology' in *Physics and Chemistry of the Earth* **6**, 133–213, 1965, provides an extensive review and an excellent summary of extant literature. Similarly 'Trace element behaviour during anatexis', by D M Shaw (1977) in *Magma Genesis* (*Oregon*

Department of Geology and Mineral Industries, Bulletin **96**, 189–213), is a useful source of references.

General accounts of the use of trace elements in igneous rock systems have also been published. This K G Cox *et al, The Interpretation of Igneous Rocks* (Allen & Unwin, 1979) contains several chapters that will be useful for the geochemist.

Low temperature geochemistry (sedimentary and aqueous geochemistry)

The last two decades have shown a sustained expansion of geochemical knowledge and information in all branches of geochemistry. However, it is probably correct to record that the most dramatic expansion to date has occurred in high temperature, especially igneous, geochemistry. There are many who believe that the emphasis has now shifted towards the geochemistry of the low temperature environment, and that sedimentary and aqueous geochemistry will show a more dramatic growth in the next few years. The literature in this field is quite comprehensive and there are several excellent texts covering substantial aspects.

Geochemistry of Sediments, by E T Degens (Prentice-Hall, 1965) is an excellent (early) review: it presents a wide-ranging survey of the geochemistry of sediments and sedimentary processes. Similarly *Principles of Chemical Sedimentology*, by R A Berner (McGraw-Hill, 1971) will be of value to those seeking information on the thermodynamics and kinetics of sediment formation. Much geochemical information will also be found in *Evolution of Sedimentary Rocks*, by R M Garrels and F T Mackenzie (W W Norton and Co., 1971). Two chapters of *Data of Geochemistry* (6th edn, *US Geological Survey Professional Paper* **440**) should also be noted, dealing with the geochemistry of specific types of sedimentary rocks: 'Chemistry of iron-rich sedimentary rocks', by H L James and 'Marine evaporites', by F H Stewart, are substantive contributions in their own fields.

The chemical processes involved in weathering are well documented, for example in *Chemical Weathering of the Silicate Minerals*, by F C Loughman (Elsevier, 1969). Two important papers must also be referenced in any discussion on the chemistry of weathering: S S Goldich's 'A study in rock weathering' published in 1938 in *Journal of Geology*, **46**, 17–58, is an early study; the influence weathering has on water geochemistry is well-documented by J H Feth, *et al* in 'Sources of mineral constituents in water from granitic rocks, Sierra Nevada' (*US Geological Survey Water Supply Paper* **1535-I**).

No discussion on the geochemistry of sedimentary rocks and processes would be complete without reference to *Solutions, Minerals, and Equilibria*, by R M Garrels and C L Christ (Harper and Row, 1965). This is a comprehensive text on the origin of sedimentary rocks which, despite its relatively early date, remains one of, if not the, most relevant books on the subject.

The geochemistry of continental and ocean waters has had a profound effect on the day-to-day activities of mankind: many important books are available on the various aspects of water chemistry. *Geochemistry of Water*, edited by Y Kitano (Dowden, Hutchinson and Ross, 1975), is a collection of papers on water geochemistry; *Aquatic Chemistry*, by W Stumm and J J Morgan (Wiley-Interscience, 1970) provides a useful account of chemical equilibria in waters; and much relevant data is given by D A Livingstone in *Chemical Composition of Rivers and Lakes* (1963) (*US Geological Survey Professional Paper* **440G**). In chemical oceanography two important works are D R Martin's *Marine Chemistry* (Marcel Dekker, 1970), and J P Riley & R Chester's *Chemical Oceanography* (Academic Press, 1975–76): the latter work is in six volumes, on a wide range of topics, by many authors. A more general work, by H D Holland *The Chemistry of the Atmosphere and Oceans* (Wiley, 1978) should also prove of considerable interest. An important paper that will be of great value to those working on the chemical evolution of the oceans is 'The geologic history of sea water', by W W Rubey, published in *Bulletin of the Geological Society of America*, **62**, 1111–1147, 1951, where much data is presented (and critically evaluated) on the role of gases from magmatic activity in the chemical evolution of the Earth's atmosphere and hydrosphere.

An excellent modern textbook in the field of aqueous geochemistry is J I Drever's *The Geochemistry of Natural Waters* (Prentice Hall, 1982). This provides a comprehensive modern coverage of the subject; it also presents several well-documented and useful summaries of "case studies".

The field of organic geochemistry should not be overlooked; it has expanded in recent years, not least because of the commercial implications of the work. There are several wide-ranging texts; for example, G Eglington and M T J Murphy *Organic Geochemistry* (Longman/Springer Verlag, 1969), or the earlier volume, edited by I A Breger, *Organic Geochemistry* (Pergamon, 1963) which includes a series of papers by workers in several branches of the subject. There are also companion volumes in the series *Advances in Organic Geochemistry* which Pergamon have published, with various editors, covering conference proceedings.

Exploration geochemistry

The extent of the literature in applied and exploration geochemistry is considerable. It is now quite impossible to do it justice
in a short section within a chapter attempting to cover the entire
literature of geochemistry. There are not only several major
standard reference books but also many entire journals devoted to
the subject. The *Journal of Geochemical Exploration* (1972–) is
now established as an important literature source for many
original publications in exploration geochemistry. Similarly the
*Transactions of the Institution of Mining and Metallurgy (Section
B; Applied Earth Sciences)* (1966–) and *Economic Geology*
(1908–) regularly publish research papers in the field, and
represent important sources of information. The *Canadian Mining
Journal* (1880–), the *Bulletin of the Geological Survey of Canada*
(1945–), the Russian publications *Doklady Akademiya Nauk SSSR*
translated into English, and *Geokhimiya* (also partly translated
into English as *Geochemistry International* (1960–) are other
publications that should be mentioned. There are many others,
including for example the *Bulletins*, *Circulars* and *Professional
Papers* of the US Geological Survey and the United Nations'
Seminars on Geochemical Exploration Methods and Techniques.

In addition to these specialized journals there are publications
on exploration geochemistry in many of the "standard" geochemical journals. There are also many textbooks, some comprehensive in approach and some on more specialized aspects (see
also Chapter 15). Two of the most useful textbooks are *Geochemistry in Mineral Exploration*, by A W Rose, *et al* (Academic
Press, 1979) and *Introduction to Exploration Geochemistry*, by A
A Levinson (Applied Publishing Co., 1974). *Geochemistry in
Mineral Exploration*, by Rose, Hawkes and Webb is a second
edition of a previous book of the same title by Hawkes and Webb
(1962). It is established as possibly the "standard" textbook in the
subject and is essential reading for all serious students of
exploration geochemistry. It offers a broad coverage of the
methods used in exploration geochemistry: this includes an
extensive discussion of the materials sampled in geochemical
surveys, soils (including residual and transported overburdens),
waters, drainage sediments (especially stream sediments), vegetation
and volatile and airborne particulates.

Mention should be made of other textbooks, for example F R
Siegel's *Applied Geochemistry* (Wiley-Interscience, 1974), which
includes a useful section on geochemical exploration methods used
in the petroleum industry. There is also an extensive literature

published in other languages, especially Russian where several of the texts have been translated into English.

There are books on the analytical aspects of exploration geochemistry: a valuable, and concise, laboratory manual is R E Stoneley's *Analytical Methods for Use in Geochemical Prospecting* (Arnold, 1976) which gives useful data, especially on rapid, field orientated, analytical methods.

Isotope geochemistry

The subject of isotope geochemistry divides conveniently into two parts, stable isotopes and radioactive isotopes. Both of these are well-covered by the literature, with many comprehensive texts available. The standard geochemical journals, and in many cases geological journals, publish isotopic papers. In some areas of geological research the incorporation of some isotopic data has become almost essential for interpretive studies. In addition, a journal *Isotope Geology*, published by Elsevier, has been introduced recently and this will provide an extensive reference source in the future.

The general geochemistry textbooks (Goldschmidt *Geochemistry*, Krauskopf *Introduction to Geochemistry* and Mason's *Principles of Geochemistry*) did not deal directly with isotope geochemistry; consequently several other texts were written. However, the importance of isotope geochemistry is now more widely appreciated and Henderson's *Inorganic Geochemistry* and Mason and Moore's 4th edition of *Principles of Geochemistry* have introductory chapters on isotope geochemistry. Two early books on isotope geochemistry were *Isotope Geology*, by K Rankama (Interscience, 1954), and *Progress in Isotope Geology*, also by K Rankama and published by Interscience in 1963. Valuable though these works were, it must be said that they are now out-of-date and have been superseded by more recent texts.

In the field of radioactive isotope geochemistry a widely used and popular book is *The Earth's Age and Geochronology*, by D York and R M Farquhar (Pergamon, 1972). This provides an excellent low cost text, with a general introduction to the uses of unstable radioactive isotopes in dating methods. It is probably the most useful book for those seeking a good account of the background theory and discussion of the methods used in radiometric dating techniques. However, it does not cover samarium-neodymium (Sm-Nd) isotopic methods and this has now become an important field of isotopic dating, notably for the older geological terrains. Sm-Nd isotopic methods and their implications are well covered in a paper in *Annual Reviews of Earth and Planetary Science*, **7**. 11–38, 1979, by R D Onions, *et al* 'Geochemical and cosmochemical applications of Nd isotope analysis'. An important reference in geochronology is *Geochronology — radiometric dating of rocks and minerals*, edited by C T Harper (Dowden, Hutchinson & Ross, 1973). This volume contains reprints of classic papers in geochronology published up to 1973.

One book that covers both radiometric and stable isotopes is *Principles of Isotope Geology*, by G Faure (Wiley, 1977) and it also contains an extensive list of references. *Stable Isotope Geochemistry*, by J Hoefs, published in 1973 by Springer-Verlag, contains a readable and useful account of stable isotopes. Many papers on the uses of oxygen and hydrogen isotopes in rocks and minerals have been published by H P Taylor: one of the most valuable of these is 'The application of oxygen and hydrogen isotope studies in problems of hydrothermal alteration and ore depostion' published in *Economic Geology*, **69**, 843–883, 1974. This paper also has a useful reference list.

Determinative (analytical) geochemistry

The introduction and refinement of the various analytical methods have provided the basic data upon which many geochemists have developed their theories. There have been substantial advances made in most of the analytical methods used in geochemistry since the 1950s. The improvements in electronics and, more recently, in computing facilities have revolutionized the quantity and quality of data available to the geochemist. The economic implications of geochemistry and the increasing competition to find exploitable mineral deposits have added further impetus to developments in analytical geochemistry.

In contrast to other fields of geochemistry the literature of analytical geochemistry is more diffuse and textbooks tend to be more specific and less comprehensive. This is probably an inevitable consequence of the specialized (in many cases compartmentalized) nature of the subject. There is certainly no shortage of books or periodical publications in the field, but almost invariably they cover only a very limited selection of topics: there is no one textbook that can be unequivocally recommended. A further complication is that many of the periodical publications have spread into the chemical literature and *The Analyst*, *Analytica Chimica Acta*, *Analytical Chemistry*, and other similar periodicals regularly provide information on analytical work relevant to the geoscientist.

A useful work on geochemical methods is *Methods in Geo-*

chemistry, by A A Smales and L R Wager (Interscience, 1960). This book reviewed numerous analytical methods and was one of the first to draw attention to the growing importance of neutron activation and radiochemical methods; however, such is the rate of progress in the field that it is now somewhat dated. The important field of X-ray fluorescence spectrometry, which is especially popular in the UK as a method for rock analysis, was covered in a relatively early book, *X-ray Emission Spectrography in Geology*, by I Adler (Elsevier, 1966). No discussion of the technique of X-ray fluorescence would be complete without reference to the book by R Jenkins and J L de Vries, *Practical X-ray Spectrometry* (2nd edn, Macmillan, 1978). Although this is not geological in orientation it has been extensively used by innumerable geochemists.

Other methods of analysis have been well covered for the geological user. *Atomic Absorption Spectrometry in Geology*, by E E Angino and G K Billings (Elsevier, 1968) covers atomic absorption spectrometry; reference should also be made here to another "non-geological" book, *Atomic Absorption Spectrometry*, edited by J Cantle (Elsevier, 1982). The inductively coupled plasma (ICP) is a method of elemental analysis which has attracted much interest and support amongst geochemists: the *Handbook of ICP Spectrometry*, by M Thompson and J N Walsh, (Blackie and Sons, 1983), covers this subject with a strong geological emphasis.

Other texts have approached analytical geochemistry in a more general manner. The early books were largely concerned with classical (gravimetric) methods of analysis for rocks and minerals. Two examples of these early texts are H S Washington's *The Chemical Analysis of Rocks* (Wiley, 1930), and A W Groves' *Silicate Analysis* (Murby & Co., 1937). *Element Analysis in Geochemistry; Volume 1, Major elements*, by A Volborth (Elsevier, 1969) was one of the first books in this field that covered a range of methods. Volborth's book includes an account of classical methods and also incorporates a section on modern instrumental methods. An excellent textbook by J A Maxwell, *Rock and Mineral Analysis*, was published in 1969 by Wiley, a second edition of which, by W M Johnson and J A Maxwell (Wiley-Interscience) was published in 1981. This book provides a rigorous and systematic treatment of geochemical methods of analysis; there are useful sections on classical methods, atomic absorption electron microprobe analysis and X-ray fluorescence. The subject of geochemical analysis was treated on an element-by-element basis in *Chemical Methods of Rock Analysis*, by P G Jeffrey (Pergamon, 1975): a second edition, by P G Jeffrey and D Hutchison was published in 1981. This is an invaluable reference book that is especially appropriate for use as a laboratory handbook, and the literature of analytical methods for the geochemist is also well covered.

New texts on geochemical analysis have continued to appear — this is clearly a growth area in the geochemical literature. Recent books include *Applied Geochemical Analysis*, by C O Ingamells and F F Pitard (Wiley-Interscience, 1986) and most notably *A Handbook of Silicate Rock Analysis*, by P J Potts (Blackie, 1987). This latter volume is a *magnum opus* covering a vast subject in a comprehensive and comprehensible manner. It is likely to remain a major reference book in the field of analytical geochemistry for years to come, and includes a detailed coverage of the literature.

Periodicals

The list of journals specifically devoted to publishing papers in the geochemical field is now too extensive to cover fully here. The journals that have established themselves as essential reading for geochemists include: *Chemical Geology* (Elsevier, 1966–) and *Isotope Geology* (Elsevier,), *Contributions to Mineralogy and Petrology* (Springer-Verlag, 1966–), *Earth and Planetary Science Letters* (Elsevier, 1966–), *Geochimica et Cosmochimica Acta* (Pergamon, 1951–), *Journal of Petrology* (Oxford Uniersity Press, 1960–), *Lithos* (Universitetsforlaget, 1968–), and *Mineralogical Magazine* (Mineralogical Society, 1876–). This short list includes only some of the very great range of journals that now contribute to the information available to the geochemist.

Several of these journals have tended to specialize in particular aspects of the subject, and some of these have been noted earlier in the various sections of this chapter.

Abstracting services are now of great importance as an information source, as the range and number of the journals and books published in geochemistry has increased. Few libraries can claim comprehensive holdings of all the relevant journals and consequently researchers have become increasingly dependent on the abstract journals, and computer-based storage systems.

Mineralogical Abstracts (1920–) is probably the most well-established of the abstracting journals and includes a comprehensive coverage of most of the geochemical literature: it has become an invaluable source of information for geochemists (as well as other areas of the geological sciences).

Some geochemical publications are abstracted into other abstracting publications, e.g. *Chemical Abstracts* (published by the

American Chemical Society, 1907–) will often provide abstracts for some of the less readily available geochemical publications; this is especially useful in the field of analytical geochemistry, much of which is published in the "chemical" rather than the "geological" literature.

Most libraries are now linked into the various computer-based abstracting systems. There are now several of these systems (see Chapter 5) operating commercially and they can be an essential source of information.

CHAPTER TWELVE

Petrology

ANTHONY HALL

Petrology is the study of rocks, and can be subdivided in two ways: firstly into the study of the major categories of rock; and secondly into petrography, which is descriptive, and petrogenesis, which concerns the origin of rocks. In the past there has tended to be a distinction between igneous and metamorphic petrology on the one hand and sedimentary petrology on the other. This distinction arose because igneous and metamorphic rocks tended to be studied by the same people, using similar techniques, whilst the study of sedimentary rocks has been carried out by specialists in other fields, such as stratigraphy or marine geology, who have no particular interest in igneous and metamorphic processes. The formation of sedimentary rocks is also more amenable to present-day observation, and this field of investigation is described as sedimentology. In recent years, igneous and metamorphic petrology have diverged from one another as each subject has grown and as separate techniques have developed for the study of these two categories of rock.

Two specialized fields related to petrology are volcanology, which overlaps to a large extent with igneous petrology, and meteoritics, the study of meteorites. Until the era of space exploration the study of meteorites was rather an abstruse branch of petrology, carried out by a very small number of people, but since the landing of men on the Moon the study of extra-terrestrial rocks has developed enormously (see also Chaper 11 and the later section in this chapter).

Organizations

The links between petrology and mineralogy are very close, as has been mentioned in the chapter on the latter subject, and there has not been a need for separate organizations in petrology: most petrologists also regard themselves as mineralogists. The existing mineralogical and geological societies cater adequately for the needs of petrologists for the organization of meetings and the production of publications. Some of the national geological societies have petrological or sedimentological sub-groups but there are no specifically petrological societies. The Meteoritical Society caters for meteorite specialists.

Rock collections are held by major national museums and by national Geological Surveys. There are only a few major meteorite collections, notably those of the British Museum (Natural History) in the UK, the Smithsonian Institution in the USA, the Natural History Museum in Vienna, and the Nininger Collection at Arizona State University, Tempe. The major repository of extra-terrestrial (i.e. lunar) rock samples is at NASA in Houston.

There are several institutes for the study of volcanology. These are situated on or near active volcanoes, and attached to nearby universities or geological surveys. Some of them have operated for only a few years or have been active only when sufficient financial support has been available, but those on Hawaii (Hawaiian Volcano Observatory) and Etna (Istituto Internazionale di Vulcanologia, Catania) are long established. The International Association of Volcanology and Chemistry of the Earth's Interior (IAVCEI) sponsors working groups and international meetings on volcanology, and is based in Rome. The Scientific Event Alert Network run by the Smithsonian Institution in Washington compiles data on both volcanic eruptions and meteorite falls, as they happen.

Textbooks and reference works

The petrological literature is very large and there have been many good textbooks on the subject. The following list therefore only contains textbooks and reference books in current use and a few older books of lasting value. In recent years, most authors have concentrated on either igneous, sedimentary, or metamorphic rocks. There are a few books which cover more than one of these fields, and these are generally elementary textbooks of petrography. Three outstanding textbooks of petrography are in world-wide use: *Petrography: an introduction to the study of rocks in thin section*, by H Williams *et al* (2nd edn, W H Freeman, 1982), *Petrology for Students*, by S R Nockolds *et al* (Cambridge University Press, 1978), and *The Study of Rocks in Thin Section*, by W W Moorhouse (Harper, 1959). At a more elementary level there are *Rocks and Rock Minerals*, by R V Dietrich and B J Skinner (Wiley, 1979), and *The Practical Study of Crystals, Minerals and Rocks*, by K G Cox *et al* (McGraw-Hill, 1974).

A more ambitious textbook, which covers the petrogenesis and petrography of all types of rock is *Petrology: igneous, sedimentary and metamorphic*, by E G Ehlers and H Blatt (W H Freeman, 1982), while a more detailed, but less comprehensive, coverage is given in *Igneous and Metamorphic Petrology*, by M G Best (W H Freeman, 1982). *Ore Petrology*, by R L Stanton (McGraw-Hill, 1972) is unique in treating ores as rocks, a particularly valuable approach since the ores are not normally described in the conventional petrological textbooks.

A reference book which is very useful to researchers in all areas of petrology is *Laboratory Handbook of Petrographic Techniques*, by C S Hutchison (Wiley-Interscience, 1974). For those who merely need to know the meaning of rock names there is the *Dictionary of Rocks*, by R S Mitchell (Van Nostrand Reinhold, 1985).

Igneous petrology

Most geological degree courses split petrology into petrography and petrogenesis, and a sound knowledge of petrography is desirable before a student can appreciate the finer points of petrogenesis.

Consequently separate textbooks are normally used for petrography and petrogenesis. For the most part, the general petrography textbooks mentioned above are quite suitable, but the very well-illustrated *Petrology of the Igneous Rocks*, by F H Hatch, A K Wells and M K Wells (13th edn, Allen & Unwin, 1973) continues to be popular. Even more beautifully illustrated is the *Atlas of Igneous Rocks and Their Textures*, by W S Mackenzie, *et al* (Longman, 1982), which is essentially a collection of colour photomicrographs with accompanying descriptions. The ultimate reference book for igneous petrography, one which is exceptionally comprehensive and detailed, although rather out-of-date, is *A Descriptive Petrography of the Igneous Rocks*, by A Johannsen, (4 vols, University of Chicago Press, reprinted 1951–1957).

Textbooks of igneous petrogenesis in current use are *Igneous Petrology*, by A Hall (Longman, 1987), *Magmas and Magmatic Rocks*, by E A K Middlemost (Longman, 1985) and *Igneous*

Petrology, by C J Hughes (Elsevier, 1982). A very good exposition of the principles of igneous petrology, which does not however contain discussions of all the individual magma types, is *The Interpretation of Igneous Rocks*, by K G Cox *et al* (Allen & Unwin, 1979). A more advanced treatment, with emphasis on physical aspects of the subject, is provided by *Principles of Igneous Petrology*, by S Maaloe (Springer-Verlag, 1985).

Experimental petrology and the meaning of phase diagrams are dealt with by a number of specialized textbooks, some of which also contain information relevant to metamorphic petrology. *Evolution of the Igneous Rocks*, by N L Bowen (Dover, reprint 1956) is still an excellent introductory account, although partly out-of-date. A commemorative volume celebrating the original publication of Bowen's book is *The Evolution of the Igneous Rocks: fiftieth anniversary perspectives*, edited by H S Yoder (Princeton University Press, 1979). This contains invited contributions by leading experts bringing up to date each chapter of Bowen's original work. A good general exposition of phase diagrams mainly relevant to igneous petrology is *The Interpretation of Geological Phase Diagrams*, by E G Ehlers (W H Freeman, 1972). A more thorough but informally written book is *Basalts and Phase Diagrams: an introduction to the quantitative use of phase diagrams in igneous petrology*, by S A Morse (Springer-Verlag, 1980); despite its title this is primarily a book on phase diagrams rather than on basalts. *Experimental Petrology: basic principles and techniques*, by A D Edgar (Oxford University Press, 1973) is mainly about experimental techniques, and *Petrologic Phase Equilibria*, by W G Ernst (W H Freeman, 1976) is less detailed and covers metamorphic as well as igneous phase relations. A number of specific aspects of igneous and metamorphic phase relations, including granites and alkaline igneous rocks are discussed in the contributions to *The Evolution of the Crystalline Rocks*, edited by D K Bailey and R Macdonald (Academic Press, 1976). Many advanced topics in igneous petrology are discussed by the contributors to *Physics of Magmatic Processes*, edited by R B Hargraves (Princeton University Press, 1980), and *Magmatic Processes: physicochemical principles*, edited by B O Mysen (The Geochemical Society, 1987).

The emplacement of igneous rocks is for many petrologists just as important as the source of the magma, and the emplacement of volcanic rocks is covered by the specialized works on volcanology. There are two outstanding textbooks in this field: *Volcanoes*, by G A Macdonald (Prentice-Hall, 1972), and *Volcanology*, by H Williams and A R McBirney (Freeman, Cooper & Co., 1979). A well-written elementary account with an emphasis on the geomorphological aspects of volcanism is *Volcanoes*, by C Ollier (MIT Press, 1969), and a more extensive elementary account is *Volcanoes of the Earth*, by F M Bullard (University of Texas Press, 1976). *Volcanoes*, by A Rittmann and L Rittmann (Orbis, 1976), is beautifully illustrated, with many colour photographs. The study of explosive volcanicity was given a tremendous impetus by the 1980 eruption of Mt. St. Helens in the North-western United States, and this is reflected in a number of books published subsequently, in particular the compilation: *Explosive Volcanism*, edited by F R Boyd (National Academy Press, Washington, 1984). The products of explosive eruptions are discussed in *Pyroclastic Rocks*, by R V Fisher and H V Schmincke (Springer-Verlag, 1984). At about the same time geologists were "celebrating" the 100th anniversary of history's most famous volcanic eruption, that of Krakatoa in 1883, and in 1983 the Smithsonian Institution Press published a beautifully illustrated book, reproducing many of the original studies of this eruption which are now difficult to find, namely *Krakatoa 1883*, by T Simkin and R S Fiske. The emplacement of intrusive igneous rocks does not hve its own specialized textbook but many aspects of the subject are discussed in *Mechanism of Igneous Intrusion*, edited by G Newall and N Rast (*Geological Journal, Special issue*, 1970).

There are many books devoted to the origin of specific igneous rock types. Basaltic rocks are described in *Basalts: the Poldervaart treatise on rocks of basaltic composition*, edited by H H Hess and A Poldervaart (2 vols, Interscience, 1967), *Generation of Basaltic Magma*, by H S Yoder (National Academy of Science, Washington, 1976) and *Spilites and Spilitic Rocks*, edited by G C Amstutz (Springer-Verlag, 1974). Ultrabasic rocks are described in *Ultramafic and Related Rocks*, edited by P J Wyllie (Wiley, 1967), *Ophiolites: ancient oceanic lithosphere?*, by R G Coleman (Springer-Verlag, 1977), and *Komatiites*, edited by N T Arndt and E G Nisbet (Allen & Unwin, 1982). Andesites are described in *Orogenic Andesites and Plate Tectonics*, by J B Gill (Springer-Verlag, 1981) and *Andesites: orogenic andesites and related rocks*, edited by R S Thorpe (Wiley, 1982). Acid igneous rocks are described in *Origin of Granite Batholiths: geochemical evidence*, edited by M P Atherton and J Tarney (Shiva, 1979), *Granites and their Enclaves*, by J Didier (Elsevier, 1973), and *Trondhjemites, Dacites and Related Rocks*, edited by F Barker (Elsevier, 1979). The most extensive survey of anorthosites is in *Origin of Anorthosite and Related Rocks*, edited by Y W Isachsen (*New York State Museum and Science Service, Memoir* **18**, 1969).

Alkaline igneous rocks including kimberlites and carbonatites have an extensive literature. The most wide-ranging survey is *The*

Alkaline Rocks, edited by H Sorensen (Wiley, 1974) while *Kimberlites and their Xenoliths*, by J B Dawson (Springer-Verlag, 1980), *Kimberlites: mineralogy, geochemistry and petrology*, by R H Mitchell (Plenum Press, 1986), and *Carbonatites*, edited by O F Tuttle and J Gittins (Interscience, 1966) provide general coverage of two important related categories. There is much detailed information on kimberlites in conference proceedings, notably *Kimberlites, Diatremes and Diamonds: their geology, petrology and geochemistry*, edited by F R Boyd and H O A Meyer (American Geophysical Union, 1979), and *The Mantle Sample: inclusions in kimberlites and other volcanoes*, also edited by F R Boyd and H O A Meyer (American Geophysical Union, 1979). *Petrology and Genesis of Leucite-bearing Rocks*, by A K Gupta and K Yagi (Springer-Verlag, 1980) is a survey of the potassium-rich group of alkaline igneous rocks. Two valuable works which concentrate on the rocks of particular alkaline provinces are *The Lovozero Alkali Massif*, by K A Vlasov *et al* (Oliver & Boyd, 1966) and *Carbonatite-nephelinite Volcanism*, edited by M J Le Bas (Wiley, 1977) which is largely on the alkaline rocks of East Africa. The *Proceedings of the First International Symposium on Carbonatites*, edited by C J Braga (Departamento Nacional de Produção Mineral, Brazil, 1980) contains a large selection of up-to-date papers on carbonatites.

Broader aspects of igneous petrology are considered in *Composition and Petrology of the Earth's Mantle*, by A E Ringwood (McGraw-Hill, 1975), *Layered Igneous Rocks*, by L R Wager and G M Brown (Oliver & Boyd, 1968) and *Petrology and Geochemistry of Continental Rifts*, edited by E R Neumann and I B Ramberg (Reidel, 1978).

Among regional studies of igneous rocks there are three recent volumes which deal with very extensive areas: *Igneous Rocks of the British Isles*, edited by D S Sutherland (Wiley, 1982), *Volcanism in Australasia*, edited by R W Johnson (Elsevier, 1976), and *Magmatism at a Plate Edge: the Peruvian Andes*, edited by W S Pitcher, *et al* (Blackie, 1985). Two detailed book-length reviews of classic areas are *The Geology of Donegal: a study of granite emplacement and unroofing*, by W S Pitcher and A R Berger (Wiley-Interscience, 1972), and *Ardnamurchan: a guide to geological excursions,* by C D Gribble, E M Durrance and J N Walsh (Geological Society of Edinburgh, 1976).

Sedimentary petrology

There are many general works on the petrography of sedimentary rocks. *Sedimentary Petrography*, edited by K B Milner, (4th edn,

2 vols, Allen & Unwin, 1962) is a long-established reference work, and *Petrology of the Sedimentary Rocks*, by J T Greensmith (6th edn, Allen & Unwin, 1978) is an equally well-established student textbook. *Microscopic Sedimentary Petrography*, by A V Carozzi (Robert E Krieger, 1972 reprint) emphasizes the study of sediments in thin section. *Sedimentary Petrology: an introduction*, by M E Tucker (Blackwell, 1981) is mainly petrographic, as is *Sedimentary Petrology*, by H Blatt (W H Freeman, 1982). The best illustrated book on sedimentary petrography, and one which is specifically written for students, is *Atlas of Sedimentary Rocks under the Microscope*, by A E Adams *et al* (Longman, 1984). Articles on the mineralogy of sedimentary rocks are contained in *Marine Minerals*, edited by R G Burns (Mineralogical Society of America, 1979) which developed from the contributions to a short course on this subject organized by the Mineralogical Society of America.

Sedimentological aspects, i.e. the deposition of sediments as opposed to the description of them in a lithified condition, are treated in: *Physical Processes of Sedimentation*, by J R L Allen (Allen & Unwin, 1970), *Principles of Sedimentology*, by G M Friedman and J E Sanders (Wiley, 1978), *Sedimentary Environments and Facies*, edited by H G Reading (Blackwell, 1978), *Depositional Sedimentary Environments*, by H E Reineck and I B Singh (2nd edn, Springer-Verlag, 1980), *An Introduction to Sedimentology*, by R C Selley (Academic Press, 1976), *Mechanics of Sedimentary Transport*, by M S Yalin (2nd edn, Pergamon Press, 1977), *Sedimentology: process and product*, by M R Leeder (Allen & Unwin, 1982), *Paleocurrents and Basin Analysis*, by P E Potter and F J Pettijohn (2nd edn, Springer-Verlag, 1977), *Cyclic Sedimentation*, by P McL D Duff *et al* (Elsevier, 1967) and *Ancient Sedimentary Environments and their Subsurface Diagenesis*, by R C Selley (3rd edn, Chapman & Hall, 1985).

Comprehensive works dealing with both sedimentology and sedimentary petrology are *Origin of Sedimentary Rocks*, by H Blatt, *et al* (2nd edn, Prentice-Hall, 1980), *Sedimentary Rocks*, by F J Pettijohn (3rd edn, Harper & Row, 1975), *Petrology of Sedimentary Rocks*, by R L Folk (2nd edn, Hemphills, 1974), *Sediments and Sedimentary Rocks*, by H Füchtbauer (E Schweizerbart'sche, 1974), and *The Origins of Sediments and Sedimentary Rocks*, by W von Engelhardt (2nd edn, E Schweizerbart'sche, 1977).

Several books are devoted to sedimentary structures, for example, *Methods for the Study of Sedimentary Structures*, by A H Bouma (Wiley, 1969), the *Atlas and Glossary of Primary Sedimentary Structures*, by F J Pettijohn and P E Potter (Springer-

Verlag, 1964), *Sedimentary Features of Flysch and Greywackes*, by S Dzulyinski and E K Walton (Elsevier, 1965), *Sedimentary Structures*, by J D Collinson and D B Thompson (Allen & Unwin, 1982), and *Sedimentary Structures: their character and physical basis*, by J R L Allen (2 vols, Elsevier, 1982).

Apart from the works on diagenesis in specific rock types mentioned in the following paragraphs, there are general accounts of diagenesis in *Diagenesis in Sediments and Sedimentary Rocks*, edited by G Larsen and G V Chilingar (Elsevier, 1979) and in *Early Diagenesis: a theoretical approach*, by R A Berner (Princeton University Press, 1980).

Milner's book on *Sedimentary Petrography*, mentioned above, has sections on techniques for the study of sedimentary rocks, and more up-to-date methods are given in *Procedures in Sedimentary Petrology*, edited by R E Carver (Wiley, 1971), *Methods in Sedimentary Petrology*, by G Müller (E Schweizerbart'sche, 1967), and *The Techniques of Sedimentary Mineralogy*, by F G Tickell (Elsevier 1965). A collection of electron microphotographs is published in *Electron Microscopy of Soils and Sediments: examples*, edited by P Smart and N K Tovey (Clarendon Press, 1981).

The standard work on arenaceous sediments is *Sand and Sandstone*, by F J Pettijohn *et al* (Springer-Verlag, 1972). An atlas of colour photographs illustrating all aspects of sandstone petrography is *Constituents, Textures, Cements and Porosities of Sandstones and Associated Rocks*, by P A Scholle (*American Association of Petroleum Geologists, Memoir* **28**, 1979). The *Atlas of Quartz Sand Surface Textures*, by D H Krinsley and J C Doornkamp (Cambridge University Press, 1974) is concerned with unconsolidated sands, and diagenesis of sandstones is covered by both *Compaction of Coarse-grained Sediments*, edited by G V Chilingarian and K H Wolf (Elsevier, 1975) and *Diagenesis as it affects Clastic Reservoirs*, edited by P R Schluger (Society of Economic Paleontologists and Mineralogists, 1979).

Many aspects of argillaceous sediments are dealt with in the books on clay mineralogy listed in Chapter 10. A great deal of additional information on clays as rocks is contained in two books written primarily for the ceramics industry: *Structural Clay Products*, by W E Brownell (Springer-Verlag, 1976) and *The Chemistry and Physics of Clays and other Ceramic Raw Materials*, by R W Grimshaw (4th edn, Benn, 1980). The best general guide to the study of argillaceous rocks is *Sedimentology of Shale: study guide and reference source*, by P E Potter *et al* (Springer-Verlag, 1980): this is short but contains extensive bibliographies. A more conventional textbook is *Geology of Clays: weathering, sedi-*

mentology, geochemistry, by G Millot (Springer-Verlag, 1970), and diagenesis is covered by *Compaction of Argillaceous Sediments*, by H H Rieke and G V Chilingarian (Elsevier, 1974). *An Introduction to Clay Colloid Chemistry*, by H van Olphen (Wiley, 1977) is relevant to the conditions of deposition of clays. The petrology of soils is described in *Fabric and Mineral Analysis of Soils*, by R Brewer (Wiley, 1964), and *Soil Components, volume 2: Inorganic components*, edited by J E Gieseking (Springer-Verlag, 1975). *Bentonites: geology, mineralogy, properties and uses*, by R E Grim and N Güven (Elsevier, 1978) is a mainly mineralogical account of this group of rocks.

A classic textbook on carbonate rocks is *Carbonate Sediments and their Diagenesis*, by R G C Bathurst (2nd edn, Elsevier, 1975). Other general works on carbonates are *Sedimentary Carbonate Minerals*, by F Lippman (Springer-Verlag, 1973), *Recent Sedimentary Carbonates: Part I, Marine carbonates*, by J D Milliman (Springer-Verlag, 1974), *A Color Illustrated Guide to Carbonate Rock Constituents, Textures, Cements and Porosities* (American Association of Petroleum Geologists, Memoir **27**, 1978) and *Carbonate Facies in Geologic History*, by J L Wilson (Springer-Verlag, 1975). Mainly carbonate fossils are described in *Introductory Petrography of Fossils*, by A S Horowitz and P E Potter (Springer-Verlag, 1971). Dolomitic rocks are dealt with in *Dolomitisation and Limestone Genesis*, by L C Pray and R C Murray (Society of Economic Paleontologists and Mineralogists, 1965) and *Concepts and Models of Dolomitisation — an introduction*, edited by D H Zenger et al (Society of Economic Paleontologists and Mineralogists, 1980). *Stromatolites*, edited by M R Walter (Elsevier, 1976) deals with an important, specialized area of limestone diagenesis.

Most of the minor sedimentary rock types have been treated by textbooks or symposia. Coal is covered by *Coal Petrology*, by E Stach et al (2nd edn, Gebrüder Borntraeger, 1975), and by *Coal Petrology: its principles, methods, and applications*, by R M Bustin et al (2nd edn, Geological Association of Canada, 1985). Chert is discussed in *The Caballos Novaculite, Marathon Region, Texas*, by E F McBride and A Thompson (*Geological Society of America, Special Paper* **122**, 1970) and in *Flint: its origin, properties and uses*, by W Shepherd (Faber & Faber, 1972). Textbooks on evaporites are *Salt Deposits*, by H Borchert and R O Muir (Van Nostrand, 1964) and *Salt Deposits: their origin and composition*, by O Braitsch (Springer-Verlag, 1971). *Marine Evaporites: origin, diagenesis and geochemistry*, edited by D W Kirkland and R Evans (Dowden, Hutchinson and Ross, 1973) is a compilation of classic papers in this field. Sedimentary deposits of aluminium, iron and manganese are described in the *Handbook of Stratabound and Stratiform Ore Deposits*, volume 7, edited by K H Wolf (Elsevier, 1976), in the proceedings of the *4th International Congress for the Study of Bauxites, Alumina and Aluminium*, edited by S S Augustithis (3 vols, National Technical University, Athens, 1978), in the symposium volume *Lateritisation Processes*, edited by M K Roy Chowdhury (Balkema, 1981), in *Manganese Nodules: research data and methods of investigation*, by R K Sorem and R H Fewkes (Plenum Press, 1979) and in *Karst Bauxites: bauxite deposits on carbonate rocks*, by G Bardossy (Elsevier, 1982). Phosphate deposits are described in *Phosphorites on the Sea Floor: origin, composition and distribution*, by G N Baturin (Elsevier, 1982).

Metamorphic petrology

There are not as many textbooks on metamorphic rocks as on igneous and sedimentary rocks. Two widely-used, comprehensive and well-balanced textbooks are *Petrology of the Metamorphic Rocks*, by R Mason (Allen & Unwin, 1978) and *Metamorphic Petrology: mineralogical, field and tectonic aspects*, by F J Turner (2nd edn, McGraw-Hill, 1981). *Metamorphism and Metamorphic Belts*, by A Miyashiro (Allen & Unwin, 1973) emphasizes the tectonic context of metamorphism and refers extensively to Japanese examples. *Petrogenesis of Metamorphic Rocks*, by H G F Winkler (4th edn, Springer-Verlag, 1976) is widely used, but its treatment is somewhat unconventional in that the author has abandoned the familiar metamorphic facies classification in the later editions. An elementary descriptive work on metamorphic rocks which emphasizes their identification and field occurrence is *The Field Description of Metamorphic Rocks*, by N Fry (Open University Press, 1984). This short volume was specifically written as a handbook for geology students, under the auspices of the Geological Society of London.

The interpretation of textures in metamorphic rocks is very difficult and has been the subject of two specialized works: *Metamorphic Textures*, by A Spry, (Pergamon Press, 1969) and *Metamorphic Processes: reactions and microstructure development*, by R H Vernon (Allen & Unwin, 1976): this field is also covered by the *Atlas of Deformational and Metamorphic Rock Fabrics*, edited by G J Borradaile *et al* (Springer-Verlag, 1982). Specialized areas of metamorphic petrology are covered in *Migmatites and the Origin of Granite Rocks*, by K R Mehnert (Elsevier, 1968) and *Theory of Metasomatic Zoning*, by D S Korzhinskii (Oxford University Press, 1970).

A four-volume treatise on metamorphism by a group of Russian workers has been translated into English under the following titles: *The Facies of Metamorphism*, edited by V S Sobolev (Australian National University Press (ANUP), 1972); *The Facies of Contact Metamorphism*, by V V Reverdatto (ANUP, 1973); *The Facies of Metamorphism at Moderate Pressures*, by N L Dobretsov *et al* (ANUP, 1973); and *The Facies of Regional Metamorphism at High Pressures*, by N L Dobretsov *et al* (ANUP, 1975). A wide range of views is also represented in the compilation of papers *Metamorphism and Plate Tectonic Regimes*, edited by W G Ernst (Dowden, Hutchinson and Ross, 1975). A useful compilation which provides up-to-date reviews of metamorphic reactions is *Characterization of Metamorphism through Mineral Equilibria*, edited by J M Ferry, published by the Mineralogical Society of America as number **11** of its series *Reviews in Mineralogy*.

Meteorites, tektites and lunar rocks

Until the era of space exploration the study of meteorites was carried out by only a very few specialists, and there was negligible information on the composition of the Moon and other planetary bodies. It is only recently that those areas of petrological investigation have expanded rapidly and have entered the earth science curriculum in universities. Consequently there are very few textbooks and much information is still only available in the original periodical literature and conference reports (see also Chapter 11).

Until recently, *Meteorites*, by B Mason (Wiley, 1962) was the standard textbook on meteorites, and, being very well written, is still a good introduction to the subject. *Meteorites: classification and properties*, by J T Wasson (Springer-Verlag, 1974) and *Meteorites: a petrological-chemical synthesis*, by R T Dodd (Cambridge University Press, 1981) are more advanced and up-to-date. Recent elementary works are *The Nature and Origin of Meteorites*, by D W Sears (Adam Hilger, 1978) and *The Search for Our Beginning*, by R Hutchison (Oxford University Press, 1983). Particular categories of meteorite are described in *Handbook of Iron Meteorites*, by V F Buchwald (3 vols, University of California Press, 1976) and *Carbonaceous Meteorites*, by B Nagy (Elsevier, 1975). Tektites are described in *Tektites and Their Origins*, by J A O'Keefe (Elsevier, 1976). *Planetary Science: a lunar perspective*, by S R Taylor (Lunar and Planetary Institute, Houston, 1982) is the best review of the petrology of lunar rocks, and describes what is known of the rocks of other planets as a result of visits by unmanned spacecraft in recent years.

A key reference work for meteorites is the *Catalogue of Meteorites, with special reference to those represented in the collection of the British Museum (Natural History), IVth edition*, by A L Graham, A W R Bevan and R Hutchison (British Museum (Natural History), 1985). This includes details of all known meteorites as well as meteorite craters and tekites.

Periodicals

Just as the professional organization of petrology is largely catered for by the existing geological and mineralogical societies, so the bulk of published petrological information is dispersed through the general geological and mineralogical literature. Descriptions of igneous, sedimentary and metamorphic rocks may appear in almost any geological publication, such as a journal or a geological survey report. There are a small number of journals which specialize in petrology and these are mostly of fairly recent origin (since 1960) and tend to concentrate on the most detailed and advanced type of petrological work. The two most important are the *Journal of Petrology* (1960–) and *Contributions to Mineralogy and Petrology* (1966–), both of which publish papers on igneous and metamorphic petrology as well as on the mineralogy of igneous and metamorphic rocks. Other journals in this field are *Lithos* (1968–), *Pétrologie* (1975–), and the *Journal of Metamorphic Geology* (started in 1983). A significant number of important petrological papers also appear in *Earth and Planetary Science Letters* (1966–) and in the *Journal of Geophysical Research* (1896–).

In the field of sedimentary petrology the most important journal is the *Journal of Sedimentary Petrology* (1931–). Papers dealing with the petrology of sedimentary rocks also appear in *Sedimentology* (1962–), *Sedimentary Geology* (1967–), the *Bulletin of the American Association of Petroleum Geologists* (1917–), *Clay Minerals* (1947–), and *Clays and Clay Minerals* (1953–). The literature of soil science (see Chapter 19) contains some articles of interest to sedimentary petrologists, although for the most part the study of soils is directed towards their importance in agriculture.

Volcanology has a specialized literature of its own. There are two major volcanological journals: *Bulletin Volcanologique* (1924–), which in 1986 changed its title to *Bulletin of Volcanology* while retaining continuity of volume numbering, and the *Journal of Volcanology and Geothermal Research* (1976–). Two other essential sources of information for volcanology are the *Catalogue of the Active Volcanoes of the World including Solfatara Fields* (1951–) published in a series of volumes by the International Association of Volcanology and Chemistry of the Earth's Interior, and *Volcanoes of the World: a regional directory, gazetteer, and chronology of volcanism during the last 10,000 years*, by T Simkin *et al* (Hutchinson Ross/Academic Press, 1981).

Meteoritics is catered for by the journal *Geochimica et Cosmochimica Acta* (1951–), which is sponsored by the Meteoritical Society as well as the Geochemical Society, and by *Meteoritics* (1953/56–). Papers on meteorites also appear in mineralogical, geochemical and astronomical journals, e.g. *Mineralogical Magazine* (1876–), *Chemie der Erde* (1926–), and *Icarus* (1962–). Most of the reports on lunar rocks and on petrological aspects of the other planets have been published in conference volumes, especially the *Proceedings of the Lunar and Planetary Science Conferences* published annually, initially by Pergamon Press, and subsequently by the American Geophysical Union as supplements to the *Journal of Geophysical Research*.

Abstracts and bibliographies

There is no abstracting service or bibliography that is specific to petrology: the services described in the chapter on mineralogy cater for petrology as well as mineralogy. *Mineralogical Abstracts* provides the most complete and rapid entry into the recent petrological literature for the new researcher or non-specialist.

CHAPTER THIRTEEN

Structural geology and tectonics

P W G TANNER

This chapter is a review of the literature relating to the study of geological structures on all scales from the microscopic to the continental. It covers such diverse aspects as the objective description of these structures, the mechanisms by which they have formed and their evolution in time.

Structural geology is concerned with the geometrical form and mode of development of structures in rocks, whereas tectonics (syn: geotectonics) is the study of the same phenomena on a regional scale, including fold belts and large portions of the earth's crust. Thus structural geology is concerned more with methods of study and mechanisms of rock deformation, whereas tectonics utilizes the data of structural geology, stratigraphy and geophysics to assemble and test large-scale models and probe into the causes of deformation of the earth's crust.

In this chapter, following details of the professional organizations available for structural geologists and a review of the primary literature sources, the historical development of structural analysis is reviewed briefly, to provide a general background. This is followed by separate sections on structural geology and tectonics in which the key textbooks, reference sources, colloquium volumes etc., are described. For ease of reference each section is subdivided into separate fields such as terminology, petrofabrics, plate tectonics, etc., and selected references to closely related disciplines such as rock mechanics, engineering and economic geology are given.

As structural geology is a field-based science a separate section on 'Collection and analysis of data' is included which is largely concerned with field techniques and methods of analyzing structural maps and data.

Finally, there is a section on sources of abstracts, bibliographies and review articles.

Organizations

A specialist group, the *Tectonic Studies Group* of the Geological Society of London, was formed in 1970 to provide a forum for structural geologists and workers in related fields such as metamorphic petrology, rock mechanics, engineering and geophysics. The Group holds an Annual Meeting in December of each year which attracts about 300 participants from Europe and North America. Conference proceedings are published in the *Journal of the Geological Society of London* and details of forthcoming meetings are advertised bimonthly in the Society's *Newsletter*. An important function of the Group is to organize thematic meetings, both in the UK and abroad, some of which generate published sets of papers; field meetings and workshops are also organized. Membership is free and open to all with an interest in structural geology or tectonics.

A sister group, the *Canadian Tectonics Group*, was formed in 1981 but attendance at the annual meeting is restricted to participants who either read a paper or present a poster. Abstracts of papers are published in the *Journal of Structural Geology*.

The *Joint Association for Petroleum Exploration Courses* (UK) (JAPEC) organizes a number of courses and workshops each year, some of which are of interest to structural geologists. The *Joint Association for Geophysics Group* and the *Metamorphic Studies Group* occasionally hold joint meetings with the *Tectonic Studies Group* and details of such activities are to be found in the Geological Society of London's *Newsletter*.

Primary literature

Structural geology

Two journals are devoted entirely to publishing papers on structural geology and tectonics — the *Journal of Structural Geology* (established in 1979 and from 1983 appearing in 6 issues per year) and *Tectonophysics* (1982–: 10 issues per year). Both journals regularly produce special issues, some of which are referenced in later sections of this chapter.

Papers on structural geology are also found in a wide range of non-specialist journals such as

Journal of the Geological Society of London (bimonthly) (1971–); formerly
   Quarterly Journal of the Geological Society of London, (1845–1970)
Bulletin of the Geological Society of America (monthly) (1890–)
American Journal of Science (monthly) (1824–)
Geological Magazine (bimonthly) (1864–)
Canadian Journal of Earth Sciences (1964–)
Transactions of the Royal Society of Edinburgh (Earth Sciences) (1788–)
Geologische Rundschau (1910–)
Geological Journal (1964–).

The *Bulletin of the American Association of Petroleum Geologists* (1917–) carries articles on tectonic structures, largely in sedimentary rocks, and on faulting and diapiric structures, while the *Scottish Journal of Geology* (1965–) includes an above-average proportion of papers on the structural geology of metamorphic rocks. Russian literature is translated in *Geotectonics* (1967–). The *Journal of African Earth Sciences* (1983–) provides a further source of structural literature.

The fields of experimental rock deformation and petrofabric analysis are covered by *Tectonophysics*, the *Journal of Structural Geology* and by occasional articles of geological importance in the *Journal of Strain Analysis* (1965–). Papers on brittle deformation may be found in the *International Journal of Rock Mechanics and Mining Sciences* (1964–), while sections on deformation, faults, fractures and tectonophysics are included in the *Journal of Geophysical Research* (1896–). Publications on mathematical aspects of structural geology are occasionally included in *Computational Geoscience* and *Mathematical Geology* (1986–); formerly *Journal of the International Association for Mathematical Geology* (1969–85).

Papers on structural geology are also published in the four-yearly proceedings of the International Geological Congress. Other sources include the *Professional Papers* series of the Geological Society of America.

Tectonics

Most of the journals referred to above carry articles on tectonics, the prime sources being *Tectonophysics, Journal of Structural Geology, Journal of the Geological Society of London*, the *Bulletin of the Geological Society of America, Geology* (1973–) and *Precambrian Research* (1974–).

Tectonics (1982–), published jointly by the American Geophysical Union and European Geophysical Society is intended for papers on the structure and evolution of the terrestrial lithosphere, including plate tectonic synthesis, and has recently been joined by the *Journal of Geodynamics* (1984–) which, although more concerned with deep-seated phenomena, does include papers on plate tectonics. Other sources include *Nature* (London) and *Transactions of the American Geophysical Union* (1920–).

Tectonic syntheses feature strongly in the *Memoirs* and *Professional Papers, Geological Society of America; Papers, Geological Survey of Canada; Special Papers, Geological Association of Canada*; and *Special Publications, Geological Society of London*.

Historical background

Tectonics began with attempts by pioneers like Steno in the seventeenth century to explain the external shape and origin of mountains. This work was largely speculative and was not placed upon a scientific basis until the first detailed geological maps were made in the nineteenth century. The early history of geotectonics is given in *The Birth and Development of the Geological Sciences*, by F D Adams (see p.463).

From around 1830 there followed a period of increasing activity, with large areas of the earth's crust being mapped geologically for the first time. Major structural syntheses resulted from this work and an appreciation of the stimulating atmosphere of that period can be gained from *Chapters on the Geology of Scotland*, by B N Peach and J Horne (Oxford University Press, 1930) and, especially, *Tectonic Essays, mainly Alpine*, by E B Bailey (Clarendon Press, 1935). Also of interest here is *Pre-cambrian Geology of North America*, by C R van Hise and C K Leith (US Geological Survey, 1909), and classical texts such as those by Suess, Heim and Collet which are referred to in the section on 'Tectonics'. The work of pioneers in field mapping, albeit with an East European bias, is recorded in *Contributions to the History of Geological Mapping*, edited by E Dudich (Akademiai Kiado, Budapest, 1984) and the pre-1945 "classical" theories of orogenesis are reviewed in lucid style by A M C Şengör in *Orogeny*, by A Miyashiro *et al* (Wiley, 1982) and by H Masson (*Eclogae Geologiae Helvetiae*, **69**, 527–575, 1976).

During the late nineteenth century inquiries into the nature of schistosity and cleavage, and the manner in which rocks deform, were also being carried out but were handicapped by a lack of knowledge of the physics of rock deformation. The development of petrofabric methods which followed, together with the more recent history, is reviewed in Chapter 1 of *Structural Analysis of Metamorphic Tectonites*, by F J Turner and L E Weiss (McGraw-

Hill, 1963) and key papers are reprinted in *Fabric of Ductile Strain*, by M R Stauffer (Hutchinson Ross, *Benchmark papers in geology*, **75**, 1983). Two valuable historical reviews of ideas and terminology in this field are 'Slaty cleavage — a review of research since 1815', by A W B Siddans (*Earth Science Reviews*, **8**, 205–232, 1972) and 'Early theories and hypotheses on pressure-solution redeposition', by D W Durney (*Geology*, **6**, 369–372, 1978).

As knowledge of the major orogenic belts was gradually compiled in such works as *Das Antlitz der Erde*, by E Suess (5 vols, Freytag, Wein, 1885–1909), the basic problems of orogenesis became apparent, but, as Adams stated in 1930: "We cannot indeed but recognize that not only is the problem of the origin of mountain ranges still unsolved, but that toward the final elucidation of this subject geological science has made a less satisfactory advance than in many, if not in most, other directions". Some of the fundamental dynamic problems were outlined by F A Donath in 'Fundamental problems in dynamic structural geology', in *The Earth Sciences, Problems and Progress in Current Research*, edited by T W Donnelly (University of Chicago Press, 1963); and progress in the various fields was reviewed by N Rast in 'Recent advances in geotectonics', (*Earth Science Reviews*, **2**, 1–46, 1967).

The advent of the plate tectonics hypothesis and its general acceptance in the late 1960s provided a mechanism for orogenesis, and fundamental advances in experimental structural geology and the development of modern methods of field analysis have greatly enhanced our understanding of rock deformation. A history of attempts (to 1979) to use the plate tectonic hypothesis to explain the development of orogenic belts has been given by K Aki in *Orogeny* (see above); the story of the evolution of new field techniques and their impact on structural geology has yet to be written.

A selection of the definitive geological literature up to 1950, and including some 24 articles largely concerned with tectonics, is given in *Source Book in Geology*, by K F Mather and S L Mason (reprinted 1964, Stechert-Hafner), and *Source Book in Geology 1900–1950*, by K F Mather (Harvard University Press, 1967).

Structural geology

Structural geology was until fairly recently largely descriptive and concerned with the geometry of structures. However, aided by both theoretical and experimental advances in the study of the mechanics of deformation, and by detailed field studies of naturally occurring structures, it has evolved over the past two decades into a true scientific discipline. The philosophical approach is reviewed by C A Anderson in 'Simplicity in structural geology', in *The Fabric of Geology* (see p.324).

Modern structural geology is concerned with the quantitative analysis of changes in shape of rock masses (strain) and of their transport from one place to another (displacement); and with identifying datum planes, both physical and temporal, which can be used in other studies involving radiometric dating and the timing of igneous and metamorphic events.

An excellent, simple introduction for those with no previous knowledge of geology is provided in Chapters 8 and 12 of *Introduction to Geology; Vol.* **1**, *Principles*, by H H Read and J Watson (2nd edn, Macmillan, 1968). Much of the same ground is covered in *Principles of Physical Geology*, by A Holmes and D Holmes (3rd edn, Van Nostrand Reinhold, 1978). A modern, concise and more detailed treatment of geological structures, rock deformation and plate tectonics can be found in Chapters 3, 9 and 13 of *The Earth; an introduction to physical geology*, by J Verhoogen *et al* (Holt, Rinehart and Winston, 1970).

Three modern introductory texts are recommended: those by Davis, Park and Suppe. *Structural Geology of Rocks and Regions*, by G H Davis (Wiley, 1984) is a thorough, up-to-date and easily read text enlivened by the occasional cartoon. *Foundations of Structural Geology*, by R G Park (Blackie, 1983) provides a concise, untrammelled introduction to modern structural geology with adequate reference material. An excellent book, clearly written and well illustrated is *Principles of Structural Geology*, by J Suppe (Prentice-Hall International, 1985). This last book is expensive and unfortunately not available in softcover at present.

Introduction to Small-scale Geological Structures, by Gilbert Wilson (Allen and Unwin, 1982) is a well-illustrated book by one of the founders of the modern approach: it provides a valuable historical perspective but, being based on his 1961 paper, has been published too late to gain the accolade it deserves. *An Outline of Structural Geology*, by B E Hobbs *et al* (Wiley, 1976) is an introductory text which, although in need of revision, is particularly useful for accounts of stress and strain and, particularly, the development of microfabrics. A more recent text for undergraduates is *Principes de Tectonique*, by A Nicolas (Masson, 1984), now available in an English translation, *Principles of Rock Deformation* (Reidel, 1987).

At a more advanced level one book has been pre-eminent since the late 1960s — *Folding and Fracturing of Rocks*, by J G Ramsay (McGraw-Hill, 1967). In clarity of exposition, illustration and mathematical treatment it was unrivalled but has been replaced by *The Techniques of Modern Structural Geology*, by J G Ramsay and M I Huber (3 vols, Academic Press, 1983–). The first volume in the series, entitled *Strain Analysis* (published 1983), sets a new standard of excellence with magnificent illustrations, a clear text, and many "real" worked examples. *Volume 1* covers the principles of rock deformation, *Volume 2* (published 1987) covers the application of these principles to folds and fractures, and *Volume 3* is planned as a specialist text.

A laboratory manual which provides a basic introduction to most aspects of structural geology at undergraduate level is *Structural Analysis and Synthesis: a laboratory course in structural geology*, by S M Rowland (Blackwell Scientific Publications, 1986).

In the field of brittle deformation, the book by N J Price, *Fault and Joint Development in Brittle and Semi-brittle Rock* (Pergamon, 1966) presents an eminently readable and thorough account of the mechanics of rock deformation. A completely revised 2nd edition of this book, including the introduction of new topics such as hydraulic fracturing and overthrusting, long-delayed, is currently planned for release in 1990. A new text in this field is *Fracture Mechanics of Rock*, edited by B K Atkinson (Academic Press, 1987).

Older textbooks on structural geology are largely descriptive and combine introductory sections on structural elements, such as folds, faults and linear and planar fabrics, with sections on regional tectonics and orogenesis. Many do not provide an adequate background in the mechanisms of rock deformation and require revision of the sections on structural mechanisms, regional synthesis and causes of orogenesis. Textbooks suitable for undergraduate students and those with a basic general knowledge of geology include *Structural and Tectonic Principles*, by P C Badgley (Harper and Row, 1965), *Structural Geology*, by L U de Sitter (2nd edn, McGraw-Hill, 1964), *Structural Geology*, by M P Billings (3rd edn, Edward Arnold, 1972) and *Elements of Structural Geology*, by E S Hills (2nd edn, Wiley, 1972). New texts in this style include *Fault and Fold Tectonics*, by W Jaroszewski (Ellis Horwood, 1985) and a thought-provoking book by two Japanese workers, *Geological Structures*, by T Uemura and S Mizutani (Wiley, 1984).

A more detailed treatment of geometry, and the terminology of folds, including a comprehensive bibliography and index, is found in *Structural Geology of Folded Rocks*, by E H T Whitten (Rand, McNally, 1966). Emphasis on the microscopic fabrics of natural and experimentally deformed rocks is given in *Structural Analysis of Metamorphic Tectonites*, by F J Turner and L E Weiss (McGraw-Hill, 1963). Somewhat neglected is an excellent text by A M Johnson, *Physical Processes in Geology* (Freeman, 1970).

Atlases, whether of field or microscopic features, are of little value unless accompanied by a comprehensive and informative text. The *Minor Structures of Deformed Rocks: a photographic atlas*, by L E Weiss (Springer-Verlag, 1972) fails this test but the *Atlas of Deformational and Metamorphic Rock Fabrics*, edited by G J Borradaile *et al* (Springer-Verlag, 1982) is a valuable reference source, though largely devoted to cleavage development and not as comprehensive in treatment as the title would suggest. A more personalized viewpoint is given in the *Atlas of the Textural Patterns of Metamorphosed (Transformed and Deformed) Rocks and their Genetic Significance*, by S S Augustithis (Theophrastus Publications, Athens, 1985).

Classical texts covering specific aspects of structural geology include:

The Dynamics of Faulting and Dyke Formation with Applications to Britain, by E M Anderson (2nd edn, Oliver and Boyd, 1951)

Earth Flexures, their Geometry and their Representation and Analysis in Geological Section with Special Reference to the Problem of Oil Finding, by H G Busk (Cambridge University Press, 1929; republished unabridged by William Trussel, New York, 1957)

Lineation, a critical review of annotated bibliography, by E Cloos (*Geological Society of America, Memoir* **18**, 1946)

Boudinage, by E Cloos (*Transactions of the American Geophysical Union*, **28**, 626–632, 1947).

Terminology

Inconsistencies and deficiencies in structural and tectonic nomenclature have led to considerable confusion and have hampered the development of structural geology. They are still an obstacle in international communication. The fundamental principle by which the terminology should be judged is that terms used to describe tectonic structures should not carry any genetic connotation whatsoever. Clear rules of nomenclature and classification, with examples of their use in the classification of faults, were proposed in 'Role of classification in geology', by M L Hill, in *The Fabric of Geology*, edited by C C Albritton, Jr. (Addison-Wesley, 1963).

A major advance achieved corporately by the International Geological Congress, the National Science Foundation and the American Association of Petroleum Geologists was the publication of the *International Tectonic Dictionary*, compiled and edited by J G Dennis (*American Association of Petroleum Geologists, Memoir* **7**, 1967). The history, and, where possible, the first use of each term is given, together with a discussion of its general usage. The dictionary represents an extremely valuable contribution to structural geology and tectonics, and includes a bibliography. A less detailed, but comprehensive, work is the *Glossary of Geology and Related Sciences with Supplement*, edited by R L Bates and J A Jackson (2nd edn, American Geological Institute, 1980) (see also Chapter 4).

Structural terms are also included in *Challinor's Dictionary of Geology*, edited by A Wyatt (6th edn, University of Wales Press, 1985); but although the historical comments are often of interest, this work suffers from a less rigorous definition of terms than is given in the works already quoted.

More detailed definitions of terms describing the morphology and geometry of minor structures are given in *Folding and Fracturing of Rocks*, by J G Ramsay (McGraw-Hill, 1967), and 'The description of folds', by M J Fleuty (*Proceedings of the Geologists' Association*, 75, 461–492, 1964). Usage of such terms is also clearly explained in the *Geological Society of London Handbook Series*. Much research in the past decade has been devoted to thrust systems and the somewhat esoteric terminology which has developed is reviewed by R W H Butler (*Journal of Structural Geology*, 4, 239–245, 1982).

A classification of fault rocks (breccia, mylonite, pseudo-tachylite, etc.) is given by R H Sibson in 'Fault rocks and fault mechanisms' (*Journal of the Geological Society of London*, **133**, 191–213, 1977). Microfabric terms are defined in *Metamorphic Textures*, by A Spry (Pergamon, 1969), and tectonic nomenclature is critically discussed by N Rast in 'Orogenic belts and their parts', in *Time and Place in Orogeny*, edited by P E Kent *et al* (*Geological Society of London, Special Publication* 3, 1969).

Collection and analysis of data

The geological map provides a basis for many structural interpretations, and structural mapping in particular has evolved into a precise scientific exercise. The philosophy which underlies the making of such maps is described in 'Nature and significance of geological maps', by J M Harrison, in *The Fabric of Geology*.

In the chapter on geological maps and remote sensing there are several textbooks mentioned which are relevant to the structural geologist. An elementary, but comprehensive, introduction to field mapping techniques is provided in *Manual of Field Geology*, by R R Compton (Wiley, 1962), and much useful information, especially for techniques of outcrop mapping, is contained in *Methods of Geological Surveying*, by E Greenly and H Williams (Murby, 1930), which was written in honour of that greatest of field geologists, C T Clough, 'to publish a few plain instructions for drawing geological boundary lines, a practical matter which seemed to have been somewhat neglected'. Some details of field procedures are also given in *Structural Methods for the Exploration Geologist*, by P C Badgley (Harper and Row, 1959), and in *Structural Geology of Folded Rocks*, by E H T Whitten (Rand McNally, 1966), with emphasis upon statistical sampling methods in the latter book.

Three of the *Geological Society of London Handbook Series* (published by the Open University Press and Halstead Press) are recommended as *vade mecums* to modern structural field mapping: *Basic Geological Mapping*, by J W Barnes (1981), and of particular value, *The Field Description of Metamorphic Rocks*, by N Fry (1984), and *The Mapping of Geological Structures*, by K R McClay (1988). An excellent guide to field procedures, with particular emphasis on the use of aerial photographs and with no pretence of being comprehensive, is *Methods in Field Geology*, by F Moseley (Freeman, 1981). A useful text but with less emphasis on structural mapping is *Field Mapping for Geology Students*, by F Ahmed and D C Almond (Allen and Unwin, 1983). Aerial photographs are often used to select an area of study, make a preliminary structural interpretation and also facilitate both the mapping and the recording of data in the field. For details of available texts reference should be made to Chapter 7.

A thorough and well-ordered guide to the study and analysis of structures seen on geological maps is *Introduction to Geological Maps and Structures*, by J L Roberts (Pergamon, 1982). This book also provides classification schemes, and compared with other texts, is particularly useful for information on structures in igneous rocks, and on unconformities. Other information on map interpretation can be found in *Analysis of Geologic Structures*, by J M Dennison (Norton, 1968), and in the appendix on 'Representation of structure' in *The Earth; an introduction to physical geology*. A newly revised, comprehensive, and well-presented laboratory manual for both map interpretation and elementary data analysis is *Structural Geology: an introduction to geometrical techniques*, by D M Ragan (3rd edn, Wiley, 1985).

Collections of elementary map exercises suitable for laboratory work include *Simple Geological Structures*, by J I Platt and J Challinor (4th edn, Thomas Murby, 1968) and *An Introduction to Geological Structures and Maps*, by G M Bennison (4th edn, Edward Arnold, 1985). By far the best and most interesting manual of this type is *Advanced Geological Map Interpretation*, by F Moseley (Edward Arnold, 1979). In *The Techniques of Modern Structural Geology*, Ramsay and Huber (see p.323) the authors move away from the classical "problem map" and present worked examples based on actual material.

Structural analysis is concerned largely with the spatial relationship of linear and planar elements and *Angular Relations of Lines and Planes*, by D V Higgs and G Tunell (2nd edn, Freeman, 1966), provides an introduction to both stereographic analysis and the use of spherical trigonometry.

Methods of stereographic analysis are described in detail in *The Use of Stereographic Projection in Structural Geology*, by F C Phillips (2nd edn, Edward Arnold, 1971); *Eléments de Tectonique Analytique*, by P Vialon *et al* (Masson, 1976); and *Manual of the Stereographic Projection for a Geometric and Kinematic Analysis of Folds and Faults*, by P J Haman (*West Canadian Research Publications of Geology and Related Sciences*, Series 1, **1**, 1961). Their use in areas of polyphase deformation is described in advanced textbooks on structural geology such as those by Ramsay and Huber (p.323) and Ragan (p.326) and in a very clear and detailed fashion in *Hemispherical Projection Methods in Rock Mechanics*, by S D Priest (Allen and Unwin, 1985). The limitations of these methods are analyzed by probability theory in *Structural Diagrams*, by A B Vistellius, a translation from the Russian edited by N L Johnson and F C Phillips (Pergamon Press, 1966).

Structural data are especially amenable to analysis by the techniques of orientation statistics, and computers are being used increasingly for the storage, analysis and plotting, of such data, and for structural simulations. One of the earliest publications in this field was *Computer Analysis of Orientation Data in Structural Geology*, by T V Loudon (*Office of Naval Research Task No. 389–135, Technical Report No. 13*, Northwestern University, 1964). A useful source of computer programs (in a field for which no book is available) which includes vector trend analyses of directional data, analysis of subsurface fold geometry and construction of π-diagrams is the series *Computer Contributions, State Geological Survey, University of Kansas* (1966–); 'structural' programs are also occasionally found in *Computers and Geosciences*. Work in this field is discussed in *Structural Geology of Folded Rocks*, by E H T Whitten (see p.323). A listing of computer programs of value to structural geologists has been prepared (1985) by the *Tectonic Studies Group* (Geological Society of London) and a register of microcomputer programs has been compiled by Dr W T C Sowerbutts, Dept of Geology, University of Manchester.

An introduction to data analysis is given in *Statistics and Data Analysis in Geology*, by J C Davis and R J Sampson (2nd edn, Wiley, 1987) and a key advanced text is *Statistics of Orientation Data*, by K V Mardia (Academic Press, 1972).

An introduction to the literature of scale-model simulation studies and finite-element analysis as used in the numerical simulation of folding in rocks is given briefly in *Computer Simulation in Geology*, by J W Harbaugh and G Bonham-Carter (Wiley/Interscience, 1970). A more complete introduction to finite-element analysis can be found in *Introduction to the Finite Element Method*, by C S Desai and J F Abel (Van Nostrand Reinhold, 1972) and in *The Finite Element Method, a Basic Introduction for Engineers*, by K C Rockey *et al* (2nd edn, Granada Technical Books, 1983).

Presentation of the results derived by data processing in structural stereograms, block diagrams and profiles or cross-sections is vital to communication with other earth scientists, and some of these methods are described in Turner and Weiss, and Ragan (see pp.320 and 326), as well as in *Block Diagrams and Other Graphic Methods Used in Geology and Geography*, by A K Lobeck (2nd edn, Emmerson-Trussell, Amherst, Mass., 1958). A classical example of presentation of structural data is 'Geology of the Tovqussap Nuna', by A Berthelsen (*Meddelelser om Grønland*, **123**, (1), 1960). The technique is described in 'Construction of block diagrams to scale in orthographic projection', by D B McIntyre and L E Weiss (*Proceedings of the Geologists' Association*, **67**, 142–155, 1956) and a revival of the use of the orthographic projection as a tool to rival the equal-area net is elegantly proposed by D G De Paor in 'Orthographic analysis of geological structures — I. Deformation theory' (*Journal of Structural Geology*, **5**, 255–277, 1983) and 'II. Practical applications' (*Journal of Structural Geology*, **8**, 87–100, 1986).

A collection of 7 papers in memory of David Elliot entitled 'Balanced cross-sections and their geological significance', edited by J R Hossack and P L Hancock (*Journal of Structural Geology, Special Issue*, **5**, No. 2, 1983) provides a valuable commentary and bibliography on the techniques of section balancing.

Tectonophysics

The study of the mechanisms of rock deformation has been approached both experimentally and by the mathematical analysis of the shapes of deformed objects occurring in rocks. The physical basis for this work was provided by the theories of stress and strain developed in physics and engineering, and an introductory account by J Handin is to be found as Chapter 11 of *Handbook of Physical Constants* revised edition, edited by S D Clark (*Geological Society of America, Memoir* **97**, 1966). A more easily read and enjoyable introduction to the subject is given by J E Gordon in *The New*

Science of Strong Materials (2nd edn, Penguin, 1976) and in *Structures* (Penguin, 1978). The basic mathematical theory, with many worked examples of its application to geological structures, is provided in an excellent book by W D Means entitled *Stress and Strain* (Springer-Verlag, 1976).

General textbooks which discuss the mechanical behaviour of rocks and describe the principles of mechanics include:

Studies in Large Plastic Flow and Fracture, by P W Bridgeman (Harvard University Press, 1964)

Theory of Flow and Fracture of Solids, by A Nadai (2 vols, 2nd edn, McGraw-Hill, 1968)

Fundamentals of Rock Mechanics, by J C Jaeger and N G W Cook (3rd edn, Chapman and Hall, 1979)

Introduction to Rock Mechanics, by R E Goodman (Wiley, 1980)

The mathematical properties of stress and strain and examples of their application to geological structures are given in the textbooks by Davis, Johnson, Ramsay and Huber, Vol 1, and Suppe (see previous pages), and in *Elasticity, Fracture and Flow, with Engineering and Geological Applications*, by J C Jaeger (corrected 2nd edn, Methuen, 1964).

More advanced mathematical treatments can be found in:

Mechanics of Incremental Deformation Theory of Elasticity and Viscoelasticity of Initially Stressed Solids and Fluids Including Thermodynamic Foundations and Application to Finite Strain, by M A Biot (Wiley, 1965)

Principles of Mechanics and Dynamics (formerly titled *Treatise on Natural Philosophy* and published in 1867), by Sir W Thomson and P G Tait (Dover, 1962)

The Classical Field Theories, by C Truesdell and R A Toupin. (*In: The Encyclopedia of Physics*, 3, edited by S Flügge) (Springer-Verlag, 1960)

Introduction to the Mechanics of a Continuous Medium, by L E Malvern (Prentice-Hall, 1969)

Tensor analysis has considerable application in this field and basic introductions to the theory are given by Ramsay (1967), also in *Methods of Mathematical Physics*, by R Courant and D Hilbert (Wiley, 1953) and in *Physical Properties of Crystals, their Representation by Tensors and Matrices*, by J F Nye (Oxford University Press, 1967). A useful text is *Applications of Tensor Analysis*, by A J McConnell (Dover, 1957).

Experimental studies

Results of fundamental importance have been obtained by the experimental deformation of rocks at elevated pressures and temperatures, notably the importance of strain rate and pore fluid pressures in deformation processes occurring over a geological time span. Much of the early work is contained in the published proceedings of conferences and symposia — for example, *Rock Deformation*, edited by D Griggs and J Handin (*Geological Society of America, Memoir* **79**, 1960), *State of Stress in the Earth's Crust: Proceedings of the International Conference, Santa Monica, California* (Elsevier, 1964), *Failure and Breakage of Rock*, edited by C Fairhurst (8th Symposium on Rock Mechanics, University of Minnesota, 1966), 'Experimental deformation of crystalline rocks', by I Borg and J Handin (*Tectonophysics*, **3**, (4), 1966), and *Flow and Fracture of Rocks*, edited by H C Heard *et al* (American Geophysical Union, 1972).

Two important monographs published recently are *Experimental Rock Deformation – the brittle field*, by M S Paterson (Springer-Verlag, *Minerals and Rocks*, **13**, 1978) and *Mechanical Behaviour of Crustal Rocks*, edited by N L Carter *et al* (*American Geophysical Union, Monograph* **24**, 1981).

Some of the earliest attempts to understand the mechanisms of folding and faulting were made with models using materials such as sand, plasticine, and putty to represent rock layers. Recent advances are described in *Gravity, Deformation and the Earth's Crust as Studied by Centrifugal Models*, by H Ramberg (2nd edn, Academic Press, 1981), and a review and bibliography of work up to 1965 is provided by 'Experimental structural geology', by J B Currie (*Earth Science Reviews*, **1**, 51–67, 1966). Experimental studies carried out during the period 1960–75 and related to the development of foliation in rocks are reviewed by W D Means (*Tectonophysics*, **39**, 329–354, 1976). A recent paper reporting results obtained by use of a simple shear machine is 'Development of sheath folds in shear regimes' by P R Cobbold and H Quinquis (*Journal of Structural Geology*, **2**, 119–126, 1980).

Petrofabric studies

Petrofabric analysis, the study of the optical and dimensional orientation of minerals in a rock, long discredited as a primary tool for understanding the evolution of tectonic structures, is able to make a positive contribution now that more is known about the mechanisms by which rocks and individual minerals deform under stress.

Pioneer work in this field was carried out in Austria in the 1920s and published in *Gefügekunde der Gesteine*, by B Sander (Springer, 1930), and in a later book which has recently been republished in translation as *An Introduction to the Study of Fabrics of Geological Bodies*, by B Sander, translation by F C Phillips and G Windsor (Pergamon, 1970). Also worthy of mention here are two American works, *Structural Petrology* (*Geological Society of America, Memoir* **6**, 1938) and *Structural Petrology of Deformed Rocks*, by H W Fairburn (2nd edn, Addison-Wesley, 1949). Two classic papers are those by M A Harker, 'On slaty cleavage and allied rock structures with special reference to the mechanical theories of their origin' (*Report of the British Association*, 55th Meeting, 1–40, 1885) and by G Voll entitled 'New work on petrofabrics' (*Liverpool and Manchester Geological Journal*, **2**, 503–567, 1960).

More recent work is reported in Turner and Weiss (see p.323), and especially in Chapter 2 of Hobbs, Means and Williams (see p.322). A clear introduction (with references) to the origin of fabrics in metamorphic rocks is given in Chapter 13 of *Igneous and Metamorphic Petrology*, by M G Best (W H Freeman, 1982). The theoretical approach is described in detail in *Crystalline Plasticity and Solid State Flow in Metamorphic Rocks*, by A Nicolas and J P Poirier (Wiley, 1976); techniques (and processes) are reviewed in *Preferred Orientation in Deformed Metals and Rocks*, by H-R Wenk (Academic Press, 1985); and some new petrofabric work is reported and reviewed in *Mountain Building Processes*, edited by K J Hsü (Academic Press, 1983).

Recent work is summarized in papers such as 'The effects of strain on the microstructures, fabrics and deformation mechanisms in quartzites', by S White (*Philosophical Transactions of the Royal Society of London*, **A283**, 69–86, 1976) and 'Fabric asymmetry and shear sense in movement zones', by J L Bouchez *et al* (*Geologische Rundschau*, **72**, 401–419, 1983). Three important thematic sets of papers which include much petrofabric work are: *Fabrics, Microstructures, and Microtectonics*, edited by G S Lister *et al* (*Tectonophysics*, **39**, Nos 1–3, 1977), *The Effect of Deformation on Rocks*, edited by G S Lister *et al* (*Tectonophysics*, **78**, 1981), and *Planar and Linear Fabrics of Deformed Rocks* (*Special Issue, Journal of Structural Geology*, **6**, No 1/3, 1984).

A related study which has come into prominence in conjunction with the discovery of the polyphase nature of deformation in orogenic belts in the study of the internal fabrics in minerals and the relationships of mineral growth to successive deformations. *Metamorphism*, by A Harker (3rd edn, Methuen, 1950), provides well-illustrated descriptions and definitions of mineral textures. The first five sections of *Controls of Metamorphism*, edited by W S Pitcher and G W Flinn (Oliver and Boyd, 1965), summarize work in this field but *Metamorphic Textures*, by A Spry (Pergamon, 1969), is unique in providing the only comprehensive account of metamorphic textures available at present; a 2nd edition of this book is in the press. A recommended specialist text is *Rotated*

Garnets in Metamorphic Rocks, by J L Rosenfeld (*Geological Society of America, Special Paper* **129**, 1970). The recent work on the interpretation of microstructures is reported in 'An evaluation of criteria to deduce the sense of movement in sheared rocks', by C Simpson and S M Schmid (*Bulletin of the Geological Society of America*, **94**, 1281–1288, 1983). A new approach to the interpretation of the fabrics in metamorphic rocks was heralded by 'Deformation partitioning and porphyroblast rotation in metamorphic rocks: a radical reinterpretation', by T H Bell (*Journal of Metamorphic Geology*, **3**, 109–118, 1985), one of a series of papers published by Bell and his co-authors in this journal.

Economic geology (see also Chapter 15)

Petroleum

The last few years have seen a revolution in the quantitative application of structural analysis to oil and gas exploration. Key papers which acted as a catalyst include: 'Some remarks on the development of sedimentary basins', by D P McKenzie (*Earth and Planetary Science Letters*, **40**, 25–32, 1978), 'On the theory of growth faulting: a geomechanical delta model based on gravity sliding', by W Crans *et al* (*Journal of Petroleum Geology*, **2**, 1980), and 'Modes of extensional tectonics', by B Wernicke and B C Burchfiel (*Journal of Structural Geology*, **4**, 105–115, 1982).

Three collections of papers provide the detail. They are *Petroleum Geology of the Continental Shelf of North-West Europe*, edited by L V Illing and G D Hobson (Institute of Petroleum, 1981); a thematic set of papers on the structural geology of the continental shelf around the UK published in the *Journal of the Geological Society of London* **181**, Part 4, 1984; and *Petroleum Geology of the North European Margin* (Graham and Trotman, 1985). Interpretation of seismic reflection data is covered in *Seismic Expression of Structural Styles, Vol. 2. Tectonics of Extensional Provinces*, edited by A W Bally (American Association of Petroleum Geologists, *Studies in Geology Series*, **15**, 1983). Of specialist interest are the first textbook on the analysis of discontinuous networks in reservoir rocks, *The Reservoir Engineering Aspects of Fractured Formations*, by L H Reiss (Editions Technip, 1980), and a collection of papers on *Deformation of Sediments and Sedimentary Rocks*, edited by M E Jones and R M F Preston (*Geological Society of London, Special Publication*, **29**, 1987).

New textbooks in this field are long overdue; the older ones include *Structural Geology for Petroleum Geologists*, by W L Russell (McGraw-Hill, 1955), and *Structural Methods for the*

Exploration Geologist, by P C Badgley (Harper and Row, 1959).
Two symposium volumes of historical interest, published by the
American Association of Petroleum Geologists, which provide
useful syntheses of oil-field information are *Possible Future
Petroleum Provinces of North America*, edited by M W Bell *et al*
(1951), and *The Habitat of Oil*, edited by L G Weeks (1958).

Finally, the plate tectonic settings in which oil deposits occur are
discussed in *Petroleum and Global Tectonics*, edited by A G
Fischer and S Judson (Princeton Unversity Press, 1975).

Ore deposits

Structural analysis is of obvious value in fields such as civil
engineering, mining and mineral exploration. However, the
structural control of the ore deposits, for example, has received
little detailed attention and the contribution which structural
geology is able to make in engineering practice has only been truly
recognized in the last decade or so. *Structural Geology with Special
Reference to Economic Deposits*, by B Stŏces and C H White
(Macmillan, 1935), although out-of-date, is superbly illustrated
and makes an ideal companion volume to advanced structural textbooks such as that by Ramsay (see p. 323). Also 'of value is *Ore
Deposits, as Related to Structural Features*, by W H Newhouse
(Princeton University Press, 1942), and symposium volumes such
as:

Stratiform Copper Deposits in Africa. Part II: Tectonics, edited by J Lombard and
P Nicolini (Association of African Geological Surveys, 1963)
Remobilization of Ores and Minerals (Associazione Mineraria Sarda, Cagliari,
Italy, 1969)
Relations of Tectonics to Ore Deposits in the Southern Cordillera, edited by W R
Dickson and W D Payne (*Arizona Geological Society Digest*, **14**, 1981)

The last is a collection of 19 papers published in memory of
James Gilluly. It is largely concerned with ore deposits in the
South West United States but the implications of the work are
broad. Many different types of mineralization and their plate
tectonic setting are described in *Mineral Deposits and Global
Tectonic Settings*, by A H Mitchell and M S Garson (Academic
Press, 1982).

Engineering geology (see also Chapter 16)

Applications of structural geology to engineering are mentioned in
Chapter 1 of *Rock Mechanics in Engineering Practice*, edited by K
G Stagg and O C Zienkiewicz (Wiley, 1968). Other useful texts
include *Elements of Engineering Geology*, by J E Richey (Pitman,
1964), *Design of Structures in Rock*, by L Obert and W I Duvell

(Wiley 1967), and *A Geology for Engineers*, by F G H Blyth and M H de Freitas (7th edn, Edward Arnold, 1984); the last provides a useful bibliography. Other texts are referred to in the chapter on engineering geology but two are worthy of mention here, *Principles of Engineering Geology*, by P B Attwell and I W Farmer (Wiley, 1976) and *Geotechnology — an introductory text for students and engineers*, by A Roberts (Pergamon, 1977).

Diapirs, glaciers and igneous intrusions

Structural geology encompasses not only phenomena such as folding and faulting but also the mechanisms of igneous intrusion and the movement of large bodies of material such as salt diapirs and glaciers.

Salt Deposits, the Origin, Metamorphism and Deformation of Evaporites,by H Borchert and R Muir (Van Nostrand, 1964) and *The Physics of Glaciers*, by W S B Paterson (2nd edn, Pergamon, 1981), deal with most of the structural aspects of salt and ice, respectively, and both provide extensive bibliographies. The book by Paterson is of particular value as structures and fabrics in ice provide sound analogues for similar features in rock. An additional work on salt deposits is *Diapirism and Diapirs; a symposium*, edited by J and G D O'Brien (*American Association of Petroleum Geologists, Memoir* **8**, 1968) and the similarity of these structures to those found in rocks is discussed in 'Salt stocks as natural analogues to Archaean gneiss diapirs', by W M Schwerdtner (*Geologische Rundschau*, **71**, 370–379, 1982). There is much of structural interest in *Dynamical Geology of Salt and Related Structures*, edited by I Lerche and J J O'Brien (Academic Press, 1987).

Structural aspects of igneous intrusions are discussed in *Structural Behaviour of Igneous Rocks*, by R Balk (J W Edwards, Ann Arbor, originally published as *Geological Society of America, Memoir* **5**, 1948), and in *Mechanism of Igneous Intrusion*, edited by G Newall and N Rast (Gallery Press, Liverpool, 1970). An important regional study is *The Geology of Donegal: a study of granite emplacement and unroofing*, by W S Pitcher and A R Berger (Wiley, 1972) and a new model, with wide implications, for the emplacement of these granites during contemporaneous simple shear is given in 'A tectonic model for the emplacement of the Main Donegal Granite, N.W. Ireland', by D H W Hutton (*Journal of the Geological Society of London*, **139**, 615–631, 1982).

Topics in structural geology

Several topics in structural geology have been of particular interest and importance since 1970 and some of these are reviewed briefly in this section.

Shear zones The study of ductile shear zones was an important research topic in the 1970s, following the paper by J G Ramsay and R H Graham, 'Strain variation in shear belts' (*Canadian Journal of Earth Sciences*, 7, 786–813, 1970). This work culminated in an international meeting in Barcelona, Spain, on many aspects of simple shear in rocks. Papers presented at the meeting were published in *Shear Zones in Rocks*, edited by J Carreras *et al* (*Journal of Structural Geology*, 2, Nos 1/2, 1980).

Folding Folding and boudinage have attracted steady attention and two contributions with an individualistic approach are a study of the morphology of folds entitled *Strain Facies*, by E Hansen, No. 2 in the monograph series *Minerals, Rocks and Inorganic Materials* (Springer, 1971), and *Styles of Folding*, by A M Johnson (Elsevier, *Developments in Geotectonics*, 11, 1977). The latter is a collection of already published papers by the author and his co-workers.

Strain analysis A considerable effort has been devoted to developing methods of measuring strain in rocks and much of this work is reviewed in Volume 1 of Ramsay and Huber, 1983 (see p. 323) and in two important collections of papers entitled *Strain Patterns in Rocks*, edited by P R Cobbold and W M Schwerdtner (*Special Issue, Journal of Structural Geology*, 5, Nos 3/4, 1983), and *Strain within Thrust Belts*, edited by G D Williams (*Special Issue, Tectonophysics*, 88, Nos. 3/4, 1982). A classical work on regional strain analysis, profusely illustrated, is *Microtectonics*, by E Cloos (John Hopkins Press, 1971). *Geological Strain Analysis*, by R J Lisle (Pergamon, 1985) is a manual for use of the Rf/ø technique, and many aspects of strain analysis as applied to igneous rocks are covered in *The Structural Analysis of Granitic Rocks*, by J Marré (North Oxford Academic and BRGM, 1986).

Structural processes A wide selection of papers is to be found in *Energetics of Geological Processes*, edited by S K Saxena and S Bhattacharji (Springer, 1977); *Deformation Processes in Tectonites (Special Issue, Tectonophysics*, 92, Nos 1/3, 1983); and in the *Philosophical Transactions of the Royal Society of London*, A283, 1976.

Thrust tectonics This topic dominated research in structural geology in the 1970s. The classic text is 'Balanced cross-sections', by C D A Dahlstrom (*Canadian Journal of Earth Sciences*, 6, 743–757, 1969) and a collection of key papers is contained in *Mechanics* *of Thrust Faults and Décollement*, by B Voight (Hutchinson, Ross, *Benchmark Papers in Geololgy*, 1976). 'Thrust systems', by S E Boyer and D E Elliott (*Bulletin of the American Association of Petroleum Geologists*, **66**, 1196–1230, 1982) provides a brilliant exposition on the subject and an extensive reference source. A useful collection of papers is contained in *Thrust and Nappe Tectonics*, edited by K R McClay and N J Price (*Geological Society of London, Special Publication* **9**, 1981) and a volume in memory of David Elliott, *Balanced Cross-sections and their Geological Significance* has been published recently (*Special Issue, Journal of Structural Geology*, **5**, No. 2, 1983). Work on ductile thrusts or slides is critically discussed in 'Tectonic slides: a review and appraisal', by D H W Hutton (*Earth Science Reviews*, **15**, 151–172, 1979).

Tectonics

The detailed results of structural mapping are compounded into regional descriptions or syntheses which deal with the geometry and tectonic development of large portions of the earth's crust. The amount of shortening across these zones can be assessed by means of structural analysis or, in some cases, palaeomagnetism. Plate tectonic models can then be erected to explain their origin, and evolution with time.

An elementary introduction to tectonics is provided in Volumes 2 and 3 (Earth History: early stages and later stages) of *Introduction to Geology*, by H H Read and J V Watson (Macmillan, 1975). *The Evolving Continents*, by B F Windley (2nd edn, Wiley, 1984) provides a more detailed but clear and readable synthesis of the evolution of the classic fold belts. In *Evolution of the Earth's Crust*, edited by D H Tarling (Academic Press, 1978) the emphasis is also upon regional synthesis but the treatment is less balanced than in Windley's book. *Tectonics*, by J Goguel, translation of the French edition, 1952, by H E Thalmann (W H Freeman, 1962) covers a wide range of topics and is 'perhaps the most perceptive and beautifully written book on many aspects of modern structural geology' (A W B Siddans, *Tectonophysics*, **104**, 258, 1984).

Eduard Suess established the study of tectonics with his publication of *Die Entstehung der Alpen* (W. Braumüller, Wien) in 1875, followed by the magnificent *Das Antlitz der Erde* (5 vols, Freytag, Wein, 1885–1909).

For the next half century, Alpine tectonics had an immense influence on tectonic hypotheses, largely through such masterly works as:

Geologie der Schweiz, by A Heim (C H Tauchnitz, Leipzig, 1919)
The Structure of the Alps, by L W Collet (Edward Arnold, 1927)
Bau und Entstehung der Alpen, by L Kober (Franz Deuticke, Vienna, 1955)

Then, as the morphology and distribution of the major orogenic belts began to be understood more clearly, a number of writers started to speculate on the causes and controls of orogenesis. These writings include largely discursive works such as:

Earthquakes and Mountains, by H Jeffreys (2nd edn, Methuen, 1929)
The Deformation of the Earth's Crust, by W H Bucher (Princeton University Press, 1933; reprinted 1941)
The Pulse of the Earth, by J H F Umbgrove (2nd edn, Martinus Nijhoff, The Hague, 1950)
Symphony of the Earth, by J H F Umbgrove (Martinus Nijhoff, The Hague, 1954)
Mountain Building; a study primarily based on Indonesia, region of the world's most active crustal deformation, by R W van Bemmelen (Martinus Nijhoff, The Hague, 1954)
Fundamentals of Geology, by S von Bubnoff (translated and edited by W T Harry; Oliver and Boyd, 1963)

Early attempts to understand the causes and mechanisms of orogenesis were severely limited by a lack both of quantitative data and the compilation of available data into a suitable form. A review of the problems involved has been published in *Time and Place in Orogeny*, edited by P E Kent *et al* (*Geological Society of London, Special Publication* **3**, 1969), and a synthesis of previous work is given by R Dearnley in 'Orogenic fold belts and a hypothesis of Earth evolution', in *Physics and Chemistry of the Earth*, **7**, 1966.

The physical problems of orogenesis are discussed in *Principles of Geodynamics*, by A E Scheidegger (2nd edn, Springer, 1963), and the relationship of orogenic events to foci of sedimentary deposition in *Geosynclines*, by J Aubouin (Elsevier, 1965).

Modern views on tectonics and orogenesis can be found in *Tectonics and Mountain Ranges* (Rodgers Volume, *American Journal of Science*, **275-A**, 1975) and *Thrust and Nappe Tectonics*, by McClay and Price (see p. 336). A W Bally in 'Thoughts on the tectonics of folded belts', a review paper in the latter volume, concludes that gravity gliding is not an important process for mountain building. A different viewpoint, emphasising the importance of vertical tectonic forces, has however been developed in Russia: *Basic Problems in Geotectonics*, by V V Belousov (McGraw-Hill, 1962), *Structural Geology*, by V V Belousov, English translation of 1961 Russian edition (Mir. Publ. Moscow, 1968), *Folded Deformations in the Earth's Crust, their Types and Origin*, by V V Belousov and A A Sorkii (Israel Program for Scientific Translations, Jerusalem, 1965) and recently expressed in *Geotectonics*, by V V Belousov (Springer-Verlag, 1980).

The part played by gravity tectonics within the tectonic belts is also emphasized in *Gravity, Deformation and the Earth's Crust, as Studied by Centrifugal Models*, by H Ramberg (2nd edn, Academic Press, 1981). *Gravity and Tectonics*, edited by K A De Jong and R Scholten (Wiley, 1973) consists of a diverse set of papers and is most often quoted for Durney and Ramsay's paper on syntectonic crystal growth.

The most up-to-date commentary on tectonics is found in *Mountain Building Processes*, edited by K J Hsü (Academic Press, 1983), being the proceedings of an International Symposium on Mountain Building held in Zürich in 1981. Geophysical aspects of mountain belts are well covered in: *Seismic Expression of Structural Styles, Vol. 3. Tectonics of Compressional Provinces/ Strike Slip Tectonics*, edited by A W Bally (*American Association of Petroleum Geologists, Studies in Geology Series*, **15**, 1983), *Continental Tectonics: structure, kinematics and dynamics*, edited by M Friedman and M N Toksöz (*Special Issue, Tectonophysics*, **100**, 1983), and in *Contributions to the Tectonics and Geophysics of Mountain Chains*, edited by R D Hatcher, Jr., *et al* (*Geological Society of America, Memoir* **158**,). Two important collections of papers on contrasting styles of tectonism have been published recently: *Continental Extensional Tectonics*, edited by M P Coward *et al* (*Geological Society of London, Special Publication* **28**, 1987) and *Collision Tectonics*, edited by M P Coward and A C Ries (*Geological Society of London, Special Publication* **19**, 1986). The field of neotectonics is reviewed in *Recent Earth Movements*, by C Vita-Finzi (Academic Press, 1986).

Regional syntheses

The works referenced here represent the essential data of tectonics: syntheses largely compiled during the past 10 years to satisfy the increasing rigour required by plate tectonic modelling. Attention here has been focussed on several well-presented orogenic belts to exemplify the general approach and indicate the present state of the art.

Caledonian and Appalachian fold belts

Early attempts to correlate events in these two contemporaneous orogenic zones now preserved on either side of the Atlantic Ocean are summarized in a number of contributions in *North Atlantic — Geology and Continental Drift; a symposium volume*, edited by M Kay (American Association of Petroleum Geologists, 1969).

Summaries and bibliographies of work in the Caledonian fold belt in the British Isles are contained in *The British Caledonides*, edited by M R W Johnson and F H Stewart (Oliver and Boyd, 1963), and *The Geology of Scotland*, edited by G Y Craig (2nd edn, Scottish Academic Press, 1983). This work is placed in a more general context in *The Geological History of the British Isles*, by G M Bennison and A F Wright (Edward Arnold, 1969) and *The Structure of the British Isles*, by J G C Anderson and T W Owen (2nd edn, Pergamon, 1980). The definitive work is *The Caledonides of the British Isles — Reviewed*, edited by A L Harris *et al* (*Geological Society of London, Special Publication* **8**, 1979) and a useful set of papers can be found in the thematic set 'Plate tectonics and the evolution of the British Isles' (*Journal of the Geological Society, London*, **139**, Part 4, 1982). A recent synthesis is published in the *Memoir of the Geological Society of London*, **9**, edited by A L Harris, entitled 'The nature and timing of orogenic activity in the Caledonian rocks of the British Isles'.

Early structural syntheses of the Appalachian fold belt in North America, include:

Appalachian Tectonics, edited by T H Clark (*Royal Society of Canada, Special Publications* **10**, 1967)
Studies in Appalachian Geology, Northern and Maritime, by E an Zen, W S White and J B Thompson, Jr. (Wiley/Interscience, 1968)
The Tectonics of the Appalachians, by J Rodgers (Wiley/Interscience, 1970)
Studies of Appalachian Geology: Central and Southern, edited by G W Fisher *et al* (Wiley, 1970)

Recent work is contained in *The Caledonides in the USA, Proceedings of IGCP Project 27*, Blacksburg, Virginia, and *Tectonic Studies in the Talladega and Carolina Slate Belts, Southern Appalachian Orogen*, edited by D N Bearce *et al* (*Geological Society America, Special Paper* **191**, 1982). Work in the Canadian portion of the Appalachian belt is reviewed in *Major Structural Zones and Faults of the Northern Appalachians*, edited by P St-Julien and J Beland (*Geological Association of Canada, Special Paper*, 1982) and, in Scandinavia, in *The Caledonide Orogen: Scandinavia and related areas*, edited by D G Gee and B A Sturt (Wiley, 1986, parts 1 and 2).

Modern work on both sides of the Atlantic is brought together in the *Caledonian-Appalachian Orogen of the North Atlantic Region* (*Geological Survey of Canada, Paper* **78–13**, 1978) and, more recently, in *Regional Trends in the Geology of the Appalachian-Caledonian-Hercynian-Mauritanide Orogen*, edited by P E Schenk (*NATO Advanced Science Institutes Series, C*, **116**, 1983). The concluding symposium of the International Geological Correlation Programme, Project 27, Caledonide Orogen, was held in Glasgow in September, 1984 and the proceedings volume for this meeting, *The Caledonian-Appalachian Orogen*, edited by A L

Harris (Blackwell Scientific Press, 1987) presents stage-by-stage reviews of the whole orogen.

Variscan fold belt

A collection of papers, *The Variscan Orogen in Europe*, edited by H J Zwart and U F Dornsiepen (*Geologie en Mijnbouw*, **60**, 1981) sets the general scene and provides a bibliography, whereas *The Variscan Fold Belt in the British Isles*, edited by P L Hancock (Adam Hilger, 1983) gives an extensive data base for the UK but fails to provide an overview. A broader perspective is provided by *The Northern Margins of the Variscides in the North Atlantic Region*, edited by D H W Hutton and D S Sanderson (*Geological Society of London, Special Publication* **14**, 1984) and a thematic set of papers on the Variscan belt of S.W. England was published in the *Journal of the Geological Society of London*, **143**, Part 1, 1986.

Himalayan fold belt

A prolific, but not widely known, documentation is available for this region which is at present in the forefront of structural and tectonic research:

Geology of the Himalayas, by A Gansser (Wiley, 1964)
Geology of the Nepal Himalaya, by S Hashimoto (Saikon Publ. Co., Japan, 1973)
Geodynamics of Pakistan, edited by A Farah and K A De Jong (Geological Survey Pakistan, 1979)
Zagros, Hindu Kush, Himalaya; geodynamic evolution, edited by H K Gupta Harsh *et al* (American Geophysics Union, *Geodynamics Series*, **3**, 1981)
Tectonic Geology of the Himalaya, edited by P S Saklani (Today and Tomorrow's Publishers, New Delhi, 1978)
Himalayan Thrusts and Associated Rocks, edited by P S Saklani (Today and Tomorrow's Publishers, New Delhi, 1986)

European Alps

The early literature has already been summarized and bibliographies of recent work are given in *Mountain Building Processes* (see p.331) and in 'An outline of the geology of Switzerland', *In: Geology of Switzerland; a guidebook*, by R Trümpy (*International Geological Congress Guidebook* **10**, Wepf and Co., 1980). Work on both the Alps and Pyrenees is referenced in *The Geology of Western Europe*, by M G Rutten (Elsevier, 1969) and *The Structure of Western Europe*, by J G C Anderson (Pergamon, 1978).

Precambrian terranes

With the help of radiometric dating the sequence of fold belts from the earliest pre-2500 ma greenstone belts to the later linear, mobile belts can be examined. The African continent provides a classic setting for such studies and this work is compiled and reviewed in *The Geochronology of Equatorial Africa*, by L Cahen and N J Snelling (2nd edn, Clarendon Press, 1984) and *Crustal Evolution of Southern Africa*, by A J Tankard *et al* (Springer-Verlag, 1982). Other work on Precambrian rocks is summarized in *The Precambrian in Mobile Zones*, edited by G Choubert and A Faure-Muret (*Earth Science Reviews*, 1980).

Other regions

Included here are books, special collections of papers, and memoirs which are of more than regional interest.

General texts include *Variations in the Tectonic Styles in Canada*, edited by R A S Price and R J W Douglas (*Geological Association of Canada, Special Paper* **11**, 1972) and *Structural Characteristics of Tectonic Zones*, edited by K L Burns and R W R Rutland (*Tectonophysics*, **47**, 1978). Metamorphic core complexes were first identified and named in the United States and the definitive work is *Cordilleran Metamorphic Core Complexes*, edited by M D Crittenden, Jr. *et al* (*Geological Society of America, Memoir* **153**, 1980) (dedicated to P Misch).

Two works on fold belts in the Southern Hemisphere are *The Phanerozoic Structure of Australia and Variations in Tectonic Style*, edited by E Scheibner (*Tectonophysics*, **48**, 1978) and *The Origin of the Southern Alps*, edited by R I Walcott and M M Cresswell (*Bulletin of the Royal Society of New Zealand*, **18**, 1979).

Finally, *Intercontinental Fold Belts*, edited by H Martin and F W Eder (Springer-Verlag, 1983) has a promising title but presents case studies from the Variscan belt in Europe and the Damara Belt in Namibia which are largely non-structural in content.

Rift systems

The major rift fault systems of the world are comparable in scale and importance with the orogenic belts. They are described and analyzed in *The World Rift System*, edited by T N Irvine (*Geological Survey of Canada, Paper* **66–14**, 1966), and in *Graben Problems (Proceedings of the International Rift Symposium, Karlsruhe, 1968)*, edited by J H Illies and St Mueller (Schweizerbart'sche, 1970). Data on the East African Rift System are compiled in *Report on the Geology and Geophysics of the East*

African Rift System and Report of the UMC/UNESCO Seminar on the East African System, both published by UNESCO in 1965. 'A discussion on the structure and evolution of the Red Sea and the nature of the Red Sea, Gulf of Aden and Ethiopia Rift Junction' organized by N L Falcon *et al* (*Philosophical Transactions of the Royal Society, London*, **A** 267, 1970), provides a bibliography and reference to aspects of rift tectonics.

Recent symposium volumes and thematic sets of papers include the following major works: *Tectonics and Geophysics of Continental Rifts*, edited by I B Ramberg and E-R Neumann (Reidel, 1978), *The Oslo Paleorift*, edited by J A Dons and B T Larsen (*Norges Geologiske Undersökelse*, **337**, 1978), and *Mechanism of Graben Formation*, edited by J H Illies (*Developments in Geotectonics*, **17**, 1981). *Processes of Continental Rifting*, edited by P Morgan and B H Baker (Elsevier, 1983) is a reprint of Volume **94** of *Tectonophysics* and contains 35 papers on the compilation and modelling of data from rift zones.

Plate tectonics

The hypothesis that the continents as we now see them were once connected to form larger crustal units, the supercontinents, was first expounded by Wegener and du Toit, based upon geological evidence presented in *The Origin of the Continents and Oceans*, by A Wegener, translated from the 4th revised German edition, 1929 by J Biram (Dover, 1966), and *Our Wandering Continents*, by A L du Toit (Oliver and Boyd, 1937). The continental drift hypothesis has only relatively recently found general acceptance, largely because of the geophysical evidence which was adduced to support it in, for example, *Continental Drift*, by S K Runcorn (Academic Press, 1962). As the geophysical results became available, a number of symposia were held which provide a testimony to the changing stage of opinion in the late 1950s and the 1960s. These include: *Continental Drift*, edited by S W Carey (Geology Department, University of Tasmania, 1958), *A Symposium on Continental Drift*, edited by P M S Blackett *et al* (Royal Society, 1965), *Continental Drift*, by G D Garland (*Royal Society of Canada, Special Publication* **9**, 1966), and *The History of the Earth's Crust, a Symposium*, edited by R A Phinney (Princeton University Press, 1968).

Geophysical results — in particular, palaeomagnetic data — led to the concept of plate tectonics, which provides a mechanism for continental drift. The value of the plate tectonics hypothesis lies in the fact that, in providing a possible mechanism for mountain building, it gave a great impetus to the compilation of tectonic data in the 1970s and an opportunity for a completely fresh look at some of the outstanding problems. The field has now become one of the most rapidly expanding parts of the earth sciences and has given rise to an extensive literature.

A general introduction is given in *Continental Drift*, by O H Tarling and M P Tarling (Bell, 1971) and early accounts of plate tectonics are presented in *The Megatectonics of Continents and Oceans*, by H Johnson and P L Smith (Rutgers University Press, 1979) and *The Sea: ideas and observations on progress in the study of the seas*, edited by A E Maxwell (Wiley/Interscience, 1970). Accounts of orogenesis and plate tectonics are given by J Sutton and E R Oxburgh, respectively, in *Understanding the Earth; a reader in the earth sciences*, edited by I G Gass *et al* (Artemis Press, 1971) and a good introduction to the geometry of plate motions can be found in Chapter 10 of Hobbs, Means and Williams (see p.322).

A concise account of major earth structures and plate tectonics is given by Park (see p.322) and *The Dynamic Earth*, a 'Scientific American' book published by Freeman (1983), is strongly recommended as a basic introductory text. Pre-eminent in this field is a fascinating new course book, *Plate Tectonics: how it works*, by A Cox and B R Hart (Blackwell Scientific Publications, 1986).

More advanced texts include *Plate Tectonics*, by X Le Pichon (Elsevier, *Developments in Geotectonics*, **6**, 1973) and especially recommended, *Plate Tectonics and Crustal Evolution*, by K C Condie (2nd edn, Pergamon, 1982). *Mechanisms of Continental Drift and Plate Tectonics*, edited by P A Davies and S K Runcorn (Academic Press, 1981) consists of 26 papers which provide an overview but not a consensus about the mechanisms responsible for plate movement.

Two bibliographies of plate tectonics literature have been prepared. *Plate Tectonics*, by J M Bird (revised edn, American Geophysical Union, 1980) consists of a selection of 69 papers published up to 1979, together with a bibliography, and the *Bibliography of Continental Drift and Plate Tectonics*, by T Kasbeer (*Geological Society of America, Special Paper* **142**, 1973) includes citations up to 1971.

At a more specialized level *Trench and Forearc Geology: sedimentology and tectonics on modern and ancient active plate margins*, edited by J K Leggett (*Geological Society of London, Special Publication* **11**, 1982) provides numerous examples of the application of the plate tectonic hypothesis, and *Plate Tectonics and the Evolution of the British Isles*, a thematic set of papers published in the *Journal of the Geological Society of London*, **139**, No. 4, 1982, illustrates the application of the hypothesis to a single region over a long period of time. An important work on the evolution of a single obliquely convergent margin is *The Geotectonic Development of California*, edited by W G Ernst (Prentice-Hall, 1981).

Examples of plate tectonic modelling with reference to basin development are included in *Structural History of the Mediterranean Basins*, edited by B Biju-Duval and L Montadert (Editions Technip, 1977) and *The Geological Evolution of the Eastern Mediterranean*, edited by J Dixon and A H F Robertson (*Geological Society of London, Special Publication 7*, 1985). The contentious application of plate tectonics to the Precambrian is given a full airing in the 28 contributions to *Precambrian Plate Tectonics*, edited by A Kröner (Elsevier, 1981).

A new era in plate tectonics modelling has been heralded by the recognition of slivers or microcontinents of exotic material accreted to plate margins. These so-called "allochthonous terranes", which can be shown to have undergone large strike-slip displacements, make up a large part of western North America. They were reported in 'Cordilleran suspect terranes' by P J Coney *et al* (*Nature*, (London) **288**, 329–333, 1980) and the first international conference on the new accretion-collision tectonics was held in Hokkaido, Japan, in 1981. The collection of papers which resulted was published in *Accretion Tectonics in the Circum-Pacific Regions*, edited by H Hashimoto and S Uyeda (Reidel, *Advances in Earth and Planetary Sciences*, 1983), whose main defect lies in poor sub-editing. This branch of plate tectonics has become one of the major growth areas in the science.

Abstracting services and reviews

The major current source for abstracts of structural geology and tectonics is the *Bibliography and Index of Geology* (see Chapter 5). Subject index headings include 'tectonics' (which is the most comprehensive), and also 'deformation', 'faults', 'folds', 'fractures', 'petrofabrics' and 'structural analysis'. Also of value is *Cahier F: Tectonique et Géophysique* of *Bibliographie des Sciences de la Terre* (see Chapter 5).

The published proceedings of conferences and symposia provide useful compilations of recent research and review papers and these have been listed quarterly since 1967 in *Geological Newsletter* (1967–76 continued as *Episodes* 1978–) (International Union of Geological Sciences).

Physical and geophysical aspects of structural geology are in *Geophysical Abstracts* (see Chapter 14) under the sub-headings of

'Geotectonics' and 'Strength and Plasticity'. In the index most of the structural topics are found under 'rock mechanics' and 'tectonics'.

Also of value is *Mineralogical Abstracts*, which includes work on petrofabrics, mineral textures, regional studies and structural geology under 'Petrology: general' and 'Petrology: metamorphism: regional, dynamic'.

Reviews of structural literature are published in *Earth Science Reviews* and *Tectonophysics*, as well as in general journals such as *Science*. Book reviews are published in *Journal of Structural Geology, Tectonophysics, Geological Journal,* and *Geological Magazine.*

Since this chapter was written in 1987 a number of books have been published in the fields of structural geology and tectonics, the most noteworthy of which are listed below.

Metamorphic Textures and Microstructures, by A J Barker (Blackie, 1989)
Interpretation of Geological Maps, by B C M Butler and J D Bell (Longman, 1988)
Introduction to Rock Mechanics, by R E Goodman (Wiley, 1988)
Geological Structures and Maps, by R J Lisle (Pergamon, 1988)
Mechanics of Tectonic Faulting, by G Mandl (Elsevier, *Developments in Structural Geology*, **1**, 1988)
Foundations of Structural Geology, by R G Park (2nd edn, Blackie, 1989)
Geological Structures and Moving Plates, by R G Park (Blackie, 1987)
Surveying and Mapping for Field Scientists, by W Ritchie, M Wood and R Wright (Longman, 1988)

CHAPTER FOURTEEN

Geophysics

JOHN MILSOM

The field of scientific activities covered by the term 'geophysics' is extraordinarily wide. The subject grades almost imperceptibly into such diverse disciplines as astronomy, meteorology, physical oceanography, and rock mechanics. Even within the range of studies of the solid earth which forms its core, two distinctly different groupings can be recognized. In the first the interest is in the physics of earth processes and the functioning of the earth as a physical system; in the second, measurements of physical phenomena are used as guides to the interpretation of geological situations in which the variations in physical properties are secondary and incidental.

The difference between these two approaches can be appreciated by contrasting the geophysicists who are concerned with understanding the mechanisms which produce and sustain the earth's magnetic field with their colleagues who use the field variations to map the distribution of certain rock formations in the upper crust. Although there is always a need for interchange of ideas between the two groups, it is obvious that the attitudes and methodologies will be quite different. 'Physical' geophysicists tend to communicate most easily with astronomers and with physicists investigating the behaviour of materials at very high temperatures and pressures, 'geological' geophysicists have more in common with conventional geologists.

The distinction between these two sides of geophysics is not quite the same as that commonly made between the 'pure' and 'applied' aspects of the science. Pure geophysicists who are concerned with the study of such things as plate-tectonics and collision processes often use techniques identical with, or at least very similar to, those used to explore for and evaluate deposits of economically useful minerals. Oil and natural gas are amongst the most important of these minerals and at the present time the majority of professional geophysicists are to be found working in the comparatively restricted field of petroleum exploration, using a single exploration tool, the reflection seismograph.

Organizations

Geophysical societies and organizations tend to observe the distinction between the pure and applied branches, rather than the more fundamental one between orientations towards physics or towards geology. The leading society in the field of pure geophysics is the American Geophysical Union (AGU), which was established in 1919 to "encompass all the physical sciences leading to a better understanding of the earth and its environment in space". It has a membership of more than 13,000, in more than 100 countries, and holds annual Spring and Fall meetings at which very large numbers of papers are read. As desribed later in this chapter, the AGU is also a major publisher of geophysical literature.

There is no direct equivalent to the AGU in the United Kingdom, but the Royal Astronomical Society and the Geological Society of London have formed a Joint Association for Geophysics which organizes a number of meetings each year for the benefit of members of both societies. They also jointly sponsor the annual UK Geophysical Assembly, the main forum for pure geophysics in the UK: in recent years there has been an increased, although still rather small, participation by applied geophysicists in this assembly.

The European Geophysical Society caters for scientists with interests mainly in pure geophysics, and works closely with the European Seismological Commission, notably in the arrangement of joint meetings and conventions.

There is also an international umbrella organization for pure geophysics, the International Union of Geodesy and Geophysics (IUGG), which is one of the eighteen scientific bodies grouped within the International Council of Scientific Unions. The IUGG has no individual members but comprises seven semi-autonomous associations, each responsible for a specific range of studies. These are the International Associations of Geodesy, Seismology and the Physics of the Earth's Interior, Volcanology and the Chemistry of the Earth's Interior, Geomagnetism and Aeronomy, Meteorology and Atmospheric Physics, Hydrological Sciences, and the Physical Sciences of the Oceans. The stated objectives of the IUGG include the promotion of the study of all the problems relating to the figure of the earth and the physics of the solid earth, oceans and atmospheres, and the coordination of the scientific activities of the 78 member countries in these areas. It sponsors, or maintains, a number of permanent services which promote standardization of measures, and collection, analysis and publication of geodetic and geophysical data. The IUGG was instrumental in setting up the first major international geophysical programme, the International Geophysical Year, in 1957–58. This outstandingly successful effort was followed by the International Year of the Quiet Sun and the International Upper Mantle Project in the 1960s, the International Geodynamics Project in the 1970s and the International Lithosphere Programme in the 1980s.

The largest organization in applied geophysics, the Society of Exploration Geophysicists (SEG), is also based in the United States. Founded in 1930, it now has about 18,000 members, of whom roughly half are classed as 'Active', with the remainder 'Associates' or 'Students' with lesser qualifications or experience. The membership is by no means confined to North America. The society organizes meetings and conventions at regional and national level and is an important publisher.

A rather similar but smaller society, the European Association of Exploration Geophysicists (EAEG) caters mainly, but not exclusively, for geophysicists working within Europe and has a membership of about 4000. There are also numerous national societies such as the Australian Society of Exploration Geophysicists, the Society of Exploration Geophysicists of Japan, and the Canadian Exploration Geophysics Society. Most of these publish journals, organize meetings and conventions and sponsor continuing professional education courses.

Textbooks and reference works

Because of the wide range of disciplines included in geophysics, even practising professional geophysicists may find themselves confronted by new terms and concepts when they move outside their own fields of expertise. In pure geophysics, however, it is possible to turn for assistance to the *International Dictionary of Geophysics* (Runcorn, 1967), a two-volume encyclopedia which, although now somewhat outdated, contains a large number of valuable introductory articles. Some information on current research in pure geophysics can be obtained from the annual *Yearbook of the IUGG*, which is available from the IUGG publications office.

The *Encyclopedic Dictionary of Exploration Geophysics* (Sheriff, 1980), published by the SEG, is a true dictionary containing concise definitions of terms in current use, as well as a number of useful tables and conversion factors. Information about users and practitioners of applied geophysics may be found in the annual issues of the *Geophysical Directory*, published by the Geophysical Directory Inc. of Houston.

A general overview and introduction to geophysics is provided by *Geophysics in the Affairs of Man* (Bates *et al*, 1982), which reviews the history and present status of all aspects of the discipline. A second historical summary, *The Road to Jaramillo* (Glenn, 1982) records the contributions of geophysics, and more particularly of palaeomagnetism, to the recent advances in understanding global tectonics. A rather different form of historical perspective is provided by the *History of Persian Earthquakes* (Ambraseys and Melville, 1982) which draws on written sources spanning a period of more than 1500 years.

Various aspects of geophysics are dealt with in the now very considerable number of textbooks available. A notable development of the last few years has been the rapid increase in the number of books which deal with the seismic reflection method and particularly with seismic processing. A selection of textbooks covering most aspects of solid-earth geophysics, but not atmospheric, oceanographic and space physics, is given below.

(a) General solid-earth and planetary geophysics

Introduction to Geophysics, by G D Garland (Saunders, 1979)
Physics of the Earth and Planetary Interiors, by A H Cook (Macmillan, 1973)
The Interior of the Earth, by M H P Bott (2nd edn, Elsevier, 1982)

(b) Gravity field of the Earth

Gravity, by C Tsuboi (Allen and Unwin, 1983)
International Gravity Standardization Net 1971, by C Morelli (*IUGG-IAG Special Publication* 4, 1973)
The Earth and its Gravity Field, by W A Heiskanen and F A Vening-Meinesz (McGraw-Hill, 1958)

(c) Earth's magnetic field

Introduction to Geomagnetism, by W D Parkinson (Scottish Academic Press, 1983)
Palaeomagnetism, by D H Tarling (Chapman and Hall, 1983)
Palaeomagnetism and Plate Tectonics, by M W McElhinney (Cambridge University Press, 1973)

(d) Earth's heat flow

Terrestrial Heat Flow in Europe, by V Cermak and L Rybach (*Inter-Union Commission on Geodynamics Special Report* 58, Springer-Verlag, 1979)

(e) General applied geophysics

Solution of the Inverse Problem in Geophysical Interpretation, by R Cassinis (Plenum Press, 1982)

Interpretation Theory in Applied Geophysics, by F S Grant and G F West (McGraw-Hill, 1965)

Time-sequence Analysis in Geophysics, by E R Kanasewich (2nd edn, University of Alberta Press, 1981)

Applied Geophysics, by W M Telford, L P Geldart, R E Sheriff and D A Keyes (Cambridge University Press, 1975)

(f) Mining geophysics

Geophysics and Geochemistry in the Search for Metallic Ores, edited by P J Hood (*Geological Survey of Canada, Economic Geology Report* **31**, 1979)

Geophysics of the Elura Orebody, edited by D W Emerson (Australian Society of Exploration Geophysicists, 1980)

Mining Geophysics (2 vols, Society of Exploration Geophysicists, 1966, 1967)

Mining Geophysics, by D S Parasnis (Elsevier, 1966)

(g) Geophysics in hydrocarbon exploration

Exploration Seismology, by R E Sheriff and L P Geldart (Cambridge University Press, 1983)

Introduction to Seismic Interpretation, by R McQuillin, M Bacon and W Barclay (2nd edn, Graham and Trotman, 1984)

Seismic Stratigraphy — application to hydrocarbon exploration, edited by C E Payton (*American Association of Petroleum Geologists, Memoir* **26**, 1977)

A First Course in Geophysical Exploration and Interpretation, by R E Sheriff (International Human Resources Development, Boston, 1978)

Seismic Data Processing, by L Hatton, M H Worthington and J Makin (Blackwell Scientific Publications, 1986)

(h) Well-logging and engineering geophysics

Applied Geophysics for Geologists and Engineers, by D H Griffiths and B C King (Pergamon Press, 1981)

Basic Well Log Analysis for Geologists, by G Asquith and C Gibson (American Association of Petroleum Geologists, 1982)

A Practical Introduction to Borehole Geophysics, by T Labo (Society of Exploration Geophysicists, 1987).

A number of hard-cover monograph series also constitute important secondary sources of information. Apart from those published by the American Geophysical Union, which have already been mentioned, the Academic Press *Advances in Geophysics* (1952–), Annual Reviews Inc's *Annual Review of Earth and Planetary Sciences* (1973–) and Elsevier's *Developments in Solid Earth Geophysics* are notable series in the pure geophysics field. In applied geophysics the Gebruder Borntraeger *Geoexploration Monographs* (which, despite their rather general title have so far been exclusively geophysical in content), the Applied Science Publishers *Developments in Geophysical Exploration Methods* and the Elsevier *Methods in Geochemistry and Geophysics* are particularly important series.

Periodicals

The primary written sources of geophysical information are research papers and survey reports published in the geophysical literature. In pure geophysics the major single publisher is the American Geophysical Union (AGU). The tripartite division of pure geophysics has been recognized since 1972 by the division of the AGU's main journal, the *Journal of Geophysical Research* (1896–), into three monthly sections, the 'blue' (Space Physics), the 'green' (Oceans and Atmospheres) and the 'red' (Solid Earth and Planetary Geophysics). The Union also publishes the monthly *Geophysical Research Letters* (1974–), the bimonthlies *Tectonics* (1982–), *Water Resources Research* (1965–) and *Radio Science* (1966–), and the weekly newsletter *Eos*, subtitled *Transactions of the American Geophysical Union* (1920–).

The AGU is a major source of geophysical literature in translation. It publishes English language versions of the *Seria Geofizicheskaia* and its successors, the *Fizika Zemli* and *Fizika Atmosferi i Okeana* sections of the *Bulletin of the Academy of Sciences of the USSR*, under the titles *Izvestiya* (1957–65), *Izvestiya — Physics of the Solid Earth* (1965–) and *Izvestiya — Atmospheric and Oceanic Physics* (1965–). It also issues complete translations of the Russian journals *Geotectonics* (1967–), *Geomagnetism and Aeronomy* (1961–), *Oceanology* (1961–), *Geodesy, Mapping and Photogrammetry* (1973– ; from 1962 to 1972 as *Geodesy and Aerophotography*) and *Soviet Hydrology* (1962–), as well as selected translations from Chinese under the title *Chinese Geophysics* (1978–).

The AGU's review journal, the quarterly *Reviews of Geophysics and Space Physics* (1970– ; as *Reviews of Geophysics* from 1963 to 1970), and the books in the various Monograph series mentioned in the previous section, can also claim to be considered as primary sources, since much of the information they contain has not been published elsewhere. The current series are the *Geophysical Monographs*, *Antarctic Research*, *Water Resources Monographs*, *Maurice Ewing Volumes* (devoted to aspects of subduction-related convergent tectonics), *Geodynamics* (which records work done under the auspices of the International Geodynamics Project) and *Coastal and Estuarine Sciences*.

No other organization even approaches this volume of publication, but one other North American organization working in the field of pure geophysics, the Seismological Society of America, publishes a bimonthly *Bulletin* (1911–) which is the main reference source for work on earthquake (as opposed to explosion) seismology.

In the UK the monthly *Geophysical Journal of the Royal Astronomical Society* (1958–) covers most of the field of pure geophysics and includes, usually in the April edition, abstracts of papers read at the UK Geophysical Assembly. The European Geophysical Society also publishes a journal, the *Annales Geophysicae* (1983–), in three sections which are roughly analogous to those of the *Journal of Geophysical Research*. This relative newcomer to the geophysical literature was created by an amalgamation of the French *Annales de Géophysique* (1944–1982) and the Italian *Annali di Geofisica* (1944-1982): papers are usually in English but are also accepted in French, German, Italian or Spanish. It differs from its American counterpart in that the 'Solid Earth' and 'Oceans and Atmospheres' sections are bound together, rather than being distributed separately.

The Deutsche Geophysikalische Gesellschaft still sponsors its own journal, the *Journal of Geophysics — Zeitschrift fur Geophysik* (1924–). The English title was added in 1974, when there was also a change from publication of papers mainly in German to papers almost exclusively in English. The East German *Geophysik und Geologie — Geophysikalischen Veroffentlichungen der Karl-Marx Universitat, Leipzig* (1975–) continues to be published in German. The *Geophysical Journal of the RAS*, the *Journal of Geophysics* and the solid-earth section of *Annales Geophysicae* were amalgamated in a new journal, known simply as the *Geophysical Journal*, in 1988.

The Osservatorio Geofisico Sperimentale in Trieste publishes a journal, the *Bollettino di Geofisica Teorica ed Applicata* (1958–), devoted mainly to earthquake seismology with papers almost entirely in English. The Society of Terrestrial Magnetism and Electricity of Japan also publishes a pure geophysics English language journal of international importance, the *Journal of Geomagnetism and Geoelectricity* (1949–).

As well as the publications associated in some way with learned societies, there are a number of commercial journals with a high content of pure geophysics. These include North Holland's *Physics of the Earth and Planetary Interiors* (1968–), the Elsevier journals *Tectonophysics* (1964–) and *Earth and Planetary Science Letters* (1966–), the Birkhaeuser Verlag *Pure and Applied Geophysics* (1964-; as *Geofisica Pura e Applicata* from 1939 to 1963), Reidel Publishing's *Marine Geophysical Research* (1970–) and also their rather oddly named *Surveys in Geophysics* (1972-; as *Geophysical Surveys* until 1986), which is devoted to surveys of advances in pure geophysics and not to "surveys" in the applied geophysics sense of the word.

The primary data gathered during the International Geophysical

Year was published by Pergamon Press as the *Annals of the IGY* in 42 volumes, of which the last is a bibliography and index. Reports on subsequent major international programmes have tended to appear either as individual texts or as special editions of established journals. Information, newsletters and reports on the current International Lithosphere Program can be obtained from the IUGG Publications Office.

The volume of published literature in the applied field is very much smaller, reflecting not a smaller degree of activity but the conditions of commercial secrecy under which much of the work is done. Both the SEG and EAEG publish regular journals, the SEG the monthly *Geophysics* (1936–) and the EAEG the slightly less frequent *Geophysical Prospecting* (1958–). The two societies have recently also begun publication of monthly, newsletter-type journals. The SEG *Leading Edge* (1982–) was designed as a magazine of both scientific and general interest; one of the issues acts as the Society's yearbook, a function formerly fulfilled by the April issues of *Geophysics*. The EAEG's *First Break* (1983–) is rather similar, its stated objectives being to constitute a means of rapid publication of results of especial interest and of articles designed to present aspects of geophysics in a simple fashion.

Other national and regional groups of applied geophysicists also publish journals which have reasonably wide circulations. These include the *Bulletin of the Australian Society of Exploration Geophysicists* (1970–; now renamed *Exploration Geophysics*) and the Scandinavian *Geoexploration* (1963–).

A number of other journals, notably those connected with the oil industry, contain a high percentage of geophysical papers. Foremost among these is the *Bulletin of the American Association of Petroleum Geologists* (1917–). The Association also publishes a variety of monographs and course notes, many with a very large geophysical content. *Marine and Petroleum Geology* (1984–), published jointly by Butterworths and the Geological Society of London, also contains a large number of articles on seismic reflection surveys.

Private companies, particularly geophysical contractors in the oil industry, are also prolific sources of primary, if somewhat partisan, information. Detailed descriptions of the latest developments in instrumentation and techniques have often been circulated by such companies for some time before any descriptions appear in the conventional published literature. Indications of what may be newly available can be gleaned from the advertising pages of *Geophysics* and *Geophysical Prospecting*.

Abstracting and indexing services

Until the end of 1971 the principle source of English language abstracts on geophysical topics was the monthly *Geophysical Abstracts*, issued by the US Geological Survey (1963–71; published in various forms by the US Bureau of Mines and US Geological Survey, 1929–1962). This journal covered world literature, the only criterion applied being that of general availability. Its publication was discontinued for economic reasons and it was thought that the gap would, to some extent, be filled by the *Bibliography and Index of Geology* (see Chapter 5). Four of the 29 sections into which the *Bibliography* is divided are exclusively geophysical, these being 17 — 'Geophysics, general', 18 — 'Geophysics, solid earth', 19 — 'Geophysics, seismology', and 20 — 'Geophysics, applied'. Publications with geophysical content are also likely to be cited in other sections.

The *Bibliography* has not proven a very satisfactory substitute for *Geophysical Abstracts*, since it contains citations only. Happily, the abstract journal was resurrected in 1977 by Geo-Abstracts Ltd. with a title change at the beginning of 1983 to *Geophysics and Tectonics Abstracts*, indicating a slight widening of scope. In 1986 there was a further title change, to *Geographical Abstracts; Geophysics and Tectonics*. Abstracts of original papers generally appear between six and nine months after their original publication. The abstracts database is now available as Geobase, for online search and retrieval through the Lockheed Corporation DIALOG database system (see also Chapter 6).

A specialist *Bibliography of Seismology* is issued monthly by the International Seismological Centre, which took over production with the start of the New Series in 1966. The publication was originally founded in 1939 by the Dominion Observatory of Canada.

Another abstracting journal that contains some geophysical material is *Physics Abstracts*, published fortnightly by the Institution of Electrical Engineers and also available online in the Inspec database. Section 9000 of each issue is devoted to geophysics, astronomy and astrophysics, 9100 to solid-earth geophysics, 9200 to hydrosphere and atmospheric geophysics, and 9300 to geophysical observatories, instruments and techniques.

In 1961 the US Government's Joint Publications Research Service commenced publication of a fortnightly guide entitled *Soviet Bloc Research in Geophysics, Astronomy and Space*. This was continued between 1975 and 1979 as *USSR and East Europe Scientific Abstracts: geophysics, astronomy and space* and thereafter as two separate journals, *USSR Report: Earth Sciences* and

USSR Report: Space. Russian literature can also be searched more directly through the *Geofizica* section of *Referativnyi Zhurnal* (see Chapter 5); the section is divided into five sub-series dealing with applied geophysics, geomagnetism and upper atmosphere studies, meteorology and climatology, oceanography and solid earth geophysics. The journal appears monthly and author and subject indexes are issued periodically.

A French journal, the *Bibliographie des Sciences de la Terre*, appeared briefly (1968–1971) as the French language counterpart of the *Bibliography and Index of Geology*, with a Section F, 'Tectonique et Géophysique' subdivided into sections on Tectonics, Global Physics and Applied Geophysics. It has now been replaced by Section 225 of the *Bulletin Signalétique*, published by the Centre Nationale de la Recherche Scientifique.

The Society of Exploration Geophysicists publishes a triennial *Cumulative Index of Geophysics* which indexes all the papers which have appeared in *Geophysics, Geophysical Prospecting* and a number of review volumes, by keyword and author.

Primary sources — data

Most geophysical data, including such things as measurements of magnetic and gravity field strengths and earthquake-wave arrival times, are originally numerical in form. Even the highly pictorial representations of subsurface geology produced by the reflection seismograph are ultimately derived from sets of voltage measurements which are now virtually always recorded digitally and processed by digital computer. Only a small fraction of this primary information is generally available but in recent years there has been a welcome tendency for this fraction to increase slightly.

One important factor in the wider availability of primary geophysical data has been the development of the network of World Data Centres. These centres were first established in 1967 specifically to disseminate the information collected during the International Geophysical Year, but their scope has expanded with time. Two universal centres, WDC-A in the USA and WDC-B in the USSR, were created to cover all the primary fields of interest, with additional centres (WDC-C1 in Europe and WDC-C2 in Japan and Australasia) where special facilities or expertise existed (Lander, 1983). The system is described in the *Guide to International Data Exchange through the World Data Centres*, issued by the International Council of Scientific Unions; this is now somewhat out-of-date and is due to be revised to incorporate new developments, including the moves being made towards online data retrieval. Although considerable effort is now going into the establishment of online facilities, most data are currently accessed and exchanged by more traditional methods.

Seismology

Interchange at an international level of basic data relating to the polarities, amplitudes and times of arrival of earthquake waves has always been of fundamental importance in studies of global seismology, since epicentres and focal depths can only be calculated by using observations from a number of widely-spaced points. The increased interest shown during the 1960s both in earthquake magnitudes and mechanisms, and in the detection of nuclear tests, resulted in the establishment of the World-Wide Standard Seismograph Network (WWSSN). Information from this net is processed by many organizations, including the British Geological Survey's Global Seismology Unit, which has a microfilm library of WWSSN data, and the UNESCO-sponsored International Seismological Centre.

Some seismological data are now being obtained and circulated in digital form, and a new term, the Global Digital Seismograph Network, is being used to denote stations operating digitally. The Digital Data Analysis Center has been established by the US Geological Survey as a service to the general scientific community (Engdahl *et al*, 1982). There is also a Historical Seismogram Microfilming Project which forms part of the activities of the WDC-A for Solid-Earth Geophysics.

As noted above, reflection seismic data obtained using artificial sources are now virtually always recorded in digital form, but only small amounts ever become generally available. Most of the work is done for specific oil companies and, unless local legislation forces their release after a certain time, the results are normally retained by the companies for later sale or exchange within the industry. Other surveys are run on a speculative basis by companies which then advertise the data for sale, at prices often measured in hundreds of thousands of dollars, through the applied geophysics journals and by direct mailing.

In recent years some of the reflection results obtained by research institutes, notably from programmes designed to probe the lower crust and upper mantle, have been made available through national data centres. The results of the surveys carried out by the US Consortium for Continental Reflection Profiling (COCORP) can be obtained for the cost of reproduction, in both raw and processed form, via the COCORP secretariat at Cornell University. Data packages acquired by the British Institutions'

Reflection Profiling Syndicate, which specializes in marine work around the British Isles, are similarly available from the British Geological Survey in Edinburgh.

Magnetism

The changes with time of the earth's magnetic field have been recorded at a number of locations for more than 100 years; such data are now routinely exchanged between observatories and are also supplied to World Data Centres. In the UK, the British Geological Survey's Geomagnetism Unit is both the National Data Centre and also World Digital Data Centre C-1, issuing WDDC catalogues and Geomagnetic Bulletins.

As a result of international cooperation, it has been possible to develop empirical formulae which define a number of International Geomagnetic Reference Fields (IGRFs). The IGRFs represent attempts to describe the large-scale variations in the earth's field in both space and time and are of prime importance in determining the regional corrections to be applied in local surveys (cf. Regan 1983).

Data on local geographical variations in field strength are much less standardized in form than observatory data, but most national and state geological institutes carry out their own surveys, usually from the air, collate information from company files and issue magnetic maps. In some instances the primary information may be available in digital form.

Gravity

Basic gravity data within their areas of responsibility are normally held by national and state geological institutes, often in computer databases. The completeness (or otherwise) of these files tends to depend on the legislation covering release of information by oil companies. In addition, the Bureau Gravimetrique of the International Association of Geodesy maintains a file of descriptions and gravity values at stations belonging to the most recent International Base Network, the IGSN71, and the US Defence Mapping Agency attempts to keep comprehensive files of results from gravity surveys in all parts of the world. The Agency actively solicits data, which may be provided under stipulations of open or restricted access; open access data can be obtained by contributors at nominal cost, on magnetic tape. Continent-wide gravity maps based on mean anomalies have been published by the Agency for Asia, Africa and South America.

References

Ambraseys, N N and Melville, C P (1982). *History of Persian Earthquakes.* Cambridge University Press

Bates, C C , Gaskell, T F and Rice, R B (1982). *Geophysics in the Affairs of Man.* Oxford: Pergamon Press

Engdahl, E R, Peterson, J and Orsini, N A (1982). 'Global digital networks — current status and future directions'. *Bulletin Seismological Society of America*, **72**, 5243–5259

Glenn, W (1982). *The Road to Jaramillo.* Stanford, Calif: Stanford University Press

Lander, J F (1983). 'World Data Center-A Activities, 1978–1982'. *Reviews of Geophysical and Space Physics*, **21**, 1539–1544

Regan, R D (1983). 'Current status of the IGRF and its relation to magnetic surveys'. *Geophysics*, **48**, 997–998

Runcorn, S K (ed.) (1967). *International Dictionary of Geophysics.* Oxford: Pergamon Press

Sheriff, R E (1980). *Encyclopaedic Dictionary of Exploration Geophysics.* Tulsa: Society of Exploration Geophysicists

APPENDIX: Addresses

The addresses of most of the societies mentioned in this chapter can be found by consulting recent issues of the journals which they produce. Data Centres are less easily located and some of the more important are therefore listed here.

Bureau Gravimetrique International, Centre National d'Etudes Spatiales/Group des Recherches Géodesie Spatiale, 31055 Toulouse Cedex, France

British Geological Survey; Marine Geophysics Research Programme (for BIRPS data); Geomagnetism Unit (WDDC-C1); Global Seismology Unit (for WWSSN data); Murchison House, West Mains Rd., Edinburgh EH9 3LA, UK

Geoabstracts Ltd, Regency House, 34 Duke St, Norwich NR3 3AP, UK

International Seismological Centre, Newbury, Berks RG13 1LZ, UK

International Union of Geodesy and Geophysics, Publications Office, 39ter Rue Gay Lussac, 75005 Paris, France

US National Geophysical Data Center/World Data Center-A for Solid Earth Geophysics, NOAA/NGDC, E/GC1, 325 Broadway, Boulder, Colorado 80303, USA

US Defense Mapping Agency, Aerospace Center, St Louis AFS, Missouri 63118, USA

World Data Centre-A for Marine Geology and Geophysics, NOAA/NGDC,E/GC3, 325 Broadway, Boulder, Colorado 80303, USA

CHAPTER FIFTEEN

Economic geology

TIM B COLMAN AND RICHARD C SELLEY

The field of economic geology embraces the study of the occurrence, exploration and exploitation of mineral and energy resources from the earth's crust. It was one of the earliest fields of human endeavour, with gold and precious stones from alluvial deposits being worked for jewellery from pre-recorded history. Flint, and later copper, tin and iron were worked for cutting implements and weapons, using some very basic form of geological knowledge.

The past two decades have seen a series of surges of interest in economic geology with iron ore, copper, nickel, uranium, and lately gold, all featuring strongly in metal mineral exploration programmes which have led to the development of new mines or extended the life of existing deposits. Oil, gas and non-metallic minerals have also experienced a sustained increase in demand whilst less conventional energy resources, such as geothermal power and oil shales, have been the subjects of increasing interest since the mid 1970s.

The growth in the exploration and exploitation of mineral and energy resources has been paralleled by the proliferation of books and journals in all aspects of economic geology.

Textbooks and monographs

The only texts which cover the entire subject area are the Open University Course texts for *The Earth's Physical Resources (S238)* (Open University Press, 1984) which provide an excellent, recent, and well-presented series on metallic and non-metallic mineral resources, energy and water resources, and future trends. However, they contain few references to other sources of information as they are intended to act as 'stand-alone' student texts. The accompanying video and audiovisual tapes can also be bought from the same source: these provide an excellent, if somewhat expensive, teaching medium with film and commentary of worldwide examples of resource exploration and exploitation.

Industrial Geology, edited by J L Knill (Oxford, 1978) also covers the full spectrum of resource geology, but in less depth, in a compilation of lectures given at Oxford University designed to make students aware of the role of the geologist in industry.

Mineral Resources, by K Warren (Penguin, 1973) and J Blunden's *Mineral Resources of Britain* (Hutchinson, 1975) provide worldwide and national accounts of mineral resources and the economic, political, and environmental problems associated with their exploitation.

Good general texts include M L Jenson and A M Bateman's *Economic Mineral Deposits* (3rd edn, Wiley, 1979) and J M Guilbert and C F Park's *The Geology of Ore Deposits* (W H Freeman, 1986). These both deal with the worldwide scene (with an emphasis on North America) and give a good general overview of metalliferous mineral deposits. The *Geology of Ore Deposits* is a completely revised and extended successor to the well known *Ore Deposits*, by C F Park and R A MacDiarmid (2nd edn, McGraw-Hill, 1975) which was somewhat dated and presented a classic magmatic hydrothermal view. The new book rectifies these shortcomings and is an attractive and reasonably priced text. A M Evans' *An Introduction to Ore Geology* (Blackwells, 1980) is a well produced and illustrated student text, with modern examples and an excellent reference section: it is based on the undergraduate course at Leicester University. *Ore Deposit Geology*, by R Edwards and K Atkinson of the Camborne School of Mines, Cornwall (Chapman & Hall, 1986) is another excellent student text. It is somewhat longer than Evans' volume and has a strong emphasis on mineral exploration: there are numerous plates from mining operations. I Smirnov's *Geology of Mineral Deposits* (MIR, 1976) presents a pre-plate tectonic development of geosynclines and leans towards the magmatic hydrothermal approach: but it is otherwise a good general text with many Russian, as well as worldwide, examples of mineral deposits.

Mining geology texts are less abundant but L J Thomas's *Introduction to Mining* (Hicks Smith 1973) is a good introductory mining text for geologists, though the examples are largely Australian based. R W LeRoy *et al*, *Subsurface Geology, Petroleum Mining Construction* (4th edn, Colorado School of Mines,

1977) provides a varied text in a North American setting but is rather compressed and dated in places. E P Pfeider's *Surface Mining* (American Institute of Mining, Metallurgical and Petroleum Engineers, 1968) contains a wealth of information. W C Lacy's *Mining Geology* (Van Nostrand, 1983) is also wide ranging, but the collection of reprints of 'classic' papers goes back to 1923 and the presentation is uneven. E H Macdonald's *Alluvial Mining* (Chapman & Hall, 1983) is a good text in this specialized area.

Metalliferous minerals

There is a large and growing interest in this field with a consequent expansion in the number and specialization of the books and journals produced. Major trends in the past 20 years have included:

(a) The development of increasingly sophisticated models for ore genesis and mineral exploration.

(b) The development of automated analysis methods such as X-ray fluorescence spectrometry (XRF) and inductively coupled plasma emission spectrometry (ICP) for the rapid, accurate and cheap analysis of large numbers of samples to increasingly low levels of detection.

(c) The development of more sophisticated computing systems to handle the large amounts of available data and facilitate the interaction of different data sets digitally, in real time.

(d) The development of remote sensing techniques. These had an important role in the selection of the areas for exploration which led to the discovery of the major Olympic Dam copper-uranium-gold deposit in South Australia in 1975.

The past 20 years have been marked by the publication of numerous texts and conference volumes on all aspects of metalliferous mineral deposits. Worldwide accounts are given in books such as P Laznicka's *Empirical Metallogeny* (Elsevier, 1985) and J D Ridge's series of *Annotated Bibliographies of Mineral Deposits* which cover the world in three massive volumes, with a fourth in preparation (Geological Society of America, 1972: Pergamon 1976 & 1984). Both of these amazing solo *tours de force* contain much of the authors' own preferences and thus the Ridge volumes are somewhat influenced by the Lindgren classification, with its emphasis on the magmatic-hydrothermal genesis of many ore deposits, and for reasons of space have no illustrations, giving a somewhat indigestible appearance. However, both these works have much to commend them as comprehensive and recent texts. The Laznicka volume, which is illustrated, is one volume of the Elsevier series *Developments in Economic Geology* which now runs to more than twenty volumes. W Lindgren's seminal *Mineral Deposits* (4th edn, McGraw-Hill, 1933) was the most important single text on the subject for many years after its first publication in 1913. It was heavily influenced by the association of vein mineralization with igneous rocks in the Western United States.

More specific areas are covered in major works such as the 1975/1976 series on *The Economic Geology of Australia and Papua New Guinea* in 4 volumes (Australian Institute of Mining and Metallurgy); which is a comprehensive compilation of the metalliferous, non-metalliferous, coal and petroleum deposits of Australia and Papua New Guinea. An earlier (1974) publication from the same source describes *The Economic Geology of New Zealand*. Full coverage of Canadian mineral deposits, together with an excellent account of Canadian geology, is given in *Geology and Economic Minerals of Canada*, edited by R J W Douglas (Geological Survey of Canada, 1970); it includes a series of maps, at 1:5,000,000, of the mineral deposits and other aspects of Canadian geology and geophysics. *Ore Deposits of the United States, 1933–1967*, by J D Ridge (American Institute of Mining, Metallurgical and Petroleum Engineers, 1968) is another massive and comprehensive survey in 2 volumes but is now a little dated, not least in its adherence to the overriding importance of the magmatic-hydrothermal genesis of ore deposits.

A compilation of Irish metalliferous ore deposits, their geology, genesis and exploration, is given in *Geology and Genesis of Mineral Deposits in Ireland*, edited by C J Andrew *et al* (Irish Association for Economic Geology, 1986). This is an important volume as it sums up the successes of Irish mineral exploration over the past 20 years and includes keynote articles on granite-related and Mississippi Valley-type ore deposits. European metalliferous ore deposits are generally well described in the series published by the Institution of Mining and Metallurgy/Mineralogical Society entitled *Mineral Deposits of Europe*. Volume 1, *Northern Europe* (1979), edited by S H U Bowie *et al* covers Finland, Sweden, Norway, Denmark, Great Britain and Ireland. Volume 2, *Southeast Europe* (1982), edited by F W Dunning *et al* covers Yugoslavia, Greece, Cyprus, Albania, Romania, Bulgaria and Hungary, countries with few modern English accounts of their mineral deposits. Volume 3 *Central Europe* (1986), edited by F W Dunning and A M Evans, covers Austria, Belgium, Czechoslovakia, East Germany, West Germany, the Netherlands, Poland and Switzerland. Two further volumes on the mineral deposits of Eastern and Southern Europe are in preparation. The only general account, in English, of Russian mineral deposits is given in the 3-volume work by V I Smirnov entitled *Ore Deposits of the USSR* (English translation of the 1974 Russian edition, published by Pitman, 1977). *India's Mineral Resources*, by S Krishnaswamy (Oxford & IBH, 1979) is the most modern account of the Indian metalliferous, non-metalliferous and constructional resources.

The mineral deposits of Africa are covered in such volumes as *Mineral Resources of the Republic of South Africa*, edited by C B Coetzee (Department of Mines, 1975) which describes the occurrence and production of all minerals, including coal, in this most important mineral-rich country. A more recent text, which describes virtually all the mineral deposits of Southern Africa, including those of Zimbabwe, Botswana and South-West Africa as well as South Africa, is *Mineral Deposits of Southern Africa*, edited by C R Anhaesser (Geological Society of South Africa, 1986). It is a massive production which is in two volumes with over 2300 pages and will form the major single source of economic geological information for this region. It consists of 180 specially written papers, which cover most aspects of the metalliferous, non-metalliferous, and energy minerals of the region. J B Wright's *Geology and Mineral Resources of West Africa* (Allen & Unwin, 1985) is a good general student text on the region.

Many national Geological Surveys have published accounts of their respective country's mineral deposits, as in the Canadian volume cited above. Others include the United States Geological Survey with its widely available series of *Professional Papers, Bulletins* and *Circulars*, which often contain information on aspects of the economic geology of the United States. A recent example is *Circular* **980**, *Prospects for Mineral Resource Assessment on Public Lands*, edited by S M Cargill and S B Green (USGS, 1986). In spite of its title this is a most useful summary of recent developments in such areas as expert systems, mineral deposit models and methods of mineral resource assessment. The British Geological Survey has published several volumes of interest to the economic geologist. These include H G Dines' *The Metalliferous Mining Region of South-West England* (HMSO, 1956). This is the only comprehensive description of this ancient, but still productive, tin and copper mining field. K C Dunham's *Geology of the Northern Pennine Orefield*, Volume **1** (HMSO, 1948) and Volume **2**, with A A Wilson, (BGS, 1985) perform a similar service for these old lead and fluorspar mining areas.

Specific metals or deposit types are the subject of numerous texts. The important porphyry copper deposits are the subject of S R Titley's *Advances in the Geology of Porphyry Copper Deposits, Southwestern North America* (University of Arizona, 1982) which is an update of the same author's classic 1966 volume on *Geology of Porphyry Copper Deposits*. They are also the subject of *Porphyry Deposits of the Canadian Cordillera*, edited by A S Brown (Canadian Institute of Mining and Metallurgy, 1976). *Geology and Metallogeny of Copper Deposits* (Springer-Verlag, 1986) edited by G Friedrich *et al* is the most recent work on this metal. The intriguing Japanese Kuroko deposits are the subject of an excellent monograph edited by H Ohmoto and B J Skinner in 1983 on *Kuroko and Related Volcanogenic Massive Sulfide Deposits* (Economic Geology Publishing Co.). The 14-volume *Handbook of Strata-Bound and Stratiform Ore Deposits*, edited by K H Wolf (Elsevier, 1976-1985) gives a massive overview, in over 6000 pages of text, of these ore deposits, in four parts: these include *Principles and general studies* and *Regional studies and specific deposits*. Each volume consists of a number of specially written articles by an authority in the field, but the series is not a comprehensive listing of all deposits. The articles vary in style and emphasis, and tend to be lengthy but the references are particularly comprehensive.

More general texts include R L Stanton's *Ore Petrology* (McGraw-Hill, 1972) which was one of the earlier works to treat ore deposits as part of the normal geological cycle. It has sections on the solubility, transport and deposition of ore minerals, and the important ore/rock associations. G C Amstutz's *Ore Genesis: the state of the art* (Springer-Verlag, 1982) was produced to celebrate Paul Ramdohr's 90th birthday and has numerous articles on genetic studies of ore deposits by Ramdohr's students and colleagues. A H G Mitchell and M S Garson's *Mineral Deposits and Global Tectonic Settings* (Academic Press, 1983) gives an excellent overview of the role played by plate tectonics and other major structural processes in the formation, localization and preservation of metalliferous ore deposits. *Economic Minerals and their Tectonic Setting*, by C S Hutchison (Macmillan, 1983) is also an excellent, well-written and modern text on the same subject. D H Tarling's smaller *Economic Geology and Geotectonics* (Blackwell, 1981) suffers by contrast, but includes sections on petroleum and coal as well as metalliferous deposits.

D S Cronan's *Underwater Minerals* (Academic Press, 1980) covers the relatively little known subject of offshore mineral deposits, such as placers and manganese nodules, and includes sections on the economic, technical and legal problems of exploiting resources from the sea bed.

The expanding series of publications from Tatsch Associates (Tatsch, 1973–) present a divergent view of mineral and energy resource formation using Tatsch's 'Tectonospheric Earth Model'. This is said to have operated from early in the earth's development, to have controlled the location of such resources and to be useful in their exploration. Current titles include volumes on gold, copper and coal.

H L Barnes' *The Geochemistry of Hydrothermal Ore Deposits* (Wiley, 1979) provides an excellent overview of many of the problems of the solution, transport and deposition of the ore-forming elements.

Mineral exploration

Mineral exploration techniques are covered in general texts such as J H Reedman's *Techniques in Mineral Exploration* (Applied Science, 1979): this work takes the student reader through the full range of activities from geological mapping to ore reserve calculation. It is almost complimentary to W C Peter's *Exploration and Mining Geology* (2nd edn, Wiley, 1987) which has more detail on mining and fieldwork and less background information on techniques and instruments: it also has particularly comprehensive references. Both books would be useful for students and in the field.

The Field Geologist's Manual, by D A Berkman (Australian Institute of Mining and Metallurgy, 1976) presents a fascinating miscellany of information, ranging from abbreviations through rock classification, surveying data and geophysical properties to the radio alphabet and metric conversions: it has irritatingly few references for the geologist with access to a library. *Mineral Prospecting Manual*, by J-B Chaussier and J Morer (North Oxford Academic Publishers, 1987) is an excellent, and very practical, handbook for the field exploration geologist. It is an updated and revised English translation of the 1981 original French text, which was written in cooperation with the Bureau de Recherches Géologiques et Minières. It is especially useful for its comprehensive descriptions of field sampling techniques, but also includes information on modern geophysical prospecting methods. However, it too has almost no references. *Structural Methods for the Exploration Geologist*, by P C Badgeley (Harper, 1959) is now dated but is still a useful reference book.

Some Geological Surveys publish papers and reports on mineral exploration, including such series as the Canadian Geological Survey's *Economic Geology Report*, for example titles such as No.**31** *Geophysics and Geochemistry in the Search for Metallic Ores*, edited by P J Hood, and the British Geological Survey's *Mineral Reconnaissance Programme Reports* (BGS, 1974–) which are a series of 90 reports giving a wealth of geological, geochemical and geophysical data about prospective metalliferous regions in Britain.

Mineral Exploration, by W C Lacy (Van Nostrand, 1983) is a collection of reprints of classic papers such as Lowell and Guilbert (1970) but dating back to 1927, in a very uneven presentation.

Exploration geochemistry

This subject is covered also in Chapter 11 but as it is very relevant to aspects of economic geology, some repetition is necessary.

A W Rose, H E Hawkes and J S Webb's *Geochemistry in Mineral Exploration* (Academic Press, 1979) is an updated version of Hawkes and Webb's classic 1962 book of the same name, but is deficient in the statistical treatment of data. A A Levinson's *Applied Geochemistry* (Applied Publishing, 1974) has been up-dated with a 300 page supplement in 1980 and is another competent and comprehensive text. F R Siegal's *Applied Geochemistry* (Wiley, 1974) is less detailed and wider ranging, with chapters on oil and marine geochemical prospecting, and geochemistry and health. A Russian viewpoint is provided by A A Beus and S V Grigorian's *Geochemical Exploration Methods for Mineral Deposits* (Applied Publishing, 1977) which is an English translation of the 1975 Russian text.

The Elsevier series of *Handbooks of Geochemical Exploration* are an expanding series of case studies in a variety of specialized subject areas. Titles published to date include *Analytical Methods in Geochemical Prospecting*, by W K Fletcher (1981), *Statistics and Data Analysis in Geochemical Prospecting*, by R J Howarth (1983) and *Rock Geochemistry in Mineral Exploration*, by G J S Govett (1983). Further titles are planned on drainage and soil geochemistry, biogeochemistry, and volatile elements. R W Boyle's *Geochemistry of Gold and its Deposits* (Geological Survey of Canada, 1979) is a comprehensive account of the noble metal, its deposits and exploration.

Exploration geophysics (see also Chapter 14)

Ninety-five percent of all applied geophysics is seismic exploration for hydrocarbons. Nevertheless there is very active research in mineral exploration geophysics based in Canada and Scandinavia. The classic text is M Dobrin's *Introduction to Geophysical Prospecting* (McGraw-Hill, 1976), now in its 3rd edition, but this has been supplanted by W M Telford *et al*, *Applied Geophysics* (Cambridge University Press, 1976). D S Parasnis' *Principles of Applied Geophysics* (Chapman & Hall, 1972) concentrates on magnetic and electrical methods, and is relatively cheap in its paperback form. *Applied Geophysics for Geologists and Engineers: the elements of geophysical prospecting*, by D H Griffiths and R F

King (2nd edn, Pergamon, 1981) attempts to remove the mystique from the subject and is a most useful book. M Brooks and P Kearey's *An Introduction to Geophysical Prospecting* (Blackwell, 1984) is also a good, modern text. The 6-volume *Developments in Geophysical Exploration Methods*, edited by A A Fitch (Applied Science Publishers, 1979–1985) provides an update in most areas of the science.

Specialized areas of applied metalliferous geology include ore mineralogy (see also Chapter 10), which is catered for by P Ramdohr's massive, 2-volume, *The Ore Minerals and their Intergrowths* (Pergamon, 1980). It attempts a full and comprehensive listing of all opaque minerals with possible parageneses. J R Craig and D J Vaughan's *Ore Microscopy and Ore Mineralogy* (Wiley, 1981) is an excellent text for students, especially for its useful summary tables. *Atlas of the Ore Minerals* (Elsevier, 1982) by P Picot and Z Johan is an English translation of a French text with excellent colour photomicrographs of many minerals. It has reflectance and X-ray diffraction tables but no VHN hardness information. P Devismes' *Atlas Photographique des Minéraux d'Alluvions* (BRGM, 1978) has a large number of colour plates from 183 detrital minerals and is a valuable reference, in French and English, for stream sediment and placer exploration.

T J Shepherd *et al*, *A Practical Guide to Fluid Inclusion Studies* (Blackie, 1985) is a most useful and practical text which covers the collection and physical preparation of samples; optical, physical and chemical analysis techniques, and data interpretation. Fluid inclusions can provide valuable information on the nature and origin of mineralizing fluids and the deposition of mineralization, and can be used as a direct exploration tool for some porphyry style deposits. *Remote Sensing and Mineral Exploration*, by W D Carter and L C Rowan (Pergamon, 1983) is a useful text devoted to this potentially important topic.

Mining Geostatistics, by A G Journel and Ch J Huijbregts (Academic Press, 1978) is a very thorough and detailed description of mathematics and statistics useful to mining geologists. M David's *Geostatistical Ore Reserve Estimation* (Elsevier, 1977) is also a very useful text and includes sections on variograms and kriging. *Mineral Resources Appraisal. Mineral endowment, resources and potential supply: concepts, methods and cases*, by D P Harris (Oxford, 1984) is a wide-ranging text on the statistical approach to actual and predicted resources from Hubbert onwards.

There are several atlases of mineral deposits. D R Derry's *World Atlas of Geology and Mineral Deposits* (Mining Journal Books, 1980) has a series of maps at the rather small scale of 1:20–30 million with major deposits indicated and a supporting text. C J

Dixon's *Atlas of Economic Mineral Deposits* (Chapman & Hall, 1979) is a series of case studies of about 50 deposits or mining areas which illustrate the setting of all the major deposit types in a group of maps and plans, with explanatory text: it is a useful teaching resource.

Metallogenic maps are also beginning to appear. The first were the 9 sheets of the *Carte Metallogenique de l'Europe* at a scale of 1:2,500,000 (UNESCO: BRGM, 1968–1970) and the single sheet *Metallogenic Map of Australia and Papua New Guinea*, with commentary by K Warren (Bureau of Mineral Resources, 1972) at a scale of 1:5,000,000. An example of more detailed mineral deposit maps can be found in the 8-volume set of *Cartes des Gîtes Minéraux de la France* produced at a scale of 1:500,000 (BRGM, 1978–1982) which list every mineral occurrence, with output and mineralogy. Computerized listings of mineral deposits are also emerging: the Canadian Mineral Occurrence Index (CANM-INDEX) of the Geological Survey of Canada, edited by D D Picklyk *et al* (Geological Survey of Canada, 1978) is a good example of these new information sources.

Industrial minerals

A wide and expanding range of non-metallic or 'industrial' minerals are consumed by society and it is now widely recognized that both their value and consumption exceeds that of the metallic minerals. They have, however, attracted a much less voluminous literature. This may be because of their often relatively simple geology and processing, though many deposits are only economic with strict geological control on their extraction. They include not only the large tonnage constructional raw materials, such as sand and gravel and crushed rock, but also a range of other minerals such as china clay, baryte and phosphate rock which also enter into international trade. Unlike metallic minerals, which are usually exploited for their metal content, the non-metallics are valued for both their physical and chemical properties; either separately or in combination.

The best general text on the non-metallic minerals is *Geology of the Nonmetallics*, by P W Harben and R L Bates (Metal Bulletin, 1984) which is an extremely readable volume, with recent references. The 50 minerals described are grouped by derivation from igneous, sedimentary and metamorphic processes, and also the surficially altered minerals. This approach can be a little confusing; for example, a significant proportion of titanium is derived from primary igneous sources such as at Tellnes, Norway, yet titanium minerals are grouped under sedimentary sources because the majority are found in placer deposits such as those of Eastern and Western Australia.

S J LeFond's 2-volume *Industrial Minerals and Rocks* (5th edn, American Institute of Mining, Metallurgical and Petroleum Engineers, 1983) is very comprehensive, with the minerals grouped by uses such as electronics, optical, and fertiliser minerals, as well as individual minerals, and there is a very useful section on sources of information; however, there has been relatively little updating of the references from the previous editions. R L Bates' *Geology of the Industrial Rocks and Minerals* (Dover, 1969) is now rather dated. M Kuzvart's *Industrial Minerals and Rocks* (Elsevier, 1984) is a useful source of information on East European and USSR deposits.

Other literature tends to specialize on one commodity or area with books such as *Gems: their sources, description and identification*, by R Webster (4th edn, Butterworths, 1983) which is a readable and useful account. The British Geological Survey series of *Mineral Dossiers* (HMSO, 1971–1984) on 26 mineral commodities, provide an overview of the selected mineral, its physical and chemical characteristics, and its geological setting and exploitation in the UK, together with a summary of its world supply and trade. Titles published include Fluorspar, Talc, Gold, Potash, Limestone, and China Clay. *Phosphate Deposits of the World, Volume 1*, by P J Cook and J H Shergold (Cambridge University Press, 1986) and R E Grim and N Guven's *Bentonites* (Elsevier, 1978) are recent examples of specialist texts on selected minerals.

The only journal which is devoted to the subject is *Industrial Minerals* (1967–) published monthly by Metal Bulletin Journals. This journal is an excellent source of information on the whole range of non-metallic minerals and also carries regular surveys of specific minerals, or countries with an important non-metallic minerals industry. It also sponsors a major biennial conference (seven to 1986) and publishes the proceedings with commendable speed in the same year as the conference.

Oil and natural gas

Geology has an important part to play in the exploration for oil and gas. Petroleum geology involves many different branches of the earth sciences ranging from remote sensing to geochemistry and from palaeontology to geophysics. There are a number of books which cover the central topic of petroleum geology, and many more which deal with particular specialities of the subject.

The classic petroleum geology textbook is A I Levorsen's

Geology of Petroleum (W H Freeman, 1967): this contains a wealth of detail but is now very dated. More up-to-date texts include G D Hobson and E N Tiratsoo's *Introduction to Petroleum Geology* (2nd edn, Scientific Press, 1981), P A Dickie's *Petroleum Reservoir Geology* (Petroleum Publishing Co., 1979) and R C Selley's *Elements of Petroleum Geology* (W H Freeman, 1985). At a more basic level Selley's *Petroleum Geology for Geophysicists and Engineers* (Prentice-Hall, 1983) is suited not only for geophysicists and engineers but also for geologists requiring a brief review of the subject.

The main exploration method for finding petroleum is seismic surveying as already mentioned in Chapter 14. There are now many books which cover this topic. Non-mathematical accounts of the seismic method will be found in A A Fitch's *Seismic Reflection Interpretation* (Gebruder Borntraeger, 1976) and *Simple Seismics*, by N A Anstey (International Human Resources Development Corporation, 1982). More detailed accounts will be found in Anstey's *Seismic Exploration for Sandstone Reservoirs* and *Seismic Interpretation: the physical aspects* (both published by International Human Resources Development Corporation in 1980 and 1977 respectively).

The application of organic geochemistry to petroleum exploration has become increasingly important in recent years. This topic is covered by three texts: *Petroleum Formation and Occurrence*, by B P Tissot and D H Welte (Springer-Verlag, 1978), *Petroleum Geochemistry and Geology*, by J M Hunt (W H Freeman, 1979) and *An Introduction to the Physics and Chemistry of Petroleum*, by R R F Kinghorn (Wiley, 1983). In the first of these three books the emphasis is more on geology than chemistry, the other two are more chemical than geological.

Geophysical well-logging, or formation evaluation, is another important tool used by petroleum geologists. Most of the wireline-logging companies give copies of their own logging manuals away free.

There are currently three books available which cover this topic. Two of these are published by the American Association of Petroleum Geologists: R H Merkel's *Well Log Formation Evaluation* (1979) is a cheap and cheerful production, while G Asquith's *Basic Well Log Analysis for Geologists* is a more lavish, and correspondingly more expensive, text. *Geological Well Log Analysis*, by S J Pirson (Gulf Publishing Co., 1983) is not a comprehensive textbook, but is an interesting, if idiosyncratic, account of the subject.

Turning from books to journals, there are a number of important ones to consider. The main journal for petroleum geology is the monthly *Bulletin of the American Association of Petroleum Geologists* (1917–). Despite its title this journal is global in outlook and catholic in subject matter, publishing papers on many branches of the subject. The American Association of Petroleum Geologists also produces a series of memoirs which contain many important papers. Membership of this association is essential for any aspiring petroleum geologist. The *Journal of Petroleum Geology* (1978–) appears quarterly and is less essential, but has a penchant for publishing stimulating and unorthodox papers. The Society of Exploration Geophysicists and the Society of Professional Well Log Analysts both publish journals which deal with their particular fields of interest. *First Break* (1983–) is an exciting new geophysical journal.

Coal and oil shale

There are several textbooks which deal with coal. A good starting point is *Coal and Coal-bearing Strata*, edited by D Murchison and T S Westoll (Oliver and Boyd, 1968). This is a symposium volume, so is rather uneven in content. It does cover coal petrography and sedimentology and deals with coalfields in Britain, Germany and the Gondwana basins.

Coal Mining Geology, by I A Williamson (Oxford University Press, 1967) is mainly written for mining engineers, but also describes British and foreign coal fields. *Coal Petrology*, by E Stach *et al* (Gebruder Borntraeger, 1975) is an English translation of the second edition of this work.

Reference to oil shales will be found in several of the petroleum geology texts already mentioned, especially those which deal with source rock geochemistry. *Oil Shale* (Elsevier, 1976), is a symposium volume edited by T F Yen and G V Chilingarian. Though it is mainly concerned with the Green River oil shale of the USA it does contain papers which review the genesis of oil shales, their distribution in time and space, and the technology of oil shale mining and processing.

Water supply

The geology of water supply is an important subject area as not only is a large part of our water resources derived from groundwater, but also the underlying rock types have a major influence on the behaviour of precipitation; whether it runs off, and can be stored in surface reservoirs, or infiltrates and recharges subsurface aquifers. The location and design of surface facilities such as dams and pipelines is also strongly influenced by the underlying rock types but this is the province of the engineering geologist.

The complete hydrological cycle is covered in W Viessmann *et al*, *Introduction to Hydrology* (Harper & Row, 1977). R Bowen's *Ground Water* (2nd edn, Applied Science, 1986) is a good general text, as is *Introducing Groundwater*, by M Price (Allen & Unwin, 1985), while H M Raghunath's *Ground Water* (Wiley Eastern, 1982) is a very practical student text, with examples drawn largely from the Indian subcontinent. A very useful series of practical examples is given in J W Lloyd's *Case Studies in Groundwater Resources Evaluation* (Clarendon, 1981). More specific texts include P T Milanovic's *Karst Hydrology* (Water Resources Publications, 1981), an English translation of the 1979 Serbo-Croat original, and aimed at the postgraduate student; many of the examples are from the classic Balkan karst areas.

The Van Nostrand *Benchmark series* includes R A Freeze and W Back's *Physical Hydrology* (1983) and Back and Freeze's *Chemical Hydrology* (1983). As with other titles in this series of reprints of classic papers, the former volume consists entirely of key, pre-1970, papers, and the latter contains numerous ones of the 1960s and earlier.

An interesting and well-produced volume is the Australian Water Resources Council publication *Groundwater Resources of Australia* (Australian Government Public Service, 1975) which describes the resources of the entire continent.

The main journals are the quarterly *Journal of Hydrology* which has been published by North-Holland, Amsterdam since 1963 and the American-based bimonthly *Ground Water*, published by the Association of Ground Water Scientists and Engineers, in association with *Water Well Journal*, also since 1963.

Geothermal energy

Geothermal energy has generally been associated with areas of recent volcanic activity such as Iceland, California, Italy and New Zealand, producing both steam for electricity generation and hot water for district heating. In the last two decades research and development has been extended to additional sources of geothermal energy. These include some sedimentary basins where aquifers containing water above 50°C may be found at depths of 1–3km and extracted for district heating. There are also areas, underlain by granite rock, with high concentrations of the radiogenic, heat-producing elements, uranium and thorium, whose radioactive decay can generate temperatures of 100–200°C at 5km depth, the so-called 'Hot Dry Rock' geothermal reservoir.

This heat can be extracted by pumping cold water into the rock, using boreholes, and extracting the hotwater or steam for power generation: this is still at a research stage.

There are an increasing number of publications relating to geothermal energy. Good general texts include *Geothermal Energy*, by H C Armstead (2nd edn, Spon, 1983) which describes its occurrence, exploration and applications, with sections on economics and pollution, and R Bowen's *Geothermal Resources* (Applied Science, 1979). The *Handbook of Geothermal Energy*, edited by L M Edwards *et al* (Gulf Publishing, 1982) is a practical account of the geology, drilling and evaluation of resources. More specialized studies include A J Ellis and W A J Mahon's *Chemistry and Geothermal Systems* (Academic Press, 1977) with examples and data from New Zealand, and J Elder's *Geothermal Systems* (Academic Press, 1981) which is more theoretical. *Geophysics of Geothermal Areas: state-of-the-art and future development*, edited by A Rapolla and G V Keller (Colorado School of Mines, 1984) is a useful European-based reference. R A Downing and D A Gray's *Geothermal Energy — the potential in the United Kingdom* (HMSO, 1986) is a recent account and includes sections on heat flow, hot dry rocks, sedimentary basins, and the geochemistry of geothermal waters.

Journals of relevance to this subject include the monthly *Geothermal Energy* (1973–87) published by Geothermal Energy Magazine, West Covina, California and the *Geothermal Resources Council Bulletin* (1972–), also monthly, published by the Council, in Davis, California. The *Geothermal World Directory* has been published annually since 1972 by Meadows, Glendora, California and lists current geothermal areas and their future development, together with companies active in this field. It also hosts conferences attracting participants from countries worldwide. Another useful source of information is *A Bibliography; geothermal resources, exploration and exploitation*, published occasionally by the National Technical Information Service of the United States Department of Commerce, with international coverage in its references: the July 1976 volume has over 600 pages.

Periodicals

The 3 main publications of metalliferous applied geology are:
(1) *Economic Geology* — a 200 page American-based journal published since 1905, which incorporates the *Bulletin of the Society of Economic Geologists* and is published 8 times a year. It covers the whole spectrum of worldwide applied metalliferous geology, with numerous special issues devoted

to a particular country, metal, or deposit type. The *75th Anniversary Volume*, edited by B J Skinner (Economic Geology Publishing Co., 1981) was a summation of the current state of the art. The Society has also begun a series of *Reviews in Economic Geology* (1984–) with three volumes published to date.

(2) The monthly *Transactions of the Institution of Mining and Metallurgy* (IMM), published since 1892, has quarterly sections on the mining industry, mineral processing and extractive metallurgy, and applied earth science respectively. The IMM also publishes the proceedings of numerous conferences on a wide variety of topics, such as the biennial *Prospecting in Areas of Glaciated Terrain* (from 1973), *Metallogeny of Basic and Ultrabasic Rocks* (1986), *High Heat Production Granites, Hydrothermal Circulation and Ore Genesis* (1986), and many other titles. The *Transactions* or *Bulletins* of the Australasian, Canadian, Indian and South African Institutes also contain articles of local and wider interest. The *Canadian Institute of Mining and Metallurgy Bulletin (CIM Bulletin)*, monthly since 1927, is the most widely available.

(3) *Mineralium Deposita* (1966–) is a newer, European oriented journal, published quarterly by Springer-Verlag. It is the *Bulletin of the Society for Geology Applied to Mineral Deposits* and tends to contain more theoretical studies of ore deposits.

The weekly, London-based, *Mining Journal*, published since 1834, and its monthly companion, *Mining Magazine*, are the main English sources of information for the mining and applied geologist. The *Mining Annual Review*, published by Mining Journal Books, is an excellent annual world summary of the minerals industry by commodity, country and profession.

The Canadian *Northern Miner*, first published in 1915, is also weekly and is mainly devoted to the worldwide activities of Canadian mining and exploration companies, and provides comprehensive listings of the mining and oil shares of the North American stock exchanges.

Other useful titles of interest to the metalliferous and mining geologist include the long-established monthly *Engineering and Mining Journal*, published by McGraw-Hill, with a useful section on markets and commodity prices, and the London-based *International Mining*, published monthly since 1984, which took over the roles of the defunct *World Mining* and *World Coal* and *therefore has a useful coal section. Marine Mining* (1977–) published by Taylor & Francis, New York, is a quarterly journal of 'sea floor minerals: their exploration, assessment and extraction'.

As already mentioned in Chapter 11 the *Journal of Geochemical*

Exploration (1972–) published in 9 issues per year, by Elsevier for the Society of Exploration Geochemists, has assumed a leading role in the dissemination of geochemical studies of metalliferous ore deposits; the Society has sponsored major biennial international conferences since 1966. *Exploration Geochemistry Bibliography*, edited by H E Hawkes (Society of Exploration Geochemists, 1982) provides a monumental compilation of worldwide literature. *Geochemistry International*, published by the American Geological Institute and Geochemical Society of Washington, (D.C.), from 1964 to date, is a translation of the Russian journal *Geokhimiya* with 9 issues per year, and it frequently has articles relating to mineral exploration in Eastern Europe and the USSR.

Geophysical exploration is catered for by *Geophysical Prospecting* (1953–), the quarterly journal of the European Association of Exploration Geophysicists and *Geophysics* (1936–) the monthly journal of the Society of Exploration Geophysicists, with its companion publication from the same source, *Geophysics: the leading edge of exploration*, published monthly since 1982. The Society also publishes a very useful comprehensive index of the three publications listed above. The latest, edited by W J Zwart, was published in 1986. The same source is responsible for the *Encyclopedic Dictionary of Exploration Geophysics*, 2nd edn, compiled by R E Sheriff in 1984. *Geophysical Journal* is a bimonthly translation of the Russian *Geofizicheskii Zhurnal* and is published by Gordon & Breach Science Publishers, New York. It carries some articles on mineral exploration in Eastern Europe and the USSR, as does its geochemical equivalent mentioned earlier.

More general references include the United States Department of the Interior, Bureau of Mines, *Minerals Yearbook* which is published annually. It has 3 sections: Metals & Minerals, Area Report (Domestic), and Area Report (International). It provides comprehensive world data on the occurrence, mining, trade and consumption of virtually all metals and minerals. The British Geological Survey has also published *World Mineral Statistics* since 1913. This is an annual publication providing statistical data on mineral production, imports and exports, by country for a range of metallic, non-metallic and energy minerals. The BGS also published the annual *United Kingdom Mineral Statistics* (HMSO) which included a full analysis of the mineral production and trade in the UK. This was replaced in 1989 by the *United Kingdom Minerals Yearbook* which includes a commentary on the UK minerals industry.

Dictionaries and abstracting services

P W Thrush's *A Dictionary of Mining, Mineral and Related Terms* (US Dept. of Interior, 1968) lists over 55000 terms. R J M Wyllie and G O Argall's *World Mining Glossary of Mining, Processing and Geological Terms* (Miller Freeman, 1975) lists over 10000 terms, with their Swedish, German, French and Spanish equivalents, while the *Rudarski Recnik*, or *Mining Dictionary* (Rudarski Institut, 1970), has entries in Serbo-Croat, Russian, English, French and German.

Abstracting services include the Institution of Mining and Metallurgy's *IMM Abstracts* which have been published bimonthly since 1948 and *Coal Abstracts*, published by IEA Coal Research, since 1977. The IMM also controls the computerized IMMAGE bibliographic database (which provides frequently updated coverage from 1979 of the full range of economic geology, mining and extraction technology). The *Bibliography of Economic Geology* (formerly *Geocom Bulletin*, 1968–1982), has been published bimonthly since 1982 by Geosystems Publications, London.

Petroleum geologists were served by the monthly *Brown's Geological Information Bulletin*, 1971–87, by Brown's Geological Information Service, London, and still have *International Petroleum Abstracts*, published by Applied Science, London, since 1973.

References

Metalliferous

Amstutz, G C *et al* (eds) (1982) *Ore Genesis: the state of the art. (Special Publication No. 2. Society for Geology Applied to Mineral Deposits)* Berlin: Springer-Verlag

Andrew, C J *et al* (eds) (1986) *Geology and Genesis of Mineral Deposits in Ireland.* Dublin: Irish Association for Economic Geology

Anhauser, C R and Maske, S (eds) (1986) *Mineral Deposits of Southern Africa.* 2 vols. Johannesburg: Geological Society of South Africa

Australasian Institute of Mining and Metallurgy (1974) *Economic Geology of New Zealand. (Monograph Series No. 4)* Melbourne: Australasian Institute of Mining and Metallurgy

Australasian Institute of Mining and Metallurgy (1975/1976) *Economic Geology of Australia and Papua New Guinea. (Monograph Series No. 5, Metals; No. 6, Coal; No. 7, Petroleum; No. 8, Industrial Minerals and Rocks)* Melbourne: Australasian Institute of Mining and Metallurgy

Badgeley, P C (1959) *Structural Methods for the Exploration Geologist.* New York: Harper

Barnes, H L (1979) *The Geochemistry of Hydrothermal Ore Deposits.* 2nd edn. New York: Wiley

Berkman, D A (1976) *The Field Geologist's Manual. (Monograph Series No. 9)* Melbourne: Australasian Institute of Mining and Metallurgy

Beus, A A and Grigorian, S V (1977) *Geochemical Exploration Methods for Mineral Deposits*. Wilmette, Illinois: Applied Publishing

Blunden, J (1975) *Mineral Resources of Britain*. London: Hutchinson

Bowie, S H U et al (eds) (1979) *Mineral Deposits of Europe, Volume 1: Northwest Europe*. London: Institution of Mining and Metallurgy/Mineralogical Society

Boyle, R W (1979) *The Geochemistry of Gold and its Deposits. (Bulletin Geological Survey of Canada* **280**). Ottawa: Department of Energy, Mines and Resources

Bureau de Recherches Géologiques et Minières (1978–1982) *Carte des Gîtes Minéraux de la France à 1:500,000*: 8 sheets with accompanying explanation. Orleans: BRGM

Brooks, M and Kearey, P (1984) *An Introduction to Geophysical Prospecting*. Oxford: Blackwells

Brown, A S (ed.) (1976) *Porphyry Deposits of the Canadian Cordillera. (Special Volume No. * **15**). Canadian Institute of Mining and Metallurgy

Cargill, S M and Green, S B (1986) *Prospects for Mineral Resource Assessments on Public Lands: proceedings of the Leesburg Conference*. (US Geological Survey, Circular **980**). Denver, Colorado: US Geological Survey

Carter, W D and Rowan, L C (1983) *Remote Sensing and Mineral Exploration*. Oxford: Pergamon

Chaussier, J-B and Morer, J (1987) *Mineral Prospecting Manual*. London: North Oxford Academic Publishers

Coetzee, C B (ed) (1976) *Mineral Resources of the Republic of South Africa*. 5th edn. Pretoria: Department of Mines, Geological Survey. The Government Printer

Craig, J R and Vaughan, D J (1981) *Ore Microscopy and Ore Petrography*. New York: Wiley

Cronan, D S (1980) *Underwater Minerals*. London: Academic Press

David, M (1977) *Geostatistical Ore Reserve Estimation. (Developments in Geomathematics*, **2**). Amsterdam: Elsevier

Derry, D R (1980) *World Atlas of Mineral Deposits*. London: Mining Journal Books

Devismes, P (1978) *Atlas Photographique des Minéraux d'Alluvions. (Bureau de Recherches Géologiques et Minières, Mémoire* **95**). Paris: BRGM

Dines, H G (1956) *The Metalliferous Mining Region of South-West England*. 2 vols. London: HMSO

Dixon, C J (1979) *Atlas of Economic Mineral Deposits*. London: Chapman & Hall

Dobrin, M (1976) *Introduction to Geophysical Prospecting*. 3rd edn. New York: McGraw-Hill

Douglas, R J W (ed.) (1970) *Geology and Economic Minerals of Canada. (Geological Survey of Canada, Economic Geology Report No.* **1**). Ottawa: Department of Energy, Mines and Resources

Dunham, K C (1948) *Geology of the Northern Pennine Orefield. Volume 1. Tyne to Stainmore. (Memoir of the Geological Survey of Great Britain)*. London: HMSO

Dunham, K C and Wilson, A A (1985) *Geology of the Northern Pennine Orefield. Volume 2. Stainmore to Craven. (Economic Memoir, British Geological Survey)*. London: HMSO

Dunning, F W et al (eds) (1982) *Mineral Deposits of Europe, Volume 2: Southeast Europe*. London: Institution of Mining and Metallurgy/Mineralogical Society

Dunning, F W and Evans, A M (eds) (1986) *Mineral deposits of Europe, Volume 3: Central Europe*. London: Institution of Mining and Metallurgy/Minerological Society

Edwards, R and Atkinson, K (1986) *Ore Deposit Geology*. London: Chapman & Hall

Evans, A M (1980) *An Introduction to Ore Geology*. Oxford: Blackwells

Fitch, A A (ed) (1979–1985) *Developments in Geophysical Exploration Methods*. 6 vols. London: Applied Science Publishers

Fletcher, W K (1981) *Analytical Methods in Geochemical Prospecting. (Handbook of Exploration Geochemistry, Volume* 1). Amsterdam: Elsevier

Friedrich, G et al (eds) (1986) *Geology and Metallogeny of Copper Deposits. (Special Publication No. 4, Society for Geology Applied to Mineral Deposits).* Berlin: Springer-Verlag

Govett, G J S (1983) *Rock Geochemistry in Mineral Exploration. (Handbook of Exploration Geochemistry. Volume* 3). Amsterdam: Elsevier

Griffiths, D H and King, R F (1981) *Applied Geophysics for Geologists and Engineers: the elements of geophysical prospecting.* 2nd edn. Oxford: Pergamon

Guilbert, J M and Park, C F (1986) *The Geology of Ore Deposits.* New York: Freeman

Harris, D P (1984) *Mineral Resources Appraisal. Mineral endowment, resources, and potential supply: concepts, methods, and cases.* Oxford: Oxford Scientific Publications

Hawkes, H E (1982) *Exploration Geochemistry Bibliography to January 1981 (Special Volume No.*11). Rexdale, Ontario: Geological Survey of Canada

Hood, P J (ed) (1979) *Geophysics and Geochemistry in the Search for Metallic Ores. (Economic Geology Report No.*31). Ottawa: Geological Survey of Canada

Howarth, R J (1983) *Statistics and Data Analysis in Geochemical Prospecting. (Handbook of Exploration Geochemistry. Volume* 2). Amsterdam: Elsevier

Hutchison, C S (1983) *Economic Deposits and their Tectonic Setting.* London: Macmillan

Jensen, M L and Bateman, A M (1979) *Economic Mineral Deposits.* 3rd edn revised. New York: Wiley

Journel, A G and Huijbregts, Ch J (1978) *Mining Geostatistics.* London: Academic Press

Knill, J L (ed) (1978) *Industrial Geology.* Oxford: Oxford University Press

Krishnaswamy, S (ed) (1979) *India's Mineral Resources.* 2nd edn. New Delhi: Oxford and IBH Publishing Co.

Lacy, W C (ed) (1983) *Mineral Exploration. (Benchmark Papers in Geology No.*70). New York: Van Nostrand Reinhold

—— (1983) *Mining Geology. (Benchmark Papers in Geology No.*69). New York: Van Nostrand Reinhold

Laznicka, P (1985) *Empirical Metallogeny: depositional environments, lithological associations and metallic ores. Volume 1. Phanerozoic environments, associations and deposits. (Developments in Economic Geology* 19). Amsterdam: Elsevier

LeRoy, R W et al (eds) (1977) *Subsurface Geology Petroleum Mining Construction.* 4th edn. Golden, Colorado: Colorado School of Mines

Levinson, A A (1974) *Introduction to Exploration Geochemistry.* Wilmette, Illinois: Applied Publishing

Levinson, A A (1980) *Introduction to Exploration Geochemistry.* 2nd edn. Wilmette, Illinois: Applied Publishing

Lindgren, W (1933) *Mineral Deposits.* New York: McGraw-Hill

MacDonald, E H (1983) *Alluvial Mining.* London: Chapman & Hall

Mitchell, A H G and Garson, M S (1981) *Mineral Deposits and Global Tectonic Settings.* London: Academic Press

Ohmoto, H and Skinner, B J (eds) (1983) *The Kuroko and Related Volcanogenic Massive Sulfide Deposits. (Economic Geology Monograph No.*5). El Paso, Texas: Economic Geology Publishing Co.

Open University (1984) *S238 The Earth's Physical Resources. Science: A Second Level Course.* Course Text in 6 Blocks. 8 vols. Milton Keynes: Open University Press

Parasnis, D S (1972) *Principles of Applied Geophysics*. 2nd edn. London: Chapman & Hall

Park, C F and MacDiarmid, R A (1975) *Ore Deposits*. 2nd edn. San Francisco: Freeman

Peters, W C (1987) *Exploration and Mining Geology*. 2nd edn. New York: Wiley

Pfeider, E P (1968) *Surface Mining*. New York: American Institute of Mining, Metallurgical, and Petroleum Engineers

Picklyk, D D *et al* (1978) *Canadian Mineral Occurrence Index (CANMINDEX) of the Geological Survey of Canada. (Geological Survey of Canada Paper* **78–8**) Ottawa: Geological Survey of Canada

Picot, P and Johan, Z (1982) *Atlas of Ore Minerals*. Amsterdam: Elsevier

Ramdohr, P (1980) *The Ore Minerals and their Intergrowths*. 2nd edn. 2 vols. (*International Series in Earth Science* **35**). Oxford: Pergamon

Reedman, J H (1979) *Techniques in Mineral Exploration*. London: Applied Science Publishers

Ridge, J D (1968) *Ore Deposits of the United States, 1933-1967. The Graton-Sales Volume*. 2 vols. New York: American Institute of Mining, Metallurgical, and Petroleum Engineers

—— (1972) *Annotated Bibliographies of Mineral Deposits in the Western Hemisphere. (Geological Society of America Memoir* **131**). Boulder, Colorado: Geological Society of America

—— (1976) *Annotated Bibliographies of Mineral Deposits in Africa, Asia (exclusive of the USSR) and Australasia*. Oxford: Pergamon

—— (1984) *Annotated Bibliographies of Mineral Deposits in Europe. Part 1: Northern Europe including examples from the USSR in both Europe and Asia*. Oxford: Pergamon

Rose, A W *et al* (1979) *Geochemistry in Mineral Exploration*. London: Academic Press

Rudarski Institut (1970) *Rudarski Recnik*. Beograd, Yugoslavia

Shepherd, T J *et al* (1985) *A Practical Guide to Fluid Inclusion Studies*. Glasgow: Blackie

Sheriff, R E (1984) *Encyclopedic Dictionary of Exploration Geophysics*. 2nd edn. Tulsa, Oklahoma: Society of Exploration Geophysicists

Siegal, F R (1974) *Applied Geochemistry*. New York: Wiley

Skinner, B J (ed) (1981) *Economic Geology: 75th Anniversary Volume*. El Paso, Texas: Economic Geology Publishing Co.

Smirnov, V I (1976) *Geology of Mineral Deposits*. Moscow: MIR Press

—— (1977) *Ore Deposits of the USSR*. 3 vols. London: Pitman

Stanton, R L (1972) *Ore Petrology*. New York: McGraw-Hill

Tarling, D H (1981) *Economic Geology and Geotectonics*. Oxford: Blackwells

Tatch, J H (1973) *Mineral Deposits*. Sudbury, Mass: Tatch Associates

—— (1976) *Gold Deposits*. Sudbury, Mass: Tatch Associates

—— (1980) *Coal Deposits*. Sudbury, Mass: Tatch Associates

Telford, W M *et al* (1976) *Applied Geophysics*. Cambridge University Press

Thomas, L J (1973) *Introduction to Mining*. Sydney: Hicks Smith

Thrush, P W (ed) (1968) *A Dictionary of Mining, Mineral and Related Terms*. Washington: United States Bureau of Mines, Department of the Interior

Titley, S R (ed) (1982) *Advances in the Geology of the Porphyry Copper Deposits, Southwestern North America*. Tucson: Ariz: University of Arizona

UNESCO (1968–1970) *Carte Métallogenique de l'Europe*. 9 sheets. 1:2,500,000. Paris: UNESCO: BRGM

Warren, K (1973) *Mineral Resources*. London: Penguin

Warren, R G (1972) *A Commentary on the Metallogenic Map of Australia and Papua New Guinea. (Bureau of Mineral Resources, Geology and Geophysics Bulletin* **145**). Canberra: BMR

Wolf, K H (ed) (1975–1985) *Handbook of Strata-Bound and Stratiform Ore Deposits*. 13 vols. Amsterdam: Elsevier
  Part 1: Principles and General Studies (1976)
  Volume 1; *Classifications and Historical Studies*: Volume 2; *Geochemical Studies*: Volume 3; *Supergene and Surficial Ore Deposits: textures and fabrics*: Volume 4; *Tectonics and Metamorphism*.
  Part 2: Regional Studies and Specific Deposits (1976)
  Volume 5; *Regional Studies*: Volume 6; *Copper, Zinc, Lead and Silver Deposits*: Volume 7; *Gold, Uranium, Iron, Manganese, Mercury, Antimony, Tungsten and Phosphorus Deposits*.
  Part 3: 1976–1981
  Volume 8; *General Studies*: Volume 9; *Regional Studies and Specific Deposits*: Volume 10; *Bibliography and Ore Occurrence Data*.
  Part 4: (1985–1986)
  Volume 11; *General Studies*: Volume 12; *General Studies*: Volume 13; *Regional Studies and Specific Deposits*: Volume 14; *Regional Studies and Specific Deposits*.
Wright, J B (1985) *Geology and Mineral Resources of West Africa*. London: Allen and Unwin
Wyllie, R J M and Argall, G O (eds) (1975) *World Mining Glossary of Mining, Processing and Geological Terms*. San Francisco: Miller Freeman
Zwart, W J (ed) (1986) *Cumulative Index. 8th edn. Supplement to October, 1986 volume*. Tulsa, Okla: Society of Exploration Geophysicists

Oil, natural gas and coal

Anstey, N A (1977) *Seismic Interpretation: the physical aspects*. Boston: IHRDC.
—— (1980) *Seismic exploration for sandstone reservoirs*. Boston: IHRDC.
—— (1982) *Simple Seismics*. Boston: IHRDC.
Asquith, G (1982) *Basic Well Log Analysis for Geologists*. Tulsa: American Association of Petroleum Geologists
Dickie, P A (1979) *Petroleum Reservoir Geology*. Tulsa: Petroleum Publishing Co.
Fitch, A A (1976) *Seismic Reflection Interpretation*. Berlin: Gebruder Borntraeger
Hobson, G D and Tiratsoo, E N (1981) *Introduction to Petroleum Geology*. 2nd edn. Scientific Press Ltd.
Hunt, J M (1979) *Petroleum Geochemistry and Geology*. San Francisco: Freeman
Kinghorn, R R F (1983) *An Introduction to the Physics and Chemistry of Petroleum*. New York: Wiley
Levorsen, A I (1967) *Geology of Petroleum*. San Francisco: Freeman
Merkel, R H (1979) *Well Log Formation Evaluation. (American Association of Petroleum Geologists, Course Note Series, No. 4)*
Murchison, D and Westoll, T S (eds) (1968) *Coal and Coal-bearing Strata*. Edinburgh: Oliver and Boyd
Pirson, S J (1983) *Geological Well Log Analysis*. 3rd edn. Houston: Gulf Publishing Co.
Selley, R C (1983) *Petroleum Geology for Geophysicists and Engineers*. Prentice-Hall
—— (1984) *Elements of Petroleum Geology*. San Francisco: Freeman
Stach, E *et al* (1975) *Coal Petrology*. 2nd edn. Berlin: Gebruder Borntraeger.
Yen, T F and Chilingarian, G V (eds) (1976) *Oil Shale*. Amsterdam: Elsevier

Non-metalliferous

Armstead, H C (1983) *Geothermal Energy*. 2nd edn. London: E & F Spon

Australian Water Resources Council (1975) *Groundwater Resources of Australia*. Canberra: Australian Government Public Service

Bach, W and Freeze, R A (1983) *Chemical Hydrology. (Benchmark Papers in Geology*, **73**). New York: Van Nostrand Reinhold

Bates, R L (1969) *Geology of the Industrial Rocks and Minerals*. New York: Dover

Bowen, R (1979) *Geothermal Resources*. London: Applied Science Publishers

—— (1986) *Groundwater*. 2nd edn. London: Elsevier Applied Science

Cook, P J and Shergold, J H (1986) *Phosphate Deposits of the World. Volume 1. Proterozoic and Cambrian Phosphorites*. Cambridge University Press

Downing, R A and Gray, D A (eds) (1986) *Geothermal Energy — The Potential in the United Kingdom*. London: HMSO

Edwards, L M *et al* (eds) (1982) *Handbook of Geothermal Energy*. Houston, Texas: Gulf Publishing

Elder, J (1981) *Geothermal Systems*. London: Academic Press

Ellis, A J and Mahon, W A J (1977) *Chemistry and Geothermal Systems*. New York: Academic Press

Freeze, R A and Bach, W (1983) *Physical Hydrology (Benchmark Papers in Geology*, **72**). New York: Van Nostrand Reinhold

Grim, R E and Guven, N (1978) *Bentonites. Geology, mineralogy, properties and uses. (Developments in Sedimentology*, **24**). Amsterdam: Elsevier

Harben, P W and Bates, R L (1984) *Geology of the Nonmetallics*. New York: Metal Bulletin

Kuzvart, M (1984) *Industrial Minerals and Rocks. (Developments in Economic Geology*, **18**). Amsterdam: Elsevier

LeFond, S J (ed) (1983) *Industrial Minerals and Rocks*. 5th edn. New York: American Institute of Mining, Metallurgical and Petroleum Engineers

Lloyd, J W (ed.) (1981) *Case Studies in Groundwater Resources Evaluation*. Oxford: Clarendon Press

Milanovic, P T (1981) *Karst Hydrology*. Littleton, Colorado: Water Resources Publications

Price, M (1985) *Introducing Groundwater*. London: Allen & Unwin

Raghunath, H M (1982) *Ground Water*. New Delhi: Wiley Eastern

Rapolla, A and Keller, G V (eds) (1984) *Geophysics of Geothermal Areas: state of the art and future development*. Golden, Colorado: Colorado School of Mines

Viessman, W *et al* (1977) *Introduction to Hydrology*. 2nd edn. New York: Harper & Row

Webster, R (1983) *Gems: their sources, description and identification*. 4th edn. London: Butterworth.

CHAPTER SIXTEEN

Engineering and environmental geology

STEPHEN FRANCIS

The prime role of the engineering geologist is to act as a source of information to the civil engineer. Therefore he must be conversant with engineering principles and concepts in addition to having an understanding of those aspects of geology which may be relevant to engineering works. Consequently, the engineering geologist needs to have knowledge of an unusually wide range of disciplines and literature sources. The central disciplines include engineering geology, soil mechanics, rock mechanics, geomorphology, hydrology, hydrogeology, remote sensing, pedology, sedimentology, and structural geology. At the other extreme, peripheral literature sources would include newspaper archives and historical records, which can sometimes provide crucial data for engineering projects.

An additional consideration, which will influence the choice of literature sources, is that the relative importance of certain aspects of engineering geology is changing; this is at least partly because of the relative youth of the discipline, which has prevented the adequate evaluation of the usefulness of certain techniques. This progressive change in emphasis is highlighted by the currently increased interest in geomorphology and remote sensing.

Environmental geology concerns the impact of geology upon civilization, consequently many aspects of engineering geology, geochemistry, geophysics and economic geology fall into this category (for instance aspects of construction, pollution, earthquakes and mineral wealth respectively). The term finds little favour in this country and many geologists are extremely unhappy with its generality and implications of 'pseudo-science'. Therefore environmental geology *sensu stricto* will not be considered in this chapter, but all geologists, either indirectly or directly, are concerned with the implications of civilization, particularly those working in the applied fields.

Due to the diversity of primary literature sources relevant to engineering geology, no attempt has been made to compile a comprehensive list of these; instead, a series of well-defined subject areas are considered in turn, with emphasis upon modern introductory texts and the most important journals.

Nearly all of the books containing introductions to engineering aspects of geology are aimed at the civil engineer rather than the geologist. An early but well-respected example is D P Krynine and W R Judd's *Principles of Engineering Geology and Geotechnics* (McGraw-Hill, 1957). Two more recent and popular works are *A Geology for Engineers*, by F G H Blyth and M H de Freitas (6th edn, Arnold, 1976) and *Geology for Engineers*, by A C McLean and C D Gribble (Allen and Unwin, 1979). A particularly relevant Appendix, 'Sources of geological information' is included in the former, and the large second section of this book can be appreciated by geologists and engineers alike. The Ministry of Defence manual *Applied Geology for Engineers* (HMSO, 1976) is also suitable as an introduction for geologists; it contains a wealth of detail, is very clearly written and contains well-selected lists of major references at the end of each chapter, together with a section on important sources of geotechnical information. A more detailed view of engineering geology is presented in *Principles of Engineering Geology*, by P B Attewell and I W Farmer (Chapman & Hall, 1976), which gradually develops an introduction to some of the analytical techniques used in soil and rock mechanics, and concludes with an extensive bibliography.

For those who intend to become seriously involved in engineering geology, the books by Blyth and de Freitas and Attewell and Farmer, together with the Ministry of Defence manual, would be useful bookshelf additions, together with some of the more specialized works discussed later.

The most important journals which specialize in engineering geology are the *Quarterly Journal of Engineering Geology* (1967–), *Engineering Geology* (1965–), and *Bulletin of the International Association of Engineering Geologists* (1970–), with a significant proportion of papers considering specific vocational problems rather than more academic aspects of the subject. Major papers are often found in *Geotechnique* (1948–), the *International Journal of Rock Mechanics, Mining Sciences and Geomechanics Abstracts* (1964–), *Ground Engineering* (1968–), *Rock Mechanics* (1969–; formerly *Rock Mechanics and Engineering Geology* (1963–68)) and the *Journal of Geotechnical Engineering* (1983–; formerly the *Journal of the Geotechnical Engineering Division of the American*

Society of Civil Engineers (ASCE) (1974–82)). The Association of Engineering Geologists in America also publishes a journal *Engineering Geology* (1964–), which should not be confused with the European journal (published by Elsevier) of the same title, quoted above. Despite all of these publications being intended as introductions to the subject, some knowledge of soil and rock mechanics is needed to appreciate much of the literature to which they refer. Those with no taste for these numerical aspects will find that most of the 'lightweight' articles are contained in the journals which do not specialize in these disciplines, such as some of the geomorphological and main stream geological periodicals (see Chapters 17 and 3 respectively).

The Engineering Group of the Geological Society of London provides a useful forum for engineering geologists, with the Society's *Newsletter* giving details of important events (also note the 'Geodiary' contained in *Ground Engineering*). Regular meetings are held at the Geological Society in London, together with occasional meetings with the British Geotechnical Society (Institution of Civil Engineers), and recently with the British Geomorphological Research Group: an annual Regional Meeting is also held. Many of the papers read at these gatherings are later published in the *Quarterly Journal of Engineering Geology* (see above).

Soil mechanics

There is now an extensive soil mechanics literature, effectively all of which is relevant to engineering geology. Much of this may initially appear forbidding to a prospective engineering geologist, but fortunately there are many introductory texts available which contain the simple principles which need to be understood. These books fall into two categories. Those in the first group discuss principles but do not consider the characteristics of soils and their influence upon soil properties: good examples of this approach are *Elements of Soil Mechanics for Civil and Mining Engineers*, by G N Smith (4th edn, Granada, 1978) and *Soil Mechanics*, by R F Craig (2nd edn, Van Nostrand Reinhold, 1978). The second category includes those works which attempt to explain the influence of soil characteristics upon soil properties and behaviour, as exemplified by T W Lambe and R V Whitman's *Soil Mechanics: S.I. version* (Wiley, 1979), R D Holtz and W D Kovacs' *An Introduction to Geotechnical Engineering* (Prentice-Hall, 1981) and I K Lee *et al*, *Geotechnical Engineering* (Pitman, 1983). Smith and Craig's books are undemanding, with the contents clearly and traditionally laid out, but this, and their relatively low cost, are the only advantages that they have over the latter three texts. Lambe and Whitmans' work is extremely well-established and of high quality, but its layout is often frustrating: Holtz and Kovacs' treatment is up-to-date and has been well received. Both of these books reflect American practice and extensively refer to mainly American literature, which is advantageous to the reader already familiar with UK practice who wishes to broaden his knowledge. The Lee *et al* text mentioned above introduces the reader to the uses of elementary statistics in data evaluation, for instance the application of probability to slope stability evaluation; this approach should become popular. This book also extensively refers to original papers and is generally clearly written.

Books of more specialized interest include laboratory and field manuals, the following of which are in common use: *British Standards, BS 1377: Methods of Testing Soils for Civil Engineering Purposes* (British Standards Institution, 1975), *BS 812: Methods for Sampling and Testing of Mineral Aggregates, Sands and Fillers* (British Standards Institution, 1975), *BS 5930: Code of Practice for Site Investigation* (British Standards Institution, 1981), *Earth Manual* (2nd edn, Water and Power Resources Service, US Dept of Interior, 1974), *Standard Specifications for Transportation Materials and Methods of Sampling and Testing, Parts I and II* (12th edn, AASHTO, 1978) and *Annual Book of ASTM Standards, Part 19, Natural Building Stones; Soil and Rock* (American Society for Testing Materials, 1982). The latter three reflect American practice, although some of these testing methods are frequently used in the UK. Any engineering geologist must be familiar with these works, particularly valuable being *BS 5930* and the *Earth Manual*; the former is already widely used, whilst the latter comprehensively considers both laboratory and field techniques.

Useful discussions of testing methods are given in T N W Akroyd's *Laboratory Testing in Soil Engineering* (Soil Mechanics Ltd, *Geotechnical Monograph No. 1*, 1957) which still remains a standard reference work, and more recently B Vickers' *Laboratory Work in Civil Engineering, Soil Mechanics* (2nd edn, Granada, 1982) which usefully complements *BS 1377*. A more detailed coverage is given in K H Head's *Manual of Soil Laboratory Testing, Volumes 1 and 2* (Pentech, 1982) and the unusual *Geotechnical Engineering Handbook*, by M Carter (Pentech, 1983) which is based upon an interesting collection of test data sheets.

Aspects of foundation types, design and construction are covered in *Foundation Engineering*, by R B Peck *et al* (2nd edn,

Wiley, 1974) which is easily followed and makes compulsive reading in places. Alternatively, M J Tomlinson's *Foundation Design and Construction* (Pitman, 1969) is a popular work.

Two works on slope stability, which are in many ways complementary, are *Slope Analysis; developments in geotechnical engineering*, by R N Chowdhury (Elsevier, 1978) and *Landslides: analysis and control*, edited by R L Schuster and R J Krizek (*Transportation Research Board Special Report* **176**, 1978).

The Elsevier book is marred by the low quality typeface used — but the contents and bibliography are worthwhile. In most respects Schuster and Krizek's work is better, and, despite each chapter being written by different authors, most aspects of slope stability are covered: this book certainly appeals to engineering geologists and would be highly recommended to readers of any background who are interested in slopes.

Interests in offshore foundation work and the hazards of the seabed have increased because of oil platform development, and two recent compendiums of papers which give an interesting introduction are *Offshore Site Investigation*, edited by D A Ardus (Graham & Trotman, 1980) and *Offshore Geologic Hazards (AAPG Education Course Note Series* **18**) (American Association of Petroleum Geologists, 1981). They primarily reflect work in the North Sea and Gulf of Mexico respectively.

Two volumes which may become trendsetters are *The Mechanics of Soils: an introduction to critical state soil mechanics*, by J H Atkinson and P L Bransby (McGraw-Hill, 1978) and *Soil Survey for Engineering*, by A B A Brink *et al* (Clarendon, 1982). The rationalization of soil mechanics that may be offered by critical state concepts has attracted considerable attention, and Atkinson and Bransby's work offers a good introduction to some of the more difficult, previously published, aspects. The book by Brink *et al* contains large sections on remote sensing of soils, terrain classification, surveys and data banking. The importance of these aspects has become increasingly accepted, and even though this book has been well received, a more detailed coverage of the use of these developments in geotechnology is still awaited.

From a historical point of view, the importance of K Terzaghi's contributions may be gauged from *Theoretical Soil Mechanics* (Wiley, 1943), which makes interesting reading for enthusiasts.

Important soil mechanics journals are listed in Appendix 1 at the end of this chapter. Papers in these journals are both vocational and academic, with the contents of *Geotechnique, Journal of the Geotechnical Engineering Division, ASCE* and the *Proceedings of the International Conferences on Soil Mechanics and Foundation Engineering* generally being regarded as state-of-the-art publications.

There is a large number of professional and research organizations concerned with civil engineering; those of interest to the engineering geologist in the UK include the Institution of Civil Engineers, the British Geotechnical Society and the International Society for Soil Mechanics and Foundation Engineering. The Building Research Establishment at Watford, and the Transport and Road Research Laboratory at Crowthorne, are the two most important research centres in the UK.

Rock mechanics

As with the soil mechanics literature, most rock mechanics publications are of importance to the engineering geologist. Rock mechanics itself covers a wide field, ranging from .the extreme limits of mining engineering to theoretical physics. A considerable amount of work in the latter area, and on the fringes of structural geology, concerns rock behaviour under conditions which are unlikely to be directly relevant to engineering geology, such as deformation under high temperatures and very high pressures. An excellent and clearly written introduction to this aspect of the subject is given by M S Paterson in *Experimental Rock Deformation: the brittle field* (Springer-Verlag, 1978) which contains a vast bibliography, which is authoritatively discussed within the text. Fortunately, Paterson's work is just one of a collection of superb books which together cover nearly all aspects of rock mechanics. These books, with the exception of one, draw the reader from simple basics to high levels of specialization with alarming ease. The exception is J C Jaeger and N G W Cook's *Fundamentals of Rock Mechanics* (3rd edn, Methuen, 1979) which treats the more theoretical aspects of this field in an unrivalled way. *Rock Slope Engineering*, by E Hoek and J W Bray (3rd edn, IMM, 1980) and *Underground Excavations in Rock*, by E Hoek and E T Brown (IMM, 1980) are essentially design manuals but with clear discussion of modern principles and techniques (the latter also contains a comprehensive bibliography on underground excavations). Both are inadequately indexed, but they should be on every engineering geologist's bookshelves. *Rock Characterisation Testing and Monitoring, ISRM Suggested Methods*, edited by E T Brown (Pergamon, 1981) contains a compilation of papers which attempt to standardize field and laboratory techniques to the approval of the International Society for Rock Mechanics. This standardization has yet to be completed and many of the methods discussed will hopefully be revised, but the volume contains essential reading. R E Goodman's *Introduction to Rock Mechanics* (Wiley, 1980) fills a gap between practice and theory, and C Jaeger's *Rock Mechanics and Engineering* (2nd edn, Cambridge University Press, 1979) successfully integrates case histories with engineering principles. All of these books are strongly recommended, with *Introduction to Rock Mechanics* which covers a wide range of topics making it the best choice as a starting point for someone new to the field.

The most important rock mechanics journals are *The International Journal of Rock Mechanics, Mining Sciences and Geomechanics Abstracts* (1964–), *Rock Mechanics* (1969–), *Mining Engineering* (1949–), *Transactions of the Institution of Mining and Metallurgy (Section A)* (1966–) and *Tunnels and Tunnelling* (1965–). The publications of the International Conferences on Rock Mechanics are also extemely valuable.

In addition to the organizations already mentioned in other sections, the International Society for Rock Mechanics is especially important in this field, particularly through the International Conferences on Rock Mechanics (see above) and its attempts at international standardization.

Finally, it should be mentioned that the non-English rock mechanics literature is considerable, of particular note being the work published in German.

Geomorphology (see also Chapter 17)

Engineering geologists have made increasing use of geomorphological techniques in recent years, which has coincided with geomorphologists becoming more interested in soil mechanics and hydraulics. The latter development, coupled with the cost effectiveness of geomorphological surveys, will ensure that this discipline will become even more relevant to engineering geology.

There is an enormous number of introductory texts to geomorphology, many of which are qualitative and simply contain descriptions of the wide variety of landforms. Specialist, quantitative, texts are of greater interest to engineering geologists, and the following list of books, whilst reflecting this author's preferences, includes both state-of-the-art works and those which can be easily appreciated by a geologist or engineer.

A S Goudie's *Duricrusts in Tropical and Subtropical Landscapes* (Oxford University Press, 1973), M F Thomas' *Tropical Geomorphology* (Macmillan, 1974), *Landform Studies from Australia and New Guinea*, edited by J N Jennings and J A Mabbutt

(Cambridge University Press, 1967), A L Washburn's *Periglacial Processes and Environments* (Arnold, 1973), which contains a notable bibliography, and D E Sugden and B S John's *Glaciers and Landscape* (Arnold, 1976), are all concerned with specific environments of interest to the engineering geologist.

More advanced texts which attempt to interpret landscape development in terms of processes include *Fluvial Processes in Geomorphology*, by L B Leopold *et al* (Freeman, 1964), M A Carson and M J Kirkby's *Hillslope Form and Process* (Cambridge University Press, 1972), *Drainage Basin Form and Process*, by K J Gregory and D E Walling (Arnold, 1973), *Beach Processes and Sedimentation*, by P D Komar (Prentice-Hall, 1976), *Geomorphology and Climate*, edited by E Derbyshire (Wiley, 1976), J B Thornes and D Brunsden's *Geomorphology and Time* (Methuen, 1977), *Timescales in Geomorphology*, edited by R A Cullingford *et al* (BGRG publication, Wiley, 1980), A J Gerrard's *Soils and Landforms: an integration of geomorphology and pedology* (Allen and Unwin, 1981) and M J Selby's *Hillslope Materials and Processes* (Oxford University Press, 1982). The themes of these books are self-explanatory, and they all contain good reading, of particular note being the works of Carson and Kirkby, Gerrard and Selby. An interesting introduction to some of the techniques useful to geomorphologists is given in *Geomorphological Techniques*, primarily edited by A S Goudie (BGRG publication, Allen and Unwin, 1981), with many of the references cited giving useful leads into a diverse literature.

Journals which contain geomorphological works of possible interest to the engineering geologist include the *Journal of Geology* (1893–) (occasional, high quality papers), *Zeitschrift für Geomorphologie* (1967–) (mainly descriptive), *Transactions of the Institution of British Geographers* (1933–) (occasional, mediocre quality papers), *Earth Surface Processes and Landforms* (1976–) (essential reading) and the British Geomorphological Research Group's *Technical Bulletins* (mainly relevant, irregularly published, available from Geoabstracts).

The British Geomorphological Research Group (BGRG) holds thematic regional meetings, promotes publications, produces a newsletter (*Geophemera*) and a journal (*Earth Surface Processes and Landforms*) (see above).

Hydrology and hydrogeology (see also Chapter 17)

Water presents major problems to many geotechnical projects consequently the engineering geologist must be capable of coping with most aspects of hydrology and hydrogeology, usually with the intention of removing or minimising the effects of water. This contrasts with the aims of the worker in the water industry who usually cannot get enough of this resource. An obvious exception to this situation occurs with the development of impounding schemes where the engineering geologist must be appreciative of catchment hydrology and hydrogeology.

Publications in these areas are dominated by the water engineer rather than the geographer or geologist. Major works which have become established over a long period include R J M de Wiest's *Hyrdology* (Wiley, 1965) and S N Davis and R J M de Wiest's *Hydrogeology* (Wiley, 1966) with D K Todd's *Ground Water Hydrology* lagging well behind until its recent second edition, (Wiley, 1980), which is virtually a new book and highly recommended, despite its occasional lapses, such as classifying limestone as an 'igneous and metamorphic rock'. Despite its age de Wiest's book is perhaps of greatest interest to the engineering geologist with its clear consideration of both ground and surface water; but both this and the other two books reflect American practice. Other good introductions to water include L Huisman's *Groundwater Recovery* (Macmillan, 1972), a very practical book with a European emphasis, J Bear's *Hydraulics of Groundwater* (McGraw-Hill, 1979) which is extremely thorough and treats the subject with mathematical enthusiasm, and W Viessman *et al, Introduction to Hydrology* (Harper and Row, 1977) which is particularly strong on the uses of statistics and modelling. Despite the engineering emphasis (with the exception of *Hydrogeology*) all of these works are well within the capabilities of most geologists: for those who waver, R Bowen's *Ground Water* and *Surface Water* (Applied Science, 1980 and 1982 respectively) have a geological emphasis but a relatively inferior standard of clarity. As most geologists have little appreciation of surface water, R C Ward's *Principles of Hydrology* (2nd edn, McGraw-Hill, 1975) can be recommended as an extremely clearly written elementary introduction to this aspect, containing comprehensive reference lists at the end of each chapter. E M Wilson's *Engineering Hydrology* (2nd edn, Macmillan, 1974) is an alternative work which is aimed at the civil engineer, but this should not be used on its own because its consideration of hydrological processes is minimal. A very popular compendium on a currently topical aspect of hydrology is *Hillslope Hydrology*, edited by M J Kirkby (Wiley, 1978). For those more interested in the practical aspects of water supply, A C Twort *et al, Water Supply* (2nd edn, Arnold, 1974) and T M Y Tebbutt's *Principles of Water Quality Control* (3rd edn, Pergamon, 1983) are good starting points.

Hydrological and hydrogeological articles of direct relevance to engineering geology will be found in the *Quarterly Journal of Engineering Geology* (1967–) (which now officially incorporates hydrogeology), *Earth Surface Processes and Landforms* (see previous section), *Ground Water* (1962–), *Journal of Geophysical Research* (1896–), *Journal of Hydrology* (1963–), *Journal of the Hydraulics Division, ASCE* (1983–), *Journal of the Institution of Water Engineers and Scientists* (1975– ; formerly *Journal of the Institution of Water Engineers* 1947–74), *Water Research Centre Technical Reports* (1975–), *Water Resources Research* (1965–), and the *Water Supply Papers* of both the USGS and BGS. In addition, virtually all of the civil and mining engineering journals carry papers on aspects of water.

Important societies in the UK include the Institution of Water Engineers, the Hydrogeological Group of the Geological Society, the embryonic British Hydrological Society and the BGRG (referred to in the previous section). The Institution of Geologists is particularly influential in the UK water industry, with virtually all of the geologists employed in the water industry being members. The Institute of Hydrology, based at Wallingford, and the Water Research Centre, based at Medmenham, are the two major research organizations in the UK, which are becoming increasingly involved in industrial work and also offer consulting services.

Remote sensing, pedology, structural geology and sedimentology

All of these fields are discussed in detail elsewhere in this book, but some comments are required on the aspects of remote sensing (including geophysics) and pedology which are relevant to engineering geology.

Remote sensing techniques are becoming increasingly important to the engineering geologist, for instance in the poorly mapped areas of the world where basemaps may have to be derived from aerial photographs. In addition multispectral, infra-red, and image enhancement methods may be useful in the preliminary evaluation of phenomena of geotechnical interest over large areas. Although geophysics is cautiously treated by most engineers, there are many instances when the use of the correct technique, together with experienced interpretation and an appreciation of the engineers' requirements, will provide data which would otherwise only be available at a higher cost. Therefore, with time, geophysical techniques may become more accepted by engineers; promising examples being borehole logging tools and ground radar. However, with the exception of some aspects of aerial photographic interpretation, remote sensing is best left to the specialist. For those wishing for background information, relevant introductions are D H Griffiths and R F King's *Applied Geophysics for Geologists and Engineers* (2nd edn, Pergamon, 1981) and the Working Party Report *Land Surface Evaluation for Engineering Practice* (Anon, 1982) both of which would be excellent starting points.

In many situations study of the soil may assist in interpreting ground conditions at more important depths, particularly in poorly exposed areas; ground water conditions may also be indicated. An elementary introduction to pedology is given in F M Courtney and S T Trudgill's *The Soil: an introduction to soil study in Britain* (Arnold, 1976) and a more interesting discussion of UK soils is contained in *Soils in the British Isles*, by L F Curtis *et al* (Longman, 1976). Particularly useful for the engineering geologist in the field is J M Hodgson's *Soil Sampling and Soil Description* (Oxford University Press, 1978), which is highly recommended. The publications of the Soil Survey of England and Wales (which are obtained directly from the Soil Survey) include some essential items, such as detailed reports on specific surveyed areas, together with soil maps; and *Technical Monographs* on methodology (such as methods of recording data in the field and laboratory physical and chemical tests). In addition, a set of 1:250,000 maps covering England and Wales is now available with an extensive, informative key (which lists, for instance, the drainage characteristics of each soil type).

Relevant journals which contain high quality articles on the physical properties of soils (in addition to the agricultural component) include the *Bulletin of the International Society of Soil Science, Journal of Soil Science, Journal of the Soil Science Society of America, Soviet Soil Sciences* and *Geoderma*. By far the most important research centre in the UK is the Soil Survey of England and Wales, based at the Rothamsted Experimental Station, which generally treats enquiries on most aspects of soil science very helpfully. (For further information on soil science literature, see Chapter 19).

Abstracting and indexing services

The engineering geologist should be aware of the major abstracting and indexing services covering mainstream geology which are discussed elsewhere in this book (see Chapter 5).

Most of the services to be considered in this section are young, but this does not cause a severe problem with literature searches because of the youthfulness of engineering geology itself, and also its associated disciplines. It is worth noting that there is a considerable overlap both in the journals which are scanned and in the articles which are listed by the abstracting services.

The Rock Mechanics Information Service is responsible for *Geomechanics Abstracts* (1973–), which is published as part of the *International Journal of Rock Mechanics, Mining Sciences and Geomechanics Abstracts*. Originally the former was known as *Rock Mechanics Abstracts* (1970–1972) which was less comprehensive in its coverage. This abstracting service now includes engineering geology, soil and rock mechanics, and hydrogeology, and is obviously strongest on aspects of rock mechanics. At a cost, the service (based at Imperial College) will conduct searches of their computer files for the period 1969 to 1977 using subject keywords only, and for 1977 to the present also searching on author's name, title, and abstract. Prior to 1969 the *KWIC Index of Rock Mechanics Literature Part 1 (1870-1968)* can be consulted. *Part 2* of the *KWIC Index* covers the period 1969–1976, and both were compiled by the Service and are available from Pergamon: as the title implies this publication does not contain abstracts. *Geotechnical Abstracts* (1970–) is the other major source of relevant abstracts. This is partly sponsored by the International Society for Soil Mechanics and Foundation Engineering, and is published by the German National Society for Soil Mechanics and Foundation Engineeing. Abstracts are also available in card form, to be used with the Geodex International Retrieval System. Soil and rock mechanics, geotechnical engineering, and engineering geology are covered in great detail, with entries under twelve principal subject groups; as would be expected, soil mechanics is especially well presented. The combination of *Geotechnical Abstracts* and *Geomechanics Abstracts* will more than adequately cover the requirements of nearly all engineering geology literature searches.

Other worthwhile, but relatively peripheral, abstracting and indexing services include the *IMM Abstracts*, from the Institution of Mining and Metallurgy (1950–) for aspects of mining engineering and rock mechanics, available also on the IMMAGE database: however a large proportion of the contents are of no geotechnical interest. As this publication does not consider literature connected with coal mining, the *NCB Abstracts C: Coal and Mining Geology* usefully fills a gap; this is available from the National Coal Board, Technical Intelligence Branch. *Geographical Abstracts A: Landforms and the Quaternary* and *Geographical Abstracts B: Climatology and Hydrology* (see also Chapter 17), started in 1960 and 1966 respectively, are sometimes valuable because they scan many publications beyond the normal reading range of geologists and engineers: both provide annual indexes, at low cost. Volume A is progressively expanding and improving in relevant content, but Volume B cannot directly compete with the excellent services provided by the Water Research Centre (WRC). Their Aqualine computerized system covers virtually all topics related to water, including pollution and biological aspects. Use of this is free to WRC members, and other benefits of membership include *WRC Technical Reports* (high quality, in-house research articles) and *WRC Information*, which is an abstracting newsletter, listing abstracts included in the Aqualine databank. Photocopies may be obtained of all papers listed in this publication, many of which are reports which have limited circulation. Combining *WRC Information* with *Geomechanical Abstracts* and *Geotechnical Abstracts* would give a comprehensive literature coverage system, with the two computerized systems also offering the bonus of WRC membership benefits, and copies of the *International Journal of Rock Mechanics and Mining Sciences* respectively.

Finally, two more general services which consider all aspects of civil engineering are the huge *Engineering Index Monthly and Author Index* (formerly *Engineering Index*, 1894–) together with the *Engineering Index Annual*, which have an American emphasis, and *ICE Abstracts* (which includes *European Civil Engineering Abstracts*).

Reference

Anon (1982) 'Land Surface Evaluation for Engineering Practice: Working Party Report'. *Quarterly Journal of Engineering Geology*, **15**, 265–316.

APPENDIX: Soil Mechanics Journals

Canadian Geotechnical Journal (1963–)
Geotechnical Testing Journal (1978–)
Geotechnique (1948–)
Ground Engineering (1968–)
Highway Engineer (1930–)
Journal of Geotechnical Engineering (1983–; formerly *Journal of the Geotechnical Engineering Division*, American Society of Civil Engineers and *Journal of the Soil Mechanics and Foundation Engineering Division*, 1956–82)
Marine Geotechnology (1977–)
Proceedings of the Institution of Civil Engineers (1952–; formerly *Journal of the Institution of Civil Engineers*, 1935–51)
Proceedings of the International Conferences on Soil Mechanics and Foundation Engineering (1936–)
Publications of the Norwegian Geotechnical Institute (1953–)
Soils and Foundations (1960–)

CHAPTER SEVENTEEN

Geomorphology and hydrogeology

KEITH M CLAYTON

This chapter covers geomorphology, glaciology, hydrology, hydro-geology, and also the Quaternary period. Of these subject areas, only hydrogeology (i.e. the study of groundwater) falls squarely within geology (see also Chapter 16). In the United States of America geomorphology is treated as part of geology, but in many countries of the world, including the United Kingdom, most of those who call themselves geomorphologists are (physical) geographers. Since this is also true of many hydrologists and glaciologists, the literature of these subject areas spills over into a wider range of journals than does most of geology *sensu stricto*. However, there is no doubt that all these subjects are earth sciences.

For each of these four subject areas there are some journals which are restricted to the single topic, but in each case the number is quite small. In addition many relevant publications are to be found in more general journals, some of which are geological in nature, some are geographic, and some are general scientific journals. The narrowest field is hydrogeology, which has a small core of specialist journals, and is covered also in geological or other more general earth science journals. Most of the literature, however, is very scattered, and insofar as these areas are covered satisfactorily by secondary services, these will be found very useful in helping towards a comprehensive coverage of relevant literature.

Another problem to be faced is the changing scope of geomorphology. Traditionally the subject has been the study of the origin of landforms, but in recent years it has extended in two ways. One has been the study, in ever greater detail, of the processes by which landforms develop, the second involves the study of that later part of the geological record when the existing landforms of the world were formed, i.e. the later part of the Tertiary and, more particularly, the Quaternary period. This period, covering the last 2 million years or so of the earth's history, includes the rapid climatic fluctuations which gave rise to the Ice Ages, and as a result, numerous different approaches to the Quaternary period have developed, many of them very loosely subsumed under the heading of geomorphology. This includes the study of the chronology of the Quaternary period using geological and geomorphological techniques, and the study of environmental change during the Quaternary in many different ways. The literature of Quaternary studies is covered in this chapter as a major subdivision of geomorphology.

The more generalized earth science and geography journals which include material relevant to the subjects of this chapter are very numerous. Indeed some are not restricted to earth science; thus *Nature* and *Science* will have a paper every few issues within the subject areas of this chapter, especially geomorphology and the Quaternary. In 1982, *Nature* had 21 papers abstracted in *Geographical Abstracts* Part A, and *Science* had eight. Journals like these, which appear frequently and circulate very widely, are used to make brief announcements of new work, and the material is often published again at greater length in another journal, perhaps a year or more later. Both also carry occasional longer articles of a review or survey nature. In addition, such general scientific journals as the *Philosophical Transactions of the Royal Society* and its equivalents in other countries, will from time to time carry relevant material. In some cases (e.g. *Comptes Rendus des Séances de l'Académie des Sciences*) the journal is divided into sections, and relevant material will normally be restricted to one or two of these.

Three major American geological journals frequently contain relevant articles: these are the *Journal of Geology*, the *American Journal of Science*, and the *Bulletin of the Geological Society of America*. The same society also publishes *Geology*, a monthly journal, presenting shorter articles appearing with less delay. Geomorphology (and Quaternary studies) are quite strongly represented in these journals, since in the USA the practitioners of these subjects are generally geologists, rather than geographers. For the same reason the official publications of the national geological surveys of many countries include much relevant material, although this is less true of the British Geological Survey. Thus the United States Geological Survey publishes a series called *Professional Papers* and this carries many papers on geomorphology, hydrology, and Quaternary topics.

In common with the earth sciences literature as a whole, much of the journal literature discussed in this chapter is tied to geographical regions. In many cases these journals are published by national, academic societies, which naturally include quite a number of articles on local topics, sometimes descriptive, sometimes more analytical. In other cases distinctive regions have encouraged the growth of a literature to the extent of supporting a specialist journal, and many universities and local societies publish material with a high proportion of the papers dealing with the local area. This pattern has been developed to a high degree in France, as will be seen later. The areal element is so highly developed in geomorphology that perhaps half the articles in any one journal will be tied to the country in which it is published. This local literature forms a convenient quarry for the home area, but can be an awkward barrier when the area of interest is in a foreign land. It means that the literature of a region is less scattered and appears in fewer journals than the literature on a systematic topic.

It is still usual for these local, and many national, journals to be published in the local language. This adds to the problems of tackling the regional literature of a foreign area. Virtually all the articles on the geomorphology of the USSR are published in Russian, (or in Georgian or another State language even less accessible than Russian) and in Russian journals. Since only a few journals are available in cover-to-cover translation (see p.36), the proportion of the geomorphological or glaciological literature available to the non-Russian reader is very small. Similar problems face anyone interested in the regional literature of France or Germany; it is necessary to be familiar with the relevant language. On the other hand, many smaller or more remote countries tend to make a greater effort to gain a wider readership by publishing at least part of their literature in English, or another major foreign language, and also by offering extended summaries in English, especially where their national language is read by few others. Thus English allows at least an introduction to the regional geomorphology of Poland, French for Romania, and German and English for Finland.

Apart from the language problem, access to the foreign literature also suffers from the limited availability of the local journals. As spending cuts bite into the purchasing power of academic libraries, foreign journals are the first to be cut. Major national collections (in the UK the British Library Document Supply Centre is outstanding) mean that copies can be located, although they are not always easy to consult. There are some universities which specialize in area studies of a particular region, but while this usually means they carry a good range of local journals in the social studies area, including geography, they are far less likely to carry some of the more technical and scientific publications. National society libraries tend to acquire foreign journals from other countries on an exchange basis (this is the basis of the collection of geographical journals from all over the world held by the Royal Geographical Society, for example) and in a similar way, the library of a national geological survey is generally a good place to locate the publications of other national geological surveys. Apart from these particular locations, that part of the literature which consists of memoirs and reports is particularly difficult to identify, and is often poorly covered in the secondary services. It forms an important part of the regional literature, but some fields (e.g. hydrogeology) have a large report literature.

While a division of articles into those of local interest and those of relevance to a wider audience may seem sensible, it is actually very disadvantageous to the geomorphologist. Interpretations of the geomorphological significance of studies of an area, whether large or small, are rarely convincing when detached from the detailed evidence on which that interpretation is based. The interpretation of field evidence is rarely entirely straightforward, and different geomorphologists will bring very different concepts to the interpretation of the landscape; in these circumstances the more that is known of the characteristics of an area from which deductions have been made about the evolution of wider areas, the better. Yet the pattern of the literature is such that this detailed material is likely to be difficult of access, for it will be in the local literature and in the local language. All too often the more general articles are unconvincing, yet access to the field evidence could well change the picture. To some extent these issues are resolved by the chance to see areas or exposures in the course of travel, or on excursions associated with an international conference or symposium, but this is rare, and confined to a very few people. For most scientists, and for most parts of the world, it is necessary to take the field evidence on trust and rely on the partial descriptions of it to be found in the literature, or acquired from maps or aerial photographs. Fortunately there are some signs that the importance of the field evidence is becoming recognized, but this is still true of too few countries, and the result is that accessible detail about the world's physical landscape remains restricted to a remarkably small part of the world.

Some insight into the scattered nature of the literature of this part of the earth sciences may be gained from Table 17.1. This analyses the literature abstracted in 13 different sections of a secondary journal (*Geographical Abstracts*) over 10 issues, from

Table 17.1

Subject Area	% non-journal items	No. of journals	Average no. items per journal	Most refs	% refs from top 10% journals
Regional physiography	25.1	133	1.53	15	33.1
Fluvial	19.9	136	1.56	7	27.9
Slopes	29.4	99	1.56	15	36.2
General hydrology	27.4	96	1.57	13	33.0
Karst	6.9	102	1.70	6	29.4
Coasts	15.1	95	1.83	22	38.4
Periglacial	13.9	97	1.92	24	45.8
Soil erosion	22.0	85	2.05	19	35.0
Glaciology	10.9	103	2.54	45	55.4
Glacial landforms and sediments	17.3	97	2.66	25	34.6
Quaternary; tropics and sub-tropics	4.4	48	2.71	34	46.3
Run-off and hydraulics	24.4	119	2.82	35	37.0
Hydrogeology	34.3	106	3.16	56	36.7
Great Britain — landforms	10.4	63	3.54	34	44.5
Average	20.1	90	2.20	25	38.7

1982 and 1983, together with the relevant regional literature on Great Britain. The total number of items per topic ranges from a little over 100 to more than 400. The topics are ordered by column 3, which divides the number of items appearing in periodicals by the number of different periodicals cited. In most cases, a comprehensive search of the journal literature would need to cover at least 150 titles, although about half of these would fail to provide an article in any one year. For example, of the 99 titles providing at least one reference on 'Slopes' in this period of 20 months, 48 provided only one relevant article, despite the fact that these 48 include a dozen of the major geomorphological journals. Of the other 36, it may be assumed that, in a sample from a different period, further journals would appear to take their place. Staying with the 'Slopes' set, it can be seen that the highest number of articles in one journal is 15, and that the top 10 journals include just over one-third of the total number of journal articles (154).

The reader may wonder why the journal literature in this field is so disorganized. So far 9 general journals have been mentioned; of these only three provided any articles on 'Slopes' in this 20 month period, yielding a total of just four articles between them. These are general journals, and despite their overall importance they cast their net so widely that they cannot contribute many papers to each sub-field of geomorphology. Only in a few journals is there any sort of concentration of articles on 'Slopes', thereby ensuring a specialist readership. The top score of 15 is a special issue of the *Japanese Geomorphological Union Transactions*, while 10 papers appeared in *Earth Surface Processes and Landforms*, 11 in *Geologia Applicata e Idrogeologia* and 17 in two international symposia. The rest are widely scattered, and it may help to understand why.

The two main factors are 1) national loyalty — there is a strong tendency for authors to publish in the journals, and often also the language, of their home country, even if they write on a systematic topic like slope evolution: 2) if, in addition, the paper is regarded by the author as a regional study, rather than a contribution to the systematic literature on slopes, then the case for local publication is increased over and above questions of national loyalty, for, as has been mentioned, most local articles get local publication. To some extent, the more significant articles are likely to gravitate towards the main journals, but this is as likely to yield an isolated article from (as in this case) the *Bulletin of the Geological Society of America*, or the *Journal of Geology*, as one of a larger group in, say, *Earth Surface Processes and Landforms*.

Given these influences and the pecking order of journals, as perceived by their contributors, it seems unlikely that much can be done to improve the situation. Cooperation among editors could move articles to more appropriate journals, but rejection of articles by inappropriate journals becomes less likely as the quality of the papers improves — yet it is just the better articles which are so annoying to lose sight of in the chaos of present publishing arrangements. Interestingly, the result of these influences is that the regional literature is less scattered than the systematic; as Table 17.1 shows, the worldwide coverage of 'regional physiography' is the most diffuse, but the more constrained scope of 'Quaternary; tropics and sub-tropics' gives an average of 2.71 papers per journal, while 'Great Britain — landforms' scores highest at 3.54. These last two also score well on the yield of the top 10% of journals. Nevertheless, they are outclassed in some respects by the unusually cohesive and organized fields of 'Glaciology' and 'Hydrogeology', with very high scores for the top journals, a high score for the top 10% for 'Glaciology', and a high average for 'Hydrogeology'.

The role of secondary sources is obvious enough in this situation, but it is worthwhile making a special comment on symposia proceedings and similar collections of papers. In many of the sub-fields covered in Table 17.1, the highest yield is not from a journal, but from a special collection of papers, or, in two or three cases, a journal with a special issue devoted to one topic. Examples include the International Association of Hydrological Sciences' Florence symposium, *Erosion and Sediment Transport Measurement (Publication* **133**, June 1981) with 18 papers on soil erosion (and a further 8 in other sub-fields), a special issue of the *Japanese Geomorphological Union Transactions* with 15 articles (out of more than 20) on slopes, and a "Festschrift" for Lewis Penny, *The Quaternary in Britain*, with 16 relevant chapters appearing both in the Great Britain group and also in 'Glacial landforms & sediments'. Such collections will rarely contain the best papers of a decade; too many are taken out from a bottom drawer for a Festschrift, or dashed off to secure financial support to attend a conference. Yet they are not to be despised, for they constitute a remarkably convenient quarry of papers which, with their references, form a useful starting point for a new research topic, or for a teacher rewriting lecture notes. This is particularly true of the more specialized symposia (rather than the major conferences), for usually these are designed to bring together researchers in an active field, and are thus likely to contribute a lot of fresh and useful material.

The final point to be stressed in these introductory remarks is that the literature has a very long life. This contrasts with the situation in most other scientific fields where much of the information published 10 years ago has already been superseded. Table 17.2 was produced for the first edition of this book, and showed the 'three-quarter lives' (period of time during which the most recent three-quarters of literature cited in recent publications was published) of the literature of some fringe earth science areas. However, in the last decade the position of geomorphology has become more complex; the long periods now apply primarily to landform studies (where regional description still has an important role), while the more fashionable study of processes prides itself on citing a relatively recent literature. Several recent issues of the *Zeitschrift für Geomorphologie* and *Earth Surface Processes and Landforms* showed a contrast between a three-quarter life of 25 years for regional articles, and 12 for process orientated papers.

Table 17.2 "Three-quarter" lives of literature (in years)

Science in general	9
Hydrology	12
Glaciology	13
Geomorphology	34

The extreme values were 60 years for one regional paper (i.e. almost one quarter of the references are to nineteenth century literature) and just 3 years for one process study. The accurate description in the literature of a long-vanished exposure remains as pertinent a piece of evidence as the most recent observation, but that observation is likely to help the understanding of chronological relationships of landforms, rather than throw light on contemporary geomorphological processes.

The continued value of the early literature in so many earth science fields creates considerable difficulties, for in practice the only way to locate earlier material is to work through bibliographies or citations in more recent articles. These are not selective, and as any research worker in this field knows, a vast store of irrelevant and outdated material must be read before the critical, factual material which is needed is discovered. There are a few selective bibliographies, but they are rarely annotated and provide no more than an introduction to this early literature. Virtually no modern bibliographic tools include any of the older literature, with the partial exception of GeoRef (see Chapter 6.), and the early volumes of the *Bibliographie Géographique Internationale*.

The more recent literature is fairly well indexed, although, in general, standards are below those of the main scientific subjects. Several of the geological and geographical abstracting and indexing systems referred to in Chapter 5 cover some of the areas of this chapter.

Alongside these international and nominally comprehensive services, there are national or regional bibliographies that can be of great value when information is being sought on a particular area. Many of these are published annually by national geological surveys or other appropriate departments. Some of the most comprehensive come from smaller countries, such as New Zealand, Bulgaria and Poland.

Geomorphology including the Quaternary

Something of the complex nature of geomorphology has already been explained. The fact that many researchers call themselves geomorphologists means that specialist groups of these scientists exist within the major national societies, such as the Geological Society of America, and recently also the Association of American Geographers. Often groups are affiliated to more than one society because of the interdisciplinary position of the subject, thus the British Geomorphological Research Group (BGRG) is affiliated to the Geological Society of London, as well as functioning as a specialized Study Group of the Institute of British Geographers. The Quaternary has its own Quaternary Research Association which publishes guides to regional Quaternary geology and the *Journal of Quaternary* Studies (1986–).

At the international level, both subject areas are now covered by organizations which hold conferences every four years and which also support a range of specialized working groups. The Quaternary has been represented by INQUA (International Quaternary Association) for 50 years, and the 12th International Conference was held in Canada in 1987. BGRG held an inaugural international conference in England (Manchester) in 1985, and the second will be in the Federal Republic of Germany in 1989. These conferences always involve the publication of a volume of abstracts which, while these are of very mixed quality and interest, remain a useful guide to current workers in the field, and often provide a preview of material due to be published at greater length elsewhere. They are also a useful indicator of national patterns of research which vary a great deal. In some cases the papers of the conference are published in full; this was true for example of the 1st International Geomorphological Conference held in Manchester which was published in two volumes called *International Geomorphology 1986* (Wiley, 1987). With almost 200 papers, such volumes represent a useful start in any search for active authors on particular topics. A perennial problem with conferences (and other less regular symposia of this type) is locating the publications, for they come from scattered publishers and are not always covered by the secondary services.

Textbooks

There are many books covering the field of geomorphology, from undergraduate texts to very specialized monographs. A F Pitty's thoughtful *Introduction to Geomorphology* has now been reissued in a revised version, with the more informative title *The Nature of Geomorphology* (Methuen, 1982). There is also a wide-ranging British-orientated review, *Geomorphology; present problems and future prospects*, edited for the British Geomorphological Research Group by C Embleton *et al* (Oxford University Press, 1978). The *Encyclopedia of Geomorphology* is a useful initial source, since it includes brief introductions to many topics, with accompanying references. Modern texts include *Geomorphology*, by A L Bloom (Prentice Hall, 1978), *Fundamentals of Geomorphology*, by R J Rice (Longman, 1977), and a series of advanced texts published by Longmans, including two by C D Ollier, *Weathering*

(2nd edn, 1983) and *Tectonics and Landforms* (1981) and *Slopes*, by A Young. In the same series *Fluvial Geomorphology*, by M Morisawa (1984) is the first full study of that topic since the now classic *Fluvial Processes in Geomorphology*, by L B Leopold *et al* (Freeman, 1964). The modern process-based approach is illustrated by *Process in Geomorphology*, edited by C E Embleton and J B Thornes (Arnold, 1979), *Geomorphological Processes*, by E Derbyshire *et al* (Dawson, 1979), and *Drainage Basin, Form and Process*, by K J Gregory and D E Walling (Arnold, 1973). Another useful text with a modern approach to the subject is *Geomorphology and Time*, by J B Thornes and D Brunsden (Methuen, 1977). An encyclopaedic survey of field techniques is contained in *Geomorphological Techniques*, edited by A Goudie (Allen and Unwin, 1981).

Glacial geomorphology is covered by many texts, some of which are really studies of Quaternary stratigraphy. The classic text, although now becoming rather dated, is R F Flint's *Glacial and Quaternary Geology* (Wiley, 1971), but *Glaciers and Landscape: a geomorphological approach*, by D E Sugden and B S John (Arnold, 1976) is a very useful text, with a strong emphasis on the explanation of glacial behaviour in the landscape. It is not concerned with areas well beyond the direct influence of the ice sheets, and here A Goudie's *Environmental Change* (Oxford University Press, 1977) will be found a very useful little book. An up-to-date review of modern stratigraphic work, emphasizing the continuous ocean record, is D Q Bowen's *Quaternary Geology* (subtitled, 'a stratigraphic framework for multidisciplinary work') (Pergamon, 1978). A useful, if fairly technical, text is *The Physics of Glaciers*, by W S B Paterson (2nd edn, Pergamon, 1981). Finally, especially in view of all that has been said about the value of the older literature, we should mention J K Charlesworth's remarkable survey of the literature of the Quaternary in his two volume work *The Quaternary Era, with special reference to its glaciation* (2 vols, Arnold, 1957).

Various specialist areas of geomorphology covered above have good texts available. Thus the standard work on periglacial topics is undoubtedly A L Washburn's *Geocryology: a survey of periglacial processes and environments* (Arnold, 1979). Karst landforms and processes are covered comprehensively in *Karst Landforms*, by M M Sweeting (Macmillan, 1972), and also in such books as *The Science of Speleology*, edited by T D Ford and C H D Cullingford (Academic Press, 1976), *Morphogenesis of Karst Regions: variants of karst evolution*, by L Jakucs (translated from the Hungarian by B Balkay) (Adam Hilger, 1977), and *Karst: important karst regions of the Northern Hemisphere*, edited by

M Herak and V T Stringfield (Elsevier, 1972). A useful text on arid areas is *Geomorphology in Deserts*, by R U Cooke and A Warren (Batsford, 1973).

Regional texts are far less common in the literature than articles with a regional emphasis. It is in the nature of geomorphology that a regional review is remarkably difficult if it is to be at all comprehensive. For larger areas regional geomorphologies tend to be closely based on geological or structural provinces, witness the classic volumes of the 1930s on the *Physiography of the Eastern (vol. 2, the Western) United States*, by N M Fenneman. For the British Isles, a series of volumes being published by Methuen is now almost complete; these include *Northern England*, by C A M King, *Scotland*, by J B Sissons, and *Southern England*, by D K C Jones: in the same series is *Ireland*, by G L Davies. Other books on whole countries are not common, as has been noted, but many books on the landscape of a region have a charm and an interest far surpassing that of an academic article. A good example is a book on the area around Wellington, New Zealand, *Rugged Landscape: the geology of central New Zealand*, by G R Stevens (A H & A W Reed, 1974).

One very useful group of books deserves particular mention; this is the series *Benchmark Papers in Geology*, published by Dowden, Hutchinson & Ross and distributed by Van Nostrand Reinhold. These facsimile (and sometimes edited) reprints of classic papers with a brief editorial introduction or commentary include a dozen volumes of direct geomorphological interest, such as the following:

Landforms and Geomorphology: concepts and history, edited by A M King (1976)
Slope Morphology, edited by S A Schumm and M P Mosley (1973)
Karst Geomorphology, edited by M M Sweeting (1981)
Glacial Deposits, edited by R P Goldthwait (1975)
Planation Surfaces, edited by G F Adams (1975)
Loess: lithology and genesis, edited by I J Smalley (1975).

The subject is fortunate in having its history well covered, notably in the still-incomplete *The History of the Study of Landforms*, by R J Chorley, *et al* (Methuen, 1964 & 1973). The first volume covers *Geomorphology before Davis*, and the second is a full and thorough study of the work of W M Davis. Interestingly, although his work has been the foundation of the teaching of geomorphology for over half a century, his only text is in German and has never been translated. Until recently students relied on the Dover reprint of his *Geographical Essays*, originally published in 1909, but now P B King and S A Schumm have reconstructed a text from the lectures attended by King, at a summer school, in 1929 (*The Physical Geography (Geomorphology) of W.M.Davis*, Geo Books, 1980). A very useful source of review essays on various topics (together with some of the fuller book reviews to be found in contemporary journals) is *Progress in Physical Geography*, published quarterly by Arnold. A recent issue also contains an analysis by C Embleton of the scope and nature of British geomorphological research and publication over the last 21 years.

Periodicals

Study of the Quaternary is necessarily interdisciplinary. The study of geomorphology is split between those who regard themselves as geographers and those who are geologists. This gives rise to a scattered periodical literature with few specialized journals in either field, although in recent years aspiring editors and ambitious publishers have started to fill in the gaps by providing journals dealiing with geomorphological processes, and with interdisciplinary study of the Quaternary. The older journals include the long-established *Zeitschrift für Geomorphologie* (1967–) from Germany (a high proportion of which is now published in English) and the French *Revue de Géomorphologie Dynamique* (1950–), still entirely in French. The title *Journal of Geomorphology* was used by Douglas Johnson for a journal founded in 1938, but which failed to survive his death (1942), largely as a result of the difficulties caused by the war. A major and long-established Scandinavian journal is *Geografiska Annaler A* (1919–). There is a Russian *Geomorfologiia* (1970–) and a Brazilian *Noticia Geomorfologica*, which uses quite a lot of its space for translations of significant English-language articles into Portuguese. Among the more modern titles, *Earth Surface Processes* (1976–80) has quickly sprung into prominence: its early restriction to process studies was relaxed when the title was amended to *Earth Surface Processes and Landforms* in 1981, but it must be admitted that the new title has made very little difference to the content. Another new journal exploiting the soils/landscape interface, and usually in English although published from Germany, is *Catena*. Like most modern journals it has an international editorial board to encourage a wide range of contributions of international scope.

   In Quaternary studies there are some long-established titles. Although of more significance for glaciology, both the *Journal of Glaciology* (1947–) and the Russian *Materialy Gliatsiologicheskikh Issledovanii Khronika Obsuzhdeniia* (1962–) include articles on Quaternary topics, covering chronology, glacial sediments, and landforms: rather similar in scope is the German journal *Zeitschrift für Gletscherkunde und Glazialgeologie* (1906–). More interdisciplinary areas are covered by a number of journals, including *Arctic & Alpine Research* (1969–), *Boreas* (1972–), *Quaternary Research* (1970–), *Quaternary Science Reviews* (1982–), and the *Journal of Quaternary Studies* (1986–) in English, and also *Eiszeitalter und Gegenwart* (1951–) and *Bulletin de l'Association française pour l'Etude du Quaternaire* (1964–).

Rather than provide a very long list of important journals which should be in any good academic library, it seems best to continue by looking in turn at some major subdivisions of the field. The subfields described are not exhaustive, but do cover the main areas of contemporary research.

Fluvial geomorphology and slopes

The two major journals are:

Earth Surface Processes and Landforms (1976–)
Zeitschrift für Geomorphologie (1967–)

For fluvial geomorphology, papers are frequently to be found in:

Water Resources Research (1965–)
Physical Geography (1980–)
US Geological Survey Professional Papers (1902–)

and for slopes in:

Geologia Applicata e Idrogeologia (1966–)
Bulletin Société Géologique de France (1830–)
Geografia Fisica e Dinamica Quaternaria, 2nd ser. (1950–)
Engineering Geology (Amsterdam) (1965–)
Catena (1974–)

Coasts

The main journals are far more 'geological' in flavour; they include

Journal of Sedimentary Petrology (1931–)
Marine Geology (1964–)
Sedimentology (1962–)
Géographie Physique et Quaternaire (1977–)
International Geographical Union, Commission on Coastal Environment, Reports
Vestnik Leningradskogo Universiteta, Seriia Geologiia-Geografiia (1956–)
Coastal Research (1969–)
Estuaries (1978–)
Estuarine, Coastal and Shelf Science (1973–)
(Die) Küste (1952–)
US Army, Coastal Engineering Research Center, Technical Reports (and other series) (1950–)
Geologie en Mijnbouw (1931–)

Karstic landforms

NSS Bulletin (1940–)
Cave Science, Transactions British Cave Research Association (1947–66) (Continued in *Journal of the British Speleological Association* (1967–))

Revue de Géographie Alpine (1920–)
Cave Geology (1975–)
Grottan
Carsologica Sinica
Helictite (1962–)
Southern Caver

Glacial landforms and sediments (those asterisked also have much periglacial material)

Boreas* (1972–)
Materialy Gliatsiologicheskikh Issledovanii Khronika Obsuzhdeniia* (1962–)
Geografiska Annaler A* (1919–)
Canadian Journal of Earth Sciences (1964–)
Quaternary Studies in Poland (1979–)
Journal of Glaciology (1947–)
Bulletin of the Geological Society of America (1890–)
Quaternary Research (1971–)
(The) Quaternary Research (Tokyo) (1957–)
Quaternary Newsletter (1970–)
Geological Journal (1964–); formerly *Liverpool & Manchester Geological Journal* (1952–63)

Periglacial geomorphology (see also those with asterisks in group above)

Biuletyn Peryglacjalny (1954–)
Engineering Geology (Amsterdam) (1965–)
Arctic and Alpine Research (1969–)
Canadian Geotechnical Journal (1963/64–)
US Cold Regions Research and Engineering Laboratory, Reports
Géographie Physique et Quaternaire (1946–)
Soviet Hydrology: Selected Papers

Quaternary of the tropics and sub-tropics

Palaeoecology of Africa and the Surrounding Islands and Antarctica (1966–)
Striae (1975–)
Modern Quaternary Research in Southeast Asia (1975–)
Quaternary Research (1971–)
Quaternaria (1954–)
Revue Tunisienne de Géografie (1977–)
Palaeogeography, Palaeoclimatology, Palaeoecology (1965–)
Geografia Fisica & Dinamica Quaternaria, 2nd ser. (1950–)
Journal of Biogeography (1974–)
Cahiers ORSTOM, Série Géologie (1969–)

A regional approach to the literature leads one to the national journals. The following lists are arranged under the main countries of Europe with a sizeable and reasonably accessible literature. For other areas, material can be traced through various geographical or regional indexes in the secondary literature.

Great Britain:

Proceedings Geologists' Association (1859–)
Quaternary Newsletter (1970–)

Scottish Journal of Geology (1965–)
Soil Survey of England and Wales, Record (1970–)
New Phytologist (1902–)
Geological Magazine (1864–)
Geological Journal (1964–) (see also under 'Glacial landforms & sediments' above)
Transactions, Institute of British Geographers (1933–)

France:

Bulletin de l'Association des Géographes Français (1924–)
Revue de Géographie Physique et de Géologie Dynamique (1928–39; 1957–)
Revue de Géomorphologie Dynamique (1950–)
Hommes et Terres du Nord (1972–)
Information Géographique (1936–)
Revue de Géographie des Pyrénées et du Sud-Ouest (1930–)
Cahiers de Géographie Physique de Lille
Annales de Géographie (1891–)
Revue d'Alsace (?1850–)
Revue Géographique de l'Est (1960–)
Mediterranée (1960–)

A feature of the organization of the French literature which is worth appreciating is that each regional (university) journal not only collects the appropriate regional literature within France, but also accepts articles on areas beyond France, according to a particular pattern. Thus *Hommes et Terres du Nord* has articles on England, while *Mediterranée* covers not only the French Mediterranean, but the area generally. This pattern is long-standing, but seems little appreciated outside France. It is interesting that no other country has adopted such a logical pattern, despite its obvious advantages for all research workers with a regional focus.

Germany:

Here again the university series dominate the local literature, including;

Mannheimer Geographische Arbeiten (1977–)
Tübinger Geographische Studien (1958–)
Frankfurter Geowissenschaftliche Arbeiten, D, Physische Geographie (1982–)
Hamburger Geographische Studien
Rhein-Mainische Forschungen
Regensburger Geographische Schriften
Heidelberger Geographische Arbeiten
Bochumer Geographische Arbeiten
Kölner Geographische Arbeiten
Berliner Geographische Abhandlungen (1964–)

Fennoscandia:

Terra (1913–)
Norsk Geografisk Tidsskrift (1926–)
Fennia (1889–)
Sveriges Geologiska Undersökning Afhandlingar och Uppsatser (1868–)
Bulletin, Geological Survey of Finland (1971–)

Poland:

There are some national journals, of which the following are useful for Polish geomorphology:

Przegląd Geograficzny (1918–)
Czasopismo Geograficzne (1923–)
Dokumentacja Geograficzna (1956–)
Geographia Polonica (1964–)
Polish Polar Research (1980–)
Biuletyn Peryglacjalny (1954–)

and the following university journals (which are remarkably successful in hiding the town where the university is located!)

Folia Geographica, Series Geographica-Physica (1968–)
Zeszyty Naukowe Uniwersytetu Jagiellonskiego, Prace Geograficzne (Warsaw) (1955–)
Annales Universitatis Mariae Curie-Sklodowska, Sectio B (Lublin) (1946–)

Other eastern Europe countries:

Acta Facultatis Rerum Naturalium Universitatis Comenianae, Geographica (Bratislava) (1961–)
Acta Universitatis Carolinae, Geographica (Prague) (1965–)
Zpravy Geografickeho Ustavu CSAV
Geograficky Casopis (Bratislava) (1953–)
Studii si Cercetari de Geologie, Geofizica, Geografie: Geografie
Geografija
Problemi na Geografijata (1974–)
Foldrajzi Ertesito (Budapest) (1952–)
Geograficky Casopis
Foldrajzi Koslemenyet (Budapest) (1873–)
Revue Roumaine de Géologie, Géophysique et Géographie: Série de Géographie (Bucharest) (1974–)
Godisnik na Sofijskija Universitet, Geologo/Geografski Fakultet; Geografija

Abstracting services and bibliographies

Three abstracting services include geomorphology, each covering its own language material best. They are *Referativnyi Zhurnal. Geografiya* (1957–); the French service published by BRGM as section 226, *Hydrologie, Géologie de l'Ingenieur, Formation Superficielles* of *Bulletin Signalétique*, and *Geographical Abstracts*, Part A *Landforms and the Quaternary* (although Part E, *Sedimentology*, covers coasts) formerly Geo Abstracts (1972–1985), *Geographical Abstracts* (1966–71) and *Geomorphological Abstracts* (1960–65). However, in all cases these publications have not existed long and are of most value for contemporary and recent material. *Geographical Abstracts*, Part A publishes over 3000 abstracts a year, and the Russian volume covering geomorphology carries a similar number, but the overlap of material between the two is remarkably small. To some extent the Russian publication includes rather a lot of ephemeral material, and many of the items included are very difficult, or impossible, to get hold of outside the USSR. The French service, in turn, is strongest in the French language literature (this includes Francophone Africa) and now co-operates with AGI's GeoRef to secure as comprehensive coverage as possible of the geological literature. Inevitably, since both services are geologically based, they are weaker on the geographical fringe of geomorphology than *Geographical Abstracts*, while GeoRef has keywords but no abstracts.

The other geographical bibliographies do not include abstracts. These are the French compilation *Bibliographie Géographique Internationale* (CNRS, 1891–), an annual volume running some time behind the current literature, and the American Geographical Society's (AGS) *Current Geographical Publications* (1938–). This is effectively a catalogue of acquisitions to the AGS library (with the separate papers from the journals all cited) and has the disadvantage for the geomorphologist that it is organized primarily on a regional basis, with systematic division within each country.

Because they are available online, two geological bibliographies will be found useful by geomorphologists. These are GeoArchive (Geosystems, London) and GeoRef (USA) (see Chapter 6). The printed version of GeoRef is the *Bibliography and Index of Geology*, and contains three divisions of 'Surficial Geology' which are of interest to geomorphologists: Geomorphology, Quaternary geology, and Soils. Some relevant items will also be found in 'Engineering geology' and 'Environmental geology'. The three Surficial Geology sections contain almost 6000 references in a year, and there are as many again in the 'Engineering geology' section, so it will be seen that this is a major source of references, albeit without accompanying abstracts. Since the total number of references in the Geo Systems bibliography is similar, it is likely that this too contains an appreciable number of references on geomorphology, Quaternary geology and soils. Also *Geographical Abstracts* (Geo Abstracts) has recently been mounted as an online database, called Geobase.

National geographical or geological bibliographies are also useful sources of geomorphology, and a number have been published: for instance the Poles publish two volumes each year (one covering geography, the other geology), providing a most comprehensive listing of their literature. There are also some older bibliographies which can be of value, for example the compilation of the UK literature assembled for the International Geographical Congress in London in 1964 — *A Bibliography of British Geomorphology*, compiled by K M Clayton (George Philip, 1964).

Glaciology

Water in the solid state occurs on earth as glaciers and ice caps, and their study has developed into a small, though remarkably distinctive, sub-discipline. In water balance terms, glaciers may be regarded as simply part of the store of freshwater on earth — indeed they include something like 98% of the freshwater on earth, and in many arid areas are an important source of irrigation water. But the distinctive behaviour of glaciers, the study of their mass balance, and also the complex matters of the way they move, the geological sediments they produce, and the landforms they form, all serve to define an independent field of study. There is no independent international association within the UNESCO family, but the International Association of Hydrological Sciences has a Snow and Ice Commission, and this holds symposia and publishes volumes based on them.

Textbooks

Books on glaciology and glacial geomorphology were covered in the geomorphology section, see page 403.

Periodicals

In recent years scientists have carried out much glaciological work in Antarctica, and the results of this have been published in a series of special journals, as well as in the normal literature. The American Geophysical Union has an *Antarctic Research Series*, the US Army's Cold Regions Research and Experimental Laboratory (CRREL) has its *Reports* and other series, and there are the publications of the Ohio State University's Institute of Polar Studies. Quick publication of results, and the recording of work in progress, is made in the *Antarctic Journal of the United States* (1966–) and the equivalents in other countries, notably the *Informatsionnyi Biulleten' Sovetskoi Antarkticheskoi Ekspeditsii* (1958–), the *Polar Record* (1931–) and the various *Bulletins* with similar names from the Japanese, South African, Australian and New Zealand contingents in the Antarctic, as well as from Chilean and Argentinian sources. The longer British papers are often to be found in the *British Antarctic Survey Bulletin* (1963–), while work on Greenland is included in the Danish publication *Meddelelser om Grønland* (1879–) which since 1980 has appeared in three series, *Bioscience, Geoscience* and *Man and Society*.

   The theoretical aspects of glaciology involve the physics of ice. Some of this fundamental work is published in the *Journal of*

Glaciology (1947–), but much occurs in *Nature, Proceedings of the Royal Society of London, Journal of Applied Physics, Physical Review Letters, Reviews of Geophysics & Space Physics* and *Low Temperature Science A, Physical Sciences.*

The geomorphological literature connected with glaciation has been covered in an earlier section of this chapter.

Abstracting services and bibliographies

Glaciology is covered in part by sections in *Geographical Abstracts, Part A* (1972–), and in *Meteorological and Geoastrophysical Abstracts* (1950–). Coverage in this latter series seems to arise from the appearance of glaciological material in journals of meteorological interest, and the number of abstracts in these sections has declined in recent years to the point where this is no longer a useful source. *Geographical Abstracts* has about 200 abstracts a year actually classified as glaciology, but many of the papers on periglacial topics, glacial landforms and deposits, and on the general climatology of the Quaternary (including the topic of climatic change) are to be found under the broader glaciological headings of other secondary sources. Similarly GeoRef (and thus the *Bibliography and Index of Geology*) has sub-sections on glacial features, glaciation, and palynology in its section 'Surficial Geology: Quaternary geology', which has about 2750 abstracts each year.

Until 1979 a very useful source, although without abstracts, was the summary of recent glaciological literature in the last few pages of each issue of the *Journal of Glaciology*. In that year this totalled about 1250 entries, if the listings of separate papers within symposia publications and conference proceedings are included. They were divided into ten sections covering snow, the physics of ice, and glaciers and ice shelves, as well as the more marginal (but equally well-represented) areas of frost action, frozen ground and permafrost. Unfortunately these items have never been indexed, so they represent a very awkward store of the older references, likely to be explored only by the most determined researcher. The older issues have a sentence or so of annotation, but the last issues carry citations only.

Another useful guide to the literature is the annual bibliography produced by CRREL under the title *Bibliography on Snow, Ice and Frozen Ground*. Between 1951, when it started, and 1969 this covered over 28,000 abstracts. The publication is available in microfiche as well as hardcopy. In 1953, the US Department of Defense started the *Arctic Bibliography* and this has appeared annually since then, although publication is now the responsibility of the Arctic Institute of North America, based in Canada. It has the feature that new material is added as it comes to light, even if it is not a current addition to the literature. This compilative aspect is also a feature of the similarly comprehensive *Antarctic Bibliography* (1951–61 *and* 1965–) published by the Library of Congress. This is now issued as a cheaply reproduced monthly publication, which is subsequently organized into annual volumes. The cover now runs from 1951; a retrospective volume, covering 1951–1961, was published in 1970. Both these bibliographies cover all material on the Arctic and Antarctic, including a good deal of biological work, medical research, ionospheric physics, political geography and international law, to give some idea of their broad scope. There is also a literature list published as a supplement to *Polar Record* (1931–) and called *Recent Polar and Glaciological Literature* (1981–).

Relevant papers are scattered through quite a number of journals, often uncovered by earth science secondary services. It is here that the *Science Citation Index* comes in handy, for by using a classic article by Nye, or more recent articles by such authors as Lliboutry or Weertman, recent articles citing these sources can be located quickly.

Hydrology and hydrogeology

Hydrology is a relatively recent addition to the various sub-disciplines associated with the earth sciences. Until well into this century much of the published work could be regarded as part of geomorphology, although an applied area concerned with irrigation, drainage and flood control was developed within civil engineering. To some extent the emergence of an independent discipline of hydrology has led to some merging of the applied and academic aspects of the subject, although the literature tends to reflect the traditional divisions. As a result, quite a large proportion of the hydrological literature is not in earth science journals.

Textbooks

A useful survey of the historical aspects of hydrology is A K Biswas' *History of Hydrology* (North Holland, 1970) which traces the history of ideas about the water cycle, including early attempts at measurement back to about 500 BC. An introductory text in wide use is E M Wilson's *Engineering Hydrology* (2nd edn, Macmillan, 1974), while the most commonly used text in departments of geography will be *Principles of Hydrology*, by R C Ward

(2nd edn, McGraw-Hill, 1975). A new text by E M Shaw *Hydrology in Practice* (Van Nostrand Reinhold, 1982) may be added to these well-established British textbooks. More advanced texts naturally tend to deal with more restricted topics, and include *Hillslope Hydrology*, edited by M J Kirkby (Wiley, 1978), *Systematic Hydrology*, by J C Rodda *et al* (Newnes-Butterworth, 1976), and *Soil and Water*, by D Hillel (Academic Press, 1971). American texts include P S Eagleson's *Dynamic Hydrology* (McGraw-Hill, 1970), *Modern Hydrology*, by R G Kaufman (2nd edn, Harper & Row, 1972), and a new edition of the long-established textbook, *Hydrology for Engineers*, by R K Linsley *et al* (3rd edn, McGraw-Hill, 1982).

Hydrogeology has produced a rich, if restricted, crop of new books in response to the growth of interest in the subject over the last decade. Without doubt the best is *Groundwater*, by R A Freeze and J A Cherry (Prentice-Hall, 1979), but the new edition of the classic text by D K Todd *Groundwater Hydrology* (2nd revised edn, Wiley, 1980) is also excellent. A useful, if less comprehensive, undergraduate text is C Fetter's *Applied Hydrogeology* (2nd edn, Merrill, 1988). Computer applications in hydrology are reviewed by G Fleming in *Computer Simulation Techniques in Hydrology* (Elsevier, 1975) and G F Pinder and W G Gray discuss numerical simulation in subsurface hydrology and hydrogeology in *Finite Element Simulation in Surface and Subsurface Hydrology* (Academic Press, 1977). *Ground Water and Wells*, by F G Driscoll (Johnson, 1986) is a much expanded version of a practically-orientated previous edition.

Finally, there is the topic of water resources. The UK water industry in particular has attracted much study, and books on it include E Porter's *Water Management in England and Wales* (Cambridge University Press, 1978), D J Parker and E C Penning-Rowsell's *Water Planning in Britain* (Allen and Unwin, 1980), and *Water in Britain: a study in applied hydrology and resource geography*, by K Smith (Macmillan, 1972). This last book has inevitably been outdated by reorganization in the water supply industry, but remains the best short account of the physical basis of water resources in Britain. A reflection of the increasing interest in the subject is the multi-volume *Water Practice Manual*, edited by T W Brandon (Institution of Water Engineers and Scientists, 1986).

Periodicals

The International Association of Scientific Hydrology was founded in 1922 and now holds regular conferences, as well as sponsoring specialist international symposia, the proceedings of which are published as special publications. It changed its name to the International Association of Hydrological Sciences (IAHS) in 1970. This society publishes a primary journal, rather awkwardly titled in two languages as *Hydrological Sciences Journal/Journal des Sciences Hydrologiques*, although often just the English part of the title is quoted. This journal was called a *Bulletin* until 1982. The increased activity of this society in recent decades has been paralleled by the emergence of two major journals, *Water Resources Research*, published since 1965 by the American Geophysical Union (with almost 1750 pages in its most recent complete volume) and the *Journal of Hydrology* (1963–) published by Elsevier and presenting much of the British material. The continuing evolution of hydrology is marked by the recent decision to found a British Hydrology Society, although it is not planning to publish a primary journal at present.

The American Society of Civil Engineers (ASCE) has a very extensive range of journals, and each of its specialist divisions has its own journal. Here we may note the *Journal of Irrigation and Drainage Engineering*, the *Journal of Hydraulic Engineering*, and the *Journal of Water Resources Planning and Management* all appearing under these titles since 1983 but having earlier titles from 1873: even the *Journal of Environmental Engineering* has an appreciable number of articles in hydrology. The British equivalent *Proceedings of the Institution of Civil Engineers* (1818–) will be found useful, and there is also the *Journal of the Institution of Water Engineers and Scientists* (1947–). A new development is the rise of journals in the engineering geology (or applied geology) field, and these include hydrology alongside soil mechanics: examples are *Engineering Geology* (Elsevier) (1965–) and the Geological Society of London's *Quarterly Journal of Engineering Geology* (1967–).

In addition to those mentioned above, significant numbers of articles will also be found in the following journals:

Transactions of the American Society of Agricultural Engineers (1906–)
Bulletin du Bureau de Recherches Géologiques et Minières, Section III (1961–)
Nordic Hydrology (1970–)
Proceedings of the Annual Eastern Snow Conference (1943–)
Civil Engineering, ASCE (1930–)
Journal of Environmental Quality (1971–)
Arctic and Alpine Research (1969–)
Earth Surface Processes and Landforms (1981–; formerly *Earth Surface Processes*, 1976–80)

Particular mention should be made of *Soviet Hydrology: Selected Papers*. This translates, in full, significant articles from a number of Soviet journals, including *Trudy Gosudarstvennogo Gidro-logicheskogo Instituta, Izvestiia Akademiia Nauk SSSR, Seriia Geograficheskaia,* and *Vestnik Moskovskogo Universiteta, Seriia Geografiia.* It also includes English abstracts of other articles and so provides an important access to the Soviet literature.

Hydrogeology is quite a precisely-defined subject area with obvious practical applications, both in the exploitation of ground-water supplies and, more recently, artificial recharge and waste disposal. There is an International Association of Hydrogeologists which meets regularly, and also on the occasions of the International Geological Congress. Some of the literature of hydrogeology is to be found in geological journals, but rather more in hydrological journals, especially *Water Resources Research* and the *Journal of Hydrology* (see previous section). There are a few specialist journals, such as *Ground Water,* and *Water Well Journal.* Some of the literature will be found in journals primarily concerned with the use to which the water might be put — thus the American Society of Civil Engineers (ASCE) *Journal of Irrigation and Drainage Engineering* will include articles on the extraction of groundwater for irrigation. The physics of groundwater movement through rocks is covered by another section of the extensive ASCE series, *Journal of Hydraulic Engineering.* However, many articles are to be found in more general journals.

Significant numbers of papers have recently appeared in the following journals:

Bulletin du Bureau de Recherches Géologiques et Minières, Section III (1961–)
Soviet Hydrology: Selected Papers
Hydrological Sciences Journal
Environmental Science and Technology (1967–)
Quarterly Journal of Engineering Geology (1967–)

In addition to these, most national geological surveys publish relevant material, often in separate series (e.g. *US Geological Survey, Water-Supply Papers*), while a recent issue of *Science of the Total Environment* (vol. **21**, 1981) was wholly devoted to groundwater.

Abstracting services and bibliographies

With regard to secondary sources, hydrology is not widely served. The coverage in *Meteorological and Geoastrophysical Abstracts* is not at all good and has declined in recent years as that journal has concentrated on the meteorological literature. Annual coverage is only about 250 abstracts, and about the same number appear each year in the 'Hydrology and Glaciology' section of *Physics Abstracts.* Fortunately the same period has seen a considerable improvement in the coverage in *Geographical Abstracts,* Part B, which now includes about 1600 abstracts a year. These are organized into the following sections: General hydrology; Observation and instrumentation; Precipitation assessment; Evaporation; Interception, soil moisture and groundwater; Runoff and hydraulics; Lakes and reservoirs; Glacial hydrology; Surface water quality; Land use effects; and Water resources. In addition, there is a section in *Geographical Abstracts*, Part F *Regional and Community Planning* on the planning of water resources and on environmental problems associated with them.

Since 1968, *Selected Water Resources Abstracts* has provided the most extensive coverage, although quite a high proportion of the items come from the NTIS database. It is published by the Office of Water Research and Technology of the US Department of the Interior, appears monthly, and includes about 6000 abstracts a year. It covers a limited range of sources thoroughly, but does not range widely through the less prolific journals. While its coverage may seem very wide compared with other sources (and there is no doubt that it is a most useful source), it includes a wide range of abstracts on such topics as water quality management and protection, (which divides between the ecology of water bodies and the treatment of wastes) and engineering works — which spreads from hydraulics into dam and spillway design. It is, as its title suggests, a general source about water, not simply a secondary source for hydrology and hydrogeology. Perhaps half its abstracts (i.e. 3000) should be regarded as falling within the field of hydrology, *sensu lato*. Those abstracts classified as hydrogeology (section 2F, groundwater) total about 500 each year.

The *Bibliography and Index of Geology*, described in Chapter 5, has a section called 'Hydrogeology and hydrology', and this has about 3300 citations (with keywords) a year. It is likely, given its sources, to provide a better coverage of hydrogeology than *Geographical Abstracts* but to do less well for hydrology. In the UK, the Water Research Centre publishes *WRC Information*, a weekly publication consisting of abstracts arranged under the following headings: Water resources and supplies; Water quality; Monitoring analysis of water and wastes; Water treatment; Underground services and water use; Sewage; Industrial effluents; and Effects of pollution. Thus it has a similar range to the *Selected Water Resources Abstracts*, though each emphasizes its national literature. The British publication covers about 4000 abstracts a year, and is paralleled by AQUALINE, a computerized retrieval system.

Some national bibliographies have been published under the auspices of the International Association for Scientific Hydrology. We may also note *Hydata* (1965–), a monthly current awareness bulletin covering both periodical and non-periodical literature. It has been published by the American Water Resources Association since 1965 and includes about 900 citations a year. The annual index is published under the title *Hydor*. The same organization publishes *Hydro-Abstracts* (1980–), which gives abstracts of the articles previously announced in *Hydata*. Abstracts are available in loose leaf or card form, and in 49 subject sections.

Hydrological data

One aspect of hydrology which should be mentioned here is the availability of run-off or water-balance data. The direct measurement of discharge in rivers involves the construction of weirs or flumes, and in most countries has been restricted, until recently, to sites being investigated for power generation or water supply. Such data as there are, have been published on an annual basis by organization, or occasionally on a national basis, and have not been collected systematically by most libraries.

An early activity of the IAHS was to publish sources for national hydrological data, and many countries are now realizing the importance of collecting information on run-off on a systematic basis. The steadily improving coverage of the *Surface Water Yearbook of Great Britain* is an example of this trend, although the recent reorganization of the Water Data Unit (now incorporated in the Institute of Hydrology) may spell the end of this series. The US has the finest records of any country, and a vast amount of data is published every year, together with periodic guides to its scope: thus the US Geological Survey has published an *Index to Catalog of Information on Water Data* in two volumes, *Surface Water Stations* and *Water Quality Stations*.

There is still no convenient source of data for larger areas than single countries, even for the major rivers of the world, and most compilations of this type appear as maps rather than statistical tables (e.g. the map *Superficial Drainage and Hydrology*, scale 1:10 million, compiled by Dr R Beno and published in 1965 by Cartographia Vallalat, Budapest). As part of the programme for the International Hydrological Decade, most countries have established centres for the collection and publication of data. These are listed in *List of Hydrological Decade Stations of the World (Studies and Reports in Hydrology*, No. **6**, UNESCO, 1969). During the International Hydrological Decade (1965–1974) UNESCO published many volumes of data, but few seem to have been continued or updated since then.

CHAPTER EIGHTEEN

Meteorology and climatology

R A S RATCLIFFE

In the broadest sense earth sciences must include the science of the earth's atmosphere and so the inclusion of a chapter on meteorology in this book is clearly appropriate. Meteorology overlaps into the earth sciences since geological phenomena such as volcanic eruptions can affect the climate of the earth: likewise the distribution of land and sea, and the position of major mountain ranges, play a large part in determining climate. The subject overlaps also into many other disciplines, e.g. atmospheric physics, atmospheric chemistry, agriculture, horticulture, forestry, etc., and there is an extensive literature on these subjects.

Oceanography is also closely linked with meteorology since climate is largely determined by ocean currents and ocean temperatures. On the longer time scale the chemistry of the oceans, by the uptake of carbon dioxide and other gases from the atmosphere, may well affect the earth's climate.

Hydrology, a subject already reviewed in Chapters 16 and 17 is closely related to meteorology. Meteorologists and hydrologists need to co-operate on such matters as flood protection and prevention, design of urban drainage systems, siting of reservoirs, etc. A knowledge of the relevant meteorological data is vital in such work.

In meteorology the boundary layer (i.e. the lowest layers of the atmosphere) is particularly important since it is here that there are the greatest fluxes of momentum, heat, and water vapour into and out of the atmosphere from the various surfaces of the globe (oceans, forests, deserts, etc.) but the high atmosphere has recently attracted much more attention than in the past. This is mainly due to the advent of meteorological satellites which now regularly orbit the earth and provide information in various forms. Not only can satellites maintain continuous photographic surveillance of the whole earth in visible light but they also monitor other radiation from the earth, notably infra-red radiation and radiation in those wave lengths which are absorbed by water vapour in the atmosphere. Infra-red pictures give information on the temperatures of the radiating bodies below and hence can be used as indicators of sea and land surface temperatures and also the temperatures of cloud tops. By measuring the absorption of radiation by atmospheric water vapour, satellites are able to give an indication of the distribution of water vapour in the atmosphere.

Climatology has also undergone a revolution in recent years: no longer is it a mere collection of statistics since it is now realized that the climate at any one place is not constant but is continually varying on different time scales. It is now known that the climate of the whole earth may well vary owing to such factors as changes in the composition of the atmosphere (notably carbon dioxide increase due to increased fossil fuel consumption), volcanic eruptions, changes in the radiation balance of the earth (due, for instance, to cutting down tropical forests).

It is clear that accurate meteorological data and knowledge of the science of meteorology is important for many purposes in the modern world. The science has developed rapidly in recent years and quotations in the literature are now commonly less than about twelve years old and there is a predominance of a few particularly important journals.

Organizations

The World Meteorological Organization

Since the global atmosphere covers all countries it is not surprising that meteorology has always been the most international of sciences. As early as 1873 nations got together to sign the Treaty of Vienna which paved the way for the standardization of instruments, observations and techniques of observing and also for the international exchange of the basic data. This Treaty led to the establishment of the International Meteorological Organization (IMO) and, after World War II, to that of the World Meteorological Organization (WMO).

Almost all countries now belong to the WMO since most realize that to forecast the weather it is necessary to co-operate with neighbouring countries in the exchange of observations. Indeed it is now realized that data from the whole globe are necessary in order to predict the weather for several days ahead at any location.

WMO, based in Geneva, is governed by Congress and an Executive Committee under which are nine technical Commissions. It is responsible for laying down international standards for instruments and observing practices, for defining the requirements for the collection, broadcasting and archiving of meteorological data, and it has a major hand in co-ordinating meteorological research, particularly when atmospheric experiments to cover large areas of the globe are planned.

WMO has prepared numerous publications for the general guidance of member states. These include:

Guide to Meteorological Instrument and Observing Practices

Weather Reporting: (a) Observing Stations
                        (b) Data Processing
                        (c) Transmissions
                        (d) Information for Shipping

Technical Regulations: (a) General
                         (b) Meteorological Service for
                             International Air Navigation
                         (c) Hydrology

Catalogue of Meteorological Training Publications and Audio-Visual Aids.

Other important WMO publications may be found by referring to *Publications of the World Meteorological Organisation* (WMO, 1980).

WMO was responsible for originating and planning the activities of the Global Atmospheric Research Programme (GARP) and a number of large scale experiments have recently taken place, including the Atlantic Tropical Experiment (GATE) and the Monsoon Experiment (MONEX). Also it has set up recently the World Climate Programme designed to study all aspects of world climate and particularly aimed at finding out the basic causes of climatic change, with a view to making reliable predictions of future world climate.

National meteorological services

In almost all countries the meteorological service is government controlled. Each country maintains a network of observing stations which operate within the guidelines laid down by WMO, as amplified by the country concerned. Observations are passed by the quickest possible means to a central collecting centre (in the UK, this is at Bracknell); the collecting centre then passes on all observations to a regional centre which exchanges its information with other regional centres. This system is very efficient and, as a result, most national meteorological services receive meteorological observations from much of the northern hemisphere within an hour or so of the observations being recorded. These data are plotted on charts and also processed as input for computerized weather forecasting in a number of countries, including the USA, UK, Germany and Sweden. These countries produce computerized forecasts in chart form for surface and various upper levels for up to five days ahead, which are disseminated over the global telecommunications system to most national meteorological services. Current meteorological information of this type is used by civil and military aviation, and participants in other outdoor activities such as sailing, mountaineering, etc.

Major outlets for weather forecasts and climatological information in Britain are the Weather Centres situated in London, Glasgow, Manchester, Newcastle, Nottingham, Southampton, Cardiff and Leeds. These centres handle many thousands of requests from the public and a few other main meteorological offices are able to accept similar enquiries; these enquiry points are listed in *Meteorological Office Leaflet No. 1, Weather Advice to the Community*, obtainable free from the Meteorological Office. Television, radio and newspaper forecasts are issued through the Weather Centres.

There is also the Weathercall. This is a service provided in co-operation with the telecommunications service whereby telephone subscribers may get the latest forecast for their local area by dialling a number. Such forecasts are updated and recorded every few hours.

The Meteorological Office provides a number of specialized advisory services and these include:

(a) *Services to builders*: A special service warning of such items as strong winds, heavy rain, frost and snow which can seriously affect many building operations.

(b) *Offshore weather services*: A service mainly for oil fields and platforms in the North Sea and other sea areas adjacent to the UK. Such concerns need to be warned of conditions which threaten the safety of oil platforms either *in situ* or when being towed to site and erected. Helicopter and diving operations related to this type of work also need meteorological information.

(c) *Services to agriculture*: The Meteorological Office maintains a number of offices which liaise with the Agricultural Advisory Service and provide information as to when conditions are conducive to the spread of various animal and plant diseases, so that farmers may be warned and may take timely preventative action. This service also undertakes to notify when spells of three days of dry weather (suitable for harvesting) are expected.

(d) *Ship routeing service*: Advice to shipping on the best routes to take in the Atlantic or Pacific to avoid bad weather and/or to save time can be obtained from the Meteorological Office at Bracknell.

(e) *Climatological services*: Climatological information is required for many purposes, e.g. the prevailing winds need to be known before siting aircraft runways, the strongest winds likely to be experienced at particular sites need to be taken into account when designing buildings, the probable heaviest falls of rain need to be known before drains and sewers are constructed, etc.

The Meteorological Office is able to supply such information both for UK and overseas locations from a study of its archives.

In the US the organization responsible for meteorology is the National Weather Service within the National Oceanic and Atmospheric Administration (NOAA). This organization maintains the meteorological observing network, collects and disseminates the weather observations and provides forecast information over periods of up to five days for various users, as does the Meteorological Office in the UK. However, meteorological information for military aviation is provided by a separate organization called the Air Weather Service (AWS) which obtains its data via NOAA. Forecasting information worldwide for the US Navy is in the hands of the Fleet Numerical Weather Facility at Monterey in California. Much of the requirement for climatological data in the US is provided by meteorological consultants and in this respect the system differs considerably from that in Britain where almost all requests are handled by the Meteorological Office. The American Meteorological Society holds a list of accredited consultants and the *Bulletin of the American Meteorological Society* frequently carries advertisements for them.

Professional associations

There are a number of non-governmental meteorological associations in the world. Probably the most important and active are:

(a) The American Meteorological Society
(b) The Royal Meteorological Society
(c) The Japanese Meteorological Society
(d) The Canadian Meteorological and Oceanographic Society.

There are other meteorological societies, notably in West Germany, India, France, Greece, New Zealand and Australia. These organizations are responsible for some key periodicals in the subject (for a list of relevant journals see Appendix I).

The American Meteorological Society (AMS) produces a number of very important scientific journals. The *Bulletin of the American Meteorological Society* is published monthly and is devoted to editorials, survey articles, professional and membership news and Society activities. The *Journal of the Atmospheric Sciences* and the *Journal of Climate and Applied Meteorology* are monthly AMS publications which publish basic and applied meteorological research respectively: they are among the most highly-rated meteorological journals. The *Monthly Weather Review*, the *Journal of Physical Oceanography* and the more popular *Weatherwise* are also published, or sponsored, by the AMS. The AMS has many branches (chapters) throughout the US and it administers and sponsors a large number of important international conferences every year on a variety of meteorological subjects. Another AMS activity is the setting of examinations for the certifying of meteorological consultants, an activity which does much to maintain the standard of meteorology in the US.

The Royal Meteorological Society (RMS) (established 1850) fulfils much the same functions in Britain as the AMS does in the US. Its major journal, the *Quarterly Journal of the Royal Meteorological Society*, is one of the most respected of all meteorological journals, carrying much of the latest research in the subject; it has been published regularly since 1871. Recently the Society has started the *Journal of Climatology* (Wiley); this new journal is international in character and publishes original research in all aspects of climatology and has been renamed the *International Journal of Climatology* from 1989. The third journal of the RMS, *Weather*, is aimed more at sixth formers and the educated lay reader, and is published monthly. In addition to these regular publications the Society occasionally arranges special publications; a recent one of these is *Topics in Dynamical Meteorology*, published by Methuen. On the occasion of the centenary of the setting up of the meteorological observatory on Ben Nevis in 1883, the Society produced a special booklet describing the way the observatory was started and giving details of its work during the 20 years it operated. The Society caters particularly for schools and has a class of membership called School Corporate Membership.

Like the AMS, the RMS co-sponsors many international conferences and symposia; one, entitled *Variations in the Global Water Budget*, held in Oxford in 1981, was particularly important. Another, held in 1984, covered the subject of meteorological education for school children, amateurs and the general public. Special publications are usually issued following these major conferences.

The Australian Branch of the RMS has recently become independent and is now the Australia Meteorological and Oceanographic Society. It deserves special mention as it has its own publication *Meteorology Australia* and, in addition to regular meetings, it has organized several international symposia. A recent successful symposium was entitled *Antarctica: weather and climate*.

Among its other activities the RMS produces lists of recommended books for different academic levels, lists of meteorological slides and filmstrips, and addresses where these may be obtained. It has also produced a comprehensive *Review of Meteorological Films World-wide*, which includes details of how to purchase or borrow the films, and also describes the contents and the academic level of them. The film review is updated regularly through the pages of *Weather*. The RMS has a number of local centres in the UK and holds many meetings and field courses. The Society has probably the best archival meteorological library in the world and this is administered with the help of the National Meteorological Library (see later). The Society is prepared to give general meteorological information to enquirers but does not handle meteorological data.

The Canadian Meteorological and Oceanographic Society was once a branch of the Royal Meteorological Society but it became independent in the 1950s and now produces its own highly respected journal *Atmosphere — Ocean*.

The Japanese Meteorological Society has been in existence for over 100 years and produces a number of journals including the *Journal of The Meteorological Society of Japan* which is published in English.

Universities

A number of British universities offer courses leading to a first degree in meteorology, although it is more usual for the subject to be studied as an option within a more general course. Consequently it is more likely that a graduate with a mathematics or physics qualification will enter the profession. For meteorologists of all grades, and personnel from overseas, the Meteorological Office runs courses at its own Training College. In the USA meteorology is more widely taught at university level and there are large meteorological departments, among the largest being Wisconsin.

European Centre for Medium Range Weather Forecasts

This organization is a unique organization to which 17 European Member States currently belong. The objectives of the Centre are:

1. To develop computerized simulation of atmospheric processes with a view to preparing weather forecasts for up to 7–10 days ahead.
2. To collect and store the appropriate data.
3. To make available some of its computing capacity to member states.
4. To assist in advanced training of scientific staff of member states in the field of numerical weather prediction.

The Centre is now operational and forecasts, in a variety of forms, are sent to the member states. The national meteorological services in the member states are free to use the information received in ways they think most suitable for their own needs and commitments.

Since September 1979 the European Centre has been archiving meteorological data, covering the globe every 6 hours at 13 levels in the atmosphere: this data set is unique and will become a most valuable source of meteorological data for research purposes. The European Centre intends to maintain the data set in perpetuity.

Trade organizations

In Britain a number of firms concentrate on the production of meteorological instruments but there are no real trade organizations.

BP Educational Services, Wetherby, Yorkshire, however, have produced two wallcharts which illustrate 'The formation of clouds' and 'The association of clouds and weather'. Each of these consists of about 50 excellent colour photographs with captions. Both are also available as sets of colour slides ideal for teaching purposes.

Diana Wyllie Limited, 3 Park Road, Baker Street, London NW1 6XP, specializes in slide sets on meteorological subjects, together with captions. Some titles in this series are: 'Stable Weather', 'Unstable Weather', 'Pollution' and, very recently, an excellent set illustrating the value of satellite pictures.

International Aeradio Limited, Bailbrook College, London Road West, Bath, organize meteorological courses to meet the requirements of individuals and overseas governments. These are tailored to meet the needs of WMO Class II, III and IV meteorologists. They also provide background familiarization courses for Air Traffic Control Officers, Briefing and Operations Officers.

Libraries and information centres

In Britain the National Meteorological Library (NML) at Bracknell is probably the finest meteorological library in the world. It is a public library and members of the public in Britain may borrow books or work in the library by arrangement. The library's author catalogue contains about half a million cards. A subject bibliography has been maintained since 1935 and volumes covering some 200 subjects are available; this subject bibliography is increasing at the rate of about 2500 entries per month. There is also a climatological bibliography: this covers most countries of the world and many countries are subdivided into regions. The climatological bibliography is subdivided into sections on synoptic meteorology, upper air meteorology, and climatology for each country, and is available on microfiche. A search for an article may be made by using author, subject, or region of the world headings.

The NML issues a monthly accessions list which is very up-to-date and contains on average some 1300 new entries. This publication is exchanged with most overseas meteorological services and is issued to libraries at research institutes and universities. Other institutes or persons may receive the accessions list by arrangement. It is planned to produce a microfiche version shortly. The NML is particularly rich in its climatological data holdings. If data have been published it is almost certain that a copy will be held, and all past observation books from weather stations operated by the Meteorological Office are archived and available for study. A rare book catalogue, listing some 600 volumes, in chronological order, and alphabetically by author has recently been produced, in cooperation with the Royal Meteorological Society: this may be purchased from the RMS in microfiche form.

The NML also holds a considerable number of items other than books and articles. It has an excellent collection of slides, usually in colour, covering many interesting meteorological phenomena. These are in frequent demand for lectures, illustrating books and even for use in commercial brochures. The library also holds some video and audio tapes. All these items may be borrowed by special arrangement.

Since 1971 the NML has been placing new acquisitions on to the computer. The system, called MOLARS (Meteorological Office

Library Accession and Retrieval System), is now available for online searching hosted by the European Space Agency Information Retrieval Service. The data base contains bibliographic details of some 130,000 books, journal articles, reports etc., including the rare book collections of the Royal Meteorological Society and the NML. Monthly updates keep abreast of current material and increase the base by some 10,000 items a year. Searching the database will be mainly by author, UDC classification, keywords and geographical location: details are available from the NML.

In the USA the Library Information Services Division (LISD) of NOAA (see p.424) at Rockville, Maryland, has developed into a highly efficient service operating largely on a sophisticated computerized basis. The collection of books, journals, reports and data amount to over 170,000 volumes and cover the field of atmospheric science since the 1870s: there is also a related collection going back to 1807 and an archival library of several hundred volumes, some dating from the 16th century. A 24-volume catalogue of all the books at Rockville (and its subsidiary collection at Silver Springs, Maryland) has been produced and is available at other major libraries throughout the USA.

All library items are now held on computer and the accessions list of new items is computerized. LISD is prepared to make searches on any subject, or combination of subjects, and will produce up to 100 citations for a standard fee. A number of computer-produced, packaged literature searches are available, free on request, for some subjects, e.g. weather modification, and tornadoes. LISD issues occasional current issue outlines: these are objective overviews of controversial subjects and include selected bibliographies, e.g. *Sea surface temperatures and climate*. LISD also helps the American Meteorological Society to produce *Meteorological and Geoastrophysical Abstracts (MGA)*, while NOAA arranges for a computerized version of *MGA* to be publicly available in the USA. It covers literature from 1970 to date and includes over 75,000 items.

LISD also houses large amounts of meteorological data including foreign data. It has, for example, climatological observations for St Petersburg (Leningrad) in the USSR since 1838, and also much data from locations in Europe.

Reference works and textbooks

Glossaries

There are a number of dictionaries or glossaries in meteorology. They include Malone's *Compendium of Meteorology* (American

Meteorological Society): Brazol's *Dictionary of Meteorological and Related Terms* (Hachette, 1955): R E Huschke's *Glossary of Meteorology* (AMS, 1959) and D H McIntosh's *Meteorological Glossary* (5th edn, HMSO, 1972). The *Glossary of Meteorology* attempts to define all important terms found in the literature at the present time with definitions that are "understandable to the generalist and yet palatable to the specialist": it does not include bibliographic references.

The *Meteorological Glossary* is less comprehensive but some of the explanations are more complete and it will be found adequate for all except the most specialist topics. WMO has published an *International Vocabulary* (1966) with lists of equivalent terms in English, French, Spanish and Russian, and definitions duplicated in English and French. This publication has recently been updated and revised.

Handbooks

A number of basic handbooks have been produced in Britain. *The Observer's Handbook* (4th rev. edn, HMSO, 1983) and the *Marine Observer's Handbook* (11th edn, HMSO, 1981) are very comprehensive books giving all the information that professional observers on land or at sea could require: they describe the necessary conditions for the acceptable exposure of meteorological instruments and how an instrument enclosure should be set up. Complementary to these texts are a handbook illustrating *Cloud Types for Observers* (HMSO, 1982) and a seven-volume publication entitled *Handbook of Meteorological Instruments* (2nd edn, HMSO, 1980–83). The latter covers instruments for measuring pressure, temperature, humidity, surface wind, precipitation and evaporation, sunshine and radiation, and visibility and cloud height respectively. The whole has been revised between 1980 and 1983. WMO has also produced an *International Cloud Atlas* (1975).

Similar handbooks have been produced for the forecaster. *Handbook of Aviation Meteorology* (2nd edn, HMSO, 1971) was reprinted in 1981 but is somewhat out-of-date in that computer forecasts have now changed many of the procedures previously followed by airfield forecasters. Similar remarks apply to P G Wickham's *Practice of Weather Forecasting* (HMSO, 1970, reprinted 1981).

Other very useful handbooks include *Tables of Temperature, Relative Humidity and Precipitation for the World*, available from the Meteorological Office. This publication is in five parts covering North America and Greenland, Central and South

America, Europe and The Azores, Africa, the Atlantic Ocean south of 35°N, and the Indian Ocean and Asia. In it are monthly averages of temperature, relative humidity and precipitation for many individual sites together with sunshine data when available: also included are extremes, number of rain days, etc. The whole makes an admirable reference source for both meteorologists and layman.

Other reference volumes of a similar calibre, published by HMSO, include the series of *Geophysical Memoirs*: one of this series, for example, is entitled *Average Temperatures, Contour Heights and Winds at 30 mb over the Northern Hemisphere*. The series of *Scientific Papers* contain the results of work done in the Meteorological Office: a useful example is *Estimation of Rainfall using Radar — a critical review*. Another very useful series are the *Naval Handbooks* produced especially for forecasters in the Royal Navy but also of great value to all sailors and ocean navigators: titles include *Weather in Home Waters: Volume I The Northern Seas, Volume II The Waters Around the British Isles and Baltic* and *Volume III The Waters Around The Azores and off SW Europe and off NW Africa*. These works include extensive tables of meteorological data such as averages and extreme values.

WMO has undertaken the preparation of a world climatic atlas indicating the principal climatic features of each continent. The first of this series, *Climatic Atlas of Europe*, consists of two sets of 13 maps showing monthly and annual mean temperature and precipitation totals respectively: a map of annual temperature amplitude is also included. A rather similar *Climatic Atlas of South America*, which uses temperature data from about 700 stations and precipitation data from about 1700 stations, has been prepared also.

Again WMO have prepared a number of handbooks, examples of which are *Guide to Agricultural Meteorology Practices* (2nd edn, 1982), *Manual of Aerodrome Meteorological Practices* (1967) and *Compendium of Meteorological Training Facilities* (5th edn, 1977).

Climatic diagrams and small scale maps of climatic regions of continents, or of the world, are to be found in many atlases. Major works of this kind include the *Soviet World Physical Atlas (Fiziko-Geograficheskii Atlas*, Mira), the German series *World Atlas of Climatic Diagrams*, and Elsevier's *Agricultural Climatic Atlas of Western Europe*.

The Meteorological Office has published a number of climatic maps of the British Isles, these include maps of average annual rainfall on two different scales, map of average duration of rainfall, maps of monthly average rainfall, maps of average duration of bright sunshine, mean number of days of snow, mean and extreme temperatures: all these relate to the period 1941–70. Maps for rainfall are available from the Climatological Services Branch.

Yearbooks

The Meteorological Office produces an annual report published by HMSO. This is a very comprehensive booklet which gives information on the work over the year of all the research branches, and of the services division which is responsible for all forecasting, including military and civil aviation and television, radio and press forecasts. It also gives details of lectures and scientific papers published by staff members during the year, and of conferences attended.

The *Annual Report of the World Meteorological Organization* includes the annual report which the Secretary-General is required to submit to members of WMO together with the annual report on the work of the organization as a specialized agency of the United Nations.

Textbooks

There are very many books about the weather, climate and various aspects of meteorology and it is difficult to single out particular volumes for special mention. However, for a general introduction to the subject, it would be hard to better D E Pedgley's *A Course in Elementary Meteorology* (2nd edn, HMSO, 1978). Climatology is comprehensively dealt with in the 15 volumes of the *World Survey of Climatology*, editor-in-chief H E Landsberg (Elsevier, 1969–). This series covers all regions of the world including the polar regions, and one volume covers the climate of the free atmosphere. A new book, covering aspects of meteorology not previously well described, is B W Atkinson's *Meso-scale Atmospheric Circulations* (Academic Press, 1981), while for those concerned with the larger scale *Large-scale Dynamical Processes in the Atmosphere*, edited by B Hoskins and R Pearce (Academic Press, 1983) is thoroughly up-to-date.

The Royal Meteorological Society has produced a list of recommended books for different age groups most of which is reproduced as Appendix II. The majority of these books are published in the UK; it does not include WMO publications or books published in the USA.

Abstracting services

The main abstracting journal for meteorologists is *Meteorological and Geoastrophysical Abstracts* (1979–), published by the American Meteorological Society. This appears monthly and covers a world-wide range of the literature.

Geographical Abstracts B — Climatology and Hydrology (1970–) appears bi-monthly and is stronger in English language material.

Periodicals

Most of the latest research is published in the recognized meteorological journals, notably *Quarterly Journal of the Royal Meteorological Society*, and *Journal of Atmospheric Sciences* (American Meteorological Society). Other journals of major importance are *Tellus*, *Monthly Weather Review*, *Journal of Climate and Applied Meteorology*, and *International Journal of Climatology*. Many countries publish a meteorological journal in their own language, some of the best known being *Zeitschrift für Meteorologie* (Berlin), *Journal de Recherches Atmosphériques* (Clermont-Ferrand), *Meteorologiya i Gidrologiya* (Moscow): a few foreign journals such as the *Journal of the Meteorological Society of Japan* are published in English.

The application of meteorology to other fields of science has become increasingly important in recent years; examples are agriculture, pollution, hydrology, and medicine. Typical journals relating to these disciplines are *International Journal of Biometeorology*, *Atmospheric Environment*, *Journal of Agricultural Meteorology*, *Rivista di Meteorologia Aeronautica*, *Climatic Change* and *Journal of Hydrology*.

Other meteorological articles of importance are occasionally found in the more general types of journals such as *Proceedings of the Royal Society*, *Nature*, *New Scientist* and *Journal of Geophysical Research*.

A more comprehensive list of important periodicals appears in Appendix I.

Report literature

The most extensive report literature is that produced by WMO. They have produced (1980) a catalogue entitled *Publications of the World Meteorological Organization*. The core of this catalogue

(Part I) is the master list of about 500 items, in three sections: the sections are (1) WMO publications listed by WMO number, (2) WMO publications listed by year of publication (for those items without WMO numbers) and (3) publications related to the Global Atmospheric Research Programme. This master list contains the complete entries for all publications still in print. Part II of the catalogue is a classification by subject of scientific and technical publications, and Part III lists all WMO reports.

Another extensive series of report literature is that produced by the United States Air Weather Service. Their *Index of Air Weather Service Technical Publications* published by Air Weather Service, Scott AFB, Illinois, lists several hundred reports. These are mainly of a forecasting nature and are intended for the guidance of operational weather forecasters in various parts of the world. However some give details of numerical models which have been developed for a variety of forecasting purposes.

In Britain unpublished research work is retained in the National Meteorological Library as a series of *Branch Memoranda*. Each research branch in the Meteorological Office maintains a series of these *Memoranda* in which full details of research projects are described.

Theses and conferences

Theses produced in universities in both the UK and the USA are often eventually published in one of the recognized journals but some universities maintain their own sets of publications, many of which may form important contributions to meteorological knowledge. Two examples are *Combined Land/Sea Surface Air Temperature Trends 1949–1972*, published by the Department of Meteorology and Physical Oceanography, Massachusetts Institute of Technology, and *Causes and Effects of Atmospheric Interannual Variability*, published by the Department of Atmospheric Science, Colorado State University.

A large number of international conferences are held on many aspects of meteorology. These are usually sponsored by WMO, the American Meteorological Society, or the Royal Meteorological Society. From most of these conferences a proceedings volume is subsequently produced but the format of these is variable. Some are published by WMO and included in their catalogue of publications referred to earlier: examples are *Proceedings of the Symposium on Long-term Climatic Fluctuations, Norwich, 1975*; *Solar Energy: proceedings of the symposium, Geneva, 1976*; *Papers Presented at the Sympsoium on the Geo-*

physical Aspects and Consequences of Changes in the Composition of the Stratosphere, Toronto, 1978; and *Proceedings of the World Climatic Conference, Geneva, 1979.*

At other conferences a symposium volume is subsequently prepared, e.g. *Meteorology over the Tropical Oceans,* a volume produced by the Royal Meteorological Society following their conference with the American Meteorological Society and the German Meteorological Society which was held in 1978 in London. A symposium volume *Variations in the Global Water Budget,* published by Reidel, followed the 1981 Oxford conference on this subject, which was sponsored by the Royal Meteorological Society and other organizations.

Data

Meteorological data differ from data in most other fields in that an enormous global communications network has been built up to exchange observations throughout the world, or at least the northern hemisphere, within a matter of hours. This network is used by national weather services to provide weather forecasts for up to about five days ahead, as described above in the section 'National weather services'. These data include temperature, humidity, rainfall amount, wind speed and direction, cloud amount and type, present and past weather, for several thousand locations worldwide. They are supplemented by upper air observations reporting temperature, humidity, and wind speed and direction at standard levels in the atmosphere, as recorded automatically by radio-sonde instruments carried aloft by balloons. For obtaining surface or upper air weather observations at a particular place or time, application should be made to the appropriate national weather service.

Other sources of primary meteorological data include satellites and weather radar. Meteorological satellites are of two types, geostationary and polar-orbiting. *Geostationary satellites* have been placed high in the atmosphere so that they rotate in orbit at the same speed as the earth itself: they therefore appear to be stationary when viewed from earth. There are five of these satellites spaced around the earth to be overhead at the equator at different longitudes, taking pictures continually and transmitting them automatically to ground stations. Such satellites always have the same field of view and hence are very useful for watching the development of clouds and weather systems.

Polar orbiting satellites are in a lower orbit and go round the earth, crossing close to both poles, while the earth rotates on its axis beneath. As a result a broad swathe of the earth from north to south is observed at each orbit, and all parts of the earth are covered twice per day.

Satellites take pictures using both visible and infra-red radiation, the latter enabling pictures to be taken in darkness. Instruments react to radiation from the earth's surface and the cloud tops, which depends on the temperature of the radiating body and so make reasonable estimates of sea and land surface temperatures and the temperature of cloud tops can be made. More sophisticated methods have been developed to measure the temperature and water vapour content throughout the atmosphere so that, as these methods develop, satellites may eventually replace the radio-sonde network.

Meteorological satellite data may be obtained in a variety of ways. It is possible for knowledgeable and enthusiastic amateurs to build their own reception station and obtain excellent pictures. Details of one such successful effort are given in *Weather*, **37** (12), 361–370. However, in Britain the best source of satellite pictures is the Department of Electrical Engineering and Electronics at the University of Dundee. Here are archived the complete sequence of polar orbiting satellite imagery since November 1978 centred approximately on the UK and covering an area roughly from Spitzbergen to North Africa, and the NE Atlantic to eastern Europe. Users can apply for any prints (for a fee): these are corrected for earth curvature distortion and may have a land edge or latitude/longitude grid superimposed if required. Dundee will select examples of particular meteorological interest if required. Sheets of 36 images illustrating a week's weather are useful for teaching purposes.

An excellent book containing 50 selected satellite images plus captions is *The Earth's Atmosphere Viewed From Space*, by R R Fotheringham (University of Dundee, c.1979). This book has an appendix outlining the characteristics of satellite systems. Dundee also provide a free information sheet giving details of their satellite image products and other useful information on orbits, sensors, etc.

Images from the geostationary satellite Meteosat can be obtained from the European Space Agency (ESA), Meteosat Data Services, ESOC, Darmstadt, Federal Republic Germany. These are full disc images, centred on 0°N 0°E, for the visible, infra-red and water vapour channels, and are available normally for every half hour. Enlargements of particular areas are available. ESA have also produced a 16 mm (silent) film *Clouds in Motion 25–27 April 1978* which lasts for 17 minutes: it is a well-produced and very instructive film which is guaranteed to stimulate an audience.

In the USA, NOAA Satellite Data Services Division, (World Weather Building, Washington DC) have available satellite imagery from both polar orbiting and geostationary satellites which cover some regions other than those dealt with at Dundee and Darmstadt. In particular, geostationary images from satellites overhead on the equator at 75°W and 135°W are available.

The Royal Meteorological Society (Bracknell, UK) produce two versions of a Satpack designed to assist the teaching of satellite meteorology at sixth form and first year university level. They also have a set of 12 excellent slides based on satellite imagery, together with a booklet giving a very full explanation of the slides. A network of weather radars, each with a range of about 100 miles, covers most of Britain: study of the radar echoes assists in the forecasting of severe storms and thunderstorms.

Other publications containing primary meteorological data included the Meteorological Office's *Daily Weather Report* and *Daily Aerological Record* but these were discontinued at the end of 1980. They contained weather observations from many stations in the British Isles and included some weather charts, notably one covering much of the northern hemisphere at 12 noon GMT. However, the Weather Centres at London, Manchester, Newcastle, Nottingham and Southampton produce local and areal weather summaries, daily, weekly or monthly, for a charge, and the London Weather Centre produces a daily weather summary for the British Isles which includes several weather charts. The Meteorological Office's *Monthly Weather Report (MWR)*, which contains charts and summaries of observations from about 620 stations in Great Britain and Northern Ireland, is still published regularly.

The Meteorological Office also updates regularly the unique series of Central England monthly mean temperatures prepared by G Manley (*Quarterly Journal of the Royal Meteorological Society*, **99**, 389–405). This series is the longest continuous temperature record in the world and covers the period from 1659 to date.

A similar series of monthly mean rainfall data over England and Wales exists from 1727 (*British Rainfall*, 299–306, 1931) which is also kept up to date by the Meteorological Office. This series has, however, recently been improved by T M L Wigley *et al* 'Spatial patterns of precipitation in England and Wales and a revised homogeneous England and Wales precipitation series', *Journal of Climatology*, 1983.

Currently the best published daily weather information is that produced by the German Meteorological Office. They publish daily data for a considerable number of European stations combined with a number of surface and upper charts which cover wide areas of the northern hemisphere.

A unique publication is *Climate Monitor* produced by the Climatic Research Unit, University of East Anglia. This is a quarterly journal containing up-to-date summaries of global climatic conditions and articles of general interest on climatology. Each issue contains a summary of global climate for the preceding season and an extensive listing of extreme events from all over the world. 500 mb circulation patterns are given for both northern and southern hemispheres and the *Daily Weather Type for the UK* (Lamb) is updated, together with the northern hemisphere mean surface air temperature. Fluctuations in high northern and southern latitude temperatures are monitored as a unique indicator of global temperature trends, and the dust veil index (DVI) appropriate to any recent volcanic eruptions is included. Short articles on a wide range of topics are included in each issue: articles of interest, not only to climatologists and meteorologists, but also to climatic historians, biologists and all natural scientists, whether professional or amateur. Apart from the quarterly issue there is a fifth issue in April each year summarizing events of the previous year.

The meteorological data holdings of the Climatic Research Unit (CRU) are unique. Unlike the raw data held by most organizations those of CRU have been meticulously screened for reliability and now represent one of the best series of verified historical data available. The material covers much of western Europe from about 800–1700 AD, with the best coverage being for the UK from 1200 AD onwards.

A large amount of unpublished primary source material, including data from medieval manorial accounts, and a comprehensive set of Icelandic historical data extending up to 1780 AD, are also held. Total holdings of instrumental meteorological data greatly exceed those in other common sources. In addition the CRU has a unique set of daily weather maps for 1781–1786 inclusive, with isolated shorter sets before and after those dates.

CRU also holds a quantity of tree ring data: these data are of two types, individual ring width measurements and data from groups of cores at particular sites which have been combined to form tree ring chronologies. The Unit has a collection of approximately 50 European chronologies from living trees and a smaller number based on archaeological material. The living tree chronologies mostly date back to the 18th century but a few go back much further. A substantial fraction of the Unit's holdings is unpublished material.

In addition to these European tree ring data, the Unit holds a selection of material from other countries. Some sea ice data are also held by CRU: limits of sea ice in the northern hemisphere from April–August 1901–1956 have been extracted from Danish Meteorological Service publications and put in computerized form. These data are available in grid point form on a 100 km grid. A variety of earlier sea ice data are also available for Iceland, Newfoundland, Davis Strait, Kara Sea, and other areas.

Both tree ring and sea ice data are valuable material for research into past climates and in the study of climatic change generally.

The USSR publishes daily upper air charts at eight levels in the atmosphere over the whole of the northern hemisphere but these are available only after about a year's delay. They also publish monthly mean surface and upper air charts, again covering the whole of the northern hemisphere, but these, too, are only available some considerable time after the event.

Various other countries have produced daily series of surface and upper air charts but these have mostly been discontinued owing to the escalating costs of publication and the fact that most charts can be produced much more quickly and easily by computer methods, making use of the daily data on the global telecommunications network.

The Meteorological Office produces a number of periodic publications. These are listed in *Meteorological Office Leaflet* No. 12 (HMSO) which also gives details of how to obtain the various items. Among the most useful are:

(a) *Weekly and Monthly Sea Ice Charts* — the former cover the North Atlantic, the Baltic, the Barents Sea, the Greenland Sea, Baffin Bay and Hudson Bay and the latter cover also the North Pacific. These charts include sea isotherms.

(b) *Charts and Information Produced under the Rainfall and Evaporation Calculation System (MORECS)*. Calculated weekly and monthly totals or averages are mapped over the UK for fourteen hydrometeorological and meteorological variables including potential and actual evaporation, potential and actual soil moisture deficit and effective precipitation for grass and also for real land use appropriate to the particular area concerned.

(c) *Monthly and Annual Totals of Rainfall for the UK*. This replaces *British Rainfall* which ceased publication after data for 1968 were published.

In the USA many periodic publications are prepared and published regularly. These are listed in *Selective Guide to Climatic Data Sources* published by NOAA, Environmental Data and Information Services. Among the most important are:

(a) *Daily Synoptic Weather Maps at Sea Level and 500 mbs* for the northern hemisphere (published monthly). This series consists of one sea level map and one 500 mb map for 1200 GMT observations each day. The sea level maps have been published for data from 1 January 1899 and the 500 mb maps began with the December 1944 issue.

(b) *Monthly Climatic Data for the World* (1948–). This publication contains monthly means of surface and upper air data for many locations throughout the world. Corrections are carried in the first issue following their receipt. Surface data include monthly mean pressure (at station level and sea level), monthly mean temperature and vapour pressure and departures from average, days of precipitation, precipitation total and departure from average and percentage of sunshine compared to the long term mean. The upper air section includes monthly mean temperature, humidity and wind data at standard levels in the atmosphere.

(c) *Weekly Weather and Crop Bulletin* (1872–). This bulletin summarizes weather and its effects on crops and farm activities for each week over the USA and other areas of the world as feasible. It began in 1872 and has continued (under various titles) with only a short break from 1881–1884. It contains a national weather summary, a national agricultural summary, and a world weather and crop update. A map and comments on the crop moisture situation are presented throughout the warm season. Weekly charts of total precipitation and departures of average temperature from normal are shown, and a chart giving snow depth on the ground is included in the winter. Other charts and special articles are included occasionally as pertinent. The *Bulletin* may be ordered from the Agricultural Weather Facility, US Department of Agriculture, South Building, Washington DC.

APPENDIX 1: Meteorological Journals

Agricultural Meteorology (Int. Journal) (Elsevier, 1964–); Bi-monthly
Archiv fur Meteorologie, Geophysik und Biolimatologie Serie A and *B* (Springer-Verlag, 1948–); Quarterly
Atmosphere — Ocean (Canadian Meteorological and Oceanic Society, 1978–); Quarterly
Atmospheric Environment (Int. Journal) (Pergamon, 1967–); Monthly
Australian Meteorological Magazine (Australian Government Publishing Service, 1954–); Quarterly
Boundary Layer Meteorology (Reidel,1970–7); Monthly
Bulletin of the American Meteorological Society (AMS, 1920–); Monthly
Climate Monitor (formerly CRUMB) (Climatic Research Unit, University of East Anglia, 1976–); Quarterly + 1
Climatic Change (Reidel, 1977–); Quarterly
Contributions to Atmospheric Physics (Deutsche Meteorologische Gesellschaft, 1956–); Quarterly
Dynamics of Atmospheres and Oceans (Elsevier, 1976–); Irregular
Geographical Abstracts B — Climatology and Hydrology (Geo. Abstracts Ltd., University of East Anglia, 1970–); Bi-monthly
Geomagnetizm i Aeronomiya (Moscow, 1963–); Bi-monthly

Geophysica (Geophysical Society of Finland, Helsinki, 1935–); Irregular
International Journal of Biometeorology (Swets and Zeitlinger, 1961–); Quarterly
International Journal of Climatology (Wiley, 1989–; formerly *Journal of Climatology* 1981–8); Quarterly)
Journal of Agricultural Meteorology (Tokyo, 1945–)
Journal of the Atmospheric Sciences (American Meteorological Society, 1962–); Monthly
Journal of Atmospheric and Terrestrial Physics (Pergamon, 1950–); Monthly
Journal of Climate and Applied Meteorology (American Meteorological Society, 1962–); Monthly
Journal of Geophysical Research (3 parts) (American Geophysical Union, 1949–); Monthly
Journal of Hydrology (Elsevier, 1963–); 8/year
Journal of Meteorological Research (Japan Meteorological Agency, 1956–); Monthly
Journal of Meteorology (Artetech, 1975–); Monthly
Journal of the Meteorological Society of Japan (Meteorological Society of Japan, 1903–); Bi-monthly
Journal of Physical Oceanography (American Meteorological Society, 1971–); Bi-monthly
Journal de Recherches Atmosphériques (Université de Clermont-Ferrand, 1963–)
Marine Observer (HMSO, 1923–); Quarterly
Meterological and Geoastrophysical Abstracts (American Meteorological Society, 1979–); Monthly
Meteorological Magazine (HMSO, 1866–); Monthly
Meteorologische Rundschau (Gebruder Borntraeger, 1947–); Bi-monthly
La Metéorolgie. (Société Metéorologique de France, 1925–); Quarterly
Meteorologiya i Gidrologiya (Moscow, 1936–)
Meteorology Australia (Royal Meteorological Society, Australian Branch, 1981–); Irregular
Monthly Weather Review (American Meteorological Society, 1874–); Monthly
Nature (Macmillan, 1869–); Weekly
New Scientist (New Science, 1956–); Weekly
Proceedings, Royal Society, London, Section A (Royal Society, 1832–); Monthly
Quarterly Journal of the Royal Meteorological Society (RMS, 1871–); Quarterly
Quaternary Research (Academic, 1974–); Bi-monthly
Rivista di Meteorologia Aeronautica (Rome) (Servizio Meteorologico Dell'Aeronautica, 1948–); Quarterly
Science (American Assn. for the Advancement of Science, 1880–); Weekly
Science Progress (Blackwell, 1894–); Quarterly
Soviet Meteorology and Hydrology (Allerton, New York, 1976–); Monthly
Tellus (Swedish Geophysical Society, 1949–); Bi-monthly
Weather (Royal Meteorological Society, 1946–); Monthly
Weatherwise (Heldref Publns, Washington, 1948–); Bi-monthly
World Meteorological Organization Bulletin (WMO Geneva, 1952–); Quarterly
Zeitschrift für Meteorologie (Meteorologische Gesellschaft, 1946–); Bi-monthly

APPENDIX 2: List of Recommended Books

6th Form, Introductory College
Barry, R G and Chorley, R J (1976) *Atmosphere, Weather and Climate* 3rd edn.
Chandler, T J (1981) *Modern Meteorology and Climatology* 2nd edn. Nelson
Gaskell, T and Morris, M (1979) *World Climate* Thames and Hudson
Harvey, J G (1976) *Atmosphere and Ocean: our fluid environments* Artemis Press

Lockwood, J G (1979) *Causes of Climate* Arnold
McIntosh, D H and Thom, A S (1969) *Essentials of Meteorology* Wykeham: Wykeham Science Series
Monteith, J L (1973) *Principles of Environmental Physics* Arnold
Perry, A H and Perry, V C (1973) *Weather Maps* 2nd edn. Oliver and Boyd
Riley, D and Spolton, L (1974) *World Weather and Climate* Cambridge University Press

General Introductory
Forsdyke, A G (1980) *The Weather Guide* 2nd edn. Hamlyn
Giles, W G (1978) *Weather Observations* EP Publishing
Holford, I (1973) *Interpreting the Weather*
Holford, I (1982) *The Guinness Book of Weather Facts and Feats* Guinness Superlatives
McAllen, P and Phillips, B (1981) *Looking at the Weather; the work of the Meteorological Office* HMSO
Pearce, R P (1980) *The Observer's Book of Weather* Frederick Warne
Pedgley, D E (1978) *A Course in Elementary Meteorology* 2nd edn. HMSO
Pedgley, D E (1979) *Mountain Weather; a practical guide for hill walkers and climbers in the British Isles* Cicerone Press

Specialist Topics
Atkinson, B W (1981) *Mesoscale Atmospheric Circulations*
Atkinson, B W ed. (1981) *Dynamical Meteorology; an introductory selection* Methuen
Barrett, E C (1974) *Climatology from Satellites* Methuen
Chandler, T J and Gregory, S, eds (1976) *The Climate of the British Isles* Longmans
Fotheringham, R R (1979) *The Earth's Atmosphere Viewed from Space* University of Dundee Press
Lockwood, J G (1974) *World Climatology* Arnold
McIntosh, D H (1972) *Meteorological Glossary (Meteorological Office 848)* HMSO
Mason, B J (1975) *Clouds, Rain and Rainmaking* 2nd edn. Cambridge University Press
Observer's Handbook (1983) 4th rev. edn (*Meteorological Office 805*) HMSO
The Marine Observer's Handbook (1977) 10th edn (*Meteorological Office 887*) HMSO
Meteorology for Mariners (1978) 3rd edn (*Meteorological Office 895*) HMSO
Oke, T R (1978) *Boundary Layer Climates* Methuen
Perry, A H and Walker, J M (1977) *The Ocean-Atmosphere System* Longman
Pickard, G L (1979) *Descriptive Physical Oceanography* 3rd edn. Pergamon
Pittock, A B et al (1976) *Climatic Change and Variability: a southern perspective* Cambridge University Press
Rosenberg, N J (1974) *Microclimate; the biological environment* Wiley
Shaw, D B ed. (1979) *Meteorology Over Tropical Oceans* Royal Meteorological Society
Tchernia, P (1980) *Descriptive Regional Oceanography* Pergamon
Tricker, R A R (1970) *Introduction to Meteorological Optics*
Wickham, P G (1970) *The Practice of Weather Forecasting* HMSO
Proceedings of the World Climate Conference (1979) (*WMO No.537*) World Meteorological Organization

CHAPTER NINETEEN

Soil science

D A JENKINS, D B JOHNSON AND W I KELSO

Soil science is unified by its subject matter rather than by any particular scientific discipline. In this respect it is a microcosm of the geological scene, in that it involves the application of a range of disciplines to the study of a natural phenomenon; however, like geology, it too has to some extent evolved its own unique philosophy. The literature relating to soils is therefore diverse in character and approach, and although some of it is concentrated in publications specific to soils, a portion of it is dispersed through the general chemical, biological, geological and civil engineering literature. As in most other branches of science, this literature is expanding rapidly. According to Jacks (1966), who presented an interesting analysis of this expansion and the trends embodied in it, the volume had more than doubled over the post-war years. More recently the number of specific soil science records processed by the Commonwealth Bureau of Soils has again roughly doubled in the last decade (B Butters, personal communication).

Historically, early investigations into the nature of soil, such as those summarized by Humphrey Davy in 1813 in his book *Elements of Agricultural Chemistry*, were essentially agricultural in nature. They involved a chemical approach to soils and plant nutrition which was carried through to the latter half of the century, receiving impetus from the publication in 1840 of Liebig's *Chemistry in its Application to Agriculture and Physiology*. In studying the nature and variability of soil as a naturally occurring material, great emphasis was placed on the geological setting until the relative importance of climate as a controlling factor was recognized by Russian scientists, such as Dokuchaev (Glinka, 1928) towards the end of the nineteenth century. Brief accounts of these early developments can be found in the introductory chapters of Wild (1988).

In this century, the study of soil has expanded rapidly to incorporate an increasing range of expertise, from mineralogy, through physical, inorganic and organic chemistry, biochemistry and microbiology, to agronomy, ecology and geomorphology. Although some of the interrelationships between these various aspects are as yet tenuous, they currently present areas of particularly rapid development, and there now exists sufficient cohesion for soil science to constitute a viable unit of study.

Owing to the importance of climate as a factor in soil development, there is a strong national or regional flavour to soil investigations and this is reflected in the resulting literature. This pattern is also seen in geomorphology but is not, of course, evident to the same degree in geology as a whole. In coping with the non-English literature the following publications are still of value: *Multilingual Dictionary of Soil Studies*, by G V Jacks *et al* (English, French, Spanish, German, Portuguese, Italian, Dutch, Swedish, Russian: Food & Agricultural Organization (FAO), Rome, 1960), *Elsevier's Dictionary of Soil Mechanics*, by A D Visser (Elsevier, 1965), together with the more specialized *Glossary of Soil Micromorphology*, edited by A Jongerius, and G K Rutherford (PUDOC, Wageningen, 1979).

The present position of soil science within the framework of geology is uneasy. Soils *per se* receive scant attention in geological circles, even with geology restyled as earth sciences — despite the fact that soils are essentially a geological material and the focus of the sedimentary geochemical cycle. The main incentive in understanding the nature and properties of soil remains agricultural, and this reflects the economic significance of soil fertility. This aspect of soils carries through into geography and ecology. It has led to the establishment of an increasing number of national Soil Surveys, although the latter have subsequently acquired varying degrees of independence from agriculture. There is yet a further aspect, and that is soil considered as a material involved in civil engineering. Here soil is grouped with other superficial deposits of the regolith, the prime considerations being its physical, mechanical or structural properties. Soil mechanics and the agricultural aspects comprise two distinct fields of study that are surprisingly independent, both in practice and in their respective literatures.

In this chapter, soil will be considered primarily in its own right as a naturally occurring material. This is, however, a geological rather than an agricultural or engineering viewpoint and conforms to the definition of 'pedology' given by Robinson (1924). Nevertheless, problems of definition in the broader context of geology still remain in that, to the soil scientist, the term 'soil' would denote a product of the interaction of the lithosphere with the atmosphere, hydrosphere and, in particular, the biosphere: its premature use for the surface deposits of extraterrestrial environments such as the moon (e.g. *Lunar Soils*, by J A Wood, 1970) is therefore to be regretted. Indeed, no apology will be given for the inclusion — and even stressing — of the organic/biological component within this review of literature in earth sciences.

Following a consideration of the relevant general texts and primary sources, additional information relating to specific aspects such as mapping and classification, mineralogy, chemistry, physics, mechanics and biology will be dealt with briefly. Finally, the secondary literature — general reviews, abstracts and bibliographies — will be discussed. Since the subject is so diverse and ramifying, this review is necessarily selective, rather than comprehensive, in nature. Its aim is simply to provide an initial point of entry for those with geological interests who are not familiar with the literature of soil science.

Textbooks

In many standard textbooks of a geological or environmental nature there will be found brief, limited accounts of the nature of soils. More comprehensive introductions are to be found in such texts as *The World of the Soil*, by E J Russell (5th edn, Collins, 1971), *An Introduction to the Scientific Study of the Soil*, by W N Townsend (5th edn, Arnold, 1973), *An Introduction to Soil Science*, by E A Fitzpatrick (2nd edn, Longman, 1986) and *Soils and Other Growth Media*, by A W Flegman and R A T George (Macmillan, 1975). An introduction to the subject which emphasizes practical aspects of soil study is given in *The Soil: an introduction to soil study in Britain*, by F M Courtney and S T Trudgill (2nd edn, Arnold, 1984). Accounts of soil from an agricultural standpoint are found in *Soils and Soil Management*, by C D Sopher and J V Baird (Reston Publishing Co., 1978) and *Soil*, by K Simpson (Longman, 1983), both of which describe chemical and physical soil problems found in the field. A totally contrasting style of introduction is found in *Soil and Civilisation*, by E Hyams (Murray, 1976) which describes the significance of soil in the context of human history.

Many textbooks which deal with soil on a more advanced level tend to have some bias in the approach which is used. The study of soils in the context of earth science is less well-covered than in, for example, their agronomic context. A geographical/geological approach is, however, used by texts such as *Formation of Soil Material*, by T R Paton (Allen & Unwin, 1978), *Soils and Landforms; an integration of geomorphology and pedology*, by A J Gerrard (Allen & Unwin, 1981), *Pedology, Weathering and Geomorphological Research*, by P W Birkeland (Oxford University Press, 1984), *Principles and Applications of Soil Geography*, by E M Bridges (Longman, 1982), and *Geology of Soils: their evolution, classification and uses*, by E B Hunt (Freeman, 1972). An agricultural context is provided in *Russell's Soil Conditions and Plant Growth*, edited by A Wild (11th edn, Longman, 1988), *Soils: an introduction to soils and plant growth*, by R L Donahue *et al* (4th edn, Prentice-Hall, 1977), *Soils and Soil Fertility*, by L M Thompson and F R Troeh (4th edn, McGraw-Hill, 1978), *Fundamentals of Soil Science*, by H D Foth (6th edn, Wiley 1978), *The Nature of Properties of Soils*, by N C Brady (9th edn, Macmillan, 1984) and *CRC Handbook of Soils and Climate in Agriculture*, edited by V J Kilmer (CRC, 1982). Soils are also of central importance to forestry, and this subject area is the theme of *Properties and Management of Forest Soils*, by W L Pritchett and R F Fisher (2nd edn, Wiley, 1987) and *Forest Soils: properties and processes*, by K A Armson (University of Toronto Press, 1977); while *Soils and Vegetation Systems*, by S T Trudgill (Clarendon Press, 1977) takes as its central theme the cycling of nutrients between plants and soils and nutrient balance within the soil system. Another important facet of soil study is the topic of *Soil Science and Archaeology*, by S Limbrey (Academic Press, 1987) whilst the broader time context is covered in *Soils and Quaternary Geology*, by J A Catt (Oxford University Press, 1986). *Introduction to Principles and Practice of Soil Science*, by R E White (2nd edn, Blackwell, 1987) is a general soil science text, suitable for undergraduate use, which displays less obvious bias in its coverage of the subject. Advanced texts which have as their central theme the processes of pedogenesis include *Pedology*, by P Duchaufour (translated by T R Paton; Allen & Unwin, 1982) and *Soils: their formation, classification and distribution*, by E A Fitzpatrick (Longman, 1982); pedogenic processes and soil forming factors are also dealt with in *Soil Genesis and Classification*, by S W Buol *et al* (2nd edn, Iowa State University Press, 1980). All three books provide descriptions, in some depth, of soil classes of the world. A complementary text to *Pedology* is *Constituents and Properties of Soils*, edited by M Bonneau and B Souchier (translation editor V C Farmer; Academic Press, 1982); originally the two texts were published in French as Volumes 1 and 2 of *Pedologie* (general editors P Duchaufour and M Bonneau). The classic theme of soil process factors is also found in *The Soil*

Resource: origin and behaviour, by H Jenny (Springer-Verlag, 1980). Finally in this section, the *Encyclopaedia of Soil Science, Part 1*, edited by R W Fairbridge and C W Finkl, jnr, (Dowden, Hutchinson and Ross, 1979) serves as a useful reference book for a number of topics in soil science.

Reference has already been made to texts which include detailed descriptions of various soil classes found throughout the world. *World Soils*, by E M Bridges (2nd edn, Cambridge University Press, 1978) serves as a well-illustrated introduction to soil genesis and diversity. Texts more specifically concerned with soils on a regional basis include *Introduction to the Study of Soils in Tropical and Subtropical Regions*, by P Buringh (3rd edn, Centre for Agricultural Publishing and Documentation, Wageningen, 1979), *West African Soils*, by P M Ahn (Volume 1 of *West African Agriculture*, Oxford University Press, 1970), *Saline and Sodic Soils: principles-dynamics-modelling*, by E Bresler *et al* (Springer-Verlag, 1982), *Soils in the British Isles*, by L F Curtis *et al* (Longman, 1976), *The Genesis and Classification of Cold Soils*, by S Rieger (Academic Press, 1983) and, on a more applied level, *Characterisation of Soils in Relation to their Classification and Management for Crop Production: examples from some areas of the humid Tropics*, edited by D J Greenland (Clarendon Press, 1981). Unfortunately, there is still a comparative dearth of textbooks in this category.

Periodicals

An indication of the range of publications covering soil science, in conjunction with fertilisers and general agronomy, is indicated by the fact that even in 1962 reference was being made to some 700 journals and periodicals in the preparation of abstracts for *Soils and Fertilisers* by the Commonwealth Bureau of Soils. However, the main bulk of the soil science literature is contained within a limited number of primary journals. Details of some of the more important journals are given in Table 1 over.

Of the English-language journals, the *Soil Science Society of America Journal* (prior to 1976 the *Proceedings of the Soil Science Society of America*) carries the largest portion of the literature. It spans soil science and agronomy, and its articles are classified under nine headings, which include: soil physics; chemistry; mineralogy; genesis; morphology; and classification. Of the other three major English-language journals, the American *Soil Science* also has an agronomic bias, whereas the British *Journal of Soil Science* and more recent *Soil Use and Management* (1985–) and the

Table 9.1 Details of some Important Soil Science Periodicals

Periodical	Language*	Country	First issue	Issues/vols p.a.	Articles p.a.	Pages p.a.
Soil Science Society of America Journal	E	USA	1936	6?/1	220	1200
Soil Science	E	USA	1916	12/2	120	800
Soviet Soil Science	E	USSR	1958	6/1	80	650
Zeitschrift fur Pflanzenernahrung, Dungung und Bodenkunde	G(E)e	Germany	1922	3/3	55	800
Anales de Edafologia y Agrobiologia	Se	Spain	1942	6/1	57	900
Journal of the Indian Society of Soil Science	E	India	1953	4/1	80	500
Geoderma	E(G)e	Internat.	1967	12/3	65	1000
Journal of Soil Science	E	U.K.	1949	4/1	40	350
Canadian Journal of Soil Science	E(F)ef	Canada	1957	4/1	75	700
Australian Journal of Soil Research	E	Australia	1963	4/1	40	400
Soil Science and Plant Nutrition	E	Japan	1955	4/1	50	400
Communications in Soil & Plant Analysis	E	USA	1970	12/1	80	1100
Soil Biology & Biochemistry	E	U.K.	1969	6/1	80	600
Pedologie	F(E)fge	Belgium	1951	3/1	20	350
Earth Surface Processes & Landforms	E	U.K.	1975	6/1	35	600
Catena	E(G,F)e	Germany	1974	2/1	25	400
Soil Use and Management	E	U.K.	1985	4/1	35	150

*E, English; G' German; F, French; S, Spanish
Letters in parentheses denote occasional articles only. Lower case denotes translated summaries.

international *Geoderma* are devoted specifically to soil science. In these last four journals the articles are unclassified.

The literature on tropical soil science is concentrated in the Indian and Pakistan journals. There is, in addition, an extensive Japanese literature, of which the main English-language journal is *Soil Science and Plant Nutrition* (formerly *Soil and Plant Food*), and this also incorporates English abstracts of Japanese articles in the current *Journal of the Science of Soil and Manure, Japan*, since 1982 called *Japanese Journal of Soil Science and Plant Nutrition*. There is a further large volume of primary literature in languages other than English (for example, from Central Europe), which space precludes dealing with here.

In 1958 a valuable addition was made to the literature available in English by the production of *Soviet Soil Science* under the auspices of the Soil Science Society of America. This is a cover-to-cover translation of the Russian *Pochvovedenie*, with a delay of the order of six months. The articles are classified under headings similar to those in the *Soil Science Society of America Journal*. It should be pointed out that the 12 numbers of this journal issued per year up to 1968 were not designated by any volume number; in that year translations of important articles from other Russian language journals were also published as a supplementary issue to *Soviet Soil Science* under the title of *Doklady Soil Science*. Since 1969, however, only a selection of the principal articles from *Pochvovedenie*, and outstandingly significant articles from other primary Soviet journals, has been published, with the remaining articles from *Pochvovedenie* included in abstract form only.

Further sources of information include the publications of a large number of government and university research stations or departments, and of regional Soil Surveys and discussion groups.

Apart from these specific journals and publications, occasional articles concerning soil science are also contained in such general journals as *Nature*, *Science*, etc.

Soil mapping and classification

Soil survey maps and publications constitute the raw data of soil science. Most of this 'literature' is obviously specific to particular areas, but it also embodies much of the philosophy and concepts current in soil science through its involvement with genesis and classification. The situation may be compared with that in geological survey, but whereas three-dimensional structural and stratigraphical complications are often dominant in the latter, the geological units involved are generally simpler, lacking the lateral heterogeneity of soils.

In Britain, mapping has been carried out by the Soil Survey of England and Wales and the Soil Survey of Scotland, at a scale of 1:25 000 (24½ inches to the mile). Earlier results were published at a scale of 1:63 360 (1 inch to the mile) accompanied by a *Memoir of the Soil Survey of Great Britain*. Subsequently attention has been concentrated on selected 10km squares of the National Grid for which maps are produced at a scale of 1:25 000, accompanied by a *Soil Survey Record* in which the characteristics, distribution and utilization of the 'Soil Series' recognized are presented, together with relevant analytical data. Special maps at other scales are occasionally published. Some 130 *Memoirs* and *Records* have now been issued, the details being available in the *Annual Reports* of the Rothamsted Experimental Station (for England and Wales) and the Macaulay Institute for Soil Research (for Scotland). A brief description of the methods employed, and classification system used, is included in each publication and further details are available in *Soil Sampling and Soil Description*, by J H Hodgson (Oxford University Press, 1978) and *Soil Classification for England and Wales (Higher Categories)*, by B N Avery (*Soil Survey Technical Monograph*, **14**, 1980). *Soil Survey and Land Evaluation*, by D Dent and A Young (Allen & Unwin, 1981) provides a more generalized account, and a journal of the same title was also introduced in 1981.

Similar surveys are carried out by other nations. For example, those published by the US Department of Agriculture (USDA) are based on individual state counties, but differ in that the area is covered not by a map but by a set of aerial photographs (scale 1:20 000) upon which the distribution of 'Soil Series' is superimposed in outline. The relevant methodology is contained within *Soil Taxonomy; a basic system of soil classification for making and interpreting soil surveys*, by the Soil Survey Staff (*USDA Agricultural Handbook*, **436**, 1975). This USDA scheme is important, since it introduced a new and systematic terminology which has now passed into general international usage, albeit with variable enthusiasm. A recent assessment of the situation has been presented by L P Wilding *et al* in *Pedogenesis and Soil Taxonomy; Part 1, Concepts and interactions; Part 2, The soil orders* (Elsevier, 1983). The situation in the tropics is discussed in *Tropical Soils and Soil Survey*, by A Young (Cambridge University Press, 1970). A publication which presents the current position in the survey and classification of soils in Australia is *Soils: an Australian viewpoint*, by the Division of Soils, CSIRO (Academic Press, 1983). An equivalent publication for New Zealand is *Soils of New Zealand*, published in 3 volumes as *Soil Bureau Bulletins, New Zealand Department of Scientific and Industrial Research*, **26**, 1968: this is summarized by H S Gibbs in *New Zealand Soils: an introduction* (Oxford University Press, 1980).

On the international scale, the production of a *Soil Map of the World*, at a scale of 1:5 000 000, was completed in 1978 under the auspices of UNESCO and FAO, in conjunction with the International Society of Soil Science. Included with the 19 Regional Maps in this major international enterprise is *Volume* 1 — *Legend* (UNESCO, 1974) which embodies a comprehensive system of nomenclature which it is hoped will become internationally acceptable. A more detailed map of the European Communities at a scale of 1: 1 000 000, in seven sheets, has also been prepared (1985). An important focus for soil data from across the world is the International Soil Reference and Information Centre (formerly the International Soil Museum) at Wageningen in the Netherlands, which is accumulating type profile monoliths, and publishes summaries of their morphological and physical/chemical characteristics.

Soil Mineralogy and Micromorphology

The composition of most soils is overwhelmingly dominated by minerals and belated appreciation of this was evident in the separation of soil mineralogy from soil chemistry as an individual section in the *Proceedings of the Soil Science Society of America* in 1969. For convenience soil mineralogy can be divided into two parts: one comprises the igneous, metamorphic and sedimentary minerals inherited from the soil's parent material, and is concentrated mainly in the sand and coarser fractions, the other part is represented by the clay fraction, comprising mostly the clay minerals themselves, hydrous oxides and amorphous materials which are mostly pedogenic, having formed within the soil. The literature for both parts overlaps with that of sedimentary mineralogy but that of the former is usually restricted to considerations of relative mineral stabilities and assessment of parentage. However, the development of clay mineralogy as a distinct field of study has to a large degree actually taken place within soil science itself (see also Chapter 10).

Brief sections on soil mineralogy are to be found in most standard geological texts though they rarely do the subject justice. In the soil science literature special emphasis is given within such texts as *Soil Parent Materials*, by T R Paton (Allen & Unwin, 1978), *The Physical Chemistry and Mineralogy of Soils, Volume* 1, by C E Marshall (Wiley, 1964), *The Chemistry of Soil Constituents*, edited by D J Greenland and M B H Hayes (Wiley, 1978), and the first part of *Fabric and Mineral Analysis of Soils*, by R Brewer (Wiley, 1964). Recently more specific and comprehensive accounts have become available in *Minerals in the Soil Environment*, edited by J B Dixon and S B Wood (Soil Science Society of America, 1977) and *Soil Components, Volume 2: inorganic components*, edited by J E Gieseking (Springer-Verlag, 1975). Clay minerals tend to dominate soil mineralogy and several texts concentrate exclusively on these minerals: supplementing the early *Clay Mineralogy*, by R E Grim (2nd edn, McGraw-Hill, 1968) have been *Geology of Clays*, by G Millot (Springer-Verlag, 1970), *The Chemistry of Clay Minerals*, by C E Weaver and L D Pollard (Elsevier, 1973) and *Clays and Clay Minerals in Natural and Synthetic Systems*, by B Velde (Elsevier, 1974).

Our knowledge of clay mineralogy is particularly dependent on advances in analytical techniques, and a valuable text with an emphasis upon this aspect is *Crystal Structures of Clay Minerals and their X-ray Identification*, edited by G W Brindley and G Brown (Mineralogical Society, 1980) to which may be added *Advanced Techniques in Clay Mineral Analysis*, edited by J J Fripiat (Elsevier, 1982). Other techniques are covered in such books as: *Differential Thermal Analysis*, by R C Mackenzie, *Volume 1: Fundamental aspects* (Academic Press, 1970), *Volume 2: Applications* (Academic Press, 1972); *The Infra-red Spectra of Minerals*, edited by V C Farmer (Mineralogical Society, 1974); *Electron Micrographs of Clay Minerals*, by T Sudo *et al* (Elsevier, 1981) and *Electron Microscopy of Soils and Sediments, volumes 1 and 2*, by P Smart and N K Tovey (Oxford University Press, 1982).

Clay mineralogy is also served by two journals, *Clay Minerals*, originally the British *Clay Minerals Bulletin* (1951–1964) but now the journal of the European Clay Groups, and *Clays and Clay Minerals*, originally the American *Proceedings, National Conference on Clays and Clay Minerals*, (1952–67). Important papers on soil mineralogy are also occasionally carried by the *Journal of Sedimentary Petrology, American Mineralogist* and similar journals described in Chapters 10 and 12.

A more recent area of study has been micromorphology. This is concerned with the spatial distribution and organization of soil components, and is of increasing importance in our understanding of pedogenic processes, classification, and detailed soil properties. In this it is analogous to petrography. The subject was initiated by the publication of *Micropedology*, by W L Kubiena (Collegiate Press, Iowa, 1938) and advanced significantly by *Fabric and Mineral Analysis of Soils*, by R Brewer (Wiley, 1964), to which may now be added *Micromorphological Features of Soil Geography*, by W L Kubiena (Rutgers University Press, 1970),

Micromorphology, by E A Fitzpatrick (Oliver & Boyd, 1984) and the new international *Handbook for Soil Thin Section Description*, by P Bullock *et al* (Waine Research Publications, Wolverhampton, 1985). The subject has also benefited from a series of proceedings of International Workshop (now Sub Commission) meetings, the last being *Soil Micromorphology, Volumes* **1** and **2**, edited by P Bullock and C P Murphy (AB Academic Publishers, 1983).

Soil chemistry and analysis

Considered as a chemical system, soil is notoriously complex. This has resulted in a concentration of the chemical literature within soil science itself, with less overlap with that of general chemistry than might be expected. The main theme has been the application of chemistry to the understanding of the behaviour of nutrients in soils, particularly of macro-nutrients such as phosphorus, potassium and nitrogen, and of phenomena such as ion exchange. This theme is in fact central to many of the general texts that have already been referred to above, and it also forms the main contribution to such journals as *Soil Science Society of America Journal* and *Soil Science*. Additional texts concerned directly with the chemical properties of soils include: *Soil Chemistry*, by H L Bohn *et al* (Wiley, 1979), which is a general introductory text; *The Chemistry of Soil Constituents* (Wiley, 1978) and *The Chemistry of Soil Processes* (Wiley, 1980), both edited by D J Greenland and M B H Hayes, which contain very clear accounts of specific topics; and some chapters in *Constituents and Properties of Soils*, edited by M Bonneau and B Souchier (Academic Press, 1982), originally published in French, which takes a more ecological approach to soil science than is usual in English textbooks.

Soil Chemistry, edited by G H Bolt and M G M Bruggenwert in two volumes *Part A: Basic elements* (Elsevier, 1976) and *Part B: Physicochemical models* (Elsevier, 1982), and *The Physical Chemistry and Mineralogy of Soil, Volume* **II**, by C E Marshall (Wiley, 1977) take a physico-chemical approach to soil problems, as do W L Lindsay's *Chemical Equilibria in Soils* (Wiley, 1979) and G Sposito's *The Thermodynamics of Soil Solutions* (Oxford University Press, 1981). The specific problems of nutrient movement are tackled in *Solute Movement in the Soil-Root System*, by P H Nye and P B Tinker (Blackwell Scientific Publications, 1977).

A number of textbooks deal specifically with organic matter in soil, for example *Humus Chemistry; genesis, composition, reactions*, by F J Stevenson (Wiley, 1982) and *Soil Organic Matter*, edited by M Schnitzer and S U Khan (2nd edn, Elsevier, 1985).

Interactions of organic matter with minerals are discussed in *Formation and Properties of Clay-polymer Complexes*, by B K G Theng (Elsevier, 1979) and there are many texts dealing with the behaviour in soils of individual organic compounds used as pesticides: *Organic Chemicals in the Soil Environment, Volume 2*, by C A I Goring and J W Hamaker (Marcel Dekker, 1972) and *Organic Compounds in Soils; absorption, degradation and persistence*, by L G Morrill *et al* (Butterworths, 1982) discuss biochemical and microbiological as well as chemical aspects.

A large number of textbooks deal with soil characteristics of importance in agriculture, and with fertilisers, drainage, etc, among them *Fertilising for Maximum Yield*, by G W Cooke (3rd edn, Granada, 1982), *Fertilisers and Manures*, by K Simpson (Longman, 1986), *Fertilisation of Dryland and Irrigated Soils*, by J Hagin and B Tucker (Springer-Verlag, 1982), *Drainage for Agriculture* (American Society of Agronomy, 1974), and *Land Drainage*, by E Farr and W C Henderson (Longman, 1986). Certain trace elements are of importance as essential micronutrients and have therefore received special attention. In addition to the standard textbooks on geochemistry, further information specific to soils can be found in *Micronutrients in Agriculture*, edited by J J Mortvedt *et al* (Soil Science Society of America, 1972), *Applied Trace Elements*, edited by B E Davies (Wiley, 1980) and *Trace Elements in the Soil-Plant-Animal System*, edited by D J D Nicholas and A R Egan (Academic Press, 1975).

A large portion of the soil chemical literature is devoted to various types of analysis. Several good manuals are available for routine soil analysis, including: *Methods of Soil Analysis*, edited by A Klute and A L Page (American Society of Agronomy, Part 1, 2nd edn, 1986; Part 2, 2nd edn, 1982), *Soil Analysis, Modern Instrumental Techniques*, by K A Smith (Marcel Dekker, 1983) and *A Textbook of Soil Chemical Analysis*, by P R Hesse (Chemical Publishing Co., 1972).

Articles concerned with soil analysis will be found in most of the primary journals already cited, and also occasionally in such journals as *Journal of the Science of Food and Agriculture* (1950–) and *Analyst* (1876/7–). Articles on soil geochemistry are also occasionally carried by *Geochimica et Cosmochimica Acta* (see also Chapter 11).

Soil physics, mechanics and conservation

Soil physics is an inherently highly complex subject, but one which has been the area of many advances over the last thirty or so years. Definable as the study of the state and transport of all forms of matter and energy in soil, it attempts to understand and predict such phenomena as the movement of the fluid and gaseous phases within the intricate spatial patterns set by the numerous, irregular solid components of soil, and to characterize soil thermal properties. The situation is further complicated by the dynamic nature of soil whereby rapidly induced changes occur which dramatically affect its physical nature, for example by wetting-drying and freeze-thaw cycles.

Fortunately, there are now available a variety of texts which approach the subject of soil physics at different levels. Most general soil science texts devote at least one chapter to soil physics, but more comprehensive introductions may be found in *Introduction to Soil Physics*, by D Hillel (Academic Press, 1982) and *Soil Physics*, by H Kohnke (McGraw-Hill, 1968). Standard texts in this area of soil science include *Soil Physics*, by L D Baver *et al* (4th edn, Wiley, 1972), *Fundamentals of Soil Physics*, by D Hillel (Academic Press, 1980) and *Soil Physics*, by T J Marshall and J W Holmes (Cambridge University Press, 1979) — the last work placing emphasis on the central role of water in soil physical processes.

The fundamental importance of soil water is emphasized in more specialized texts. E C Child's *An Introduction to the Physical Basis of Soil Water Phenomena* (Wiley, 1969) has been a standard text on this topic for some years. More recent works include *Soil and Water; physical principles and processes*, by D Hillel (Academic Press, 1971) and *Physical Aspects of Soil Water and Salts in Ecosystems*, edited by A Hadas *et al* (Chapman & Hall; Springer-Verlag, 1973). The field water cycle and its control form the basis of *Applications of Soil Physics*, by D Hillel (Academic Press, 1980), while the relevance of theoretical knowledge of soil water and temperature to actual practical problems is described in *Applied Soil Physics*, by R J Hanks and G L Ashcroft (Springer-Verlag, 1980). Other books dealing with specific areas of soil physics include *Modification of Soil Structure*, edited by W W Emerson *et al* (Wiley, 1978) and the English translation of *Soil Physics, Selected Topics*, by A Kezdi (Elsevier, 1979), the latter text concentrating more on mechanical aspects of soil physics. The specific physical problems and properties of tropical soils is the subject of *Soil Physical Properties and Crop Production in the Tropics*, edited by R Lal and D J Greenwood (Wiley, 1979).

The study of the mechanical properties of soils forms the interface between soil physics and civil engineering, a field with its own considerable literature (see also Chapter 16). Texts which pertain more specifically to soil mechanics include *Elements of Soil Mechanics for Civil and Mining Engineers*, by G N Smith (4th edn,

Granada, 1978), and *Agricultural Soil Mechanics*, by A J Koolen and H Kuipers (Springer-Verlag, 1983); a different approach to the subject, based on the mathematical solving of theoretical problems, is found in *Solution of Problems in Soil Mechanics*, by B H C Sutton (Pitman, 1979).

Two of the most important aspects of applied soil physics involve the methods by which soil water contents are artificially modified by irrigation and drainage. A number of texts are specifically devoted to these topics. The *Drainage Manual*, published by the US Department of the Interior, Bureau of Reclamation (US Government Printing Office, 1978), and *Drainage Principles and Applications*, published by the International Institute for Land Reclamation and Improvement, Wageningen (*Publication* **16**, Vols. **1–4**, 1974) describe the principles of design and installation of drainage systems. Techniques and problems of irrigation encountered in arid zones are documented in *Arid Zone Irrigation*, edited by B Yaron *et al* (Chapman & Hall; Springer-Verlag, 1973) and *Arid Land Irrigation in Developing Countries — environmental problems and effects*, edited by E B Worthington (Pergamon, 1977), while a text published by the British Agricultural Development and Advisory Service (ADAS) entitled *Irrigation* (5th edn, HMSO, 1982) is more relevant to developed, temperate lands. A general text which places irrigation in the general field of soil physics is *Physical Edaphology: the physics of irrigated and non-irrigated soils*, by S A Taylor and G L Ashcroft (Freeman, 1972).

The importance of soil conservation in the light of the constant problem of soil erosion induced by intensive agricultural and forestry practices has continued to be highlighted in the literature. Conservation is specifically covered in *Soil Conservation*, by N Hudson (Batsford, 1971) and *Soil Conservation and Management in the Humid Tropics*, edited by D J Greenland and R Lal (Wiley, 1977), while texts more specifically concerned with the processes of soil erosion include *Soil Erosion and Conservation*, by R P C Morgan (Longman, 1986) and *Soil Erosion*, by D Zachor (Elsevier, 1982), the latter book providing a more expanded coverage of the subject.

The primary literature relating to the various aspects of soil physics is mostly to be found in the general soils journals already referred to. Articles are also carried by *Soil and Tillage Research* (1983–), *Journal of Soil and Water Conservation* (1946–), *Journal of Hydrology* (1963–) and *Journal of Agricultural Engineering Research* (1956–). In the field of civil engineering, articles on soil mechanics and general physical properties can be found in a number of journals, for example *Géotechnique* (1948–) and two publications of the American Society of Civil Engineers — *Journal of Geotechnical Engineering* and *Journal of Irrigation and Drainage Engineering* (1983–; formerly *Journal of the Irrigation and Drainage Division, Proceedings of the American Society of Civil Engineers*).

Soil biology and biochemistry

The soil environment is the permanent habitat for many macro- and microorganisms, and a temporary habitat for numerous others. The size and nature of the biological component of a soil is largely dependent on the physicochemical conditions prevailing, but at the same time soil organisms are able to modify these conditions to a greater or lesser extent. The turnover and transformation of organic and inorganic materials in soils are brought about mainly by soil microorganisms, and chemically and biologically active substances are continually being produced by them. The roots and other subterranean structures of plants are also important in modifying soil, since they both add to (e.g. various organic substances) and extract from (e.g. water, nutrients) the medium in which they grow. The living component of soil is thus highly important in maintaining its overall fertility, and also in the processes of pedogenesis themselves. Although the biological component of soil presents a size continuum, the traditional practice has been to divide it between the soil microbial population (viruses, bacteria, actinomycetes, fungi, algae and protozoa) and soil meso- and macrofauna (animals larger than protozoa). Plants are normally dealt with separately, though increasing attention continues to be given to the vital association of plant roots and soil microorganisms found in the rhizosphere.

General textbooks on meso- and macrofauna inevitably include fauna whose main habitat is the soil environment. More specific texts include *The Distribution and Diversity of Soil Fauna*, by J A Wallwork (Academic Press, 1976) and *Soil Biology*, by W Kuhnelt (2nd edn, Faber & Faber, 1976). Specific organisms of particular relevance to soils are described, for example, in *Termites and Soils*, by K E Lee and T G Wood (Academic Press, 1971) and *Earthworms: their ecology and relationships with soils and land-use*, by K E Lee (Academic Press, 1985). On a more concise and general basis *Life in the Soil*, by R M Jackson and F Raw (Arnold, 1966) provides an introduction to soil biology as a whole, and includes a chapter on methods of study. A more detailed description of methods used in the study of soil macro- and mesofauna is presented in *Methods of Study in Quantitative Soil Ecology: population, production and energy flow*, edited by J

Phillipson (Blackwell, 1971). *Introduction to the Soil Ecosystem*, by B N Richards (Longman, 1974) places emphasis on the role of soil organisms in the balancing of terrestrial ecosystems and, whilst the total sphere of soil biology is discussed, the text does concentrate predominantly on the role of soil microorganisms.

Increasing awareness of the crucial importance of the soil microbial population has resulted in a rapid increase in publications on the subject in recent years. The soil environment features strongly in general environmental microbiology texts, such as *Microorganisms in Action; concepts and applications in microbial ecology*, edited by J M Lynch and J E Hobbs (Blackwell, 1988) and *Environmental Microbiology*, by W D Grant and P E Long (Blackie, 1981). Books dealing exclusively with the topic of soil microbiology include *Soil Microorganisms*, by T R G Gray and S T Williams (Longman, 1971) and *Introduction to Soil Microbiology*, by M Alexander (2nd edn, Wiley, 1977). Both may serve as introductory texts, and give comprehensive coverage of the subject, whereas *Soil Micro-biology*, edited by N Walker (Butterworths, 1975) takes selected topics and covers them in some depth. A detailed account of common techniques used for the study of soil microbiology is given in *Methods for Studying the Ecology of Soil Microorganisms*, by N Parkinson *et al* (Blackwell, 1971), whereas the possibilities which exist for manipulating soil microorganisms so as to optimize crop productivity are discussed in a recent text by J M Lynch *Soil Biotechnology — microbiological factors in crop productivity* (Blackwell, 1983). Present knowledge of various economically important aspects of soil microbiology, and future prospects for developing certain areas are also described in *Advances in Agricultural Microbiology*, a collection of reviews, edited by S Subba Rao (Butterworths, 1982).

Bacteria and fungi always dominate the soil microbial biomass, and in many soils the fungal population is the more significant of these two groups. Interest in soil fungi also arises from the fact that a number of them are plant pathogens, or parasites, and some form important symbiotic associations with higher plants. Texts have therefore been produced which are entirely devoted to the subject of soil fungi: these include *Soil Fungi and Soil Fertility*, by S D Garrett (2nd edn, Pergamon, 1981), the English translation of *Fungi in Agricultrual Soils*, by K H Domsch and W Gams (Longman, 1972) and the reference work *Compendium of Soil Fungi, Volumes* I and II, by K M Domsch *et al* (Academic Press, 1981). Soil fungi also predominate in *Ecology of Root Pathogens*, edited by Y R Dommergues and S H Krupa (Elsevier, 1979). A parallel text, edited by the same two scientists is *Interactions between Non-pathogenic Soil Microorganisms and Plants*

(Elsevier, 1978). Interactions between soil microorganisms and plant roots are of great significance and are described in all general soil microbiology texts. *Plant Root Systems: their function and interaction with the soil*, by R Scott Russell (McGraw-Hill, 1977) describes the interactions between soil and plant roots, including those involving the rhizosphere microbial population.

Soil biochemistry is a subject area which encompasses aspects of soil microbiology, but goes beyond to include, for example, the study of cell-free biological molecules which exist in the soil environment, and the behaviour and fate of pesticides in soils. A series of texts entitled *Soil Biochemistry* has been published intermittently over a number of years and presents topical reviews in each volume. The series, published by Marcel Dekker, has had a number of editors over the years; these are A D McLaren and G H Peterson (Vol.1, 1967), A D McLaren and J Skujins (Vol.II, 1972), E A Paul and A D McLaren (Vols.III and IV, both 1975) and E A Paul and J N Ladd (Vol.V, 1981). Texts which deal in depth with specific aspects of soil biochemistry include *Soil Enzymes*, edited by R G Burns (Academic Press, 1978) and *Nature and Origin of Carbohydrates in Soil*, by M V Cheshire (Academic Press, 1979).

Articles on soil biology and biochemistry appear occasionally in many of the mainstream soil science and microbiological journals, and also in those with an environmental bias, such as *Applied and Environmental Microbiology* (1953–) and *Microbial Ecology* (1974–). Other periodicals cater more specifically for this subject area, however, and include *Soil Biology and Biochemistry* (1969–), *Pedobiologia* (1961–) and *Revue d'Ecologie et de Biologie du Sol* (1964–). All of the latter three journals publish articles in a number of European languages.

Reviews and symposia

At the present time general review articles are regrettably sparse and sporadic, and there are unfortunately no review journals actually centred on soil science. However, journals such as *Soil Science* and *Soil Science Society of America Journal* occasionally carry invited review articles and there were often such articles preceding the abstract section in individual numbers of *Soils and Fertilisers* (1938–). The only current review journal to incorporate regular contributions on soil science is *Advances in Agronomy*, published annually since 1949 by Academic Press. This generally contains from five to ten articles, each of some 25–100 pages in length, and although the overall emphasis is agronomic, one or two articles per issue are usually on some aspect of pedology, soil chemistry or soil mineralogy. A specific annual review journal, *Advances in Soil Science* is published by Springer-Verlag (1985–). *Earth Science Reviews* also occasionally carries articles relating to soil science.

The major international meeting, whose contributed papers are subsequently published, is the International Congress of Soil Science. A full congress is held every four years, the thirteenth being that in Hamburg, West Germany, in 1986. The papers presented are grouped under seven commissions, namely: soil physics; soil chemistry; soil biology; fertility and plant nutrition; genesis, classification and cartography; technology; and mineralogy. The *Transactions* are usually obtainable via the International Society of Soil Science, International Soil Reference and Information Centre, PO Box 353, 9, Duivendal, 6700 A.J. Wageningen, The Netherlands. Conferences on selected themes are also held by individual commissions in the intervening years.

Other symposia are held by various universities or institutions on a wide range of topics, some regularly, some sporadically, the proceedings being published as monographs or reports.

A number of individual topics have been covered at regular intervals by international conferences and symposia. The resultant publications are typically of the order of 300–500 pages, containing 40–50 articles. Some papers delivered at European meetings are in French or German, but there are usually summaries in English. These books form valuable sources of data, particularly in those branches of soil science which are developing rapidly. Some examples are the *Proceedings of the Bangkok Symposium on Acid Sulphate Soils*, edited by H Dost and N van Breeman (International Institute for Land Reclamation and Improvement (IILRI), Wageningen, 1982) and *Acid Sulphate Weathering (Soil Science Society of America, Special Publication* **10**, 1982). *Soils with Variable Charge*, edited by B K G Theng (New Zealand Society of Soil Science, 1980) reports a conference held in New Zealand. *Pseudogley and Gley*, edited by E Schlichtling and U Schwertmann (Verlag Chemie, 1973) deals with waterlogged soils and *Soil Micromorphology, Vols.* **1** and **2**, edited by P Bullock and C P Murphy (AB Academic Publishers, 1983), the proceedings of the international working meeting held in 1981, provides a recent survey of research in a specialized field, as does *The Soil-Root Interface*, edited by J L Harley and R S Russell (Academic Press, 1979), the proceedings of an international symposium held in Oxford in 1978.

Abstracting services

As described in Chapter 6, literature searches are now often accomplished with the aid of computer data banks, and such systems as the DIALOG service are likely to become even more important and more widely available in the near future. Recent developments have been discussed in conference proceedings such as *Developments in Soil Information Systems; proceedings of the second meeting of the ISSS working group on soil information systems, Bulgaria, June 1977*, edited by A Sadowski and S W Bie (PUDOC, 1978). However, without doubt the most valuable and comprehensive system of abstracting in soil science is that operated by the Commonwealth Bureau of Soils. Their publications include the abstract journal *Soils and Fertilisers*, and *Irrigation and Drainage Abstracts* (1975–), as well as annotated bibliographies on selected topics.

Soils and Fertilisers was initiated in 1938 replacing various earlier publications covering this field, and it is issued every month. Each issue now contains some 1000 abstracts which are designed to be informative rather than indicative, and which, according to Jacks (1966), represented approximately 60 per cent of the total number of articles actually reviewed by the Bureau. The abstracts are arranged in groups by subject matter according to a simplified version of the Universal Decimal Classification system, and fall into one of ten main classes, of which the first, soils, is relevant in this context. This class comprises pedology, soil chemistry, analysis, physics, classification and biology, and it accounts for over half of the abstracts in each issue (i.e. approximately 600); the remainder cover the cultivation, fertilization and management of soils, plant diseases and pesticides. The time lag between publication in a primary journal and appearance of the abstract varies from under a year for articles in the major English language journals, to two or three years in the case of the more obscure journals in other languages. In addition to the abstracts, each issue also contains an author index. All these data are rationalized and annual indexes of subjects and authors are published. Prior to 1962 indexes were produced trienially.

Another valuable form of publication produced by the Commonwealth Bureau of Soils is the *Annotated Bibliography*. This is a duplicated typed series of abstracts of the literature on selected topics. The bibliographies cover a specified period, which ranges from 4 to 30 years, and on important topics supplements are issued to keep the abstracts up to date. A bibliography may contain from 10–100 abstracts and vary in price from £1 to £17. Over 2000 such bibliographies have now been issued, some of the 1000 currently available being listed occasionally in the front of *Soils and Fertilisers*. They are obtainable direct from the Commonwealth Bureau of Soils (Rothamsted Experimental Station, Harpenden, Hertfordshire, England).

Soil science literature is also covered by several other abstracting journals, although none is as comprehensive or convenient as *Soils and Fertilisers*, nor is the rationale behind their selection systems so evident. *Bibliography of Agriculture* (Onyx Press, 1942–) subdivides material only into sections for 'Soil science', 'Soil improvement materials' and 'Soil resources and management'. *Biological Abstracts* (1926–) contains a total of over 170,000 abstracts per year, of which 1,800 come under the section headed 'Soils' — that is, around 70 abstracts per issue. The section, however, is subdivided into three parts only — 'general', 'inorganic' and 'organic': abstracts on soil microbiology are included in a separate section. *Chemical Abstracts* (1907–) operates on a similar scale and contains some 25 abstracts per issue relating to soils in the section headed 'Mineralogical and geological chemistry' and some 60 abstracts in the section 'Fertilisers, soils and plant nutrition': in this case there is no subdivision of the soil abstracts. *Mineralogical Abstracts* (see also Chapter 10) operates on a smaller scale but is particularly valuable for the section on 'Clay minerals' introduced in 1960. This section is in turn subdivided into two parts — 'Techniques; Structures; Properties' and 'Petrology; Weathering; Soils'. Both of these parts average some 20 abstracts each per issue, and other relevant abstracts may also be found in the sections on 'Apparatus and techniques' and 'Geochemistry'. Similarly, *Geographical Abstracts*, Part A, which bears the title 'Landforms and the Quaternary' contains a section on soils, each issue having about 60 abstracts.

References

Glinka, K D (1928) Dokuchaev's ideas on the development of pedology and the cognate sciences. *Proceedings and Papers of the 1st International Congress of Soil Science*, **1**, 116

Jacks, G V (1966) The literature of soil science. *Soils and Fertilisers*, **29** (3), 227–230

Kellog, C E (1940) Reading for soil scientists, together with a library. *Journal of the American Society of Agronomy*, **32**, 867–876

Robinson, G W (1924). Pedology as a branch of geology. *Geological Magazine*, **61** (9), 444–455

Wild, A (1988) ed. *Russell's Soil Conditions and Plant Growth*, 11th edn. Longmans

Wood, J A (1970). The lunar soil. *Scientific American*, **233**, (2), 14–23

CHAPTER TWENTY

History of geology

DOUGLAS A BASSETT

The standard works on the history of geology for the first half of this century were: *The Founders of Geology*, by Sir Archibald Geikie (Macmillan, 1897; 2nd enlarged edition, 1905; Dover Reprint, 1962), *Geschichte der Geologie und Paläontologie bis Ende des 19 Jahrhunderts*, by Karl von Zittel (Oldenbourg, 1899; Johnson Reprint, 1965), which was translated into English as *History of Geology and Palaeontology to the End of the Nineteenth Century* (Walter Scott, 1901; facsimile reprint, Wheldon & Wesley, 1962), and *The Birth and Development of the Geological Sciences*, by Frank Dawson Adams (Baillière, Tindall & Cox, 1938; Dover Reprint, 1954). Fuller details of the contents and editions of these volumes — all by practising geologists — and the way they complement one another, as well as bibliographical details of other general works, both continental and British, are given in the first version of this chapter (p.400–402 *in* D N Wood, (ed.) *Use of Earth Sciences Literature*, Butterworths, 1973).

During the nineteen-fifties a number of volumes by professional historians threw new light on the history of the subject and in part complemented, in part superseded, the earlier volumes. They included: *Genesis and Geology: a study in the relations of scientific thought, natural theology, and social opinion in Great Britain, 1790–1850*, by C C Gillispie (Harvard University Press, 1951; reprinted 1969; Harper Torchbook, 1959), *The Death of Adam: evolution and its impact on western thought*, by J C Greene (Iowa State University Press, 1959; Mentor Paperback, 1961), and *Natural Law and Divine Miracle. A historical-critical study on the 'principle of uniformity' in geology, biology, and theology*, by R Hooykaas (Brill, 1959), re-named *The Principle of Uniformity* in a new edition (Brill, 1963), with a French translation *Continuité et Discontinuité en Géologie et Biologie*, (Éditions de Seuil) in 1970.

During the nineteen-sixties the study of the history of geology, as of all the sciences, underwent an even greater change in perspective. This followed the combined work of scholars such as Michael Polanyi (e.g. *Personal Knowledge. Towards a post-critical philosophy*, Routledge & Kegan Paul, 1958), Norman Hanson (e.g. *Patterns of Discovery: an inquiry into the conceptual foundation of science*, Cambridge University Press, 1961), and particularly, Thomas S Kuhn in *The Structure of Scientific Revolutions* (1962; 2nd enlarged edn, 1970, *International Encyclopedia of Unified Science*, **2**(2), University of Chicago Press), which, in mapping out the intricate interrelations of fact and theory and of science and society, seriously questioned the pre-suppositions of the philosopher of science and the routine assumptions of the historian of science. In essence they presented scientific work as skilled craftsmanship, practised within a shared tradition, maintained within a social community of scientists.

The various changes in perspective are considered in such publications as: 'A revolution in historiography of science', by G Buchdahl (*History of Science*, **4**, 55–69, 1965, in which the discussion centres on T S Kuhn's volume and Joseph Agassi's *Towards an Historiography of Science*, Mouton & Co, 1963); 'Historiography of science: its aims and methods', by R Hooykaas (*Organon*, **1**, 37–49, 1970); *The History of the Natural Sciences as Cultural History*, by M J S Rudwick (Free University, Amsterdam, 1975); 'Sir Archibald Geikie (1835–1924), geologist, romantic, aesthete and historian of geology: the problem of Whig historiography of science', by D R Oldroyd (*Annals of Science*, **37**(4), 441–462, 1980); 'Science, scientists, and historians of science', by N Reingold (*History of Science*, **19**, 274–283, 1981); 'Pitfalls in the historiography of geological science', by R Hooykaas (*Histoire et Nature*, **19–20**, 21–34, 1983); and 'On Whiggism', by A R Hall (*History of Science*, **21**, 45–59, 1983).

The changes of perspective are also reflected either explicitly or implicitly in the following major works:

The Meaning of Fossils: episodes in the history of palaeontology, by M J S Rudwick (Macdonald & American Elsevier, 1972), which was the first explicit 'textbook' of the application of the new historiography to a geological subject. It records five 'revolutions' in palaeontology and is above all a history of argument. A second edition of the book, in 1976 (and the re-issue by the University of Chicago Press in 1985) are facsimile reprints of the first with a new preface which contains the author's defence of the episodic nature of the work.

The Great Devonian Controversy: the shaping of scientific knowledge among gentlemanly specialists, also by M J S Rudwick (University of Chicago Press, 1985) is another 'textbook' of the new historiography. It contains a summary of the changes in perspective that have taken place.

The Earth in Decay: a history of British geomorphology 1578–1878, by G L Herries Davies (Macdonald, 1969), which does for the study of the history of geomorphology what Rudwick has done for palaeontology.

The Making of Geology: earth science in Britain 1660–1815, by R S Porter (Cambridge University Press, 1977). This work 'sees science as an integral part of the spectrum of man's intellectual and social activities' and is designed specifically to demonstrate that there existed in Britain, by 1815, a science of geology, a discipline recognized as such and one with clearly formulated methods and concepts, a defined programme of research, its own institutions and an extremely enthusiastic public. The author extends certain aspects of his view of 'The making of geology' in 'Creation and credence: the career of theories of the Earth in Britain 1660–1820' (*in* B Barnes and S Shapin (eds): *Natural Order: historical studies of scientific culture*, Sage Publications, 1979) and in a chapter on 'The terraqueous globe' in *The Ferment of Knowledge: studies in historiography of eighteenth century science*, edited by G S Rousseau and R S Porter (Cambridge University Press, 1980).

Toward a History of Geology, edited by C J Schneer (MIT Press, 1969), which contains papers read at one of the very first inter-disciplinary and international conferences on the subject of the history of geology before the publication of Darwin's *Origin of Species*.

Images of the Earth: essays in the history of the environmental sciences, edited by L J Jordanova and R S Porter (British Society for the History of Science, *Monograph No.*1, 1979), which includes papers read at another inter-disciplinary conference designed 'to move away from the biographical mode which has been prominent in such collections as *Toward a History of Geology*. And, perhaps above all, . . . to stimulate thought about the earth broadly considered as an object of scientific investigation, transcending the conventional straight-jacket of the "history of geology".' Unlike the American collection, however, ten of the eleven essays in *Images of the Earth* deal primarily, or entirely, with British geology and nine concentrate on the period 1750–1830. The introduction surveys the developments in the historiography of the earth sciences.

Science in Culture: the early Victorian period, by Susan F Cannon (Dawson & Science History Publications, 1978), which, although not specifically concerned with geology, is of particular relevance for three reasons. First, it illustrates the author's continuing attempts to develop analytical tools into the study of the history of science by means of revised editions of earlier and important papers; second, it is relevant in arguing that the approach to science in early nineteenth century Britain should be Humboldtian (the interplay between geology, geophysics and chemistry) rather than Baconian; and third, in highlighting the singular importance of the period up to 1859 for science as a whole.

The Naturalist in Britain: a social history, by D E Allen (Allen Lane, 1976), the first ever social history of naturalists in Britain. Aspects of the main theme are taken further in two papers in *The Journal of the Society for the Bibliography of Natural History*: 'Natural history and social history' (**7**, 509–516, 1976) and 'Naturalists in Britain: some tasks for the historian' (**8**, 91–107, 1977).

The frequent references to these eight important volumes in the remainder of the text are to their short titles only.

Organizations

The increase in interest shown by professional historians of science has been matched by an upsurge of interest among practising geologists in the history of their subject. This is expressed particularly in the establishment of a committee under the aegis of the International Union of the Geological Sciences in the mid-nineteen-sixties and, in the early nineteen-eighties, the formation of the first society and the first journal concerned specifically with the history of the earth sciences.

International Committee on the History of the Geological Sciences (INHIGEO)

The concept of an International Committee on the History of the Geological Sciences (INHIGEO) was discussed at the 22nd International Geological Congress in 1964 and the USSR delegation given the responsibility of bringing such a body into being as a committee of IUGS. As a result, the first meeting of INHIGEO was held in Yerevan, Armenia, in 1967 and the Committee has since organized a number of meetings through its own national groups and in association with other national societies. Among its main scientific symposia (with the resulting publications) have been: (i) 'History of concepts in mineral deposits', Freiberg, 1970 (Special issues of *Geologie: Zeitschrift fur das Gesamtgebeit der Geologischen Wissenschaften*, **20** (4/5), (6/7), 1971); (ii) 'History of concepts in Precambrian geology', Montreal, 1972 (*Geological Association of Canada Special Paper* **19**, edited by W O Kupsch and W A S Sarjeant, 1979); (iii) 'The Charles Lyell Centenary Symposium', London, 1975 (Special issue of the *British Journal for the History of Science*, **9**(2), 1976); (iv) 'Regional influences on the origin and development of geological theories (especially in the 18th and 19th centuries)', Munster-Bonn, 1978 (Special issue of *Munsterche Forschungen zür Geologie und Paläontologie*, 1983); (v) 'The development of geology up to the death of Cuvier: work in the French language in the international exchange of ideas', Paris, 1980 (Special issue of *Histoire et Nature*, **19–20** (1981–82), 1983); (vi) 'Development of geological mapping and cartography in connection with the progress of geological thought', Budapest, 1982 (*Contributions to the History of Geological Mapping*, edited by E Dudich, Akademiai Kiado, Budapest, 1984).

The INHIGEO *Newsletter* (1967–) contains reports on the work of the national groups, the scientific symposia and on other symposia of interest, and includes reviews of works on the history of geology and lists of other relevant material. The aims of the Committee are outlined by R Hooykaas (*Newsletter* No. 14, 15–20, 1980) and the complete sequence of INHIGEO Symposia (from 1967 to 1983) described by M Guntau (in No. 17, 4–11, 1983).

History of Earth Sciences Society

On the initiative of a number of geologists with an interest in the history of their subject and of some historians of science, a History of Earth Sciences Society was formed in 1981. Since 1982 it has issued a journal, *Earth Sciences History*, which contains articles, reviews of books and lists of other relevant publications, as well as a calendar of events. The journal has included a number of special thematic issues, for example: History of geology and geological concepts in the North Eastern United States (**2**(2), 1983); The 19th century history of geology in Kansas (**3**(2), 1984); History of geology and geological education in the Southern and Border

States (**4**(1), 1985); Plate tectonics and biogeography (**4**(2), 1985); as well as special European (**3**(1), 1984) and Australian/New Zealand (**5**(1), 1986) issues, and one to celebrate George W White's contributions to geology (**2**(1), 1983).

References arranged chronologically

The arrangement adopted in the first version of this chapter, (1973, 400–436) based on George Sarton's three co-ordinates ('time', 'subject' and 'location'), is maintained. The chronological sequence has been completely rewritten to incorporate new material; the references cited in the first edition of the other two sections are not repeated, but a selection of new references is provided. The sections on bibliography and biography are also completely rewritten and virtually all of the references are additional to those in the first version. A new section ('Other works of reference') has been added.

Classical times and the dark and middle ages

The most accessible works which deal with these early periods are: *The Birth and Development of Geological Sciences*, by F D Adams (which has chapters on 'Geological sciences in classical times' and 'The conception of the universe in the Middle Ages'); Archibald Geikie's *Founders of Geology* (with a chapter on 'Geological ideas among the Greeks and Romans'); *A Source Book in Greek Science*, compiled by M R Cohen and I E Drabkin (McGraw-Hill, 1948) which contains 600 pages of extracts on geography, geology, meteorology, physics, etc.; the chapter on 'The development of geological thought and the scientific study of scenery', by R W Clayton (*in* John Brierley, ed. *Science in its Context: a symposium with special reference to sixth-form studies*, Heinemann, 1964), which contains a list of classical works, classified according to their subject matter, and a very useful chart illustrating the growth of geology, and the names of the major authorities; and *Science in the Early Roman Empire: Pliny the Elder, his sources and influence* (Croom Helm, 1986), a series of essays edited by R French and F Greenaway.

Articles include: J W Harrington's 'The first, first principles of geology' (*American Journal of Science*, **265**, 449–461, 1967) which discusses Herodotus' understanding of geological principles; 'Pliny's *Historia Naturalis*: the most popular natural history ever published', by E W Gudger (*Isis*, **6**, 269–281, 1924); 'Ptolemy's *Geography, Book VII*, Chapters 6 and 7', by O E H Neugebauer (*Isis*, **50**, 22–29, 1959); Rushdi Said's 'Geology in tenth century

Arabic literature' (*American Journal of Science*, **248**, 63–66, 1950); and 'Geology in embryo (up to 1600 A.D.)', by C E N Bromehead (*Proceedings of the Geologists' Association*, **56**, 89–134, 1945) which contains a useful bibliography of works by contemporary and modern authors.

The three volumes of George Sarton's *Introduction to the History of Science* (Williams and Wilkins, for the Carnegie Institution, 1927–48) provide a very important reference series for the period up to the fourteenth century, being rich in biography and bibliography. The titles of the volumes are: *From Homer to Omar Khayyam* (1927); *From Rabbi Ben Ezra to Roger Bacon* (1931: in two parts); and *Science and Learning in the Fourteenth Century* (1948). Three other earlier works which are of particular importance are: *Della Storia Naturale delle Gemme, delle Pietre è di tutti i Minerali, ovvero della Fisica Sotteranea*, by D Giacinto Gimma (2 vols, Naples, 1730), which provides one of the best guide-books for the student who wishes to explore the maze of the ancient literature; *The Failure of Geological Attempts Made by the Greeks from the Earliest Ages Down to the Epoch of Alexander*, by J Schvarcz (1868), which reviews the whole body of ancient Greek literature prior to the time of the expedition of Alexander the Great (334 B.C.); and W Whewell's *History of the Inductive Sciences, from the Earliest to the Present Time* (3 vols, Parker, 1837; facsimile reprint of 3rd edn, Cass, 1967). References to other early commentaries are given by Adams.

The works of Aristotle (384–322 B.C.), *Historia Naturalis*, by Pliny (A.D. 23–79), *De Rerum Natura*, by Lucretius (99–55 B.C.), *On Architecture*, by Vitruvius (?37 B.C.), and the *History* of Herodotus, are available in English translation in the Loeb Classical Library and in the *Great Works of the Western World* series (Encylopaedia Britannica, 1952).

Secondary sources for geographical ideas are contained in *The Geographical Lore of the Time of the Crusades. A study in the history of medieval science and tradition in Western Europe (1095–1270)*, by J K Wright (American Geographical Society, 1925; Dover Reprint, 1965) and *Geography in the Middle Ages*, by G H T Kimble (Methuen, 1938), which contains a chapter on 'Physical geography in the Middle Ages'. The Dover edition of the former contains a new introduction by C J Glacken which attempts to bring the reader up to date.

Fifteenth and sixteenth centuries

An introduction to the study of geological phenomena in the fifteenth century is given by E Stokes in 'Fifteenth century earth science' (*Earth Science Journal*, **1**, 130–148. 1967), in which the parts dealing with earth science phenomena in two late mediaeval encyclopaedias — *Mirrour of the World* and Ranulph Higden's *Polychronicon* (both printed by William Caxton in the 1480s) — are examined in relation to fifteenth-century ideas about the physical nature of the earth and the universe. Topics such as the four elements, the earth and the spheres, gravity, the arrangement of the continents and oceans, the unity of waters, earthquakes and volcanoes, erosion, fossils, mountain building, and climatic zones are summarized and reference is made to the Biblical and classical Greek sources of these ideas.

Reference to the work of the oustanding figure of the period, the universal genius Leonardo da Vinci (1452–1519), is made in several works: 'Some early references to geology from the sixteenth century onwards', by W P D Stebbing (*Proceedings of the Geologists' Association*, **54**, 49–63, 1943); *Geografia e Geologia Negli Scritti di Leonardo da Vinci*, by Agostino Gianotti (Museo Nazionale Scientia, Milan, 1953?); and 'Leonardo da Vinci's geologische studien', by R Weyl (*Natur und Volk Senckenbergische Naturforschende Gesellschaft*, **79**, 2–10, 1949). *Leonardo da Vinci's Note Books, Arranged and Rendered into English*, by E McCurdy (1906), should also be consulted.

The extensive literature on the work and influence of arguably the most contentious figure, Phillipus Aureolus Theophrastus Bombastus von Hohenheim (1493–1541) — usually referred to as Paracelsus — can be readily traced in general works on the history of science and of medicine. Three books of particular interest are *Paracelsus. An introduction to philosophical medicine in the era of the Renaissance*, by W Pagel (Karger, 1958), a thoroughly documented study which is a reliable guide to the writings and biographical studies of Paracelsus, *The English Paracelsians* (Oldbourne Press, 1965), and *The Chemical Dream of the Renaissance* (Heffer, 1968), both books by A G Debus. These, and other works, are assessed by W Pagel in 'Recent Paracelsian studies' (*History of Science*, **12**, 200–211, 1974).

The 'chemical, medical, vitalistic and religious backdrop' against which to view the work of the English physician-chemist Edward Jorden (1569–1632), who sought a new explanation of metals and the source of the internal heat of the earth, are considered by Debus in *Toward a History of Geology* ('Edward Jorden and the fermentation of metals: an iatrochemical study of terrestrial phenomena').

Theories of the earth in the cosmologies of the period approximately delimited by Nicolas of Cusa (1401–64) on the one hand and Nathanael Carpenter (1589–1628) on the other, are outlined by Suzanne Kelly in *Toward a History of Geology*. Of the many topics surveyed, three are discussed in some detail, namely, the origin and future destruction of the earth, the structure of the earth, involving the material or materials of which the earth was made and their arrangement, and the changes in the surface of the earth, both minor and major. Other topics such as the origin of metals, rivers and springs are discussed in F D Adams' *Birth and Development of the Geological Sciences*.

Three of the main names in sixteenth-century geology are included in the title of P Brunet's article 'Les premiers linéaments de la science géologique: Agricola, Palissy, George Owen' (*Revue d'Histoire des Sciences et de leurs Applications*, **3**, 67–79, 1950), which forms a useful starting point for the study of the period.

The best-known reference to the work of Georgius Agricola (1494–1555) is the translation by H C Hoover and L H Hoover of *De Re Metallica* (The Mining Magazine, 1912; Dover reprint, 1950), which includes important biographical and bibliographical material as well as extensive discussions of the early history of geology and the mineral industry. *Agricola on Metals. . . .*, by B Dibner (Norwalk, Conn.; Burndy Library, 1958), on the other hand, is a slim book which contains an historical introduction, summaries of the contents of each chapter of the original placed in their proper historical setting and explained in language intelligible to the non-technical reader, and an explanation of the relation of the work of Agricola to that of the later geologists Abraham Werner and James Hutton. Another important translation is that from the first Latin edition (1546) of *De Natura Fossilium* [Textbook of Mineralogy] by M C and J A Bandy for the Mineralogical Society of America (*Geological Society of America Special Paper* **63**, 1955). Attention to Agricola's *De Ortu et Causis Subterraneorum* (1546) is drawn by Joan Eyles (*Nature*, **176**, 949–950, 1955). Two other essential references are the series of seven books under the collective title *Georgius Agricola — Ausgewählte Werke*, edited by H Prescher and sponsored by the Staatliches Museum für Mineralogie und Geologie, Dresden, and the commemorative volume *Georgius Agricola, 1494–1555, zu Seinem 400. Todestag 21* (1955), issued by the Central Agricola Commission of the German Democratic Republic.

A translation of the best known work by 'Palissy the Potter' (1520?–1590), *Discours Admirables, de la Nature des Eaux et Fontaines, tant Naturelles qu'Artificielles, des Métaux des Sels et Salines, des Pierres, des Terres*, etc. (Paris, 1580), under the title *The Admirable Discourses* (University of Illinois Press, 1957) is given by Aurèle La Rocque. The sources of Palissy's work and the important influences in his life are outlined by the same author, in

Toward a History of Geology, in which he also catalogues the comments by subsequent authorities on Palissy's work and thus provides a useful bibliographic key. A supplement to La Rocque's work is 'The geographical and geological observations of Bernard Palissy the Potter', by H R Thompson (*Annals of Science*, **10**, 149–165, 1954). The rare edition of Palissy's *Oeuvres Complètes* (J J Dubochet et Cie, 1844) was reprinted in 1961 (Librairie Scientifique et Technique Albert Blanchard). The original notes and historical notice by Paul-Antoine Cap are reproduced and the new preface by Jean Orcel is designed as a popular introduction.

The best reference to the geological work of George Owen (1552–1613) in south-west Wales, is still F J North's 'From Giraldus Cambrensis to the geological map. The evolution of a science' (*Transactions of the Cardiff Naturalists' Society*, **64**, 20–97, 1933); the best general study is *George Owen of Henllys: a Welsh Elizabethan*, by B G Charles (National Library of Wales Press, 1973); and the most reliable transcription of the ms *The Description of Pembrokeshire*, is in the Cymmrodorion Record Series No. 1, edited by H Owen (1892) with quotations from it in John Challinor's paper (*Annals of Science*, **9**, 127–129, 1953).

A name which should have been included in the title of Brunet's article is that of Conrad Gesner (1516–65), see *Conrad Gesner, physician, scholar and scientist; a quatercentenary exhibit*, by Richard Durling (National Library of Medicine, 1968). Gesner's work on minerals is considered in Adams' *Birth and Development*, and on fossils by Rudwick in *The Meaning of Fossils*. As the latter demonstrates, *De Rerum Fossilium* [On Fossil Subjects] (Tiguri, 1565) marked a crucial moment in the emergence of the science of palaeontology because it incorporated three innovations of outstanding importance for the future and yet, at the same time, its form and content epitomized the scientific and social matrix within which that emergence took place. The innovations were: the use of illustrations to supplement verbal descriptions, the establishment of collections of specimens, and the formation of a scholarly community co-operating by correspondence.

Details of the geological studies of a fifth significant figure, the Portuguese Joao de Castro (1500–48), are given in R Hooykaas' comprehensive study of his work, *Science in Manueline Style* (Academia Internacional da Cultura Portugesa, 1980 — in English).

Secondary sources of more general interest for the fifteenth and sixteenth centuries include: *Tudor Geography 1485–1583*, by E G R Taylor (Methuen, 1930), which contains a 'Catalogue of English geographical or kindred works (printed books and mss) to

1583', *Science and Thought in the Fifteenth Century*, by Lynn Thorndyke (Columbia University Press, 1929), and *The Origins of Museums: the cabinet of curiosities in sixteenth and seventeenth century Europe*, edited by O Impey and A MacGregor (Clarendon Press, 1985). The last was the product of an international symposium held at Oxford as part of the Tercentenary Celebrations of the founding of the Ashmolean Museum: it contains a wide-ranging chapter on 'Early collecting in the field of geology' by Hugh Torrens.

Seventeenth century

During the first half of the seventeenth century geological subjects received comparatively little attention, but in the second half discoveries were deliberately undertaken, collections of objects were made and the seeds of some of the basic principles underlying modern stratigraphy and palaeontology sown.

C J Schneer discusses 'The rise of historical geology in the seventeenth century' (*Isis*, **45**, 256–268, 1954) and stresses that the investigators were commonly interested in both man-made and natural antiquities. J M Levine in *Dr Woodward's Shield: history, science and satire in Augustan England* (University of California Press, 1977) states that for John Woodward natural history and civil history were the same kinds of activity, dealing with two halves of the same story; F V Emery in *Edward Lhuyd FRS, 1660–1709* (University of Wales Press, 1971 — bilingual) describes the relationships between Lhuyd's work in geology, botany, archaeology and philology; and M J S Rudwick's 'Natural antiquities' in *The Meaning of Fossils* (1972) discusses the difficult contemporary problem of distinguishing organic and inorganic material in the context of seventeenth century science, and the allied consideration of using fossils as one of the sources of evidence for reconstructing and explaining the history of the earth, as distinct from the history of man.

The contributions of Nicholas Steno (Niels Stensen) (1638–86) and Robert Hooke (1653–1703) are the subject of an extensive literature. Examples include: *Nicolaus Steno and his Indice*, edited by Gustav Scherz (Munksgaard, 1958) which contains numerous articles on Steno as well as the text of his *Indice di Cose Naturali, Forse Dettato da Niccolo Stenone*; and *The Earliest Geological Treatise (1667) by Nicolaus Steno* (Macmillan, 1958), translated by Alex Garboe (with an introduction and notes) from *Canis Carchariae Dissectum Caput*. Steno's other geological treatises are submitted in their original texts, in translation with annotations, and with critical reviews, in 'Steno: geological papers', edited by G Scherz and translated by A J Pollock (Odense University Press, 1969) (*Acta Historica Scientiarum Naturalium et Medicinalium*, **20**). Steno's relations with the Royal Society are outlined in 'Nicolas Stenon et la Royal Society of London', by Remacle Rome (*Osiris*, **12**, 244–268, 1956).

Robert Hooke's *Micrographia: or some physiological descriptions of minute bodies made by magnifying glasses with observations and inquiries thereupon* (Martyn and Allestry for the Royal Society, 1665) is reproduced in R T Gunther's series *Early Science in Oxford*, Volume **13** (Oxford University Press for the Subscribers, 1938; reprinted 1968). Attention to Hooke's geological views is drawn in 'The first English geologist', by A P Rossiter (*Durham University Journal*, **29**, 172–181, 1935), 'Robert Hooke and his conception of earth history', by G L Herries Davies (*Proceedings of the Geologists' Association*, **75**, 493–498, 1964), and, together with the biography *Robert Hooke*, by Margaret 'Espinasse (Heinemann and University of California Press, 1956), they provide much additional material and useful bibliographies of both primary and secondary sources.

Hooke's major geological contribution *A Discourse of Earthquakes* has been studied by a number of historians, for example: 'Robert Hooke, Rudolf Erich Raspe and the concept of earthquakes', by A V Carozzi (*Isis*, **61**, 85–91, 1970), 'Robert Hooke's methodology of science as exemplified in his *Discourse of Earthquakes*', by D R Oldroyd (*British Journal for the History of Science*, **6**, 109–130, 1972), and 'Hooke on earthquakes: lectures, strategy and audience', by R Rappaport (*British Journal for the History of Science*, **19**, 129–146, 1986).

During the late seventeenth century, Britain witnessed a series of systematic attempts to explain the nature, history and future of the earth. The best known of these, by Thomas Burnet, William Whiston, Edmund Halley, Erasmus Warren, John Arbuthnott, John Harris, Thomas Robinson, John Woodward, Robert Hooke and John Ray and the various responses to them are discussed in a number of publications. These include *Cosmogonies of our Fathers: some theories of the seventeenth and eighteenth centuries*, by K B Collier (Columbia University Press and P S King & Son, 1934), Chapter III in *The Earth in Decay* (1969) by G L Herries Davies, Part 1 of *The Making of Geology* (1977) by Roy Porter, 'La théorie de la terre au XVIIe siècle', by Jacques Roger (*Revue d'Histoire des Sciences*, **26**, 23–48, 1973) and *Time's Arrow; time's cycle. Myth and metaphor in the discovery of geological time*, by S J Gould (Harvard University Press, 1987).

Bibliographic data relating to various theories of the earth published in the seventeenth century, with special reference to contemporary discussion and published commentaries in the light of their significance for the study of the history of geology, are given in 'Bibliography and the history of science', by V A Eyles (*Journal of the Society for the Bibliography of Natural History*, **3**, 63–71, 1955).

Eighteenth century (the first three quarters)

V A Eyles' contribution in *Toward a History of Geology* ('The extent of geological knowledge in the eighteenth century, and the methods by which it was diffused') provides a comprehensive guide to the primary literature of this period. He maintains that, contrary to popular opinion, a great deal of geological work was achieved during the century, and his contention is confirmed in the other eight papers in this symposium volume, which deal with various aspects of eighteenth century geology in America, France, Germany, Russia and Sweden. Eyles makes little reference to secondary sources, however, and for this reason the papers by R Rappaport 'Problems and sources in the history of geology, 1749–1810' (*History of Science*, **3**, 60–77, 1964), Eyles himself 'The history of geology: suggestions for further research' (*History of Science*, **5**, 77–86, 1966) and R Porter and K Poulton 'Research in British geology 1660–1800, a survey and thematic bibliography' (*Annals of Science*, **34**, 33–42, 1977) are useful supplements.

Two important aspects of eighteenth century geology not considered by Eyles are the growth of geomorphology and the search for order. The former is dealt with in a comprehensive way in *The Earth in Decay* (1969) by G L Herries Davies; the latter was a search shared by most of the sciences. One aspect is discussed in a short chapter 'The eighteenth century and the idea of order' in Jacob Bronowski's *The Common Sense of Science* (Heinemann, 1951; Penguin Books, 1960), another in *The Great Chain of Being; a study of the history of an idea*, by A O Lovejoy (Harvard University Press, 1936; reprinted 1964; Harper Torchbook, 1960), the classic study of the *Scala Naturae* or 'Ladder of being'. Lovejoy's monograph is reassessed in 'The great chain of being after forty years', by W F Bynum (*History of Science*, **13**, 1–28, 1975) and by L Formigari in *Dictionary of the History of Ideas*, edited by P Wiener (**1**, 325–335, 1968). See also W D I Rolfe's contribution 'William and John Hunter: breaking the great chain of being' in W F Bynum and Roy Porter (eds), *William Hunter and the Eighteenth Century Medical World* (Cambridge University Press, 1985).

The most recent survey of geology in the century as a whole is that provided by R Porter in *The Making of Geology* (1977). The book is strongest for the period 1660–1710 and for the end of the eighteenth century, particularly on Hutton's place in the history of geology. A work which provides a wider geographical context is 'Historicism and the rise of historical geology', by D R Oldroyd (*History of Science*, **17**, 191–213, 227–257, 1979) in which the author surveys the work of Giovanni Arduino in Italy, J G Lehmann, G C Fuchsel and A G Werner in Germany, G L Buffon, Nicolas Desmarest, J A De Luc and others in France and Switzerland, as well as of James Hutton, William Smith and Charles Lyell in Britain.

A selection of other relevant works includes the following studies: of the crucial role of deluges in many of the theories of the earth prevalent in France, R Rappaport's paper 'Geology and orthodoxy: the case of Noah's Flood in eighteenth century thought' (*British Journal for the History of Science*, **11**, 1–18, 1978); of three terms ('monuments', 'revolutions', 'accidents') from the technical vocabulary of early geology borrowed from common usage, R Rappaport, 'Borrowed words: problems of vocabulary in eighteenth century geology' (*British Journal for the History of Science*, **15** (1), 27–44, 1982); the place of Alexander Catcott, a follower of John Hutchinson, and an accomplished fieldworker, M Neve and R Porter, 'Alexander Catcott: glory and geology' (*British Journal for the History of Science*, **10**, 37–60, 1977); and *Religion and the Rise of Modern Science* (Scottish Academic Press, 1972), by R Hooykaas.

The industrial revolution and geology

Studies of the direct impact of the industrial revolution on geology are few. They include: 'Die geologische Wissenschaften an der Bergakademie Freiberg in der Periode der industriellen Revolution', by M Guntau (NTM-Schriftenr.Gesch., Naturwiss., Technik. Med. 11, 16–23, 1974); 'The Industrial Revolution and the rise of the science of geology', by R Porter *in* M Teich and R Young (eds), *Changing Perspectives in the History of Science* (Heinemann, 1975); *Geologie und Industrielle Revolution*, edited by E Wachter and G-R Engewald (Bergakademie, 1979), a series of essays on the relations between geology and industrialization; and 'The history of coal prospecting in Britain 1650–1900', by H S Torrens in *Energie in der Geschichte: zur actualitatder technikgeschichte, Dusseldorf, 1984*, (p.88–95 [11th International Commission on the History of Technology Symposium]). The detailed study of the history of coal mining in one of the first parts of the South Wales coalfield to be developed, M V S Symons's *Coal Mining in the*

Llanelli Area. Volume One, 16th century to 1829 (Llanelli Borough Council, 1979) is of interest because it discusses the level of geological understanding achieved by the coal viewers and their relationship with the early geological surveyors, including William Logan and Henry De la Beche.

Bibliographical details of a number of facsimile reprints and/or translations of major publications (other than those by Hutton and Werner), which are mentioned later, are given in the first edition of this chapter (p.411–412, 1973); they include works by R E Raspe, Dr J B A Beringer, B de Maillet, and J W von Goethe.

A wider context

Publications which set the seventeenth and eighteenth century studies of geology in a wider context (and additions to those cited on pages 410 and 412 of the original version of this chapter) include: *The Dark Abyss of Time: the history of the earth and the history of nations from Hooke to Vico*, by P Rossi (University of Chicago Press, 1985); a translation by L G Cochrane of *I Segni del Tempo* (Feltrinelli, 1979) which argues that the discovery of time combines the insights of those we would now call theologian, historian, linguist, as well as geologist; *The Abyss of Time. Changing concepts of the earth's antiquity after the sixteenth century*, by C C Albritton (Freeman, Cooper & Co, 1980); and the wide-ranging survey *Traces on the Rhodian Shore: nature and culture in western thought from ancient times to the end of the eighteenth century*, by C J Glacken (University of California Press, 1967) which is summarized by the author as 'Environment and culture' in P Wiener (ed.), *Dictionary of the History of Ideas*, **2**, 127–157, 1973.

The 'Heroic Age of Geology' (1770s–1820s)

The last quarter of the eighteenth and the first of the nineteenth century are commonly recognized as being of critical importance in the history of geology and are sometimes referred to as the 'Heroic Age'. The term was first coined by Zittel (*History of Geology . . .*, 1901) to denote the period 1790–1820 but was later extended by G Sarton in 'La synthèse géologique de 1775 à 1918' (*Isis*, **11**, 357–394, 1919) to cover the period from 1775 to the 1820s.

Formative figures

The activities of this short but revolutionary period can be highlighted by referring to some of the many publications dealing with the lives and works of three of the formative figures — James

Hutton (1726–97), Abraham Gottlob Werner (1749–1817) and William Smith (1769–1839).

Hutton's works are probably more accessible today than at any time in the past. The two volumes of *Theory of the Earth, with Proofs and Illustrations* (Cadell and Davies, 1795) are available in facsimile reprint (Wheldon and Wesley, 1959), as is John Playfair's *Illustrations of the Huttonian Theory of the Earth* (Cadell and Davies/William Creech, 1802), with an introduction by G W White (University of Illinois Press, 1956; Dover, 1964). Other examples of Hutton's geological writings, not previously reprinted, are brought together in *James Hutton, System of the Earth, 1785; Theory of the Earth, 1788; Observations on Granite, 1794;* together with Playfair's biography of Hutton in *Contributions to the History of Geology Series* Vol. 5, 1970, published by Hafner. The first published account of Hutton's *Theory* was issued separately for the first time in 1987 (Scottish Academic Press). Extensive quotations from Hutton's works, with explanatory notes, are given in such works as *James Hutton — The Founder of Modern Geology*, by E B Bailey (Elsevier, 1967), and in the anthology (with suggestions for its use by teachers) by D A Bassett (*Geology, Journal of the Association of Teachers of Geology*, **2**, 55–76, 1970). Twenty-six facsimiles of the 'lost' illustrations (almost all by John Clerk of Eldin (1728–1812)) which were to have accompanied the projected third volume of Hutton's *Theory*, and which were discovered at Penicuik in 1968, are reproduced in *James Hutton's Theory of the Earth: the lost drawings*, by G Y Craig *et al* (Scottish Academic Press, 1978). Much new material unearthed since that discovery is incorporated by Jean Jones in her chapter in D Daiches, *et al* (eds), *A Hotbed of Genius: the Scottish Enlightenment* (Edinburgh University Press, 1986), one of the contributions to the celebrations commemorating the Scottish Enlightenment and which sets Hutton effectively into his regional, temporal and social context.

The most readily accessible studies to Hutton's *Theory of the Earth* are: the 1949 volume of the *Proceedings of the Royal Society of Edinburgh*, which commemorates the 150th anniversary of Hutton's death; 'The Huttonian Earth Machine 1785–1802', by G L Herries Davies in *The Earth in Decay* (1969); 'James Hutton and the Concept of a Dynamic Earth', by R H Dott in *Toward a History of Geology* (1969); 'James Hutton und die Ewigkeit der Welt', by R Hooykaas (*Gesnerus*, **23**, 55-66, 1966); 'James Hutton on Religion and Geology: the unpublished preface to his "Theory of the Earth"', by D R Dean (*Annals of Science*, **32**, 187–193, 1975); 'A comparison of James Hutton's *Principles of Knowledge* and *Theory of the Earth*', by J E O'Rourke (*Isis*, **69**, 4–20, 1978);

'Hutton's *Theory of the Earth*', by R Grant in *Images of the Earth*, 1978, 23–38); 'James Hutton and the Study of Landforms', by G L Herries Davies (*Progress in Physical Geography*, **9** (3), 383–389, 1985); and 'An Outline of the Philosophy of James Hutton (1726–97)', by P Jones, in V Hope (ed) *Philosophers of the Scottish Enlightenment*, Edinburgh (1984).

Contrasting views of the relationship of the work of G H Toulmin to that of Hutton are presented by G L Herries Davies in his paper 'George Hoggart Toulmin and the Huttonian Theory of the Earth' (*Bulletin of the Geological Society of America*, **76**, 121–124, 1967) and R Porter in 'George Hoggart Toulmin and James Hutton: a fresh look' (*Bulletin of the Geological Society of America*, **89**, 1256–1258, 1978), and 'George Hoggart Toulmin's Theory of Man and the Earth in the Light of the Development of British Geology' (*Annals of Science*, **35**, 339–352, 1978).

An introduction to Werner's work is given in V A Eyles' essay review of A V Carozzi's translation of Werner's first book, *Von den Ausserlichen Kennzeichen der Fossilien* (Leipzig, 1774) published in *History of Science*, **3**, 103–115, 1964. The translation, *On the External Characters of Minerals* (University of Illinois Press, 1962), is based on Werner's personal and annotated copy of the original edition and is therefore a rendering of what could have been the 'second edition' of the work. Werner's second book, the 28-page pamphlet *Kurze Klassifikation und Beschreibung der Verschiedenen Gebirgsarten* (1785), is translated as *Brief Classification and Description of the Different Rocks*, by A M Ospovat (Hafner, 1970) and its contents discussed by him in *Toward a History of Geology*, and in *Isis*, **58**, 91–95, 1967.

A wide range of papers was published to celebrate the 150th anniversary of Werner's death. Fifteen of these, including several in English, are contained in 'Abraham Gottlob Werner: Gedenkschrift aus Anlass der Wiederkehr seines Todestages nach 150 Jahren am 30 Juni 1967', edited by H J Rösler *et al* (*Freiberger Forschungshefte*, Reihe C, No.223, 1967).

Werner's life and work is described in *Abraham Gottlob Werner* (Teubner Verlagsgesellschaft, 1984) by M Guntau, which contains a chronology of Werner's life and a comprehensive list of references. There is also 'Abraham Gottlob Werner: history and folk-history', by G Seddon (*Journal of the Geological Society of Australia*, **20**, 381–395, 1973).

The imposing list of Werner's students includes Jean D'Aubuisson de Voisins (1769–1841), author of one of the three most trustworthy reports on Werner's 'geognosy', *Traité de Geognosie* (Strasburg and Paris, 1819); and Robert Jameson (1774–1854), Werner's British advocate and author of another of the reports on Werner's 'geognosy', *Elements of Geognosy* (Edinburgh, 1808). The facsimile reprint of the latter (Hafner, 1976) contains an assessment of the author's influence on popularizing Werner's geology, by J M Sweet.

The much publicized dispute between Hutton the vulcanist (or plutonist) and Werner the neptunist is a complicated matter and needs much further study. Among the relevant studies are 'The University of Edinburgh's Natural History Museum and the Huttonian-Wernerian debate', by A C Chitnis (*Annals of Science*, **26**, 85–94, 1970), 'Hutton and Werner compared: George Greenough's geological tour of Scotland in 1805', by M J S Rudwick (*British Journal for the History of Science*, **1**, 117–135, 1962), 'Neptunismus und Plutonismus', by W von Engelhardt (*Fortschrift der Mineralogie*, **60**, 21–43, 1982), and 'Neptunists, vulcanists and plutonists' in A Hallam's *Great Geological Controversies* (Oxford University Press, 1983).

The extensive list of papers issued on the Werner anniversary dwarfs the equivalent one for the bicentenary in 1969 of William Smith's birth. Compared to the many studying Werner, one historian only, Joan Eyles, was studying the works of the 'Father of English Geology and Stratigraphy' at that time. Among her contributions are: 'William Smith: some aspects of his life and work' in *Toward a History of Geology*; 'William Smith (1769–1839). A bibliography of the published writings, maps and geological sections, printed and lithographed' (*Journal of the Society for the Bibliography of Natural History*, **5**, 87-109, 1969); and two papers on the links between Smith and American and French geologists, respectively (*Earth Sciences History*, **3** (1), 54–57, 1984; and in *From Linnaeus to Darwin. Commentaries on the history of biology and geology*, edited by A Wheeler and J H Price, 1985). Work on a biography of Smith that Joan Eyles was compiling at her death in 1986 has been taken over by H S Torrens.

A synoptic table of the British stratigraphical column (as interpreted by Smith between 1797 and 1817, in the context of other interpretations proposed between 1725 and 1833) is provided by J Challinor in his paper 'The progress of British geology during the early part of the nineteenth century' (*Annals of Science*, **26** (3), 177–234, 1970); the question 'if Bath *was* the cradle of modern British geology, was Smith the father or merely the midwife to previously conceived ideas?' is posed by H S Torrens in 'Geological communication in the Bath area in the last half of the eighteenth century' in *Images of the Earth*, 1979; Smith's use of fossils is questioned by R Laudan in 'William Smith: stratigraphy without palaeontology' (*Centaurus*, **20**, 210–226, 1976); and an anthology of writings on and by Smith, compiled by D A Bassett (*Geology*, **1**, 38–51, 1969).

Two of the three major figures just mentioned have been described as 'father of geology'. Another is the slightly earlier Russian polymath, Mikhail Vasilievich Lomonosov (1711–65). The most accessible general portraits of him are: the chapter 'Mikhail Lomonosov. One-man university' in Albert Parry's *The Russian Scientist* (Collier-Macmillan [Russia Old and New Series], 1972); and the item by B M Kedrov in the *Dictionary of Scientific Biography*, edited by C C Gillispie. Lomonosov's geological work is described by V V Tikhomirov in 'The development of the geological sciences in the USSR . . . ' in *Toward a History of Geology*, and 'Lomonosov und die geologie in Russland des 18. Jahrhunderts' (*Zeitschrift für geologische Wissenschaft*, **8**, 107–113, 1980), and by D I Gordeev in *M.V. Lomonosov — osnovopolozhnik geologicheskoi nauki* (Moscow, 1961).

A fifth formative figure is Georges-Louis Leclerc, Comte de Buffon (1707–88). His work is considered by Jacques Roger in the critical edition of 'Les Époques de la Nature' (*Mémoires du Muséum National d'Histoire* Série C: *Sciences de la Terre* **10**), which contains an introduction, a reproduction of the manuscript, notes, scientific vocabulary and bibliography, and also in *Les sciences de la vie dans la pensée française du 18ᵉ siecle* (Paris, 1964, 527–584), also by Roger.

The work of these five formative figures is considered by M Guntau in 'The emergence of geology as a scientific discipline' (*History of Science*, **16**, 280–290, 1978).

Geology: the word and the concept

The use of the word chosen for the new science is considered by D R Dean, 'The word "Geology"' (*Annals of Science*, **36** (1), 35–43, 1979); some of the tensions in the new concept are outlined by R Laudan, 'Tensions in the concept of geology: natural history or natural philosophy' (*Earth Sciences History*, **1**, 7–13, 1982); and conflicts in the nature of geology considered by R Rappaport, 'The geological atlas of Guettard, Lavoisier and Monnet' in *Toward a History of Geology*, 1969.

The usefulness of a major contemporary encyclopaedia in throwing light on the various conflicts is described by H R Pestana in 'Rees's *Cyclopaedia* (1802–1820) as a sourcebook for the history of geology' (*Journal of the Society of the Bibliography of Natural History*, **9**, 353–361, 1979).

Nineteenth century (second, third and fourth quarters)

The period 1820 to 1840 has been referred to as that of the 'Great Masters', and as the 'Golden Age of Geology'. Others maintain that the limit should be extended to 1875 so as to include the geological syntheses of Charles Lyell, Leopold von Buch, Alexander von Humboldt and Elie de Beaumont, as well as the evolutionary synthesis of Charles Darwin. Contemporary reviews of developments during this time by W D Conybeare (1833), W H Fitton (1832–33) and particularly by Le Vicomte E J Adolphe D'Archiac (1847–60), are cited in the first edition of this chapter (1973, p. 415).

The particular interest of the first half of the nineteenth century for the general historian, as pointed out by Cannon in 'History in depth' (*History of Science*, **3**, 20–38, 1964 and *Science in Culture* (1978)) is threefold: first, the documentary evidence is possibly better than at any other period before or since because crucial informal exchanges were often recorded in long scientific letters; second, the period may well have been the watershed for general history and that the history of science is the clearest and most direct vehicle for demonstrating this; third, it was not so much a late summer of the amateur but the spring time of the professional. See, for example, 'Gentlemen and geology: the emergence of a scientific career, 1660–1920', by R Porter (*Historical Journal*, **4**, 809–836, 1978) and 'Specialization and professionalization in British geology', by J G O'Connor and A J Meadows (*Social Studies of Science*, **6**, 77–89, 1976).

Assessments of the status, popularity and relative standing of the subject during these years are given in G L Herries Davies' *The Earth in Decay* (1969), D E Allen's *The Naturalist in Britain* (1976) and S Cannon's *Science in Culture* (1978).

Geological societies

Societies devoted exclusively to geology appeared for the first time in the early decades of the century. The work of the first, founded in 1807, is described in *The History of The Geological Society of London*, by H B Woodward (Geological Society, 1907) and 'The foundation of the Geological Society of London: its scheme for co-operative research and its struggle for independence', by M J S Rudwick (*British Journal for the History of Science*, **1**, 325–355, 1963). The latter, based in part on the private papers and correspondence of G B Greenough (1778–1856), the first President of the Society, also provides an introduction to the state of geological study, to the conflict between practical and theoretical geologist and, consequently, to the influence of Francis Bacon.

The 'prehistory' of the Society and of its forerunner (the short-lived but energetic and productive British Mineralogical Society: 1798–1806) are described by Paul Weindling in *Images of the Earth* (p.248–271) and in *Metropolis and Province: science in British culture 1780–1850*, edited by I Inkster and J Morrell (Hutchinson, 1983). Other discussions of the Geological Society are given in the papers 'London institutions and Lyell's career 1820–1841', by J B Morrell (*British Journal for the History of Science*, **9**, 132–146, 1976), 'Ideas and organization in British geology: a case study in institutional history', by R Laudan (*Isis*, **68**, 527–538, 1977), and in *The Making of Geology*, by R Porter.

The work of another British society of the period (and one in which geologists played a prominent part) is described in considerable detail in two sesquicentennial publications: *Gentlemen of Science: early years of the British Association for the Advancement of Science*, by J Morrell and A Thackray (Oxford University Press, 1981), a definitive and 'magisterial' history providing a deep analysis of the structure of the early Victorian scientific community (1831–1844); and *The Parliament of Science*, a series of essays by eleven scholars edited by R MacLeod and P Collins (Science Reviews Ltd, 1981). The former is complemented by, *Gentlemen of Science: early correspondence of the British Association for the Advancement of Science*, also edited by Morrell and Thackray (Royal Historical Society, 1984), which contains 293 letters judged by the editors to be 'of great significance to the student of early Victorian science and culture'.

Bibliographic details of histories of other national societies· (France, Germany, Ireland, the USA, etc.) are given in the first edition of this chapter (1973, p.415, 416).

Geological surveys

The story of the development of the Geological Survey of Great Britain, founded in 1835, is told by two of its Directors: an 'official' history, *The First Hundred Years of the Geological Survey of Great Britain*, by Sir John Flett (HMSO, 1937), and a much more personal account, *Geological Survey of Great Britain*, by Sir Edward Bailey (Murby, 1952). A detailed study of the Survey as a 'research school', from 1839 to 1855, is given by J A Secord (*History of Science*, **24** 223–275, 1986); and the first study, in book form, of its founder and first Director by P J McCartney, *Henry De la Beche; observations on an observer* (Friends of the National Museum of Wales, 1977). The overlapping history of the Geological Survey of Ireland is incorporated in *Sheets of Many Colours: the mapping of Ireland's rocks*, by G L Herries Davies (Royal Dublin Society, 1983).

The success of the British Survey resulted in the initiation of a large number of national surveys. Recent studies of these include: *Government in Science. The US Geological Survey, 1867–1894*, by T G Manning (University of Kentucky Press, 1967) and the comprehensive official history *Minerals, Lands and Geology for the Common Defence and General Welfare*, by M C Rabbitt (3 vols, US Government Printing Office, 1979, 1980, 1986); *Reading the Rocks; the story of the Geological Survey of Canada 1842–1872*, by M Zaslow (Macmillan, 1975); and *History and the Role of Government Geological Surveys in Australia*, by R K Johns (Adelaide, 1976).

A visual language

One of the most striking features of the majority of late eighteenth century books and journals on topics relevant to the future science of geology is the scarcity and poor quality of their illustrations. A similar survey of material published in the 1830s shows a remarkable change: the texts are now complemented by a wide range of maps, sections, landscapes and diagrams of other kinds. The emergence of this visual language for the science is considered by M J S Rudwick (*History of Science*, **14**, 149–165, 1967). In this article, he attributes the spectacular change in part to the development of aquatints, wood engravings, steel engravings and lithographs in the decades around 1800, and draws attention to the seminal volume *Prints and Visual Communication*, by W M Ivins Jr (Routledge & Kegan Paul, 1953: reprint Da Capo Press, 1969).

Examples of studies of aspects of this visual language include: 'Geology and landscape painting in nineteenth century England', by M Pointon (p.84–118 in *Images of the Earth*, 1979); *Ben Peach's Scotland: landscape sketches by a Victorian geologist*, by A Anderson (Institute of Geological Sciences, 1980); and '"A philosophical landscape": Susanna Drury and the Giant's Causeway', by M Anglesea and J Preston (*Art History*, **3** (3), 252–273, 1980). A catalogue of 86 separately published prints of fossils in nineteenth century Britain is provided by J Thackray (*Archives of Natural History*, **12** (2), 175–199, 1985).

Surprisingly, there is no single comprehensive study of the most significant expression of this visual language — the geological map. A useful outline 'History of the development of geologic maps' (*Bulletin of the Geological Society of America*, **54**, 1227–1280, 1943) is given by H A Ireland; a brief history of geological maps and sections from 1644 to the present, 'Zur Geschichte der Geologischen Karte' (*Zeitschrift für Angewandte Geologie*, **3**, 417–424, 1957) by W Steiner; and D A Bassett prefaces *A Source-*

book of Geological, Geomorphological and Soil Maps for Wales and the Welsh Borders (1800–1966) (National Museum of Wales, 1967) with a general essay on the development of the geological map and its derivatives in Britain, with elaborate bibliographical references and many quotations.

More recent analyses include: 'The early development of British geological maps', by R C Boud (*Imago Mundi*, **27**, 73–96, 1975), *Sheets of Many Colours: the mapping of Ireland's rocks 1750–1890*, by G L Herries Davies (Royal Dublin Society, 1983), 'Recherches et réflexions sur la naissance de la cartographie géologique, en Europe et plus particulièrement en France', by F Ellenberger (*Histoire et Nature*, **22/23**, 5–54, 1985), and the papers presented at the 10th INHIGEO Symposium at Budapest, *Contributions to the History of Geological Mapping*, edited by E Dudich (1984). The 25 papers by Russian authors are also published separately in V V Tikhomirov (ed.) *The Story of [the] Geological Map; essays on the history of geological knowledge*, volume **21** (USSR Academy of Sciences, 1982).

Studies of individual maps and atlases include: 'William Maclure's geological map of the United States', by C J Schneer (*Journal of Geological Education*, **29**, 241–245, 1981 — with a colour reproduction [NAGT History of Science Series]), 'John Macculloch, M.D., F.R.S., and his geological map of Scotland: his years in the Ordnance 1795–1826', by D Flinn (*Notes and Records of the Royal Society of London*, **36**, 83–101, 1981), 'The geological atlas of Guettard, Lavoisier and Monnet', by R Rappaport, in *Toward a History of Geology* (1969).

The construction of early geological sections is discussed by T D Ford in 'The first detailed geological sections across England — by John Farey 1806–8' (*Mercian Geologist*, **2**, 41–49, 1967) and in his introduction to the facsimile reproduction of White Watson's *The Strata of Derbyshire* (1811) (Moorland Reprints, 1973). The three-dimensional aspect of geology is further considered in papers on geological models by S Turner and W R Dearman in 'The early history of geological models' (*Bulletin of the International Association of Engineering Geology*, **21**, 202–210, 1980), 'Discovery of working drawings for the Sopwith models of 1841 at the Hancock Museum' (*Geological Curator*, **2** (8), 467–495, 1980), 'Thomas Sopwith's large geological models' (*Proceedings of the Yorkshire Geological Society*, **44** (1), 1–28, 1982), and 'Models illustrating John Farey's figures of stratified masses' (*Proceedings of the Geologists' Association*, **94**, 97–104, 1983).

A more specialized aspect of the visual language is considered in 'Caricatures as a source for the history of science: De la Beche's anti-Lyell sketches of 1831', by M J S Rudwick (*Isis*, **66**, 534–560,

1985) and in *Henry De la Beche: observations on an observer*, by P J McCartney (Friends of the National Museum of Wales, 1977).

The extent of the change brought about by the use of maps and sections is very well illustrated in the *Transactions of the Geological Society*. A comparison of the first and second series of the *Transactions* reflects another striking change: from a clear emphasis on crystallography and mineralogy in the former (1811–1821) to that on stratigraphy and regional geology in the latter (*see* J W Judd: *Quarterly Journal of the Geological Society*, 1887).

A comparison of science in the metropolis with that in the provinces is given in *Metropolis and Province; science in British culture 1780–1850*, edited by I Inkster and J Morrell (Hutchinson, 1983). In one sense the volume is a successor to R E Schofield's *The Lunar Society of Birmingham: a social history of provincial science and industry in eighteenth century England* (Clarendon Press, 1963) and *Science and Technology in the Industrial Revolution*, edited by A E Musson and E Robinson (Manchester University Press, 1969) in which the various authors argue that it was in the regions, especially Birmingham and Manchester, that science was advancing most spectacularly.

Charles Lyell and the 'Principles of Geology'

In spite of the emphasis on empirical work, theory inevitably played a major role in the development of the subject, as demonstrated in studies of the work of Charles Lyell. The publication of Lyell's *Principles of Geology: being an attempt to explain the former changes of the Earth's surface by reference to causes now in operation* (3 vols, Murray, 1830–33) was one of the major events of the century and one which had an effect on geological thought similar to that of Charles Darwin's *Origin of the Species* on biological thought some three decades later.

The most important re-assessment of Lyell's work in the context of that of the whole group of scientists who were working in the early nineteenth century in Britain is the selection of papers from the Lyell Centenary Symposium issued in *The British Journal for the History of Science*, **9**, 91–242, 1976. The papers consider: Lyell and the principles of the history of Geology (R Porter); Lyell, radical actualism, and theory (W F Cannon); Lyell and the philosophers of science (M Ruse); London institutions and Lyell's career: 1820–41 (J B Morrell); Charles Lyell speaks in the Lecture Theatre (M J S Rudwick); The rivalry between Lyell and Roderick Murchison (L Page); The non-progress of non-progression: two responses to Lyell's doctrine (M Bartholomew); Lyell and G B Brocchi: a study in comparative historiography (P J McCartney);

The distortion of Werner in Lyell's *Principles* (A M Ospovat); Darwin, Lyell, and the geological significance of coral reefs (D R Stoddart); The first Spanish translation of Lyell's *Principles* (J Ordaz).

The different approach of two of the most prolific writers on Lyell, M J S Rudwick and L G Wilson, is clearly illustrated in their respective contributions to *Toward a History of Geology* (1969). Part of this difference is explained in Wilson's criticism of the basic premises of Rudwick's interpretation in 'Geology on the eve of Charles Lyell's first visit to America, 1841' (*Proceedings of the American Philosophical Society*, **124**, 168–202, 1980). Other aspects are clear from a comparison of their other contributions. Those by Rudwick include: the introduction to the facsimile of Lyell's *Principles* (Johnson Reprint Corporation), *Sources of Science Series* No. 48, 1970); 'The strategy of Lyell's *Principles of Geology*' (*Isis*, **61**, 5–33, 1969); 'Uniformity and progression: reflections on the structure of geological theory in the age of Lyell' (p.209–227 in D H D Roller (ed) *Perspectives in the History of Science and Technology*, Norman, Oklahoma, 1971); 'Transported concepts from the human sciences in the early work of Charles Lyell' (p.67–83 in *Images of the Earth*, 1979); 'Historical analogies in the early geological work of Charles Lyell' (*Janus*, **64**, 89–107, 1977); and 'Charles Lyell's dream of a statistical palaeontology' (*Palaeontology*, **21**, 225–244, 1978). Those by Wilson include: the first volume of the proposed three-volume biography *Charles Lyell. The years to 1841: the revolution in geology* (Yale University Press, 1972); and *Sir Charles Lyell's Scientific Journals on the Species Question* (Yale University Press, 1970).

Uniformitarianism v. catastrophism

The majority of references to Lyell's work, in both geological and historical textbooks, imply that it completely, or almost completely, resolved the controversy between the proponents of the doctrine of catastrophism on the one hand and of uniformitarianism, and the closely related actualism, on the other. That the controversy was not resolved is clear, however, from the following: the remarkable defence of Cuverian catastrophism in the two-volume *Synthetische Artbildung*, by N Heribert-Nilsson (Lund, Gleerup, 1953); the demonstration by Stephen Toulmin in 'Historical inference in science: geology as a model of cosmology' (*Monist*, **47**, 142–158, 1962) that the controversy is still part of contemporary cosmological thinking; *Catastrophes and Earth History; the new uniformitarianism*, edited by W A Berggren and J A Couvering (Princeton University Press, 1984, from the proceedings of two symposia held in the USA in 1977); and 'The return of catastrophism and saltatory development', by R Hooykaas (*INHIGEO Newsletter*, No. 5, 11–17, 1981).

The dampening of the swing of the pendulum from catastrophism to uniformitarianism following the formulation of Darwin's theory and of the second law of thermodynamics was the subject matter of T H Huxley's famous presidential address to the Geological Society of London in 1869 ('Catastrophism, uniformitarianism, evolutionsim') and M King Hubbert's article 'Critique of the principle of uniformity', in C C Albritton Jr (ed.) 'Uniformity and Simplicity. A symposium on the principle of uniformity of nature' (*Geological Society of America, Special Paper* **89**, 1967).

The main analyses of the controversy are: the major volume *The Principle of Uniformity*, by R Hooykaas (Brill, 1963) and its sequel 'Catastrophism in Geology: its scientific character in relation to actualism and uniformitarianism' (*Mededelingen der Koninklijke Nederlandse Akademie van Wetenschappen. Afd. Letterkunde* **33**, no. 7, 1970); 'The uniformitarian-catastrophist debate' (*Isis*, **51**, 38–55, 1960) and 'The impact of uniformitarianism: two letters from John Herschel to Charles Lyell, 1836–1837' (*Proceedings of the American Philosophical Society*, **105**, 301–314, 1961), both by W F Cannon; 'Uniformitarianism and catastrophism', by L G Wilson (in P Wiener (ed.) *Dictionary of the History of Ideas*, **4**, 417–423, 1973); the chapter 'Catastrophists and uniformitarians' in *Great Geological Controversies*, by A Hallam (Oxford University Press, 1983); the chapters 'Life's revolutions' and 'Uniformity and progress' in Rudwick's *The Meaning of Fossils* (1972); the close examination of Cuvier's classic catastrophist text, *Discours sur les Révolutions du Globe* (1812) by A V Carozzi in his paper 'Une nouvelle interprétation du soi-disant catastrophism de Cuvier' (*Archives des Sciences*, **24**, 367–377, 1971); and *Time's Arrow, Time's Cycle*, by S J Gould (Harvard University Press, 1987) which also cites many examples of oversimplification and misrepresentation of the concepts in modern geological textbooks.

Charles Darwin and geology

It is generally suggested that the study of Lyell's *Principles* was a seminal factor in Darwin's intellectual development. This link is strongly upheld by M T Ghiselin in *The Evolution of Charles Darwin: the triumph of the Darwinian method* (University of California Press, 1969). Papers that describe Darwin's direct involvement in geology and with geologists include: 'The Sedgwick-Darwin geological tour in North Wales [1831]', by P H

Barrett (*Proceedings of the American Philosophical Society*, **118**, 146–164, 1974); 'The eye of reason: Darwin's development during the *Beagle* voyage, by H E Gruber and V Gruber (*Isis*, **53**, 186–200, 1962); 'Remembering Charles Darwin as a geologist' in R G Chapman (ed.) *Charles Darwin 1809–1882; a centennial commemorative* (Croom Helm, 1983) and 'Darwin as a geologist' (*Scientific American*, **254**, 94–101, 1986), both by S Herbert; the chapter 'Life's ancestry' in Rudwick's *The Meaning of Fossils* (1972); 'Darwin and Glen Roy: a "great failure" in scientific method?', by M J S Rudwick (*Studies in the History and Philosophy of Science*, **5**, 96–185, 1974); 'Darwin and the dilemma of geological time', by J D Burchfield (*Isis*, **65**, 301–321, 1974); and the facsimile of Darwin's *The Structure and Distribution of Coral Reefs* (University of California Press, 1962 and University of Arizona Press, 1984), with a foreword by M T Ghiselin.

The centenary of the publication of *On the Origin of Species by Means of Natural Selection* (John Murray, 1859) and of Darwin's death both led to a spate of biographical and other studies. An impression of the extent of this 'Darwin Industry' is provided by D R Oldroyd in 'How did Darwin arrive at his theory; the secondary literature to 1982' (*History of Science*, **22**, 325–374, 1984) and in *Darwinian Impacts: an introduction to the Darwinian revolution* (Open University Press and New South Wales Press Ltd., 1980) which contains a comprehensive bibliographical essay.

A work which undoubtedly did great service in familiarizing the reading public with the idea of evolution and thus prepared them for the more complete and more efficient *Theory* presented by Darwin was the anonymous *Vestiges of the Natural History of Creation* (Churchill, 1844), written by the essayist, publisher and amateur geologist, Robert Chambers. A facsimile of the book, with an introduction by Sir Gavin de Beer, was issued in 1969 (Leicester University Press) and its appeal to the 'ordinary' reader, is considered in 'The universal gestation of nature: Chambers' *Vestiges* and explanations', by M J S Hodge (*Journal of the History of Biology*, **5**, 127–151, 1972).

Other major controversies

The spectacular success of stratigraphical studies during the eighteen twenties, thirties and forties (reflected in the fact that all but three of the major systems were named between 1819 and 1841) almost inevitably led to conflicts and controversies. Two of these were on a major scale and both have been studied in considerable detail.

The first is the controversy surrounding the establishment of the

Devonian System and which Lyell considered to be one of the most important theoretical issues ever to be discussed at the Geological Society of London. The stratigraphical, historical, personal and institutional aspects of this conflict have been studied by M J S Rudwick for over 20 years and there have been a number of 'progress reports' which have confirmed the title 'Great Devonian Controversy', given to it by G B Greenough and others in the late 1830s. The shape of the controversy from the standpoint of modern geologists and palaeontologists was the basis for an address at the International Symposium on the Devonian System, at Bristol, in 1978, 'The Devonian: a System born from conflict', in 'The Devonian System', edited by M R House and M G Bassett (*Special Papers in Palaeontology*, No.23, 1979). The completed analysis, published as *The Great Devonian Controversy: the shaping of scientific knowledge among gentlemanly specialists* (University of Chicago Prerss, 1985) is the major work of its kind.

The second is the controversy over the limits of the Cambrian and Silurian systems as originally defined by Adam Sedgwick and Roderick Murchison, respectively, and the consequent splitting of stratigraphers throughout the world into two camps. Aspects of the controversy were discussed at a meeting of the Geological Society of London in 1976 devoted to geological controversies (e.g. 'The Murchison-Sedgwick controversy', by J C Thackray (*Journal of the Geological Society of London*, **137**, 367–372, 1976) and the full extent described and analysed in *Controversy in Victorian Geology: the Cambrian-Silurian dispute*, by J A Secord (Princeton University Press, 1986).

A third major controversy surrounded the identification of what are now known as glacial phenomena. There is no single all-embracing study but different aspects are considered in: 'Centenary of the glacial theory', by F J North (*Proceedings of the Geologists' Association*, **54**, 1–28, 1943); 'The Ice Age' (in A Hallam, *Great Geological Controversies*, Oxford University Press, 1983); 'Darwin and Glen Roy: a "great failure" in scientific method?', by M J S Rudwick, referred to in the previous section; and 'The tour of the British Isles made by Louis Agassiz in 1840', by G L Herries Davies (*Annals of Science*, **24**(2), 131–146, 1968). Rudwick's contribution is of particular interest because it emphasizes the importance of fieldwork in this kind of historical research.

Societies and museums

The eighteen fifties and sixties was a period in which a number of geological and natural history societies were inaugurated and in which many of them launched *Transactions* and established museums. The various trends in creating societies are outlined by D E Allen in *The Naturalist in Britain* (1976) and the history of journals by A P Harvey in his paper 'A history of geological publications [from 17th century onwards] in the United Kingdom' (*Earth and Life Science Editing*, No.6, 7–12, 1978).

The statement that 'Museums are probably the aristocrats among sources for the history of science' is of particular relevance to geology because of the nature of the material. The prevalence of geological collections in British museums of the late nineteenth century is reflected in *Museums and Galleries*, by T Greenwood (Simkin Marshall & Co., 1888), and the current situation assessed in detail by P S Doughty in *The State and Status of Geology in United Kingdom Museums: Report of a survey conducted on behalf of the Geological Curators' Group (Geological Society [of London] Miscellaneous Paper* **13**, 1981). The range of works on museums is illustrated in T Sharpe's *Geology in Museums: a bibliography and index* (National Museum of Wales, 1983), which contains 77 references to museum history and 190 to collections and collectors. The latter contain many references to the series 'Geological collections and collectors of note' in *The Geological Curator*, initiated in 1974 and still continuing.

Advances in petrography, palaeontology and geomorphology

The 1850s and 1860s have also been described as significant decades for the fragmentation of the sciences into specialized disciplines, each with their own methods, techniques, approaches and subject boundaries, and also for the budding, branching and institutionalization of interdisciplinary sciences. One example is discussed in W H Brock's 'Chemical geology or geological chemistry' (*Images of the Earth*, 147–170), which concentrates on the debate in the 1860s between the Manx geologist David Forbes and the North American chemist and geologist Thomas Sterry Hunt.

The 1850s was also the period when Henry Clifton Sorby pioneered the use of the microscope for the study of rocks in thin sections. Sorby's own recollections of the story are given in 'Fifty years of scientific research' (Sheffield Literary and Philosophical Society, 1897), and the primary literature brought together in *Sorby on Geology: a collection of papers from 1853 to 1906 by Henry Clifton Sorby*, edited by C H Summerson (*Geological Milestones*, No.3, Miami, 1978). 'New biographical data on Sorby', by M J Bishop (*Earth Sciences History*, **3** (1), 69–81, 1984) records the discovery of the diary of his father covering the years 1845–1846 and lists all earlier biographical and bibliographical works, including the standard biography by N Higham (Pergamon Press, 1963) and the comprehensive bibliography by D W Humphries in *The Sorby Symposium on the History of Metallurgy* (The Metallurgical Society Conference, **27**, 43–58).

The flowering of Sorby's technique came in the period from the 1870s onward, as described by B M Hamilton in 'The influence of the polarising microscope on late 19th century geology' (*Janus*, **69**, 51–68, 1982). An index of the resulting successful developments in petrography is provided by the two editions of Ferdinand Zirkel's *Lehrbuch der Petrographie*, the first two volumes (Bonn, 1866) being concerned primarily with the macroscopic methods of the older teaching, and the second three volumes (Leipzig, 1893–94) a frank and full acknowledgement of the petrographical reform necessitated by microscopic and microchemical methods. The latter are discussed by B M Hamilton in 'The development of chemical geology in the nineteenth century with special reference to the situation in Britain' (*Earth Sciences History*, **5** (2), 114–123, 1986).

This was also the time when one of the major confrontations in the history of establishing a time scale for the history of the earth took place. As J D Burchfield points out in *Lord Kelvin and the Age of the Earth* (Macmillan, 1975), it involved Kelvin, the acknowledged leader of British physical science and the archetypal example of the successful Victorian scientist on the one hand, and T H Huxley, acting on behalf of the biologists, and particularly Darwin, on the other. The author concentrates: (i) on the problem of establishing a quantitative limit to the probable age of the earth, and not on the problem of determining the relative ages of strata, a problem discussed by W B N Berry in *Growth of a Prehistoric Timescale Based on Organic Evolution* (Freeman, 1968); (ii) on the interaction between two branches of science only — geology and physics; and (iii) deals almost exclusively with British and American scientists from 1860 to 1930.

Two other striking developments of the second half of the nineteenth century were the advances in palaeontology and geomorphology which resulted in large measure from the opening up of the American Far West by pioneers such as E D Cope (1840–97), G K Gilbert (1843–1918), Clarence King (1842–1901), O C Marsh (1831–99), J W Powell (1834–1902) and others. The book *Great Surveys of the American West*, by R A Bartlett (University of Oklahoma Press, 1962) provides an introduction to the surveys and includes a very good bibliography of published and unpublished government documents as well as of secondary sources.

Exploration and Empire: the explorer and the scientist in the winning of the American West, by W H Goetzmann (Alfred Knopf, 1956) is equally useful. The first two volumes of the 'official' history of the USGS by M Rabbit, mentioned earlier, and covering the period up to 1904, effectively complements these studies. The subtitle reads: *A History of Public Lands, Federal Science and Mapping Policy, and Development of Mineral Resources in the United States.*

An introduction to the geomorphological story is given in *The History of the Study of Landforms or the Development of Geomorphology*. Vol.1, *Geomorphology before Davis*. Vol.2, *The Life and Work of William Morris Davis*, by R J Chorley *et al.* (Methuen, 1964 and 1973). It was Davis, to quote G L Herries Davies' review of the second volume (*History of Science*, **13**, 139– 145, 1975) who 'created the science of geomorphology'. What Werner was to geology around 1800, W M Davis was to geomorphology a century later — 'an infallible oracle to whose followers every utterance seemed sacred and immutable. Werner's influence long survived his own demise; so too did that of Davis.'

The palaeontological story is outlined in Jean Piveteau's contri-bution to *La Science Contemporaine* (Presses Universitaire de France, edited by René Taton, 1961: see also the English translation, *A General History of the Sciences*, Thames & Hudson, 1963–66). The critical years that saw the general adoption of an evolution perspective (and the emergence of fully professional careers in the earth sciences) are covered in *Archetypes and Ancestors: palaeontology in Victorian London, 1850–1875*, by A Desmond (Blond and Briggs, 1982). The book's main figures are Richard Owen and T H Huxley and their respective supporters and disciples; the main topic, their interpretation of dinosaurs and the relations between the classes of vertebrates. A more detailed assessment of developments in Central Europe is given in *Colloque International sur Joachim Barrande à Prague et à Liblice le 17 à 20 Mai 1969*, edited by M M B Bouček and L Marek (*Casopis pro mineralogii a geologii*, Prague, 1970).

International co-operation

During the 1870s geologists began discussing the need for an international gathering to bring order to the language and literature of geology. The Société Géologique de France was approached with a view to organizing a congress in connection with the Paris Exposition of 1878. They accepted and the meeting proved to be the beginning of the International Geological Congress. The other early congresses were at Bologna (1881),

Berlin (1885), London (1888), Washington (1891) and Zurich (1894). The activities at the first of these are described briefly by W Thurston 'The First International Geological Congress' (*Geotimes*, **13**, 16–17, 1968), and by F Ellenberger 'The first International Geological Congress, Paris, 1878' (*Episodes*, **2**, 20–24, 1978).

The lost limb

The major change in the general popularity of geology in Britain in the second half of the century as opposed to the first is analysed by D Allen in his paper 'The lost limb: geology and natural history' in *Images of the Earth* (1979), whereas the successful developments in the more technical aspects of the subject are reflected in J Challinor's *The History of British Geology: a bibliographical study* (David & Charles, 1971). Aspects of the state of university geology are considered by R Porter in 'The Natural Sciences Tripos and the "Cambridge School of Geology" 1850–1914' (*History of Universities*, **3**, 193–216, 1982), but little has been published on geology in the many 'civic' universities established in the 1870s and early 1880s, other than in the various centenary volumes of individual institutions. Little also is published on geology in general education, other than such works as *T H Huxley: Scientist, Humanist and Educator*, by C Bibby (Watts, 1959).

Twentieth century

Among the most striking developments in the twentieth century have been the concepts of continental drift and plate tectonics. Historical accounts of the controversies created by these new theories are provided in: *Debate about the Earth; approach to geophysics through analysis of continental drift*, by H Takeuchi *et al* (Freeman, Cooper, 1970); *Continental Drift: the evolution of a concept*, by U B Marvin (Smithsonian Institution Press, 1973); *The Road to Jaramillo: critical years of the revolution in earth sciences*, by W Glen (Stanford University Press, 1982); *A Revolution in the Earth Sciences*, by A Hallam (Oxford University Press, 1973); in the longest chapter in *Great Geological Controversies*, by A Hallam (Oxford University Press, 1983) and 'Plate tectonics and biogeography', a selection of the papers read at Symposia at Washington 1982 and New York 1984 (*Earth Sciences History*, **4** (2), 1985).

Four contributions by H Frankel deal with 'Alfred Wegener and the specialists' (*Centaurus*, **20**, 305–323, 1976), 'Arthur Holmes and Continental Drift' (*British Journal for the History of Science*, **11**, 130–150, 1978), 'The reception and acceptance of

Continental Drift as a rational episode in the history of science' (in
S H Manskopf (ed.) *The Reception of Unconventional Science*,
West-view Press, 1979), and 'The career of Continental Drift
theory: an application of Imre Lakatos' analysis of scientific
growth in the rise of the drift theory' (*Studies in the History and
Philosophy of Science*, **10**, 21–166, 1979).

Alfred Wegener's *The Origin of Continents and Oceans* (translated from the 4th revised German edition of 1929 by J Biram) was
published by Methuen in 1966 and the most recent biographical
studies are: *Alfred Wegener und die Drift der Kontinente* (Wissen-
schaftliche Verlagsgesellschaft, Stuttgart, 1980), which presents an
outline of the history of the continental drift theory within a
biographical framework; and *Alfred Wegener (Biographien
hervorragender Naturwissenschaftler, Techniker und Mediziner*,
vol.**46**, BSG Teubner Verlagsgessellschaft, 1980) by H G Körber,
which provides detailed descriptions of Wegener's four expedi-
tions to Greenland, and analyses his comprehensive scientific
work in various fields.

Reviews of the development of particular branches of the
subject by practising geologists are obviously numerous. Some of
these are cited in the sections dealing with individual subjects and
individual countries in the original version of this essay (1973,
p.420–436). The main review journals are discussed in Chapter 4.
Bibliographical details of a selection of works containing extracts
from some of the important publications of the century are given
in the first version of this chapter (1973, p.420).

References arranged by subject

The following selection of books and papers is an attempt to
reflect the range and diversity of the publications of the last ten
years or so. It is additional to the relevant section in the previous
edition of this chapter (1973, p.420–431). Items marked with an
asterisk(*) are from the didactic *History of Geology Series* in the
Journal of Geological Education, sponsored by the National
Association of Teachers of Geology (NAGT). They are well
illustrated and contain helpful lists of further reading. *Benchmark
Papers* refer to volumes in the Dowden, Hutchinson & Ross series
Benchmark Papers in Geology.

Methods and philosophy.   Publications dealing with the methods
and philosophy of geology include: *Philosophie und Geologie*,
edited by E Fabian *et al* (*Schriften reihe für Geologische
Wissenschaften*, Heft 24, 1985), the contributions to the third

GDR-USSR Symposium on the History of Geological Sciences (Grieswald, 1983); *Philosophy of Geohistory, 1785–1970*, edited by C C Albritton (*Benchmark Paper*, No.14, 1975); *The Structure of Geology*, by D B Kitts (Southern University Press, 1977) [a collection of Kitts' studies on the theoretical and philosophical questions of geology]; *The Scientific Ideas of G K Gilbert*, edited by E L Yochelson (*Geological Society of America, Special Paper*, **183**, 1980); and refer also to the bibliography of 213 studies by 100 authors on the philosophy of the earth sciences published in the GDR between 1945 and 1975, with an introduction by M Guntau (Bibliothek der Bergakademie DDR, 92).

Palaeontology (and evolution). *From Linnaeus to Darwin: commentaries on the history of biology and geology*, edited by A Wheeler and J H Price (*Society for the History of Natural History, Special Publication*, **3**, 1985); 'History of palaeontology: before Darwin', by M J S Rudwick (p.375–384 in R W Fairbridge and D Jablonski (eds) *The Encyclopedia of Palaeontology*, Dowden, Hutchinson and Ross, 1979); *The Fossil Hunters: in search of ancient plants*, by H Andrews (Cornell University Press, 1980); *Fossils and Progress: paleontology and the idea of progressive evolution in the nineteenth century*, by P J Bowler (Science History Publications, 1976); *The Hot-blooded Dinosaurs: a revolution in palaeontology*, by A J Desmond (Blond & Briggs, 1975).

The Role of Palaeontology in the Development of Geology in the Soviet Union, by M V Kulikov and V S Belenkov (Proceedings of the 27th Session of the All-Union Palaeontological Society, Nauka, Leningrad, 1985); 'The birth of palaeontology in France: 1700–1750', by K B Bork (*Journal of the Scientific Laboratories of Denison University*, **54**, 65–78, 1973); 'Dutch palaeontologists of the eighteenth century', by R Visser (*Janus*, **62**, 125–149, 1975).

'A l'aurore de la stratigraphie paléontologique: Jean-Andre de Luc, son influence sur Cuvier', by F Ellenberger and G Gohau (*Revue d'Histoire des Sciences*, **34**, 217–257, 1981); 'The relevance of Cuvier's *Lois Zoologiques* for his palaeontological work', by B Thennissen (*Annals of Science*, **43**, 1543–1556, 1986); 'Saleswoman to a new science: Mary Anning and the fossil fish *Squaloraja* from the Lias of Lyme Regis', by M A Taylor and H S Torrens (*Proceedings of the Dorset Natural History and Archaeological Society*, **108** (for 1986), 135–148, 1987); 'Robert E Grant's later views on organic development: the Swiney lectures on 'Palaeozoology' 1853–1857', by A J Desmond (*Archives of Natural History*, **11**, 395–413, 1984).

'*Formed Stones', Folklore and Fossils*, by M G Bassett (National Museum of Wales, 1982); 'Fossil sales', by W D I Rolfe (p.32–38 in J M Chalmers-Hunt (ed.) *Natural History Auctions 1700–1972. A register of sales in the British Isles* (Sotheby Parke Bernet Publications, 1976). 'Changing concepts of the nature and significance of fossils', by J Gregory* (*Journal of Geological Education*, **32**, 108–118, 1984).

Stratigraphy and historical geology. 'Facies in stratigraphy from "Terrains" to "Terranes"', by C M Nelson* (*Journal of Geological Eduction*, **33**, 175–187, 1985); 'The concept of biotic succession', by K B Bork* (*Journal of Geological Education*, **32**, 213–225, 1984); '"Transition Rocks and Grauwacké" — The Silurian and Cambrian Systems, through 150 years', by M G Bassett (*Episodes*, **8**, 231–235, 1985); 'The Great Taconic Controversy', by C J Schneer (*Isis*, **69**, 173–191, 1978); 'The rise and fall of the Deluge', by D R Dean* (*Journal of Geological Education*, **33**, 84–92, 1985); *Sedimentary Rocks: concepts and history*, edited by A V Carozzi (*Benchmark Paper*, No.12, 1974).

Geochronology. 'Geologic time', by C C Albritton* (*Journal of Geological Education*, **32**, 29–37, 1984); 'The age of the earth controversy: beginnings to Hutton', by D R Dean (*Annals of Science*, **38**, 435–456, 1981).

Mineralogy. *Studien zur Geschichte der Mineralnamen in Pharmazie, Chemie und Medizin von den Anfängen bis auf Paracelsus*, by D Goltz (*Sudhofts Archiv, Beiheft* **xiv**, 1972), and Walter Pagel's essay review 'The history of mineral terminology' (*History of Science*, **12**, 70–76, 1974); *Albertus Magnus and the Sciences; commemorative essays, 1980*, edited by J A Weisheipl (Pontifical Institute of Medieval Studies, Toronto, 1980); *Jacob Berzelius: the emergence of his chemical system*, by E M Melhado (Almqvist and Wiksell and University of Wisconsin Press, 1981); 'The science of minerals in the age of Jefferson', by J C Greene and J G Burke (*Transactions of the American Philosophical Society*, **68** (4), 1–113, 1978); *An Essay about the Origin and Virtues of Gems*, by Robert Boyle, a facsimile of the 1672 edition with an introduction by A F Hagner, in Hafner's *Contributions to the History of Geology Series*, vol.7 (1972); *Manual of the Mineralogy of Great Britain and Ireland*, by R P Greg and W G Lettsom (1858), reprint with a foreword by D G Embrey (Lapidary Publications, Broadstairs, 1977); 'Early mineralogy in Great Britain and Ireland', by W C Smith (*Bulletin of the British Museum (Natural History)* Historical Series, **6** (3), 49–74, 1978); 'Mineralogy: a historical review', by R M Hazen* (*Journal of Geological Education*, **32**, 288–298, 1984) which includes a synoptic table of mineral classification systems from Theophrastus (*c*.314 BC) to R W G Wycoff 1948; 'Meteorites, the Moon and the history of geology', by U B Marvin* (*Journal of Geological Education*, **34**, 140–165, 1986).

Crystallography. *Istorija Kristallografii*, by I I Shafranovskij (Nauka, Leningrad) in three parts — Part I: *From Ancient Times to the Beginning of the 19th Century* (1978); Part II: *The 19th Century* (1980); Part III: in preparation; 'An overview of crystallography in North America', by C Frondel (in D McLachlan and J P Glusker (eds) *Crystallography in North America,* American Crystallographic Association, 1983); 'The Renaissance background to crystallography', by C J Schneer (*American Scientist*, **71**, 243–263, 1983); *Crystal Form and Structure*, edited by C J Schneer (*Benchmark Paper*, No.34, 1977).

Geomorphology and glacial geology. *A Short History of Geomorphology*, by K J Tinkler (Barnes & Noble, 1985); *The Development of Landform Studies in Australia*, by H I Scott (Artarmon, 1977); *Landforms and Geomorphology: concepts and history*, edited by C A King (*Benchmark Paper*, No.28, 1976).

Upptacken ar Istaden: studier i den moderna Geolognius Framvaxt [The Discovery of the Ice Age], by T Frängsmyr (*Lychnos-Bibliotek*, **29**, Almqvist & Wiksell, 1976); 'James Geikie, James Croll and the eventful Ice Age', by C Hamlin (*Annals of Science*, **39**, 565–583, 1982); 'Glaciology and the Ice Age', by A V Carozzi* (*Journal of Geological Education*, **32**, 158–170, 1984).

A large number of references to publications dealing with the overlap between geography and geomorphology are given in 'The history of science and the history of geography: interactions and implications', by D N Livingstone (*History of Science*, **22**, 271–301, 1984); for example, *Empiricism and Geographical Thought: from Francis Bacon to Alexander von Humboldt*, by M J Bowen (Cambridge University Press, 1981).

Tectonics, geotechnics and geophysics. *History of Seismology, Seismics and Earth Tides Research* — 22 papers dealing with problems covering the period from antiquity to the present, edited by H Kantzleben (*Publications of the Zentralinstitut für Physik der Erde der Akademie der Wissenschaften der DDR*, No.64, 1981); *History of Geophysics*, a diverse collection of studies and accounts edited by C S Gillmor (American Geophysical Union, 1985); *History of Geophysical Research in the Netherlands and its Former Overseas Territories*, a more elaborate version of an earlier work in Dutch by J Veldkamp (*Verhandelingen Koninklijke Nederlandse Akademie van Wetenschappen Afd Natuurkunde*, 1st series, vol. **32**, 1984); *Geophysics in the Affairs of Man: a personalised history of exploration geophysics and its allied sciences of seismology and oceanography*, by C C Bates *et al* (Pergamon, 1982); *Geosynclines*, edited by F Schwab (*Benchmark Paper*, No.64, 1982); *Orogeny*, edited by J Dennis (*Benchmark Paper*, No.62, 1982); 'Nineteenth-century debates about the inside of the earth: solid, liquid or gas?', by S G Brush (*Annals of Science*, **36**, 225–254, 1979); 'John Flamsteed's letter concerning the natural causes of earthquakes', by F Willmoth (*Annals of Science*, **44**, 23–70, 1987); 'Terrestrial magnetism and the development of international collaboration in the early nineteenth century', by J Cawood (*Annals of Science*, **34**, 551–587, 1977).

References arranged by country

The following section is additional to the section in the previous edition of this chapter (1973, pp.431–434).

Britain. 'Robert Plot: Britain's "Genial Father of County Natural Histories"', by S Mendyk (*Notes and Records of the Royal Society of London*, **39**, 159–177, 1985); *Down to Earth: one hundred and fifty years of the British Geological Survey*, by H E Wilson (Scottish Academic Press, 1985), with the emphasis on development during the last half century. *Humphry Davy on Geology: the 1805 lectures for the general audience*, by R Siegfried and R H Dott (Wisconsin University Press, 1980); *The Great Chain of History: William Buckland and the English school of geology, 1814–1849*, by N A Rupke (Oxford University Press, 1983); 'Arthur Aikin's mineralogical survey of Shropshire 1796–1816 and the contemporary audience for geological publication', by H S Torrens (*British Journal for the History of Science*, **16**, 111–153, 1983); 'John Farey's mineral survey of south-east Sutherland and the age of the Brora coalfield', by C D Waterston (*Annals of Science*, **39**, 175– 185, 1982).

Ireland. 'The history of Irish geology', by G L Herries Davies (p.303–315 in C H Holland (ed.) *A Geology of Ireland*, Scottish Academic Press, 1981); 'Geology in Ireland before 1812: a bibliographical outline', by G L Herries Davies (*Western Naturalist*, **7**, 79–89, 1978).

Continental Europe. 'La géologie en Suisse des debuts jusqu'à 1882: digression sur l'histoire de la géologie suisse depuis Konrad

Gesner (1565) jusqu'à Heinrich Wettstein (1880)', by A V Carozzi (*Eclogae Geologicae Helvetiae*, **76**, 1–33, 1983). *Leben und Wirken deutscher Geologen im 18 und 19 Jahrhundret*, edited by H Prescher (VEB Deutscher Verl. für Grundstoff industrie 1985). *Geologen der Geothezeit*, by H Prescher (*Abhandlungen des Staatlichen Museums für Mineralogie und Geologie zu Dresden*, Bd **29**, 1979). *History of the Geological Institute of the USSR Academy of Science. Development of the Institute, its scientific schools, bibliography of works*, by V V Tikhomirov *et al* (Nauka, Moscow, 1980).

Extensive lists of European publications are given in the INHIGEO *Newsletter*, particularly in the most recent issues. See also: *Bibliographie in der DDR zür Geschichte der Geologie, Mineralogie, Geophysik und Paläontologie Vergeligten Arbeiten*, by P Schmidt (Bergakademie, 1978); and 'Bibliographie sommaire sur le développement de la géologie de langue française jusqu'en 1832', by K L Taylor (*Histoire et Nature*, **19/20** (1981–82), 141–147, 1983).

North America. *Two Hundred Years of Geology in America: Proceedings of the New Hampshire Bicentennial Conference on the History of Geology*, edited by C J Schneer (University of New England Press, 1979); *Geologists and Ideas: a history of North American geology*, edited by E T Drake and W M Jordan (Geological Society of America [The Decade of North America Project Series], 1985); *The Geological Sciences in the Antebellum South*, edited by J X Corgan (University of Alabama Press, 1982); *Frontiers of Geological Exploration of Western North America*, edited by A E Leviton *et al* (California Academy of Sciences, 1982); *North American Geology; early writings*, edited by R M Hazen (*Benchmark Paper*, No.51, 1979).

American Science in the Age of Jefferson, by J C Greene (Iowa State University Press, 1984). *Wealth Inexhaustible: a history of America's mineral industries to 1850*, by M H and R M Hazen (Van Nostrand Reinhold, 1985). *Charles Lyell on North America* (Arno Press, 1977) consists of reproductions of lectures presented by Lyell during two visits to the United States and of 40 of his short articles on American geology. 'History of geology', by M T Greene (in S G Kohlstedt and M W Rossiter (eds) 'Historical Writing on American Science', *Osiris*, 2nd Series, no.1). It is also worthwhile to see the special issues on aspects of American geology in *Earth Sciences History* (1982–).

Bibliographies

During the period since the compilation of the first version of this chapter a number of important bibliographic tools for the historian of natural history, and of geology in particular, have been published.

(i) *Geology and the History of Geology; an international bibliography from the origins to 1978*, by W A S Serjeant (vols. 1–5, Macmillan, 1980; reissued Kreiger Publishing Co., 1986. Supplement for 1979–1984, vols. 6, 7, 1987). The first volume lists books and articles treating the history of geology in general and dealing with its specialist subdisciplines, together with related sciences (e.g. botany). It also details histories of geological societies, of the petroleum industry, and of voyages and travels. Volumes 2 and 3 are organized bibliographically, alphabetically by geologists: an outline life is given indicating areas of expertise and publications, followed by a listing of secondary works relating to the geologist's life and career. Volume 4 lists these same geologists separately, according to nationality, describing their geological activity and scientific speciality. The fifth volume is an author index of all the works cited in the first three. A *Supplement* for the years 1979–1984, with additions to earlier volumes is provided in two further volumes (Kreiger Publishing Co., 1987).

These volumes provide an almost perfect complement to the *Isis Cumulative Bibliographies* (1913–65), edited by M Whitrow (5 vols, Mansell and the History of Science Society, 1971; and 1966–75, edited by J Neu: Vol.1, *Personalities and Institutions*, Mansell and the History of Science Society, 1980).

(ii) *The History of British Geology: a bibliographical study*, by J Challinor (David & Charles, Barnes & Noble, 1971) which provides a selection of primary publications arranged chronologically (1538–1969), and a number of short, thematic essays.

(iii) *The Earth Sciences: an annotated bibliography*, by R Porter (Garland, 1983) (*Bibliographies of the History of Science and Technology*, vol.3, edited by R Multhauf and E Wells) — a selection of 808 publications, arranged into the following categories: Bibliography and reference works (26); General histories — of science and of geology (47); Specialist histories — including a section on geological philosophies and methods (192); Cognate sciences — including geophysics, geography, oceanography, natural history and crystallography (143); Studies by area (100); Biographical studies (282); Institutional histories (62); The Social dimension (11); Geology and religion (16); Geology, culture and the arts (23). The great majority of the entries include a short descriptive, and sometimes critical, paragraph. There is also a short, helpful introduction and a selection of geological illustrations.

(iv) *Annual Bibliography of the History of Natural History* (1985–) (British Museum (Natural History)). Volume 1 of the series (1985) covers the literature for 1982, and Volume 2 (1986) for 1983.

A selection of bio-bibliographies is cited in the accompanying section on Biographies.

Biographies

In much the same way that the situation regarding bibliographic works of reference has improved almost out of recognition, so has the situation as regards biography. This is due in very large measure to one enormous (and enormously helpful) publication: *viz, Dictionary of Scientific Biography*, editor-in-chief C C Gillispie (14 vols, Scribner, 1970–80). All periods of science from classical antiquity to modern times are represented, with the exception that there are no articles on the careers of living persons. In many instances the articles are either the first or the most considerable yet made of an individual body of work, for the purpose of the *Dictionary* is described as 'not only to draw upon existing scholarship but to constitute scholarship where none exists'. Each item includes a brief bibliography of primary and secondary sources, and the elaborate index volume contains locality, subject and author items.

Among recent bio-bibliographies of well-known figures in the history of geology, the following are worth mentioning: *The Letters of Georges Cuvier: a summary calendar of manuscript and printed materials preserved in Europe, the United States of America and Australasia*, compiled by D Outram (British Society for the History of Science, 1979); 'Edward Forbes (1815–1854) — an annotated list of published and unpublished writings', by P F Rehbock (*Journal of the Society for the Bibliography of Natural Science*, **9** (2), 171–218, 1979); *A List of the Papers and Correspondence of George Bellas Greenough (1778–1855) held in the Manuscript Room, University College London Library*, by Jacqueline Golden (University College London, 1981); *A Bibliography of George Poulett Scrope, Geologist, Economist and Local Historian*, by Paul Sturges (Baker Library, Harvard, 1984); and *Guide to the Charles D. Walcott Collection, 1851–1940*, by W R Massa Jnr (Smithsonian Institution, 1984).

Aspects of the éloges of the Paris Academy of Sciences are considered in *Science and Immortality* by C B Paul (University of California Press, 1981), and of Cuvier's éloges — including an extensive bibliography of the biographical writing of Cuvier, by D Outram (*History of Science*, **16**, 153–178, 1978). Eighteen biographical portraits of leading Russian geologists, from A V Lomonosov onwards, is provided in *Biographen bedeutender Geowissenschaften der Sowjetunion: 19 biographische Darstellungen zu bedeutenden Gelehrten der russischen und sowjetischen Geologiegeschichte*, edited by M Guntau (Akademie Verlag, 1979).

Examples of recent biographical studies (not mentioned earlier) include: *Georges Cuvier: vocation, science and authority in postrevolutionary France*, by D Outram (Manchester University Press, 1984); *Grove Karl Gilbert: a great engine of research*, by S J Pyne (University of Texas Press, 1980); *Richard Griffith 1784–1878*, edited by G L Herries Davies and R C Mollan (Royal Dublin Society [*Historical Studies in Irish Sciences* . . . No.1], 1980); *Johann Heinrich Merk Werke und Briefe*, by A Henkel and H Kraft (Frankfurt a-Main, 1968) — a little known biography of the man whose influence on Goethe was considerable; *Lamark*, by L J Jordanova (Oxford University Press [*Past Masters Series*], 1984); *Linnaeus: the man and his work*, four essays (including one on Linnaeus as geologist), edited by T Frängsmyr (University of California Press, 1983) — a translation from an earlier edition in Swedish; *Hugh Miller: outrage and order. A biography and selected writings*, by G Rosie (Mainstream, Edinburgh, 1982); *John Milne: father of modern seismology*, by A L Herbert-Gustar and P A Nott (Tenterden, Kent, 1980); *Adam Sedgwick: geologist and Dalesman, 1785–1873. A biography in twelve themes*, by C Speakman (Broad Oak Press, The Geological Society and Trinity College, Cambridge, 1982); *Edward Suess, 1831–1914*, a collection of seven articles (three with English abstracts), edited by A Tollman and E Kristan-Tollman (Oesterrichische Geologische Gessellschaft, 1981); and *John Taylor: mining entrepreneur and engineer, 1779–1863*, by R Burt (Buxton, 1977).

The series 'Biographies of Outstanding Scientists, Technicians and Physicians', in German (B G Teubner Verlagsgesellschaft) includes Alexander von Humboldt (K-R Biermann, 1980), Lyell (G Zirnstein, 1980), Paracelsus (I Kästner, 1985), V I Vernadskij (P Krüger, 1981) and Werner (M Guntau, 1984).

Other works of reference

Other works of reference of relevance to the history of geology are:
Dictionary of the History of Ideas: studies of selected pivotal ideas, P P Wiener (Editor-in-Chief) (5 vols, Scribner, 1973–74) — a collection of just over 300 essays which neatly complements the *Dictionary of Scientific Biography* (1970–80).

Macmillan Dictionary of the History of Science, edited by W F Bynum *et al* (Princeton University Press and Macmillan, 1981; paperback 1983) — the first work of its kind. It has an analytical table of contents, in which 55 items are listed under earth sciences (including actualism, catastrophism, continental drift, denudation and decay, fossils, geology, geophysics, stratigraphy); 42 items under historiography and sociology of science (including metaphor in science, paradigm, revolutionary science, science and religion, scientific institutions, and Whig history); and 131 items under philosophy of science.

A Dictionary of Geology, compiled by J Challinor (University of Wales Press, 1961) was the first dictionary or glossary of geological terms to be compiled on historical principles. The 6th edition (1986, edited by A Wyatt) has appropriately been re-named *Challinor's Dictionary of Geology*.

United Kingdom Research on the History of Geological Sciences: a directory 1981 (The Royal Society, 1983) is a volume which lists work in progress and recent publications (mostly 1976–1981) by 86 individuals.

Milestones in the History of Geology: a chronological list of important events in the development of geology (Council on Education in the Geological Sciences (CEGS) *Short Review Number* **25**) by A La Rocque *et al* (*Journal of Geological Education*, **22** (5), 195–203, 1974).

Natural History Manuscript Resources in the British Isles, compiled by G D R Bridson *et al* (Mansell & R R Bowker, 1980) is a comprehensive, detailed and authoritative guide (begun in 1968 and based on questionnaires and on personal visits) to the material for the period 1600 to 1900.

British Archives, compiled by J Foster and J Sheppard (Macmillan, 1982) is a full listing of the chief archive repositories, their opening times and facilities, and the main kinds of records and archives deposited there.

Index